TUNNELS: Planning, Design, Construction

International Edition

ELLIS HORWOOD SERIES IN ENGINEERING SCIENCE

STRENGTH OF MATERIALS
J. M. ALEXANDER, University College of Swansea.
TECHNOLOGY OF ENGINEERING MANUFACTURE
J. M. ALEXANDER, R. C. BREWER, Imperial College of Science and Technology, University of London, J. R. CROOKALL, Cranfield Institute of Technology.
VIBRATION ANALYSIS AND CONTROL SYSTEM DYNAMICS
CHRISTOPHER BEARDS, Imperial College of Science and Technology, University of London.
COMPUTER AIDED DESIGN AND MANUFACTURE
C. B. BESANT, Imperial College of Science and Technology, University of London.
STRUCTURAL DESIGN AND SAFETY
D. I. BLOCKLEY, University of Bristol.
BASIC LUBRICATION THEORY 3rd Edition
ALASTAIR CAMERON, Imperial College of Science and Technology, University of London.
STRUCTURAL MODELLING AND OPTIMIZATION
D. G. CARMICHAEL, University of Western Australia
ADVANCED MECHANICS OF MATERIALS 2nd Edition
Sir HUGH FORD, F.R.S., Imperial College of Science and Technology, University of London and J. M. ALEXANDER, University College of Swansea.
ELASTICITY AND PLASTICITY IN ENGINEERING
Sir HUGH FORD, F.R.S. and R. T. FENNER, Imperial College of Science and Technology, University of London.
INTRODUCTION TO LOADBEARING BRICKWORK
A. W. HENDRY, B. A. SINHA and S. R. DAVIES, University of Edinburgh
ANALYSIS AND DESIGN OF CONNECTIONS BETWEEN STRUCTURAL JOINTS
M. HOLMES and L. H. MARTIN, University of Aston in Birmingham
TECHNIQUES OF FINITE ELEMENTS
BRUCE M. IRONS, University of Calgary, and S. AHMAD, Bangladesh University of Engineering and Technology, Dacca.
FINITE ELEMENT PRIMER
BRUCE IRONS and N. SHRIVE, University of Calgary
PROBABILITY FOR ENGINEERING DECISIONS: A Bayesian Approach
I. J. JORDAAN, University of Calgary
STRUCTURAL DESIGN OF CABLE-SUSPENDED ROOFS
L. KOLLAR, City Planning Office, Budapest and K. SZABO, Budapest Technical University.
CONTROL OF FLUID POWER, 2nd Edition
D. McCLOY, The Northern Ireland Polytechnic and H. R. MARTIN, University of Waterloo, Ontario, Canada.
TUNNELS: Planning, Design, Construction
T. M. MEGAW and JOHN BARTLETT, Mott, Hay and Anderson, International Consulting Engineers
UNSTEADY FLUID FLOW
R. PARKER, University College, Swansea
DYNAMICS OF MECHANICAL SYSTEMS 2nd Edition
J. M. PRENTIS, University of Cambridge.
ENERGY METHODS IN VIBRATION ANALYSIS
T. H. RICHARDS, University of Aston, Birmingham.
ENERGY METHODS IN STRESS ANALYSIS: With an Introduction to Finite Element Techniques
T. H. RICHARDS, University of Aston, Birmingham.
ROBOTICS AND TELECHIRICS
M. W. THRING, Queen Mary College, University of London
STRESS ANALYSIS OF POLYMERS 2nd Edition
J. G. WILLIAMS, Imperial College of Science and Technology, University of London.

TUNNELS:
Planning, Design, Construction
International Edition

T. M. MEGAW, M.Sc., Hon.D.Sc., F.I.C.E.
Consultant and Former Senior Partner

and

J. V. BARTLETT, C.B.E., M.A., F.Eng., F.I.C.E., F.I.E.Aust., F.A.S.C.E.
Director and Senior Partner
Mott, Hay & Anderson, Consulting Civil Engineers
Croydon, Surrey

ELLIS HORWOOD LIMITED
Publishers · Chichester

Halsted Press: a division of
JOHN WILEY & SONS
New York · Brisbane · Chichester · Toronto

First published in International Edition in 1983 by

ELLIS HORWOOD LIMITED
Market Cross House, Cooper Street, Chichester, West Sussex, PO19 1EB, England

Originally published in two volumes
Volume 1 (1981): Volume 2 (1982)

The publisher's colophon is reproduced from James Gillison's drawing of the ancient Market Cross, Chichester.

Distributors:

Australia, New Zealand, South-east Asia:
Jacaranda-Wiley Ltd., Jacaranda Press,
JOHN WILEY & SONS INC.,
G.P.O. Box 859, Brisbane, Queensland 40001, Australia

Canada:
JOHN WILEY & SONS CANADA LIMITED
22 Worcester Road, Rexdale, Ontario, Canada.

Europe, Africa:
JOHN WILEY & SONS LIMITED
Baffins Lane, Chichester, West Sussex, England.

North and South America and the rest of the world:
Halsted Press: a division of
JOHN WILEY & SONS
605 Third Avenue, New York, N.Y. 10016, U.S.A.

British Library Cataloguing in Publication Data
Megaw, T. M.
Tunnels. – (Ellis Horwood series in engineering science).
1. Tunnels
I. Title II. Bartlett, J. V.
624.1'93 TA805

Library of Congress Card No. 83-162

ISBN 0-85312-611-9 (Ellis Horwood Ltd., Publishers – International Edn.)
ISBN 0-470-27217-1 (Halsted Press)

Typeset in Press Roman by Ellis Horwood Ltd.
Printed in Great Britain by Butler & Tanner, Frome, Somerset.

Table of Contents

Chapter 3 Investigations

Chapter 4 Tunnelling in Soft Ground

Chapter 5 Compressed Air Working

Chapter 6 Rock Tunnelling

Chapter 7 Tunnelling Machines for Soft Ground and Rock

Chapter 8 Permanent Linings

Chapter 9 Survey and Control

Authors' Preface

This book aims to explain the principles and practice of tunnelling, and to show the part which tunnels may play in solving some of the problems encountered by man, as his increasing numbers and activities cause congestion on the surface of the earth and a need for better communications. It is written in the hope that it will prove of use, not merely to those already involved in tunnelling but to engineers and others who may be drawn into a tunnelling project, or who should give consideration to tunnelling as contributing to the achievement of some desirable objective. It should also be of help to the tunnel specialist in its wide scope, and perhaps in giving him an insight into the problems of his colleagues and of the tunnel users.

The very question of what to include within the definition of a tunnel poses a problem. There are three aspects to be considered: form, function, and method of construction. In form it is characteristically an underground cavity, long in relation to its cross-section, and horizontal rather than vertical in alignment. In function a tunnel is typically for passage and conveyance of men and materials below ground. Method of construction might, in a narrow definition, be restricted to boring, by excavation and lining from within. In a wider sense, building or assembling the structure in a trench, under land or water, does produce a tunnel, and whether or not the operation can be called 'tunnelling' the means and the end are included and described in this book. Conversely, caverns or other underground storage may be tunnelled but the end product might not be categorised as a 'tunnel' and the particular features of such structures are not described in any detail. A formal definition of a tunnel might be thus: a passage constructed below ground, or water, of essentially cylindrical form and having its axial alignment not greatly differing from the horizontal; of dimensions sufficient for the passage of a man; excavated and lined from within, or by other means.

This book is not intended to be a catalogue of the latest tunnelling techniques or machines, nor does it contain any novel or advanced theoretical analysis; the extensive bibliography will enable readers to pursue such matters in books, periodicals and professional papers. There is no attempt to deal with costs, any figures for which become so quickly out of date.

In writing each chapter it was apparent that its subject could well fill a whole book by itself, but limitations of time and space had to be respected. It is to be hoped that a reasonable balance has been attained in providing two volumes. In the interests of readability, and usefulness to those seeking guidance on a particular subject, each chapter is largely self-contained. The authors are conscious that some overlapping and repetition has resulted, but believe that to be the lesser of two evils.

The plan of the book is to deal in the first volume with the history and scope of tunnelling and the techniques of bored tunnel construction, with a chapter describing and comparing two major undersea projects, the Channel Tunnel and the Japanese Seikan Tunnel.

The second volume describes other techniques – for tunnels in trench, under land and water – and the ancillary operations of shaft sinking and ground treatment. The relevance of geology is discussed, and there are chapters on the special requirements of tunnels for highways, metros and railways and their ventilation.

Each chapter has a brief bibliography dealing with its subject matter, while a more comprehensive and detailed bibliography is provided at the end of the book. In the text very few detailed references have been given because it is intended that it should be freely readable without interruption, to provide a general grasp of the subject.

Acknowledgements

In collecting material for the book and in writing it we have had invaluable assistance from some London Transport engineers and from many colleagues in Mott, Hay & Anderson. We are most grateful for all the help we have received and in particular thank the following for their very substantial contributions to specialist sections:

Peter Bishop, Arthur Cairncross, Harry Chanter, Steve Gunn, Harry Hadaway, Ken Henderson, David Henson, Ian Hill, Terry Hulme, Jeff Marshall, Frank Turner.

Over a period of several years John Finch, as librarian, has given invaluable help as have Sue Mill and her willing colleagues in typing and retyping the manuscripts.

Illustrations used in the text have been derived from many sources, named where possible in the captions, and we acknowledge with thanks the assistance of those contributing.

1

History

1.1 FUNDAMENTAL OPERATIONS

The fundamental operations of tunnelling are:
(1) Survey
(2) Excavation of ground
(3) Immediate support of ground
(4) Permanent support of ground
(5) Management of water

Where all conditions are favourable, construction becomes relatively simple, but the history of tunnelling shows how often unexpected physical conditions have made a project impossible as originally conceived, and ultimately only achievable, after years of immense effort, by virtue of new methods.

1.1.1 Relationship to mining

Tunnelling as an art has its origins in the mists of the past. Its relationship to mining is very close, but there are important distinctions to be made. The primary objective in mining is to dig out valuable minerals: after they are worked out the access passages and other voids are no longer of use, and in that sense all mining is temporary. The time may be very long—salt is still mined at Hallstatt where the deposits were worked as early as 2500 B.C. In contrast, the primary objective in tunnelling is to provide a passage or other space for permanent use, which must remain safely available for its purpose indefinitely. In addition to provision of passages for men or materials tunnels are constructed for storage, for installation of plant and, throughout the ages, for attack and defence. In size, transport tunnels must be as large as necessary for traffic, whereas mining passages will rarely exceed dimensions of about 3 m.

Another distinction to be drawn is that the organisation for tunnel construction is usually temporary; there is no settled community, and no continuity of construction. A team is built up, changing as the phases of construction develop, and is finally dispersed. Achievement of the objective is the end;

operation of the tunnel calls for entirely different men and skills. In mining, by contrast, a permanent community is established for the life of the mine, which may extend over several generations.

Most of the technology is the same for tunnelling and for mining and much is to be gained by interchange of information and ideas, but the differences are important. Attempts to employ coal miners in constructing London's underground in London Clay in the 1930's were said to be unsatisfactory because the coal mining tradition was to put in props but let the weight of the ground come down on to them, whereas the tunnelling technique was to wedge up to take the maximum possible weight immediately so as to minimise settlement and disturbances of the overlying strata.

It is also relevant that in tunnel construction the geometry and in particular the line must be pre-determined and followed closely from portal to portal, whereas in some types of mining the whole area of a stratum is to be excavated leaving only the necessary support to control subsidence, and in others an ore-bearing lode is followed wherever it may lead, excavating no more barren rock than essential for access. In all cases accurate survey is important.

The difference may be summarised paradoxically as permanent work by a temporary organisation contrasted with evanescent work by a permanent community.

1.2 ANCIENT MINES AND TUNNELS

1.2.1 Prehistoric mining

Despite the attempt to discriminate between tunnelling and mining, the earliest history must be traced through mining operations, first in extension of surface working for flint and later in the search for copper and other metals. The first tunnels in the narrower sense, excluding mining, may have been for supply of water to towns and other communities. Flint mines, such as Grimes Graves in Norfolk, were developed also in Belgium, northern France, Portugal, and elsewhere, dating back beyond 2000 B.C.; picks made from deer antlers were used, with flint axes and bone shovels. Salt mining at Halstatt from 2500 B.C. has already been mentioned and about 1000 B.C. those mines supported a very prosperous Bronze Age and early Iron Age community. Bronze picks and wooden mallets and shovels were initially used, but the use of iron was developed.

Copper, not merely native, but reduced by smelting from oxides was used by the Sumerians in the 4th millennium B.C., the ores being mined in the mountains of Armenia. Some time in that era adits were driven to follow the outcrops into the mountain-sides. Egypt of the Pharaohs obtained copper from Sinai and adjacent areas in the 3rd millennium B.C. Stone tools and wooden wedges expanded by wetting were used to drive up to 50 metres into the rock. Later, Egypt obtained its copper from Cyprus, named after its copper, and Armenia, and mined gold from deep shafts by slave and prisoner labour, mainly in the Nubian desert. 'Fire setting' seems to have been employed to mine hard rock.

Tin was essential for the production of bronze, which is an alloy of copper and tin and which is so much harder than natural copper that it competed with flint for tools and weapons and superseded it for many purposes. Phoenician sailors and traders produced supplies from Cadiz and from Cornwall before 1000 B.C. Cartagena was an important centre for silver mining.

Iron working seems to have originated about 1400 B.C. in Armenia, under Hittite domination, but it was not until about 1000 B.C. that knowledge spread over a wide area, effecting the transformation of the Bronze Age into the Iron Age.

1.2.2 Ancient tunnels

Water supplies provide further early examples of tunnels, and perhaps may be considered to mark the beginning of tunnelling as distinct from mining. Rock hewn temples and underground burial chambers in many parts of the world also exemplify methods and uses of underground excavation.

Extensive canal building was carried out in the 'Fertile Crescent' (Egypt, Palestine, Syria, Iraq) in ancient times for water supply and irrigation. A canal must follow a level course, and the need to pass a ridge entails either a long diversion round the end of the ridge, or a deep cutting, or a tunnel through the ridge. Sometimes the tunnel is the only possible solution if water from a mountain is to be conducted to a city in the plains. Where the supply of pure water is limited there are added advantages in carrying it underground which reduces evaporation and preserves it from surface contamination. Another reason for water tunnels was defensive: to preserve the water supply to a walled city under siege.

An early example from Greece of tunnelling through a ridge is the work of Eupalinus of Megara in 687 B.C. on the island of Samos. The tunnel, which was lined, was about 1000 m long and carried water through clay pipes laid in a trench cut in the floor.

The 'qanaats' of Iran and surrounding areas are small tunnels carrying water underground from hillside springs to towns or villages. Excavated normally from shafts at centres of about 50 m, their course across the desert is marked by the mounds of spoil at the shafts. Some of the shafts are gently inclined to provide easy access and the qanaats may terminate in deep wells.

Ancient strategic aqueducts are typified by the Siloam tunnel of Jerusalem excavated about 700 B.C. by King Hezekiah when an Assyrian siege was in prospect. The bible record in II Kings 20.20 reads, 'And the rest of the acts of Hezekiah, and all his might, and how he made a pool, and a conduit, and brought water into the city, are they not written in the book of the chronicles of the kings of Judah'.

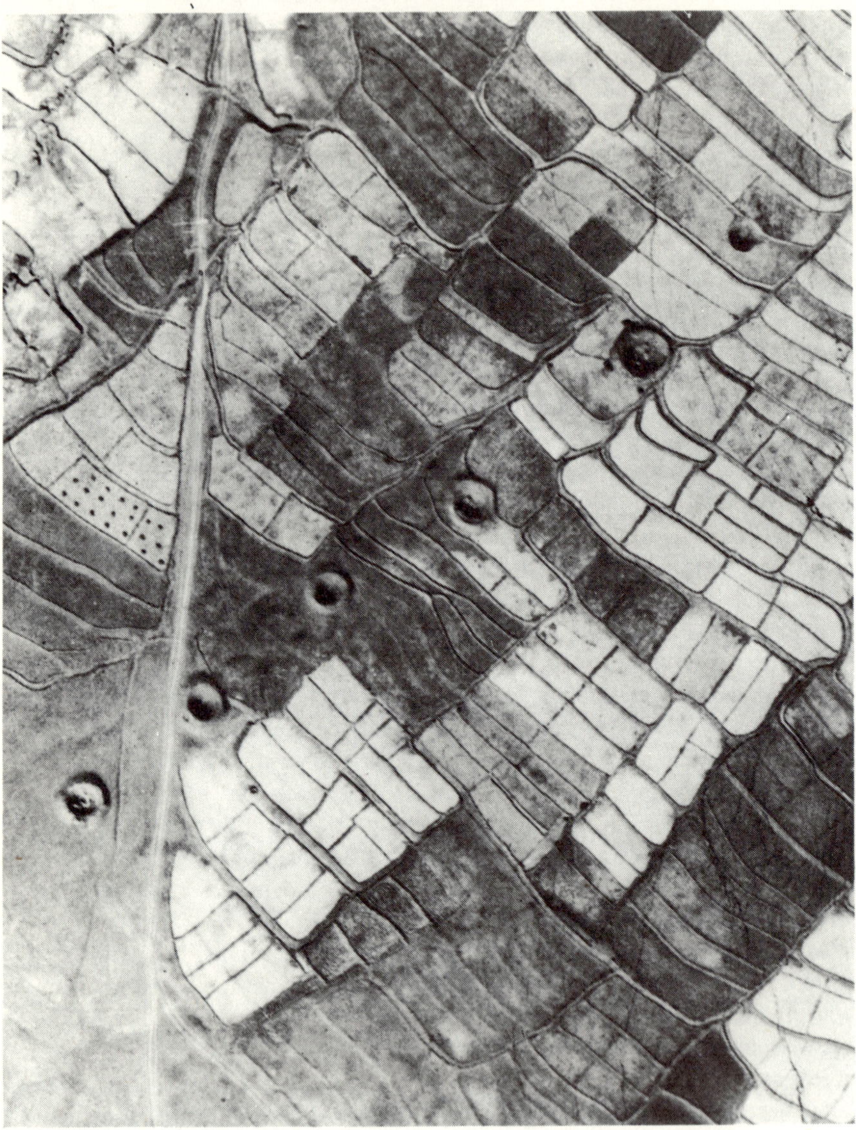

Fig. 1.1 – Air photograph of shafts marking the line of a qanaat (Prof. Ambraseys)

Examples of tunnelling from the classical periods of Greece and Rome are numerous, perhaps mostly for aqueducts. In Rome, water supplies were brought into the city from springs in the surrounding hills; over the period from 312 B.C. until about A.D. 52 aqueducts totalling some 350 km were built, almost wholly in tunnel, the system being based on gravity flow.

The Romans also used tunnelling for drainage and reclamation of marshes, and occasionally for roads, of which an outstanding example was the Pausilippo tunnel built in 36 B.C. between Naples and Pozzuoli which was about 8 m wide x 10 m high x 1470 m in length.

After the collapse of Roman power few major works were undertaken for many centuries. Tunnelling in mediaeval times became largely the art of the sappers in attacking walled fortresses, used also perhaps for the romantic underground passages reported to exist in many castles and monasteries. Civil engineering was very much in abeyance until stimulated by the needs of the canal builders.

1.3 CANALS

Except in very flat alluvial plains, canal engineers always face two problems to which tunnelling is relevant: changes of level and sufficient supply of water. Canal locks, by which barges can be transferred from one level to another, were the usual, though not the only solution of the first problem, but at the expense of demanding a supply of water for their operation. If a canal has to pass across a ridge of high ground water supply at the summit is likely to be scarce or non-existent unless pumps are used, while a flight of locks on each side becomes necessary if the canal is to follow the ground surface.

This is the situation where tunnelling may offer a solution which at once saves construction of locks with their demand for water and delays in transport and makes water supply much easier; by tunnelling, the ‘summit level’ may be lowered within reach of streams, and indeed the tunnel itself may tap underground streams. Passage through the tunnel, although it may be slow, is still likely to save much time and water compared with passage up and down again through flights of locks.

The pioneer tunnel of the Canal Age was constructed on the Canal du Midi, built in the years 1666–81 to connect the Atlantic to the Mediterranean and to save long, costly, and hazardous journeys round the Iberian peninsula and through the Straits of Gibraltar. It was a vast enterprise comprising 240 km of canal with over one hundred locks and with a summit level at nearly 200 m above sea level. Water was supplied from reservoirs built in the Montagnes Noires 20 km away. The project was promoted and executed by Pierre-Paul Riquet with the support of Louis XIV and his Minister of Finance, Colbert. The tunnel was not on the summit level but under a ridge over 100 km further east, near Béziers. The tunnel, 6.5 m x 8 m x 157 m long, was constructed in the years 1679–81. It was left unlined until 1691. Gunpowder was used in its construction, the first recorded use in tunnelling, although it seems probable that sappers, in their military capacity, had experimented earlier.

Before the days of gunpowder the nearest thing to explosives was fire setting, where the rock was heated up by burning quantities of wood and was then quenched with cold water, which opened up cracks and split off lumps. This method seems to have been used in ancient Egypt.

In England water transport became of increasing importance in the 16th century and was developed, first by way of improvement of navigable rivers, and then in the 18th century by canal construction, which indeed largely created civil engineering as a profession. Transport of coal and all heavy cargoes was almost entirely by water and the value of the river systems of the Severn, Thames, Mersey, Trent and others was enormously enhanced by construction of connecting canals, which of course had the typical summit problems, frequently resolved by tunnelling.

The first canal tunnels were on the Worsley–Manchester canal built by Brindley for the Duke of Bridgewater to carry coal direct from the mines at Worsley, some six miles west of Manchester, to the city. The canal was tunnelled into the sandstone at the colliery for about a mile, where later some forty miles of tunnels on various levels were developed. It was carried across the river Irwell on an aqueduct 12 m above the river and terminated again in a tunnel and vertical shaft under Castle Hill in Manchester. The canal, on a level throughout, and without locks, was opened in 1761 and resulted in the price of coal in Manchester being halved.

Brindley also built the Trent–Mersey, or Grand Trunk, canal 146 km long, which linked these rivers, and the Staffordshire–Worcestershire canal, 75 km long, connecting to the Severn at Stourport. This Grand Trunk canal was sponsored and strongly supported by Josiah Wedgewood of pottery fame. The Harecastle tunnel, 2090 m in length, carrying the canal at its summit through the ridge north of Stoke-upon-Trent, was a major undertaking. There were four other tunnels on the route including one of 1135 m at Preston-on-the-Hill, near Runcorn; there the canal joined the Duke of Bridgewater's canal leading down to the Mersey by a flight of locks. The Harecastle tunnel proved difficult. It was driven, with the use of gunpowder, from a number of shafts; springs were struck which flooded the works, and a drainage heading in advance of the work proved necessary. In contrast to the other four tunnels, which were 4.1 m wide x 5.3 m high, this tunnel was only 2.2 m x 3.6 m, and in operation a one way system had to be worked, the barges being 'legged' through, which took two hours. In legging, men lay on their backs on the barge deck and pushed the barge through by their feet against the tunnel roof. The canal took eleven years to build, largely because of this tunnel, and was not completed until 1777, five years after Brindley's death. The canal was a great success in reducing transport costs between Liverpool and the Potteries, but this 'single lane' tunnel remained a bottleneck and was not duplicated until 50 years later, when Telford engineered a new tunnel, also as a one way tunnel, but 4.3 m x 4.9 m, including a 1.5 m towpath. Fifteen shafts were sunk for the construction and sixteen cross headings to the old tunnel. Difficulties were again encountered with running sand and with extremely hard Millstone Grit, but the work was completed in three years, including brick lining, and without any loss of life. The tunnel is still open, but principally for recreational use; extensive repair work including concrete lining and roof bolting has been necessary in 1974.

Following Brindley's example, canals proliferated throughout England with their quota of tunnels. There was a boom in promotion of schemes with its peak in the years 1793-4. Development continued until the Railway Age. A notable tunnel was at Standedge about five miles south west of Huddersfield where the Pennine ridge was pierced by a tunnel 5000 m in length and at a height of 194 m above sea level. Work started in 1794 under Outram by sinking shafts, one of which was 70 m deep; the tunnel was 3.0 m wide x 5.5 m high and had four 'wides' within its length to provide passing places. Much of the excavation was through hard Millstone Grit and it took seventeen years to complete, being opened in 1811. Many lives were lost.

Mention may also be made of the Sapperton tunnel (1789), 3 km long, which was the summit level of the Thames–Severn canal, driven through the Oolite limestone strata into which it lost water. Telford's work on the second Harecastle tunnel has already been described. He had earlier been responsible for the Chirk and Whitehouses tunnels on the Ellesmere canal, in the vicinity of his famous Pont Cysyllte aqueduct–'the stream in the sky'. Late in his life, in 1830, the Cowley tunnel, driven under his direction on the Birmingham–Liverpool Junction canal, ran into such trouble that most of its length had to be opened out into a deep cutting and the tunnel finished up 74 m in length instead of the planned 630 m.

On the continent of Europe canal construction also demanded tunnels. Napoleon's St. Quentin canal had three major tunnels in its 100 km length. The Tronquoy tunnel, 1075 m long and 8 m wide, was driven through sandy gravel in 1803 by construction of the brick arch in a sequence of headings working upwards prior to removal of the 'dumpling'. This technique later developed into the 'German System'. The Riqueval tunnel, 5550 m long, was constructed simultaneously, in less difficult ground. A section of the canal was opened in 1810 and it was extended in 1822 by driving of the Noirieu tunnel, 12000 m long. The Canal de Bourgogne, started in 1775 and completed in 1825, is 245 km long and includes a 3 km tunnel. The Canal de la Marne au Rhin, 300 km long, built 1838–53, includes the 4 km Mauvages tunnel and four others.

In Belgium in 1828 the Charleroi Canal had to be tunnelled through a hill with quicksand, for which was developed the 'Belgian System', in which a small top heading is driven first and then enlarged progressively to allow the top half to be timbered and the arched lining built in brick or masonry, followed by underpinning of the sides in trench and final excavation of the 'dumpling' and construction of the invert. The canal has been modernised and the tunnel opened out and an inclined plane with a rise of 70 m has been provided.

American canal building also developed at the end of the 18th century. The first tunnel was driven in 1818 on the Schuylkill canal, its dimensions being 6 m x 5.5 m x 137 m long. It was opened out in 1855. The second American canal tunnel was the Lebanon tunnel at the summit of the Union canal, excavated in 1824–26 to dimensions 5.5 m x 4.6 m x 220 m.

After the coming of the railways few canal tunnels were built, but the latest and largest is the Rove canal tunnel providing a connection between Marseilles and the Rhone. It is one of the largest of all bores being 22 m x 15 m x 7.3 km through limestone. It was started in 1911 but delayed by the First World War and it was not completed and opened until 1928.

1.4 BRUNEL'S THAMES TUNNEL

Between the Canal Age and the Railway Age came the construction of Brunel's great Thames Tunnel, the first to pass under a tidal river and the first shield driven tunnel. It was in some respects far ahead of its time and its successful completion demanded so much time and effort and personal supervision that no similar enterprise was undertaken for nearly half a century.

Brunel's concept of a shield, in certain aspects inspired by the boring methods of the shipworm, *Teredo Navalis*, was covered by two patents in 1818 and was applied to the proposed dual carriage-way Thames crossing between Rotherhithe and Wapping, started in 1825 by the Thames Tunnel Company. An earlier attempt in this vicinity, first by Vazie and then by Trevithick in 1807 to drive a heading, or drift, as a preliminary to enlargement to a 5 m diameter tunnel had failed in the soft ground.

The fundamental problems of excavation and temporary support outlined at the beginning of this chapter were described by Brunel in his patent specification: 'I shall premise by observing that the chief difficulties to be overcome in the extension of tunnels under the beds of great rivers lie in the insufficiency of the means of forming the excavation. The great desideratum, therefore, consists of finding efficacious means of opening the ground in such a manner that no more earth shall be displaced than is to be filled by the shell or body of the tunnel, and that the work shall be effected with certainty'. . . . 'In the formation of a drift under the bed of a river, too much attention cannot be paid to the mode of securing the excavation against the breaking down of the earth . . . I propose to resort to the use of a casing or a cell, intended to be forced forward . . .'.

Brunel's shield was designed to provide:

(1) A skin, covering the ground on all sides. This was done by means of iron staves sliding forward as the excavation was advanced, and supported by iron frames.

(2) Means to support the face and yet provide access for excavation. The iron frames carried horizontal oak poling boards 150 mm x 75 mm in section, supported by screw jacks which could be moved forward one at a time as the small area of face open was excavated.
(3) Means to move the shield forward into the excavated space and then to build the permanent lining at the rear. The framework of the shield consisted of twelve 3-tier frames, supporting the staves at sides and top and providing working cells, each occupied by one miner; these frames were propelled forward one at a time by jacking from the completed arch brickwork.

The whole shield was 11.4 m wide x 6.8 m high x 2.7 m long. It was first assembled at the bottom of a 15 m diameter shaft, and then its somewhat complex mechanisms had to be operated by initially unskilled labour, in dirty, wet and dangerous conditions. As well as troubles in ensuring proper use of the various devices there were difficulties in keeping the shield on line, but in good clay it functioned satisfactorily. The hope of at least a roof of sound clay was not realised, and management of water became the main tunnelling problem. This is of course the fifth fundamental operation noted at the beginning of this chapter. The water softened the silt and clay and came in ever increasing volume: it was accompanied by methane causing minor explosions and by sulphureted hydrogen and other pollutants causing illness and death to the

Fig. 1.2(a) – Brunel's Shield. Longitudinal section of Brunel's Thames Tunnel showing: (i) excavation in progress in the shield at the face, (ii) brick lining built close behind shield, (iii) screw jacks at top and bottom thrusting the shield forward from the brickwork.

Fig. 1.2(b) – Brunel's Shield
One of the twelve divisions of the shield, showing: (i) the poling boards at the face supported by screw jacks, (ii) the top staves, providing roof support, (iii) the bottom shoes, resting on timbers.

face workers. Five times the whole tunnel was inundated from the river above and five times was reclaimed by dumping clay and gravel in the river bed and pumping out the water and digging out the debris. Tunnelling had started at the end of 1825, but after the second inundation, which was in January 1828 when 184 m out of 366 m had been completed, work was stopped for lack of capital and was not resumed until a government loan was granted at the end of 1834. A year was spent in replacing the shield with an improved structure and another five and a half years from March 1836 until November 1841 was required to finish the tunnelling. The tunnel was opened in 1842 but only to pedestrians as no money was available for carriageway access. It was ultimately sold in 1865 to the East London Railway Company who constructed the necessary approaches to make it a railway link through which London Transport trains still run regularly.

Sir Marc Isambard Brunel himself took a day-to-day part in the construction and controlled every detail. His even more famous son, Isambard Kingdom Brunel, was Resident Engineer in the first half up to the stoppage in 1828, but when work was resumed in 1835 he was no longer available as he had other interests, including the Box tunnel on the Great Western Railway.

1.5 RAILWAYS

Modern railway construction started with the Liverpool and Manchester Railway, opened in 1830, and involved tunnelling from the very beginning. Loco haulage terminated at Edge Hill in Liverpool where the railway faced a formidable barrier of high ground, and from that point two tunnels were driven: (1) a short tunnel of 265 m rising to Crown Street, up which passenger coaches were rope hauled to the surface, and (2) the long Wapping tunnel for goods traffic descending for 1930 m down a gradient of 1 in 48 to the docks. The tunnel was 6.7 m x 4.9 m in section. Shortly after opening, the railway was extended in tunnel from Edge Hill to its present terminus at Lime Street. This tunnel was 1852 m long and ran down a gradient of 1 in 97. It was worked by rope haulage until 1870 when loco working combined with artificial ventilation by an extract fan was introduced. About 1866 the tunnel was opened out into a cutting and widened to take four tracks. These tunnels were mainly in Triassic sandstone which stood up well, only requiring to be supported over parts of their length.

The enormous spread of railway construction in Great Britain by competing lines for much of the 19th century brought its quota of tunnels, mostly to minimise gradients and to obtain direct routes. Between 1830 and 1890 over fifty railway tunnels exceeding one mile (1.61 km) in length were completed, 25 of them between 1838 and 1850. Most were through hills, and in many water was the greatest enemy, inundating shafts, washing out sand, or softening shale. Exceptionally, the Mersey Railway Tunnel (1879–88) and the Severn Tunnel (1873–86) were beneath tidal rivers.

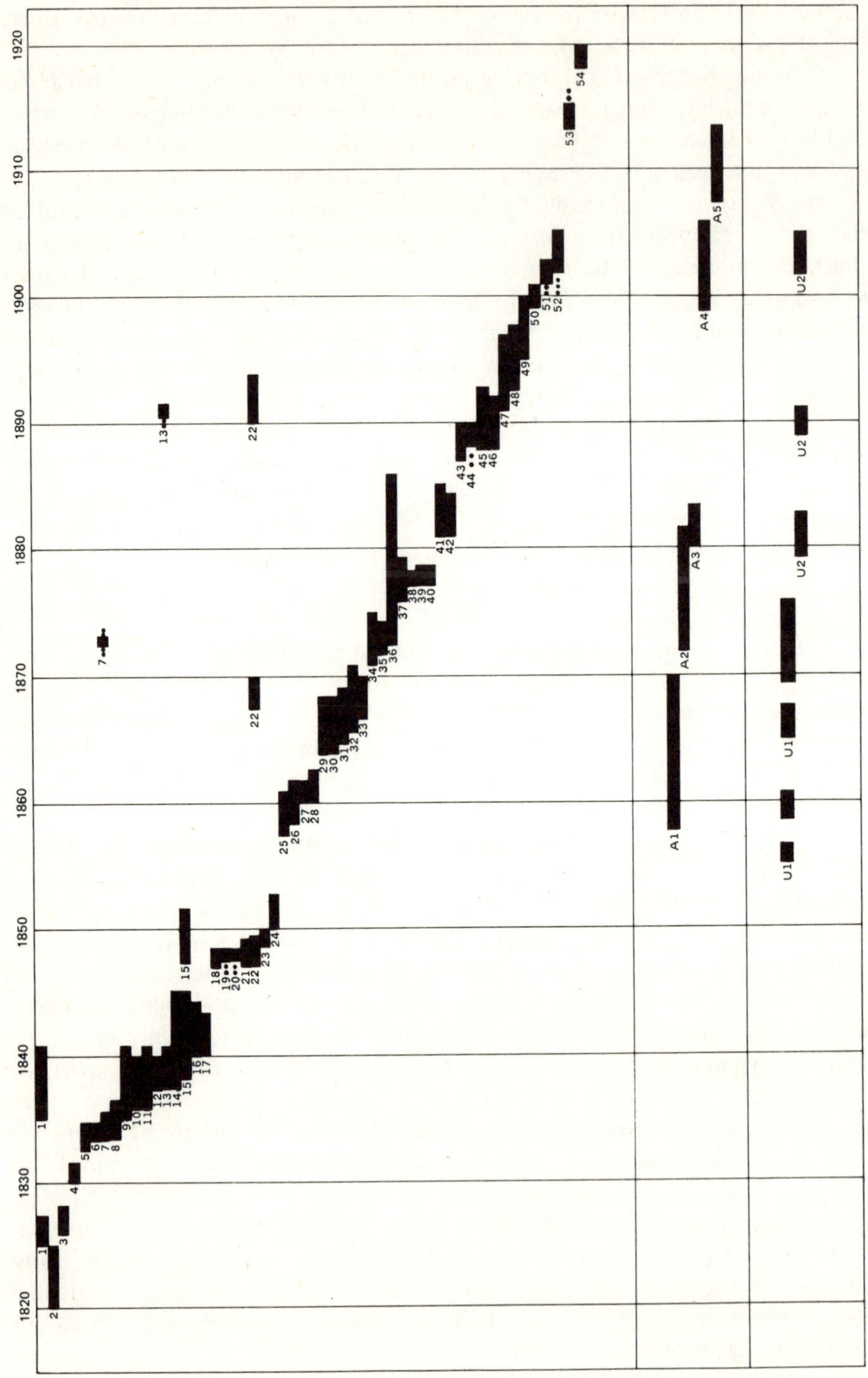
1820
1830
1840
1850
1860
1870
1880
1890
1900
1910
1920

	Dates	Name	L'gth m.	Notes
1	1825–43	Thames Tunnel	459	Brunel's tunnel. Converted to rail
2	1820–24	Strood	2130	Canal tunnel converted to rail
3	1826–29	Wapping	2057	Liverpool terminus. Rope haul
4	1830–32	Glenfield	1633	Leicester. Robert Stephenson
5	1833–36	Lime Street	1852	Liverpool terminus. Later opened out
6	1834–36	Primrose Hill	1081	London to Birmingham. George Stephenson Watford duplicated about 1873 (6–8)
7	1834–37	Watford	1660	
8	1834–38	Kilsby	2218	
9	1836–41	Box	2937	London to Bristol. I. K. Brunel. Up to 4000 men, 300 horses
10	1837–40	Summit	2638	Manchester to Leeds
11	1837–41	Clayton	2066	London to Brighton
12	1838–40	Clay Cross	1631	Derby to Leeds
13	1838–41	Merstham	1674	London to Brighton. Up to 6000 men, 500 horses. Adjacent Quarry tunnel 1892
14	1838–45	Sapperton	1701	London to Bristol. I. K. Brunel
15	1839–45	Woodhead	4848	Trans Pennine. Followed by down line 1847–52. Replacement 1949–53.
16	1840–44	Abbotscliff	1776	Near Dover. Also Shakespear Cliff 1273 m and Martello 485 m
17	1840–43	Morley	3080	Leeds
18	1846–48	Sough	1843	Blackburn
19	1846?–48	Harecastle	1615	Midlands. Near Brindley's canal tunnel
20	1846?–48	Victoria	2474	Liverpool, to docks
21	1846–49	Bramhope	3439	Airedale — Wharfedaile
22	1846–49	Standedge	4885	Also canal tunnel. Rail duplicated 1868–70 Double track duplication 1890–94
23	1848–50	Halton	1770	Birkenhead line
24	1850?–53	Merthyr	2283	Great Western
25	1857–61	Penge	1951	South London
26	1858–61	Lydden	2166	Near Dover
27	1860–62	Greenock	1930	Scotland
28	1860–63	Dove Holes	2729	Derby to Manchester
29	1864–68	Sevenoaks	3157	London to south coast (29–31)
30	1864–68	Polhill	2387	
31	1865–69	Oxted	2067	
32	1866–71	Caerphilly	1767	Wales
33	1867–70	Bradway	1853	Sheffield
34	1871–75	Blea Moor	2404	Settle to Carlisle
35	1872–74	Combe Down	1672	Bath to Templecombe. 6 tunnels in 26 miles
36	1873–86	Severn	7012	Subaqueous. T.H. Walker, contractor
37	1876–79	Ffestiniog	3528	Wales
38	1877–78	Queensbury	2287	Bradford — Halifax. 14 tunnels in 19 miles
39	1877–79	Corby	1761	Kettering to Nottingham 10 tunnels in 48 miles (39–40)
40	1877–79	Glaston	1684	
41	1881–86	Mersey Railway	—	Subaqueous. Sir Francis Fox
42	1881–85	Drewton	1933	Hull — Barnsely. Closed 1959
43	1887–90	Rhondda Valley	3148	Wales
44	1887?–90	Sharnbrook	1701	Leicester — Bradford
45	1888–93	Totley	5697	Sheffield — Manchester (45–46)
46	1888–92	Cowburn	3385	
47	1891–96	Wenvoe	1707	Pontypridd
48	1893–97	Catesby	2743	South of Rugby
49	1895–00	Gildersome	2131	Leeds
50	1899–01	Disley	3535	Midlands
51	1900–03	Chipping Sodbury	4063	Near Bristol
52	1901?–05	Bolsover	2400	Chesterfield — Lincoln
53	1913–?	Llangyfelach	1785	Wales
54	1918–20	Ponsbourne	2454	North of London

ALPINE RAILWAY TUNNELS

	Dates	Name	Length
A1	1857–70	Frejus	12.2 km
A2	1872–82	St. Gotthard	14.9 km
A3	1879–84	Arlberg	10.5 km
A4	1899–06	Simplon	19.5 km
A5	1907–13	Lotschberg	14.6 km

CLASSIC U.S. RAILWAY TUNNELS

	Dates	Name	Length
U1	1855–76	Hoosac	7.4 km
U2	1879–05	Hudson River	—

The table shows construction dates of British railway tunnels, generally exceeding one mile (1609 m) in length, and also of the major Alpine railway tunnels and the American Hoosac and Hudson River tunnels. These are also displayed in the figure above, from which can be appreciated the relationship in time, and the periods of intense activity and relative inactivity.

Fig. 1.3 – A century of Railway tunnels.

Stephenson's Liverpool to Manchester line was quickly followed by his London to Birmingham line, and in the pursuit of good alignment and easy gradients five tunnels were necessary, the most difficult being those at Primrose Hill, Watford and Kilsby. Primrose Hill tunnel (1081 m) was driven in London Clay, which swelled and developed heavy pressures, cracking the brickwork lining in places. The contractor abandoned the work, which was completed by direct labour. Watford Tunnel (1660 m) was driven through the Chalk ridge which forms the north rim of the London Basin, but fissures in the chalk, filled with gravel, sand and clay and penetrated by water, caused much trouble and many lives were lost. Kilsby tunnel (2218 m) proved the most difficult, through a ridge of Liassic age south of Rugby. The ridge had already been pierced by a canal tunnel which had been routed to avoid quicksands shown by borings. The railway route was chosen in expectation of missing the quicksands, but failed to do so. The work was done from eighteen shafts, and ever-increasing demands on pumps forced the contractor to abandon his contract, the tunnel being successfully completed by direct labour in 1838.

I. K. Brunel was appointed Engineer to the newly formed Great Western Railway in 1833, at the age of 26, and surveyed and planned the route from London to Bristol and designed and supervised the construction. Its major tunnel is Box tunnel (2937 m), 8 km north east of Bath, where the line pierces a ridge of Jurassic rocks, including the Great Oolite limestone, the Fullers Earth, and the Inferior Oolite. Water entering through fissures flooded sections of the tunnel twice during construction, but it was completed successfully and opened, a year late, in 1841. Opponents had expressed alarm at its length, although not greatly in excess of Kilsby, and its gradient of 1 in 100, the whole proposal having been described as 'monstrous and extraordinary, most dangerous and impracticable'.

Another major early railway tunnel was Woodhead (4828 m) through the Pennine ridge at the summit of the Manchester line. The first single track tunnel was built in 1838–45, and the duplicate in 1847–52. These were high on the bleak moors and were built by tough gangs of railway navvies ('navigators' from the canal construction days), living in their own community in the roughest conditions and sometimes terrorising the neighbourhood. Both tunnels gave trouble from lack of ventilation and general maintenance and were replaced in 1954 by a new Woodhead tunnel taking both tracks and incorporating overhead electrification. The new tunnel took over four years to build, also suffering from falls of rock; nature's obstacles remained very real.

1.5.1 Mersey and Severn Railway Tunnels

Two subaqueous railway tunnels were built almost at the same time; first opened was the Mersey Railway tunnel, 1879–86, followed by the longer and more difficult Severn tunnel, which took 13 years, from 1873 to 1886. The Mersey tunnel, engineered by Sir Francis Fox, was through the Bunter Sandstone of the Triassic period, under the tidal River Mersey. It was driven from shafts 1618 m apart and 52 m deep on each side of the river, which allowed drainage headings to be driven on a rising gradient to mid-river. Colonel Beaumont's Tunnelling Machine, also used at this time in Channel Tunnel experiments, was used for much of the heading excavation, cutting a circular bore of 2.2 m diameter. It was also used for ventilation headings on the Liverpool shore. The main tunnel was 8 m x 7 m, lined with 0.7 m thick brickwork. The provision of forced ventilation to the stations, at the rate of 109 m^3/s, calculated to dilute engine fumes 500 times, is of interest. Much water was and still is pumped from this tunnel.

The Severn tunnel crosses the tidal river a little south of the modern suspension bridge. Its purpose was to replace the ferrying of passengers and shorten the journey from London or Bristol to Newport and Cardiff in South Wales. Parliamentary authority was obtained in 1872 and work started in 1873 with shaft sinking and a 2 m x 2 m pilot heading under the river. In July 1877, after 1500 m of pilot had been successfully completed, tenders for the main tunnel were invited, but rejected as too high; the Railway Company proceeded with the work themselves. Progress was satisfactory, almost to the meeting of the headings under the river; until in October 1879 the landward drive on the Welsh side struck the 'Great Spring' at the junction of the Millstone Grit and Carboniferous Limestone. The tunnel was flooded and work was held up for over a year while new arrangements were made. With Hawkshaw as Engineer and Walker as contractor work was resumed and the world's largest pump was installed. The junction was effected in 1881 and 3600 m of the bore had been enlarged and lined when in 1883 the Great Spring again broke in and flooded the tunnel, followed by breakdown of the largest pump and then by flooding of the working shafts and boiler plant by an abnormal tidal flood. The tidal range on the Severn is 15 m at maximum spring tides and its onset as the 'Severn Bore' is famous, but this tide was something quite exceptional. The tunnel was soon reclaimed, however, and in 1885 a train was passed through, but water pressure dangers appeared at the Great Spring and a new drainage shaft had to be sunk before the tunnel could be opened to traffic in 1886. The tunnel continues to carry main line traffic. Pumping from the Sudbrook shaft to which the flow of the 'Great Spring' was directed continues at a rate of 9–20 million gallons per day. In the years 1929–31 extensive and effective grouting was carried out where the tunnel runs through marl near the English side, and also round the adjacent shaft and the vent shaft on the Welsh side, because leakage through the brickwork had become excessive and cavities were being formed by erosion under and round the brick lining.

Another Severn tunnel has been driven in recent years to carry 400 kV power cables. It crosses at the site of the suspension bridge. The tunnel was lined with concrete segments. Part of the excavation was by drilling and blasting and part by machine, but some of the rock was too hard for economical machine cutting. Water was again troublesome but not catastrophic.

1.5.2 Alpine tunnels

The barrier interposed by the Alps between northern Europe and Italy was a challenge to the railway builders of the 19th century which could only be met by tunnelling. The major tunnels started in that century were the Frejus (Mont Cenis), St. Gotthard, Arlberg, Simplon, followed in 1907 by the Lötschberg. Across the Atlantic, in Massachussetts, the Hoosac tunnel, whose construction was spread over twenty-one years, was the proving ground for valuable developments in compressed air rock drills.

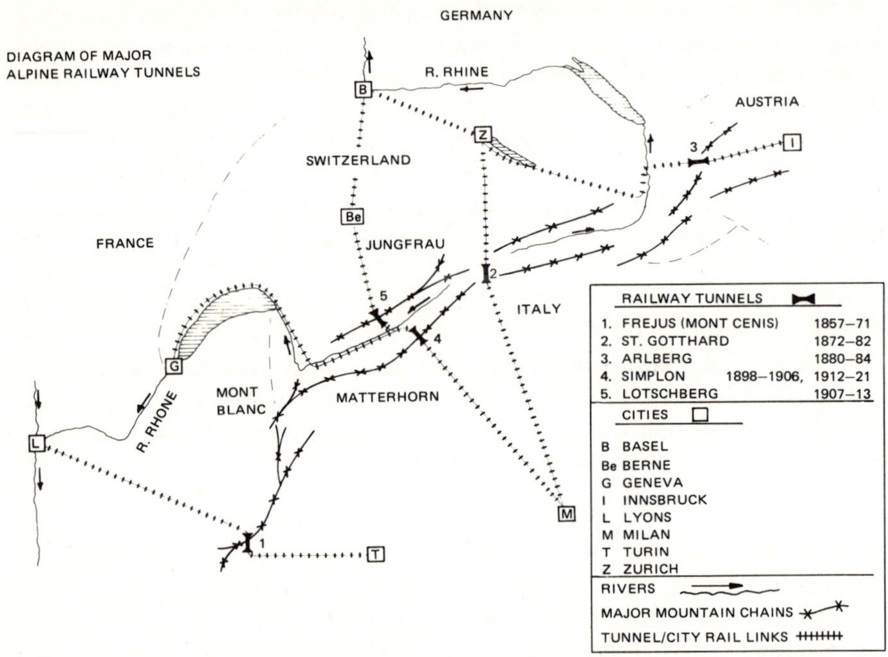

Fig. 1.4 – Diagram of Alpine railway tunnels

For each tunnel a choice had to be made between a short tunnel at high level approached by severe gradients and sharp curves and vulnerable to winter snow, and a much longer tunnel at a lower level. The operating advantages of the lower and longer tunnel were generally decisive, and the long tunnels deep below the mountain surface were chosen.

These tunnels involved new developments in drilling equipment and explosives and the builders faced almost insoluble problems of ventilation of the works, enhanced by high temperatures of the rock, and hot springs up to 60°C. Very severe pressures from rock deteriorating after exposure demanded quite exceptional methods of construction. All the tunnels were eventually completed.

The col de Frejus, or Mont Cenis, tunnel to carry the railway from France to Turin was the first great Alpine tunnel, 1340 m above sea level. Its bored length appears to have been 12.23 km but it is now recorded as 13.7 km. Mountain streams were utilised to compress air to operate drills and provide ventilation. Housing with schools, hospitals, etc., was provided to accommodate a work force of 2000 men on each side. Work started in 1857 and the tunnel was opened fourteen years later in 1871, shortly after the death of its builder, Germain Sommeiler.

During the same period the Hoosac tunnel, connecting Boston to the Hudson Valley, was being built in America, but its construction was so protracted, from 1855 to 1876, that it became known as 'the Great Bore'. Contractor after contractor took it on and abandoned it. After various periods of direct labour under a State Commission a fifth contractor effected completion. The tunnel was 7.44 km long, mainly through schist and gneiss. Even in 1865 progress was only at the rate of 0.32 m per day, improved the following year to 0.46 m when machine drills were introduced, but by 1873 the rate had improved to about 1.65 m.

In the Alps, completion of the Frejus tunnel was followed by construction of the St. Gotthard tunnel over the years 1872–82. It provided the railway link from the Rhineland through Zurich to Milan. It is 14.9 km long and at a height of 1164 m. Louis Favre accepted an eight year contract, but died in 1879 when struggling to overcome appalling conditions of water, high temperatures and bad ventilation. The worst rock was decomposed felspar and gypsum which became plastic in moist air and developed pressures which were eventually held by a granite lining comprising arch, walls and invert with respective thicknesses of 0.6 m, 2.5 m, 1.4m. The bad conditions and accidents resulted in many deaths —about 25 a year—and much sickness and injury. Nearly 4000 men were employed at the peak of the work.

The Arlberg tunnel, 10.5 km long, linking the Austrian railways to Switzerland, was built during the period 1880–84 without any major setbacks. It was marked by competition between hydraulic and compressed air drills without showing any very clear balance of advantage in either system.

The Simplon tunnel between Berne and Milan is 19.8 km long, the longest of its kind, and was started in 1898 after geological forecasts of good rock. Great emphasis was laid on safety and welfare precautions, and as a result, despite some very difficult conditions, particularly from springs of water as hot as 60°C, the tunnel was completed and opened in 1906 with a far better record of health and safety than St. Gotthard: nevertheless, 39 lives were lost. The design was for twin single-track tunnels, both headings being driven concurrently from both ends, which produced great advantages in ventilation and also in drainage and escape routes. In addition to the high rock temperatures and hot water, major trouble was experienced with a zone of micaceous limestone and gypsum which softened when exposed and developed very high pressures in the 3.5 m x 3.2 m heading. Over a 42 m length no timber was strong enough to withstand the squeeze, and what was virtually a box heading of 74 steel frames built of joists 356 mm deep was constructed. Enlargement to full size had to be executed by excavating under and round the frames in short lengths, filling in with heavy masonry as quickly as possible. As so often in those enterprises the man carrying the greatest responsibility, Alfred Brandt, died in the course of the work in 1898. Enlargement and completion of the duplicate heading to provide the second track was not done until the years 1912–21.

The Lötschberg tunnel which is 14.6 km in length at a height of 1245 m is complementary on the north side of the Rhone valley to the Simplon tunnel on the south side, and its construction followed accordingly in the years 1907–13. The whole rail link was 58 km long including the summit tunnel and thirty-three other tunnels. The special tunnelling hazard of this work was that the heading unexpectedly ran out of rock and into alluvial filling more than 186 m deep below the Gastern valley. The tunnel filled with waterborne sand and gravel, burying 25 men. It had to be blocked off and the tunnel driven anew on a series of curves so as to cross under the buried valley higher up its course. The tunnellers also suffered another catastrophe when the air blast from an avalanche swept away the works canteen and its occupants.

The next generation of great Alpine tunnels was for highways, not for railways, but a deep level St. Gotthard railway tunnel 49 km long has been the subject of studies.

1.5.3 Rock tunnelling development

The development of rock tunnelling has been largely dependent on three things: rock drilling machines, drill bits, and explosives. Gunpowder gave way to the much more powerful nitroglycerine, quickly followed by dynamite, introduced by Nobel in 1867, and by gelignite, which are much safer to store and handle and more controllable in use. There have, of course, been improvements and refinements, but no change of scale or new principle.

For rock drills compressed air became the accepted motive power, although hydraulic power was preferred for a time in Europe and is again coming into favour. Much ingenuity went into the design of compressors, from the falling water devices of the Frejus tunnel to steam driven reciprocating engines and modern reciprocating or rotary machines driven by electricity or diesel power. The great problem with drills themselves lies in their self-destructive characteristics; they hammer themselves to pieces. The wear and tear on the points was first met by making them as detachable bits of the toughest possible steel, but then in turn the shanks had to be improved to endure the increased fatigue stresses, while the whole mechanism, subjected to hammer blows and vibrations, and to dust, grit and water, had to be kept within strict limits of weight so that it might be handled by men for horizontal and overhead drilling. The Alpine tunnels and, perhaps most of all, the Hoosac tunnel were the cradle of these developments, taking advantage as time went on of every improvement in steel technology. One of the great advances, as late as the 1950's, was the development of the tungsten carbide bit. It was in Sweden, where the abundant rock is mostly ancient and very hard, that ways and means of incorporating this material in long-lasting bits were found and brought into practical use.

1.6 NATIONAL SYSTEMS

During the 19th century various systems of tunnelling, principally for canals and railways, were given national labels from the countries in which they were developed. Every type of ground and tunnel section has its own problems so that each system was modified to suit the conditions and to overcome defects that appeared. Much more space would be required to describe each system and its development fully, but the details are hardly relevant now, when timber is so largely replaced by steel and concrete and when so many new techniques of excavation and ground control are available. In any case, there was always modification to adapt any system to the ground and the circumstances of the particular tunnel. The principles, however, remain of interest and are relevant when considering methods of construction. Apart from the structural adequacy of any system, old or new, its economy of materials can be vital, as can speed of construction which may depend on clear access to the working areas. Additional headings were frequently incorporated in order to assist transport of materials and so open out additional working points.

The English system comprised a top (crown bar) heading 3-6 m in length driven ahead of the main face. In this, two principal crown bars were set longitudinally, resting on the completed lining at the back and on props ahead. This heading was then enlarged laterally down to springing, the top being supported by additional crown bars ultimately propped from a substantial sill. The lower half of the tunnel was then excavated, the main sill being undercut progressively and propped down to the tunnel floor. This left a clear chamber within which a complete length of lining was built, roof timbers being removed progressively if possible but built in when necessary.

The advantage of this system is the clear building space, but it is dependent on the stand-up time and the ground loading being within the capacity of the crown bars.

The Austrian (cross bar) system was essentially based on a short top heading enlarged laterally to the full size of the arch, the roof being supported on timber segmental arch ribs at relatively short centres. The ribs usually had to be provided with radial props from sills which had to be changed as excavation progressed.

After excavation down to springing level the lower half was excavated in much the same manner as in the English system.

Fig. 1.5 – Two drawings reproduced from Simm's Practical Tunnelling (1896). The English System of Tunnelling, as used in the Bletchingley Tunnel (1840) in clay. (a) Longitudinal section showing top crown bar heading in advance of brickwork, and bottom heading. (b) Face timbering, ahead of brickwork.

The merit of the system lay in the ability to build the ribs at as close centres as necessary to resist high ground pressures, but successive repropping could result in relaxation of the ground causing increased loading. The lengths available for construction of the lining were liable to be rather short and obstructed by timber.

The Belgian (underpinning) system initially resembled the English system in the excavation of the arch section down to springing, but at that stage, before excavation of the lower half, the building of the arch was completed. The arch had then to be undercut and underpinned, which could be done in short lengths in small pits if necessary.

The principal advantage lay in the early completion of the roof structure, provided that the underpinning operations did not result in undue relaxation of the arch and consequent heavy loading.

The German (centre core) system was also claimed as a French system, having been used about 1803 in the Tronquoy canal tunnel. In this system the bases of the tunnel walls were first built in separate small bottom headings, followed by a succession of upper headings until the crown was reached and the arch completed.

The core of ground was left in as a 'dumpling' until completion of the lining and gave support to vertical and radial props. The advantages of working in small headings in bad ground are apparent, but could be nullified by squeezing and distortion of the core and also by difficulties of securing sound workmanship in the building of the lining in small units in confined spaces.

The Italian (invert arch) system was developed about 1870 for the very difficult Cristina railway tunnel through the Appenines. Very severe clay pressures were encountered, crushing timbering and masonry. In the system finally successful the invert was first built in solid masonry from a bottom heading, so as to provide a secure base for props and struts. This masonry was extended laterally and upwards in small units, including a thick temporary masonry arch over the working heading, so that the crown excavation could finally be timbered and the arch lining built with the temporary arch as centering.

The American system was not a standard national pattern in the same sense, but there was a common pattern resembling the Austrian system. A top heading was driven from which excavation was enlarged laterally to accommodate a polygonal timber arch rib down to springing level. Support at springing might then be by vertical side posts.

Alternatively, 'wall plate' headings might be driven first, the legs and cap of the arch rib being built up from the longitudinal sill comprising the wall plate. Having secured the top, excavation and timbering progressed downwards for the 'length' under construction.

It is clear that every one of these systems had its merits and disadvantages and the safe use of each depended on familiarity with the system in all its details and with the ground being tunnelled. Adaptation and modification in any particular tunnel was necessarily continuous.

1.7 SHIELDS AND COMPRESSED AIR

1.7.1 Tower subway

In the field of subaqueous tunnelling 1869 was an important year because it marked the successful construction of the Tower subway in London using a shield and cast iron lining. The tunnel was used for a cable-hauled car and then for pedestrians until 1894 and now carries water mains. The scheme was Barlow's, but his pupil, Greathead, became responsible for it and in particular the design and development of the shield, which is the ancestor of almost all subsequent shields. The subway was 402 m long lined with cast iron segments within a 2.18 m diameter circular excavation. The tunnel was driven through London Clay 19 m below high water, with 7 m cover below the river bed.

The shield comprised a cylinder formed of 13 mm thick wrought iron skin plates 1.45 m long, stiffened with cast iron rings and having a stiffening diaphragm incorporating a rectangular doorway, which could be closed off by dropping boards across it. Behind the diaphragm were six screw jacks, thrusting the shield forward from the completed cast iron lining, which was erected ring by ring within the tail of the shield. This permanent lining comprised three segments of a circle plus a short key segment, all flanged and bolted together. Because of the thickness of the shield's tail skin and working clearances the cast iron lining did not entirely fill the excavated void, and the injection of cement grout was adopted to minimise ground movement, but a hand syringe was not very effective. Progress of nearly 3 m a day demonstrated the efficiency of the system, albeit in good ground. The importance of this small tunnelling operation lies in its pioneering and developing and proving in practice of so many devices essential to shield tunnelling in soft ground, namely: rigid and simple shield, protective diaphragm, tail skin protecting building of lining, cast iron lining providing, in combination with the principle of grouting, immediate ground support.

In 1940 the tunnel was damaged by an adjacent bomb crater in the foreshore near the low tide line. A length of about 50 m was repaired by enlarging to 3 m diameter with cast iron rings using tunnelling methods. It is of interest that the effect of the forces was to fracture the cast iron rings and crush the tunnel, at one point to a diameter of about 1.3 m, but when exposed in the re-excavation the clay outside had flowed plastically to the new diameter, and appeared unchanged except for a skin of about 50–75 mm.

Barlow in 1864 had patented most of these ideas, but Greathead receives the credit for putting them into practical shape. Working possibly from Barlow's 1864 patent, Beach in New York built a shield for a pneumatic subway under Broadway, brick lined 2.4 m diameter. The body of his shield was of timber with iron cutting edge and tail, and he used hydraulic rams. His tunnel was successfully driven through sandy soil in 1869-70, and a shield of his design was again used for a 2.4 m sewer in Cincinnati in 1871-2.

Although the Mersey and Severn tunnels were major subaqueous railway tunnels in rock, the techniques were not fundamentally different from the other railway and canal tunnels. Water, of course, was an even more dangerous enemy there than on land because any breach to the river opened up an unlimited source of supply, but the distinctive problems of soft ground subaqueous tunnelling, exemplified in Brunel's Thames tunnel, were not faced again on that scale until the attempt to tunnel under the Hudson River between Jersey City and New York started in 1879.

1.7.2 Hudson River Tunnel

New York had from an early day railway connections from the north leading eventually to the terminus at Grand Central Station but no direct links from the west and south where the Hudson river was a formidable barrier dependent on ferry transport for the crossing until 1908. That was the date of completion of the Hudson River Tunnel for which a preliminary shaft was sunk in 1874. At the chosen site the tunnel is almost entirely in silt but the use of compressed air† appeared to offer a means of excluding water and holding the ground, initially for an iron plate lining, followed by a brick lining. Work started in earnest in 1879. After installation of an air lock and breaking out from the base of the shaft it was thought that 'the great problem that the silt would hold the air and the air would hold the silt was practically solved'. Unfortunately, the balance was too unstable and either drying out of the silt or intrusion of water caused falls of ground. In 1880 a serious accident, in which 20 lives were lost, occurred when trying to form the junction at the shaft. The tunnel was inundated, but was recovered and work proceeded until 1882, when funds ran out and the tunnel was sealed off. This was the first large scale use of compressed air† for tunnelling, though a small tunnel in dock works in Antwerp had been driven under compressed air a few months earlier. The use of compressed air for shaft or caisson sinking for bridge piers was well known, but no earlier use in tunnels is recorded. A patent for working in compressed air had been taken out in 1830 by Lord Cochrane. It included shaft sinking and also tunnelling with the use of an air lock for entry. Brunel's Thames tunnel would have benefited from employment of compressed air; it is said that the idea was brought to his notice in 1828 by a Dr. Colladon. It may be worth noting here that there is

† Footnote:
Previous references to compressed air have been to the use of high pressure air (at pressures of about 7 atmospheres) as the driving power for machines such as rock drills. The current reference is to the compression of the whole atmosphere within the tunnel in which the miners work and breathe at a pressure of the order of one or two atmospheres, calculated to balance in some degree the water pressure acting at the face of the tunnel. The sense in which the term is used is usually clear from the context.

Fig. 1.6 – Hudson River Tunnel. Excavation in silt in compressed air, with use of Greathead shield and cast iron lining (1890).

an inherent instability in this use of compressed air to exclude free water at the vertical face of a tunnel because the water pressure is so much greater at the bottom of the face than at the top while the air pressure is the same all over, and so must either be inadequate at the bottom or excessive at the top.

The combination of shield and compressed air together with cast iron lining was first proved on the City and South London Railway tunnelling as described later. With advice from Baker and Greathead the techniques were adopted successfully in the Hudson tunnel when work was resumed there in 1889, under a contract with Pearsons. The silt was so soft in places as to be extruded into the tunnel through apertures in the shield when this was jacked forward. The tunnel was not, however, completed under this contract, which was suspended for lack of funds in 1891; another contract was let in 1902 and tunnelling by the same method was completed in 1905. The length driven by shield was 1070 m and the external diameter of the cast iron lining 5.9 m.

An important contribution to compressed air working was the introduction at the Hudson tunnel in 1889 by Moir, Pearson's agent, of the 'medical lock' for treatment of caisson disease, whereby complete relief was given in ordinary cases, and fatalities were reduced from the appalling figure of 25% of the men employed in a year to two deaths in 15 months out of 120 men.

1.7.3 London tube lines

The construction of the City and South London Railway 1886-90 (now part of the Northern Line in the London underground system) provided the first large-scale use of the Greathead shield system, and also the first use of shield, compressed air and segmental cast-iron lining in combination. This was also virtually the first electric railway. The tunnel was driven as twin cast-iron lined tubes, starting on the north bank of the Thames near London Bridge and crossing under the river and running south for 5.07 km to Stockwell. The first third of the tunnel was only 3.100 m in diameter and the remainder only 3.200 m. Most of the drive was in London Clay but water-bearing gravel was encountered for a length of 180 m near Stockwell. It was for this work that compressed air was introduced. The unsatisfactory grouting technique of the Tower Subway was replaced by Greathead's device of the grouting pan, using high pressure compressed air to inject the mixture through holes provided in the cast-iron segments.

The original Metropolitan and District lines had been constructed by cut and cover, but disturbances to the surface and to property were unacceptable. The success of this scheme therefore pointed the way first of all to Glasgow to build its 6½ mile loop subway (1897), passing under the river Clyde in difficult conditions, and then to the expansion of London's underground system. The same tunnelling principles were followed in construction of the Bank–Waterloo railway (1898) where clay-pocketing was introduced by Dalrymple Hay; the Central London railway (1900); the Bakerloo line (1906); the Piccadilly line (1906); and the modern Victoria line (1968) and Fleet line (1979). All the lines have been extended, partly in tunnel but also, in the outer suburbs, by running on surface lines.

A notable development of the original City and South London railway was its enlargement to standard 3.56 m diameter and lengthening of stations in the early 1920's, some of the work being carried out with trains running through annular tunnelling shields, and some in compressed air. The Eastern Extension of the Central line (1935–39) was marked by some very difficult compressed air working in bad ground and the use of chemical ground treatment, and also by the introduction of pre-cast concrete segments for the tunnel lining where the ground was good enough and cast-iron was not deemed essential. The construction of the Victoria line introduced digger shields, and flexible joined segments. The projected route of the Fleet line ran through difficult water-bearing gravels in the south-east and, in anticipation, trial tunnelling to develop the 'Bentonite shield' was carried out successfully. Another aspect of such a railway system is the very complex tunnelling for interchange stations, such as

Euston, where the main line terminus at the surface is served by stations on six deep tube tracks and the two shallower tracks of the Inner Circle. These, with connecting passages and escalator tunnels on a slope of 30°, make a very intricate three-dimensional pattern.

1.7.4 General adoption

Following the success of the City and South London railway the combination of shield and compressed air was adopted before the end of the century for a number of tunnels. In America—apart from the second instalment of the Hudson river tunnel—the earliest and most notable was the 6.4 m diameter cast-iron lined Sarnia tunnel for the Canadian Grand Trunk Railway, passing from Ontario to Michigan under the St. Clair river, and completed in 1890. Air pressures ranged from 0.7 to 1.9 bar* and the shield was provided with an erector arm to handle the C.I. segments. In 1892–94 a gas main tunnel under the East river, New York, was started in good rock but ran into trouble with bands of decomposing felspar which washed out when acted on by water. A shield, compressed air up to 3.6 bar, and a cast-iron segmental lining were necessary to complete the troublesome sections. The use of timber for initial lining of some other small pipe tunnels is of interest.

In Great Britain, apart from tube railways and the Blackwall tunnel, described more fully below, this period saw the Vyrnwy aqueduct under the Mersey rescued by use of a shield and compressed air, while in Scotland three 5.2 m diameter tunnels 210 m long were driven under the river Clyde for the Glasgow Harbour authority in 1890-93, providing an early example of a subaqueous highway tunnel, although it may be noted that access to the tunnels for vehicles and pedestrians was by lift only. Edinburgh also in 1893–4 constructed twin 5.3 m highway tunnels in compressed air under the Mound, compressed air being used to give support rather than to expel water.

In France the method was used at the end of the century for a number of sewer tunnels of diameters of 2 m to 3 m under the rivers Seine and Oise in the vicinity of Paris.

In Germany a 4 m diameter tramway tunnel 373 m long was driven under the river Spree in 1896–99.

In Australia, the City of Melbourne built some 11 km of tunnelled sewers with diameters ranging from 1.5 m to 3.5 m; seven lengths totalling 1620 m were constructed in compressed air, and cast iron lining and concrete was used for six of these lengths and concrete blocks for one. For the free air tunnels, cast-iron, concrete, bluestone, and even timber segments were used.

1.7.5 Blackwall Tunnel

The Blackwall Tunnel (1892–97) under the river Thames may perhaps be considered as the achievement of Brunel's vision of a highway under the

* 1 bar in the metric (SI) system is $10^5 N/m^2$ and equals 14.50 psi, nearly 1 atmosphere.

river, built by a public authority and free for public use. It was in many ways a bold pioneering work. The tunnel was deliberately located at a level where cover at the deepest part of the river was no more than 1.6 m of gravel. Reliance was placed on the combination of shield and compressed air, which at the date of the decision had not been proved elsewhere, and on the protection of a clay blanket dumped in the river bed. In fixing the level the limiting depth was decided as 24 m below high water, to limit maximum air pressure to 35 lbs/in^2. Near the critical section the tunnel was twice inundated to a depth of 2½ m, fortunately without casualties.

The advances since Brunel's day included the rigid structural steel circular Greathead shield with stiffening girders and two safety diaphragms, cast-iron segmental lining bolted together within the tail of the shield and pressure grouted, and the use of compressed air to keep the water under control. Other features were the use of hydraulic power for the main rams, which were 28 in number, subsequently increased to 34, capable of a total forward thrust of 5000 tonnes, and the provision of a segment erector arm working in the tail of the shield to handle the heavy cast-iron segments.

Two shafts were sunk on each side of the river and shield tunnelling started from No. 4 on the south bank. The shield, weighing 220 T, was assembled in a shallow pit near the shaft and then floated along a channel to the shaft and lowered into place by pumping out the water with which the system had been filled. From shaft No. 4 the shield was driven through shaft No. 3, then under the river to No. 2 and on to No. 1. At each end a length of tunnel was built as cut-and-cover and beyond that a ramp in open cut up to surface level.

The ground comprised river gravels overlying London Clay on both banks, with the variable Woolwich and Reading beds below the clay. Under the river the clay thinned out altogether so that the tunnellers had a face of gravel entirely open to the river, except for the protection afforded by depositing a 3 m thick 'clay blanket' over the worst section. The shield face when necessary was kept closed up by a system of horizontal steel planks held by jacking screws, much as in Brunel's shield, and released one by one as each move forward was made. For the most critical part only a small 'window' in each plank was opened. In good ground, where sand and shale underlay the clay, progress up to 4 m in 24 hours was made, but in mid-river no more than 1½ m in a week was possible. Other points of interest in this tunnel were the use of electric lighting in the tunnel, the provision of hydraulic power by pipe line from surface pumps, the repair of damage to the cutting edge of the shield in the invert caused by striking rock in the early stages of the drive. Very great precautions were taken in subjecting men to the compressed air working and in the event no lives were lost while only three cases of permanent illness were recorded.

The internal diameter of the tunnel was only 7.5 m allowing a 4.9 m roadway with restricted headroom. The alignment was straight between shafts but with sharp bends at the shafts. This was acceptable in the days of horse-drawn dock traffic, but became intolerable for modern vehicles. A new tunnel was built adjacent in 1960–68 in which again shield, compressed air and cast-iron lining were used, but with the additional device of extensive ground treatment by injection of complex grouts from top and bottom pilot tunnels, to make the ground coherent and watertight. After completion of this new tunnel some improvements in the old tunnel were made, in particular to improve clearances at the bends and each of the pair of tunnels now carries one-way traffic only.

Other highway tunnels under the tidal Thames, also shield driven in compressed air, are at Rotherhithe (1904–08), very close to Brunel's tunnel, and at Dartford, 22 km down river from Blackwall tunnel, where a pilot tunnel was driven before 1939 but work on the main tunnel was not resumed until 1956, the tunnel being opened in 1963. The design of shield closely resembled that for Blackwall, including segment erectors, and hydraulic pipe lines, but the bore under the river was mostly in chalk and the more difficult strata were near the river banks where the top of the face was in gravel on one side and in soft silty clay on the other. Duplication of this tunnel was completed in 1980.

1.8 20TH CENTURY TECHNIQUES

By the beginning of the 20th century the basic techniques for bored tunnels had been devised and proved the hard way, so that most ground could be tunnelled if the promoters were prepared to pay the price. The new century therefore saw a steady expansion of tunnelling throughout the world to serve many and varied purposes. A system of tunnelling widely adopted in recent years has been the submerged tube for river crossings where the tunnel is prefabricated in long lengths in a dry dock, floated out into the river and sunk into a pre-dredged trench where the lengths are connected up. Much work by this system has been done in Holland, and also in the U.S.A. Excavating machines working within a tunnel have been developed and have been combined with shields in very powerful tunnelling machines. Ground treatment by injection of a wide range of chemical grouts has proved a valuable technique in difficult ground, both for reducing its permeability to influx of water and improving its cohesive strength. Freezing whether by brine circulation or by use of liquid nitrogen can also provide temporary cohesion and impermeability. In rock, preservation of the integrity of a rock mass by rock bolting is a method of growing importance. Sprayed concrete linings immediately after exposure, can prevent progressive loosening of the rock.

For small tunnels and large pipes, pipe jacking has become a useful method. It may be considered as shield tunnelling in which the tunnel lining itself constitutes the shield. All of these techniques are further described in later chapters.

1.8.1 Railways

Tunnels for main line railway systems have continued to be built throughout the world, with many notable achievements on the Alpine scale. New Zealand built the Otisa tunnel (1923), 9 km long through crumbling shale, and the Rimutaka tunnel (1955) of similar length, and others. Italy's Bologna–Florence Diretissima (1934) included thirty tunnels, of which the Great Appenine tunnel, 18½ km long through soft clay and running sand with methane gas, was the most formidable. In the U.S.A. and Canada old summit tunnels have been and are being replaced by longer tunnels at a lower level and with easier gradients and curves to handle heavy goods traffic more efficiently. Japan has engaged in major railway development in recent years to improve communications in their mountainous country, and has built many notable tunnels, both for the general railway system and for the new high speed lines.

The Channel tunnel project, and the Seikan tunnel, are described in Chapter 10.

1.8.2 Metro systems

Most cities having a population of the order of a million or more, and some smaller, are finding the problems of surface transport so difficult that they are using the third dimension, depth, to tunnel underground for railway construction. The various names–Underground, Metro, Subway, U-bahn, Rapid Transit–can perhaps best be generalised under the term Metro, whether this be thought of as deriving from Paris or from London's first system, the Metropolitan Railway.

The varieties of system are many but almost all have in common tunnelling for at least some sections of their city routes. Some were planned as self-contained lines usually radial but sometimes as a loop like the Glasgow Subway. Others develop from city tramway or street car routes, which are run below the surface to clear the streets of congestion. Others again arise from the extension of main line or suburban railways into the heart of a city or from interconnection of existing termini.

Insofar as generalisation is possible the independent lines tend to be deep level bored tunnels, while the extensions and tramways tend to be shallow, constructed largely by cut-and-cover methods.

Before the end of the 19th century, in addition to London, lines were constructed in Glasgow (1896), Budapest (1896), Boston (1897) and Vienna (1898). The Mersey Railway into Liverpool (1886), now extended into a city loop, might also be put in this category. After London, the next major system was the Paris Metro, whose first line was opened in 1900, followed by Berlin in 1902. New York had then an elevated railway but also went underground progressively from 1904, with Jersey City following in 1908 and also Philadelphia.

New York's problems of crossing the Hudson river, and more so the East river, were solved for main line and metro railways by driving numerous tunnels. Six railway tunnels in all were built under the Hudson river in the years 1902–08; eight were built under the East river from 1903–09 and eight more from 1914–20. All are single track tunnels lined with cast-iron 4.6 to 5.0 m diameter, shield driven in compressed air through varying strata of silt, sand and rock.

Additional tunnels were driven from time to time and more recently a tunnel to carry four tracks under the East river has been constructed, the river section being a submerged tube approximately 12 m x 11 m in section with rail tracks at two levels, but connecting into bored tunnels.

The first quarter of the 20th century saw underground city lines opened in Hamburg (1912), Buenos Aires (1913), and Madrid (1919). In 1935 the first section of the Moscow Metro was opened, since when it has been steadily expanding. It has been followed in the U.S.S.R. by Leningrad, Kiev, Tbilisi and Baku.

Major world wide expansion began about 1950 and in the following twenty years eighteen cities began operating new systems, large or small, as well as many additions and extensions to existing systems, and planning and construction in progress on many others.

Toronto Subway (1954) was the first in Canada, followed by Montreal Metro (1965). In Japan, Tokyo opened its first line in 1927, then Osaka (1933), Nagoya (1957), and Kobe (1968). In the U.S.A. the earlier Metros in Boston, New York and New Jersey, Philadelphia and Chicago (1943) grew out of overhead railways, but the new Bay Area Rapid Transit (BART), opened in stages from 1972, is a landmark of careful planning for a very high standard of comfort and speed in an attempt to attract traffic away from the private car. Its 10 km underwater tunnel across the bay was constructed in prefabricated lengths sunk into a dredged trench by submerged tube methods. A noteworthy feature is the design of tunnel end joints to meet the contingency of earthquake movements.

Rotterdam's line across the river Maas (1968) was also built as a prefabricated submerged tube in concrete, and the method was even adopted there for the city streets in which canals were first dug so that the units might be floated and sunk.

Almost every method of tunnelling has been used in these Metros: cut-and-cover, shield tunnelling with or without compressed air, digger shields, rock tunnelling with explosives or machines, and as described, submerged tubes. They vary in size of tunnel from the Glasgow Subway which in cross-section is a circle of 3.35 m internal diameter, up to a circle of 8.70 m diameter for the Paris express line, while the rectangular section adopted for cut-and-cover construction is usually for double track with a span of about 7½ m. The largest span is 34 m for the Brussels main line link. Depth below surface may exceptionally be as much as 60 m.

Although a city may initially build a shallow subsurface line, later lines are forced deeper at crossings and cut-and-cover methods have to be supplemented by deep tunnelling techniques in increasing degree. Access to deep stations by escalators or lifts provides a new range of tunnelling problems.

Melbourne is currently completing city loops with main line capacity linking up the various suburban lines and providing through running.

So many new Metros are now planned and under construction that further enumeration or description is impracticable.

1.8.3 Highway tunnels

Tunnelling for highways did not become a common practice until much later than for railways. Because of gradient and curve limitations railways needed tunnels early and often, but much steeper gradients and very sharp bends were at that time acceptable on highways when they did not carry heavy long-distance traffic and were not the subject of large capital investment. Short road tunnels through rocky spurs and ridges are numerous in mountain areas and on coastal roads, but the longer highway tunnels are now either subaqueous or carry the modern highway with its increasingly heavy vehicles and loads through ridges and mountains for exactly the same reasons as formerly applied principally to railways–gradients, curves and large capital investment. There is also now a third situation demanding highway tunnels, namely, city congestion, eased by underpasses short or long.

The Thames tunnel (1843) was, as described, the frustrated forerunner of highway tunnels and although hill tunnels were built in the 19th century it is significant that the major early road tunnels were subaqueous tunnels in port and dock areas. Blackwall (1897; 1361 m) fulfilled Brunel's dream and was followed by its near neighbour Rotherhithe (1908; 1482 m), both shield driven in compressed air and lined with cast-iron rings. In Hamburg a tunnel under the river Elbe (449 m) was opened in 1911, and in Pittsburg the Liberty tunnel (1800 m) in 1924, but the first of the modern generation for motor vehicles was really the Holland tunnel (2600 m) in New York, opened in 1927 to provide the first road access other than by ferry, to the city from the west.

The Holland tunnel was planned and designed for modern motor traffic which is characterised by emission of poisonous carbon monoxide. The problem of ventilation became for the first time an issue of vital importance and was very fully studied and so successfully provided for that the system has worked satisfactorily without major change for nearly fifty years. The tunnel, under the Hudson river, was shield driven in compressed air and lined with cast-iron (8.33 m int. dia.) through the same ground that had proved so difficult in the 19th century. Six hooded shields each 9.19 m diameter were employed to drive the twin tubes: the tunnel face had to be kept close timbered, and in some sections clay pocketing round the hood was found necessary. Over some lengths the shield face, which was divided into 19 compartments, was kept closed except for small areas in the bottom and the silt was extruded through these slots into the invert and thence removed. For a short section on the New York side rock in the invert had first to be excavated in a pilot heading and replaced by a concrete cradle.

The success of this tunnel led to the construction of the Lincoln tunnel, also under the Hudson river between New Jersey and Manhattan Island. Three two-lane tunnels of lengths 2280, 2510 and 2260 m were built, also shield driven with cast-iron lining, and were opened in 1937, 1945 and 1957. Under the East river the Queen's Midtown tunnel was opened in 1940.

The Mersey Queensway tunnel, opened in 1934, was another major highway tunnel. It was driven essentially by rock tunnelling methods–drilling and blasting–under the river Mersey to connect Liverpool with Birkenhead. It was lined with cast-iron under the river, with an internal diameter of no less than 13.4 m in order to provide for four lanes of traffic between its main entrances in the city centres. The lane width is only 2.7 m. Two two-lane branch tunnels were driven direct to the dock areas, but because of the problems of crossing traffic they are not now much used. The main tunnel is 3237 m long and the branches 504 m and 478 m. Ventilation was a major problem, more costly than originally foreseen, but satisfactorily solved. A second tunnel, Mersey Kingsway (2487 m) opened in 1971 and duplicated in 1973, was built about a mile downstream, through the same Triassic sandstone, by use of a full face excavating machine, again with a circular segmental lining but of concrete faced with steel. In this case a pair of two-lane tunnels 9.67 m diameter each providing two 3.7 m lanes was preferred to the single large diameter tunnel. The Dartford tunnel (1963; 1429 m) and second Blackwall tunnel (1968); 1174 m) have already been mentioned. Similar subaqueous tunnels, driven in compressed air and cast-iron lined were completed under the Clyde (1963; 762 m) and under the Tyne (1967; 1684 m).

A new pattern in highway tunnel construction was set in Oakland, California by the Posey tunnel (1081 m) opened in 1928, which was probably the first highway tunnel constructed by the submerged unit method, although a railway had already in 1910 been carried under the Detroit river by similar means. Circular steel shells in 62 m lengths were encased in concrete and lowered into a dredged trench in the bed of the river, then joined up to form the tunnel, the trench being backfilled over the tunnel. This tunnel was duplicated in 1962 by the Webster Street tunnel.

The submerged tunnel has perhaps now become the commonest method for subaqueous tunnels in the U.S.A., some of the longer examples being: Detroit–Windsor (1930; 1191 m), Virginia–Elizabeth river (1952; 1020 m, 1962; 1280 m), Baltimore (1957; 2331 m), Virginia, Hampton Roads (1957; 2280 m), Chesapeake Bay (1963; 1749 m and 1661 m), together with many others under 1 km.

In Europe, Holland has been the leader in this method, but with substantial differences in construction technique, using precast concrete units of rectangular

Fig. 1.7(a) – First Mersey highway tunnel. Mersey Queensway Tunnel (1930) in sandstone. Excavation and erection of 13.4 m dia. cast iron lining for upper half, using pilot heading.

Fig. 1.7(b) – First Mersey highway tunnel. Mersey Queensway Tunnel (1930) in sandstone. Completion of excavation of lower half, with cast iron lining underpinning upper half

section for such tunnels as: Rotterdam, Maas (1942; 1075 m), Amsterdam, Ij (1967; 1037 m), Amsterdam, Coen (1967; 587 m). In Belgium also the method was used for the J. F. Kennedy tunnel (1968; 690 m), and France, Sweden and Denmark have completed similar tunnels in recent years.

Japan has built a number of long highway tunnels, mostly by rock tunnelling methods, such as the Kanmon tunnel (1958; 3461 m) connecting the islands of Honshu and Kyushu, and the twin Teenosan tunnels (1963; 1390 m and 1453 m) for the Nagoya–Kobe expressway. Under construction is the Enasan tunnel (8450 m) on the Tokyo–Nagoya highway which was briefly second only to the Mont Blanc tunnel in length. Submerged tube methods have also been used in Japan but principally for short Metro system tunnels.

The Alpine railway tunnels have been succeeded by a generation of highway tunnels of which Mont Blanc (1965; 12650 m), Grand St. Bernard (1964; 5800 m), San Bernadino (1967; 6596 m) and St. Gotthard (1980; 16320 m) are the longest. The same problems of water, cold or hot, and swelling rock producing extremely heavy pressures, and rock bursts, as well as avalanches, have had to be faced. Although so much more is known and equipment and tools are so much improved, the difficulties remain formidable.

1.8.4 Water

Water tunnels may be looked at broadly in two major divisions, namely: aqueducts and hydro-electric schemes. A third, of lesser magnitude, is for cooling water systems in thermal power stations.

Aqueducts follow the ancient tradition of the Persians, Greeks, Romans, and others in carrying water supplies for long distances from mountain areas not only to cities but also for irrigation.

The Great Lakes in America were the scene of some early subaqueous tunnelling. Chicago drove a brick lined 1.5 m x 1.5 m tunnel 3221 m long out under Lake Michigan in 1867 to ensure ample clean water, and has gone on extending the system ever since. Cleveland in a similar enterprise under Lake Erie in 1874 had much difficulty with methane gas, but also has added to its tunnels subsequently. In southern California a number of cities have combined to draw water from the Colorado river in 370 km of aqueduct of which 148 km is in bored tunnel and 88 km in cut-and-cover. The work was started in 1932, and includes 29 tunnels of 4.9 m diameter, the longest being the East Coachello (29 km) and the San Jacinto (21 km) at depths up to 200 m. The New York City–Delaware water tunnel of 4 m diameter is said to be the world's longest at 137 km.

London's Metropolitan Water Board in 1955–60 drove a 2 m tunnel 30 km long to transfer water from west to east into the system of reservoirs based on the river Lea. The long lengths carefully sited as far as possible in the London Clay stratum stimulated development of soft ground tunnelling machines together with expanded concrete linings which are assembled without bolting immediately behind the digger shield and wedged out so as to fit tight against the machined surface of the clay without the usual need for grouting. The success of this first tunnel was followed by further similar tunnelling in London and elsewhere and the adoption of the idea of the expanded lining also for Metro construction in the Victoria line.

In South Africa the Orange River Project for diversion of water from the Orange river into the basin of the Fish river for irrigation and also for supply to Port Elizabeth, has as its key link one of the world's very long tunnels, the Orange–Fish tunnel, 82.4 km of 5.34 m finished diameter concrete lined tunnel, driven through rock by drilling and blasting, working from seven shafts up to 278 m deep.

Hydro-electric power schemes are typically in mountain areas and depend on storage of water at high level and utilisation of its energy of descent to turbines situated as far as possible below storage level. Tunnels are normally required to feed the turbines and frequently also for the discharge, but also through the mountains to supplement the supply available in one valley with that from an adjacent collecting area. With improvement of rock tunnelling techniques it has also become advantageous in some schemes to site the turbines deep in the heart of the rock in caverns excavated for that purpose. The advantages lie in the shortness of the pressure tunnels leading to the station, and also in protection of the generating stations.

An early example of tunnelling for water power was at Niagara in 1890, where water was drawn off above the falls and returned to the river lower down after utilisation of its energy. Further power and control schemes have followed to make use of as much of Niagara's water as possible without too serious interference with the spectacular flow over the falls.

Sweden built a small underground power station in 1910 at Mockfjord, and followed with the first such major station in 1910–14 at Porjus, within the Arctic Circle. This has a 530 m supply tunnel feeding five shafts 50 m deep and a 1152 m tailrace. The machinery chamber is 12 m x 20 m x 90 m long at a depth of 51 m. Power was required for a railway serving iron ore mines, and underground construction was adopted partly to ensure freedom from freezing up in winter. The construction techniques have been continuously improved, sizes increased and costs reduced, so that the Stornorrfors station opened in 1958 has a generator chamber 18.5 m x 29 m x 124 m and a tailrace 16 m x 26 m in section and 4000 m long. The power generated is 375 MW. More than half of Sweden's power is now produced in such stations. Similar underground stations have been constructed in most countries with suitable topography.

Apart from complete underground stations, there has been much tunnelling done to feed turbines above ground and to link up sources of supply as well as for pumped storage schemes. Scotland has many miles of such tunnels. Australia's Snowy Mountains scheme has required extensive tunnelling. Generation combined with river control and irrigation is typified on the Indus river by the Mangla Dam, where five 11 m tunnels each 500 m long were driven by means of a Robbins Mole, the same machine being subsequently adapted to drive the twin tubes of the Mersey Kingsway tunnel. The hazards still to be encountered are exemplified by the collapse of ground into the tunnels at the Tarbela Dam, also on the Indus river.

In hydro-electric water tunnels there is an important distinction between those in which water is contained at high pressure and those in which it flows as in a covered channel. In high pressure tunnels the rock is most unlikely to be sufficiently watertight and the relations between stresses in the tunnel lining and in the surrounding rock are of vital importance if cracking and leakage are to be avoided. The science of rock mechanics has been developed in recent years to deal with this problem among others.

1.8.5 Sewers

The growth of cities, often astride rivers, makes sewerage an ever-increasing problem, if the discharge is to be collected and disposed of efficiently. The volume to be handled is similar to the volume of water supplied, and there are special problems of additional surface water flows in rainstorms. The critical requirement of small but steady gradients leaves little choice as to pipe depths so that tunnelling soon becomes necessary, even for pipes, and much more so for the larger trunk sewers. The largest sewers do not often exceed a few metres in diameter but they have to be driven in conditions that are often very difficult, because the level can rarely be determined to suit easily tunnelled strata and the sewer must be driven typically through very variable alluvial deposits and even through made ground, in the vicinity of building foundations and other subsurface structures peculiar to cities.

London was the first great city to provide comprehensive sewerage. Before 1817 contamination of drains was a penal offence but in 1847 it first became compulsory to drain houses into sewers, which had previously provided only for surface water drainage. Eight independent bodies were superseded by a series of Metropolitan Commissions of Sewers and eventually in 1856 by the Metropolitan Board of Works. Abolition of cesspools was followed by shocking pollution of the river Thames from the direct discharge into the tidal river. Bazalgette, as Engineer to the new Board, planned and executed a major scheme of some 70 miles of interception sewers up to almost 3 m diameter, discharging into the river 12 miles downriver. Numerous tunnels were necessary under streets, rivers, canals, railways, and house property. While many were through clay, others were through water-bearing sand and gravel and through ‘quicksands’. Bazalgette developed a very satisfactory method of ground water lowering by sinking pumping wells to 2 m below excavation level, providing a shingle filter at the bottom and laying earthenware drain pipes below the invert of the proposed sewer. The whole scheme was substantially completed by 1865.

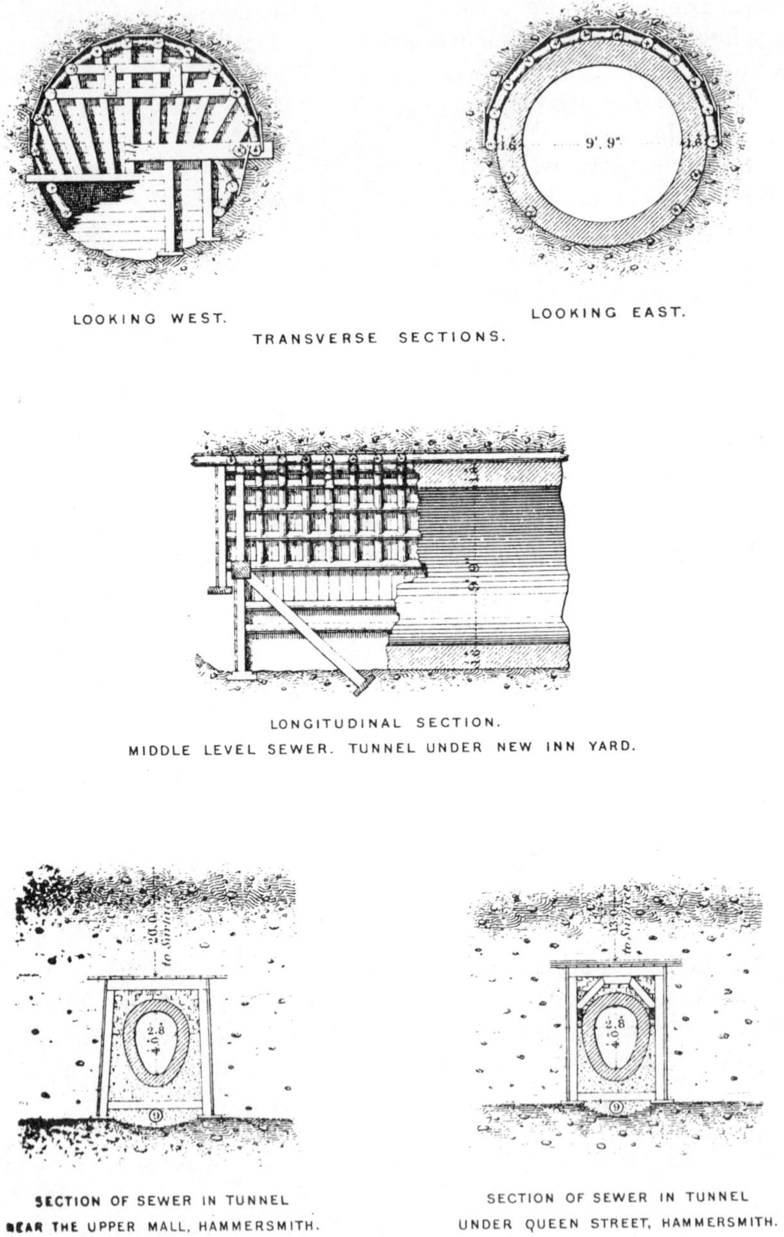

Fig. 1.8 – Sections of sewers, illustrating Bazalgette's paper of 1865 to the Institution of Civil Engineers.

No doubt the varied tunnelling in progress at this time stimulated thought and experiment, and contributed to Greathead's development of the shield for the Tower Subway in 1869.

Bazalgette's great plan for London was substantially enlarged in the years 1898–1912 when a further 75 km of main sewers were built, about 57 km being tunnelled. Cast-iron linings were used for some 25 km, about half with the aid of a shield and compressed air.

The brickwork in these sewers was built to very high standards of workmanship, and generally they still continue in use.

Shields and compressed air were widely used before the end of the century for sewer tunnelling in such diverse cities as Cincinnati, Melbourne and Paris.

In many cities now the greater demands of growing population and industry, and higher standards for rivers makes continuous improvement necessary, including large tunnelled intercepting sewers to eliminate direct discharge into rivers or sea. Extensions of existing outfalls into the sea to much deeper levels further offshore are being demanded. A notable example is that of Los Angeles, where a 4 m tube has been extended 8 km into the Pacific to a depth of 60 m of water; this was laid by submerged tube methods—perhaps more akin to pipe laying than to tunnelling—in exposed oceanic conditions to this great depth. Sydney, Australia, now proposes sea outfalls 4 km long.

Another city with special problems is Mexico City, built on deep lacustrine deposits which continue to subside steadily under the weight of the city's buildings through the years, so that a surrounding canal which used to drain the city is now left high above it.

1.8.6 Other tunnels

Tunnels to carry cables or pipes are numerous and important but have not usually been of large dimensions or otherwise remarkable in construction. The cable tunnel under the river Severn has already been mentioned. A similar cable tunnel was driven through chalk under the river Thames near Tilbury. The London Post Office tube railway runs for some 11 km between Paddington and Liverpool Street and carries mails in automatic trains without drivers. It was tunnelled over the years 1914–17 but not equipped and commissioned until 1924–27.

Tunnels for storage of oil and other materials have been constructed widely, as well as tunnels for shelter and defence purposes, as for example in Gibraltar where provision was made for all requirements for a garrison to remain underground for a long period. Another aspect of underground construction is for car parking in cities, but not necessarily by tunnelling techniques.

Many new methods of excavation have been proposed including rock fusion and hydraulic jetting, but none has yet been sufficiently developed to supersede some form of digging tool for soft ground, and drilling and blasting or mechanical excavations for hard rock–merely developments of the primitive antler pick and fire setting.

1.9 FUTURE PROGRESS

Perhaps the most fitting conclusion for this chapter is to record an international beginning and continuing developments. An O.E.C.D. Advisory Conference was held in Washington in 1970 and attended by representatives of nineteen countries. The main aims of the conference were:

(1) to assess the most important strengths and weaknesses of present-day tunnelling technology;
(2) to identify the principal needs for improvement; and
(3) to recommend government policies–including policies for research, development and demonstration–designed to stimulate rapid progress in tunnelling technology.

According to a survey some 13000 km of tunnels were constructed in the decade 1960-69, and twice as much was expected in the succeeding decade. This excludes mining tunnels.

The conference agreed on five recommendations. They are too lengthy to be quoted in full but their subjects may be indicated as follows:

(1) That each country should designate a national agency for coordination and stimulation of improvement.
(2) That in every urban area a single agency should maintain records and a master plan.
(3) That in evaluation account be taken of indirect as well as direct costs and benefits.
(4) That each country should promote application of technical advances in tunnelling.
(5) That working ties be established between national agencies as well as through an international commission.

Since that conference, the professional aspects of tunnelling have progressed steadily. The International Tunnelling Association was formed in 1974 and meets annually, principally these days to review the progress of its various working parties each covering some specialised aspect. Its twenty six member nations include almost all the major countries to whom tunnels are important.

Tunnels have been the subject of an increased amount of research at universities and laboratories (such as the Transport and Road Research Laboratory in Britain), and there are now regular conferences on tunnelling in many countries and a vastly increased exchange of information internationally.

All this activity has helped to bring the potential value of tunnels to the attention of politicians and planners, and although many projects have been delayed recently by the widespread recession, it seems certain that tunnels will play an increasing role in the future.

BIBLIOGRAPHY

A short bibliography appears at the end of each Chapter. A comprehensive bibliography will be included in the second volume.

There are a few books which specifically deal with the history of tunnelling, and much additional historical material is to be found in descriptions of methods and projects, in biographies and incidentally in recorded discussions at meetings of the Institution of Civil Engineers and elsewhere.

The appended list first gives some early classic books and papers *in order of date of publication* followed by books primarily historic in their content. It is far from exhaustive, and reference should be made to other chapters, although some titles may be repeated.

Simms, F. W., *Practical Tunnelling,* Crosby Lockwood, 1844 (revised and enlarged 1859, 1877, 1896).

Bazalgette, J. W., 'On the main drainage of London, and the interception of the sewage from the River Thames', *Min. Proc. Instn. Civ. Engs.* 1865, **24.**

Brunel, I., *The Life of Isambard Kingdom Brunel, Civil Engineer,* Longmans, 1870.

Drinker, H. S., *Tunneling, explosive compounds and rock drills,* J. Wiley, 1878.

Gripper, C. F., *Railway tunnelling in heavy ground,* Spon, 1879.

Baker, B, 'The Metropolitan and Metropolitan District Railways,' *Min. Proc. Instn. Civil Engineers,* 1885, **81**.

Wolfe Barry, J., 'The City lines and extensions (Inner Circle Completion) of the Metropolitan and District Railways , *ibid,* 1885, **81**.

Burr, S. D. V., *Tunneling under the Hudson River,* J. Wiley, 1885 (reprinted A. A. Mathews, Arcadia, 1968).

Fox, F., 'The Mersey Railway', *Min. Proc. Instn. Civ. Engs.* 1886, **86.**

Walker, T. A., *The Severn Tunnel,* Bentley, 1888 (reprinted Kingsmead, 1969).

Greathead, J. H., 'The City and South London Railway; with some remarks upon subaqueous tunnelling by shield and compressed air', *Min. Proc. Instn. Civ. Engs.* 1895, **123**.

Leitch, W. O., 'Iron Tunnels', *ibid,* 1896, **125**.

Hay, D. and Fitzmaurice, M., 'The Blackwall Tunnel', *ibid*, 1897, **130**.

Dalrymple-Hay, H. H., 'The Waterloo and City Railway', *ibid*, 1899, **139**.

Halcrow, W. T., 'A Century of Tunnelling', *Proc. Instn. Mech. Engrs.,* 1941, **146**.

Rolt, L. T. C., *Isambard Kingdom Brunel,* Longmans, 1957.

Lampe, D., *The Tunnel; the story of the world's first tunnel under a navigable river, dug beneath the Thames, 1824–42*, Harrap, 1963.

Sandstrom, G. E., *The History of Tunnelling*, Barrie & Rockliff, 1963.

Coleman, T., *The Railway Navvies,* Hutchinson, 1965.

Clements, P., *Marc Isambard Brunel*, Longmans, 1970.

Beaver, P., *A History of Tunnels,* P. Davies, 1972.

2

Functions and Requirements

In any systematic consideration of tunnels and the appropriate methods of construction, the function of the finished tunnel is of fundamental importance. The purposes of tunnels–and consequently their size, shape and lining–are diverse. In planning and construction important factors to be examined in relation to function include: situation, ground, dimensions and geometry, structural form, construction method and permanent equipment. The object of this chapter is to review the range of uses and the interactions between function and these other factors. Particular functions, structures and methods are dealt with in more detail in later chapters.

Costs are not discussed in any precise terms. Too much space and time would be needed to abstract comparable figures and to present them in any useful form, even apart from making proper allowances for inflation and other changing conditions. The resources demanded by practicable alternatives may be the most significant measure. Another aspect is cost/benefit analysis which takes account of indirect social and environmental costs as fairly as possible.

2.1 FACTORS RELATED TO FUNCTION

The factors referred to above may here be explained a little more fully in general terms. They are mutually interdependent and are not separate aspects for decision.

1. 'Situation' for a tunnel may be through a mountain or hill, or subaqueous, or urban.
2. 'Ground' may be anything from soft silt to hard uniform rock covering a very wide range of behaviour in excavation; water may play an important part. Any choice of ground implies changes in geometry, structural form and construction method.
3. 'Dimensions and geometry' are those of the finished tunnel: width, height and length, together with levels and gradients and curves. Specified limits may be very narrow or offer a wide range of possibilities.

4. 'Structural form' may be circle, horseshoe, rectangle or other shape incorporating cast-iron, concrete, brickwork, sprayed concrete, etc., to carry the loads imposed. The nature of the ground and method of construction influence strongly the structural form.
5. 'Construction methods' range from boring by drilling and blasting, or by tunnelling machine, or with or without a shield, to cut-and-cover in various sequences, and submerged prefabricated tunnels. The choice of method is limited by the ground conditions, but also by available resources in the widest sense.
6. 'Equipment' of the completed tunnel includes such features as roadway or rail track, lighting, ventilation, decorative and functional finishes, control systems. The special needs of each category of tunnel will be further referred to.

All of these things should be taken fully into account in the planning and design of the project. It is usually most unsatisfactory and inefficient to make substantial additions or alterations at a later stage, except where proper provision has been made in the original planning.

2.2 PRINCIPAL FUNCTIONS

The principal functions for which tunnels, other than mining tunnels and military sappers' tunnels, are built include:

A. Transportation

1. People and goods:	pedestrian and cycle subways
	railways and metros
	highways
2. Water:	canals
	city supplies
	irrigation
	hydroelectric power
	cooling water
3. Sewers:	
4. Cables and piped services:	

B. Storage and Plant
   1. Car parks
   2. Cavern storage of oil
   3. Underground power stations
   4. Military stocks
   5. Disposal of radioactive waste

C. Protection of People
   1. Shelters
   2. Control posts

2.3 SITUATION

The reason for tunnelling for transportation is to pass underneath obstacles, which may be hill or mountain ridges, rivers or seas, city streets and buildings or other structures.

2.3.1 Mountain Range

If the obstacle is a mountain range passage through it by tunnel can effect a great saving in time and energy and, indeed, may make practicable a crossing which would otherwise demand gradients and curves quite unacceptable for the purpose. The progression from canals, which must be level except for steps at locks, through railways, where gradients greater than about 1 or 2% are unsatisfactory, to highways with acceptable gradients of 3 or 4% marks also the historic sequence of tunnel development.

If the attempt is made to plan the layout of a surface route up a valley having a col at the head to be crossed, progressive difficulty is to be expected as the col is approached and the valley becomes narrower and steeper. The ruling gradient of railway or highway can only be worked to by attempting to climb as quickly as permissible from the starting point, utilising the sides of the valley and cutting steadily upwards across the contour lines. With increasing steepness and reducing width for deviations the col may be unattainable, and a summit tunnel becomes inevitable. Such a procedure offers the minimum length of tunnel, but there is usually the alternative of driving a longer tunnel at a lower level with shorter or easier gradients and easier curves, effecting operational savings in distance, climb, fuel consumption and time, but at greater capital cost. This is the typical problem that becomes extreme in the Alpine type of tunnel, whether for rail or road. The long costly tunnel justifies the capital investment if sufficient traffic can be attracted.

2.3.2 Subaqueous

Examples of passage under waterways have been described in their historical context. The choice of a tunnel rather than a bridge for the crossing of a river or estuary depends on the factors peculiar to each situation. Where numerous traffic lanes are wanted and where foundation conditions make possible moderate spans, the bridge may well be the cheaper alternative but the cost of very long bridge spans increases disproportionately, and if foundation conditions are unfavourable and if shipping requires long spans and high clearances, the tunnel becomes the better choice. However aesthetically exciting great bridges may be in their main spans, long approaches at high level through a city are as a rule far from pleasing and can form a serious obstruction to traffic circulation and land use at ground level.

Another advantage of tunnelling may arise from construction in stages. An ultimate requirement for a six-lane river crossing as traffic develops over a period of years can be met by a series of two-lane tunnels as needed, whereas virtually the whole capital cost of a six-lane bridge would be incurred in the initial construction and it would remain under-employed for a long period. The successive construction of two-lane tunnels in this pattern of development is illustrated by the Lincoln tunnels in New York, opened in 1937, 1945 and 1957, and in England by the Mersey Kingsway tunnels opened in 1971 and 1974 and the Dartford tunnels opened in 1963 and 1980.

In the case of surface railways the subaqueous tunnel is handicapped by the need to descend far below ground level and rise to the surface again while providing a gradient much flatter than is acceptable for road transport. This does not apply in the same degree to metro systems which generally remain well below ground level.

2.3.3 Urban

Tunnelling under city streets has now become a common activity for nearly all the types of transportation listed except canals, but long urban tunnels are principally for metros, water supply, and sewers. Much shorter tunnels are required for highway underpasses and pedestrian subways, most usually constructed in cut-and-cover. Longer highway tunnels are sometimes suggested but problems of access and ventilation add greatly to the difficulty and cost, against which preservation of surface land and gains in environmental amenity may prove a quite inadequate offset.

2.4 PEDESTRIAN AND CYCLE SUBWAYS

2.4.1 Pedestrian subways

Pedestrian subways are in many ways the least demanding of tunnels—perhaps they are also the most primitive—because the pedestrian is able to descend and ascend steps and quite steep gradients, and to turn corners sharply. The absolute limitations on tunnelling are therefore few, but most modern pedestrian subways are intended to offer an attractive alternative to the use of the surface, as in crossing a street, and therefore most go far beyond minimum standards. The old gloomy passage, brick-lined, damp and poorly lit is very different from the modern subway of the city centre—spacious and bright, lined with decorative tiling and having easy approach ramps.

The shallow subway will probably be constructed in cut-and-cover, but for connecting passages in metro stations at deeper level, bored tunnelling will be necessary. Minimum space requirements are: headroom of about 2.3 m, width for passing of at least 2 m, and preferably more, and moderate gradients which should be less than 10%. Where steps are acceptable the rise can be as steep as 1 in 2, but the total rise of each flight should be no more than about 3 m.

Drainage, adequate for the worst rainstorm, must be provided, utilising storage and pumping if necessary. Emergencies such as tidal flooding and bursting of water mains must be examined at the design stage and provided for, at least to the extent necessary to prevent major catastrophes.

Lighting should be of the highest standard practicable and of robust construction, vandal-proof if possible in exposed situations.

The usefulness of a subway can be greatly enhanced by good decorative finishes which attract users. Ceramic tiles or terrazzo provide such a finish: vitreous mosaic tiling is perhaps even better and less easily damaged, but there are many other suitable finishes depending on situation and usage.

It is usually important also that a subway should be as shallow as possible so as not to discourage the user by long descents and ascents. This implies minimum roof construction depth and may make it difficult to incorporate ceiling lighting.

2.4.2 Cycle subways

There is in some highway systems a need to provide separate accommodation for cyclists. Their use of the main carriageways may be dangerous and obstructive. A subway similar to that for pedestrians may be provided at street crossings, but with gradients not exceeding about 1 in 10 and, of course, without steps.

A special requirement appears where a subaqueous tunnel is built to replace a ferry in an industrial area where cycle transport is important. For the Tyne crossing a separate cycle tunnel (and also a pedestrian tunnel) served by an escalator was built (see Fig. 2.1). In the Clyde tunnel a cycle track and pedestrian path were incorporated in the invert of the main tunnel beneath the road deck.

2.5 RAILWAYS

Tunnels for railways are discussed at greater length in later chapters and therefore the references here are only intended to relate their special features to those of other tunnels. The typical situations for main line railways are mountain ranges, hills and subaqueous crossings. Every kind of ground is liable to be encountered. In mountain ranges, complex patterns of folding and faulting present various strata dipping at any angle and including belts of shattered rock and squeezing rock. Water is usually encountered, sometimes at high pressure and temperature. In undulating country, the conditions are likely to be less severe, but clays may involve particular difficulties.

For subaqueous crossings, tunnels have been cut through anything from the near liquid silts of the Hudson river to the volcanic and other beds of the Seikan tunnel, and water is always a hazard.

The typical railway tunnel is about 5 m wide x 7 m high for a single-track tunnel and about 8.5 m x 7 m for twin tracks, the choice depending on volume of traffic, and also on problems of rock support in a wide tunnel. What governs the longitudinal profile of all railway tunnels is an acceptable gradient. Less than 1% is preferred but steeper gradients may have to be adopted in many

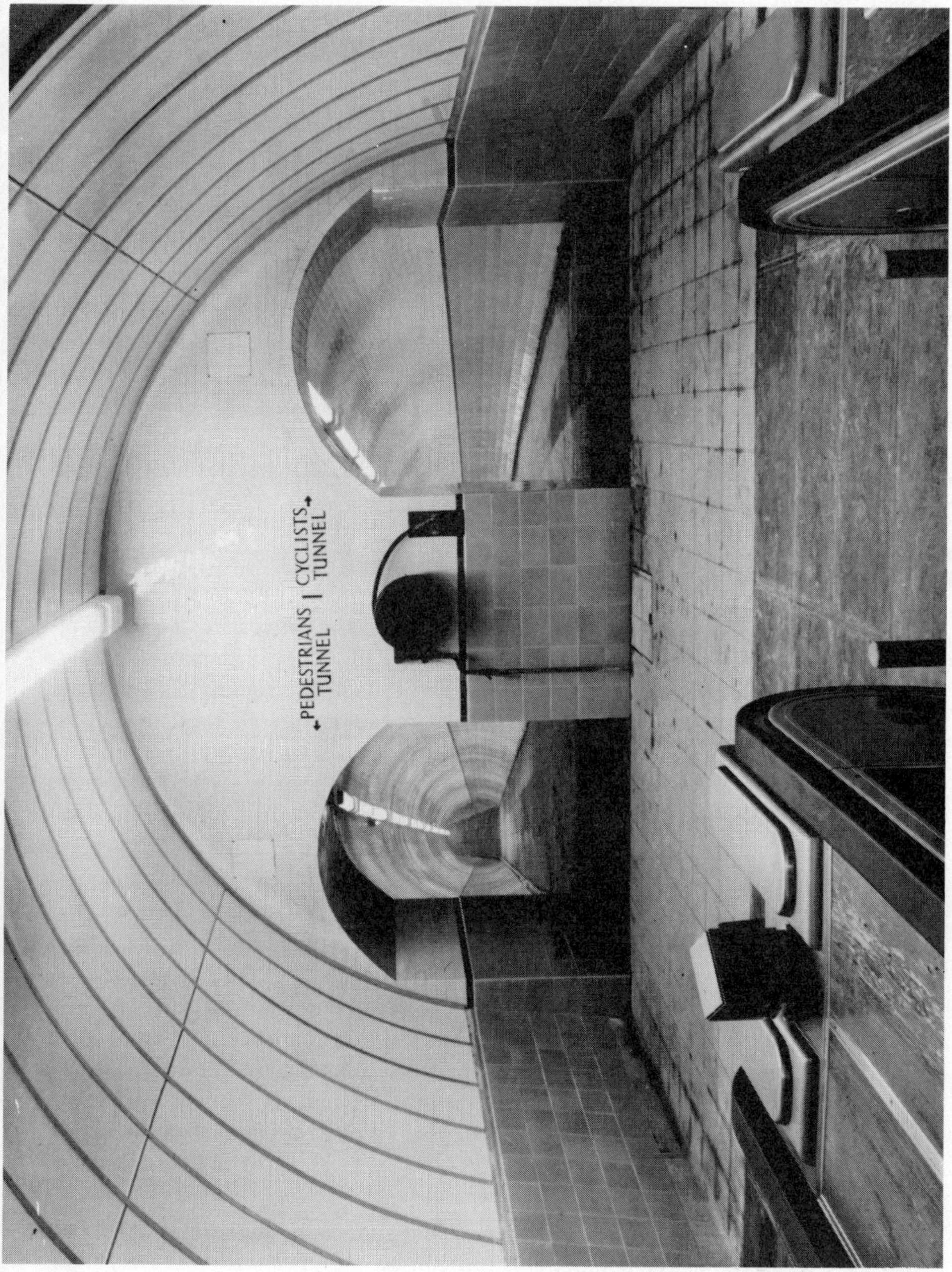

Fig. 2.1 – Pedestrian and Cycle Tunnels under river Tyne.

circumstances. The ability to restart a stopped train on a rising gradient is an operational necessity. The ultimate controlling factors are loco power and rail adhesion. The other principal geometrical factor is curvature, which is governed by much the same rules as on the open line in relation to speed.

The commonest structural form for railway tunnels is probably the horseshoe, the ground being protected and supported by a lining wherever there is any looseness or uncertainty about the exposed rock. In the past brickwork and masonry from local sources were often used. In modern tunnels sprayed concrete, steel arch ribs and *in situ* concrete are most usual in sound rock. Where the tunnel is through clay slow plastic movement is to be feared and the bottom must be protected with an invert arch or at least a strut across the base of the side walls. A complete circular form with segmental concrete or even cast iron lining has been constructed in some cases.

Construction methods are not peculiar to railway work except where rail rather than road access and transport are adopted.

Equipment comprising track, signalling and other features are dealt with in Volume 2.

2.6 METRO SYSTEMS

As with main line railway tunnels the requirements for metros are dealt with in later chapters in more detail, but are outlined here. They differ significantly from main line railways in many respects. The tunnel, the rolling stock and all the equipment are, or should be, designed as an integrated system if it is reasonably self-contained. Sometimes it is necessary to accommodate main line or suburban rolling stock, but this is unlikely to lead to overall economy unless the metro line is essentially a subsurface section of a primary surface railway.

Metro tunnels are intrinsically urban, but will frequently include subaqueous sections, because few major cities are without rivers to cross, and many are sea ports. Hill tunnelling on a minor scale is also encountered, but not mountain tunnelling in its full sense. In general, metro tunnelling is as shallow as practicable because of the importance of easy and rapid access to and from the surface.

Many major cities are situated on rivers and built on alluvial ground, with the consequence that if their metro tunnels are immediately sub-surface they must be through soft, and usually water-bearing ground. In many other cities, however, rock is present at quite shallow depth or outcropping at the surface in hilly areas. Typically, metro tunnels present a wide range of soils and usually considerable variability along any line.

Gradients may be steeper than with main line tunnels, because the system does not have to accept heavy goods trains but can be equipped with relatively light coaches with multiple motor drives. In contrast to some main line tunnels, watertight linings are usually considered essential and rails are therefore dry enough to give good adhesion. Gradients of 3.5% are acceptable, and, indeed, were unavoidable in some cities such as, for example, Toronto where the whole land slopes up at about that rate from the shores of Lake Ontario. In Montreal, steeper ground required gradients up to 6.3%, and pneumatic-tyred drive was adopted in order to improve adhesion, and hence acceleration, as well as for smooth riding.

There is sometimes merit in a deliberate switchback profile, whereby trains climb up to stations, the gradient providing assistance in braking, and reabsorbing kinetic energy, while a corresponding descent aids acceleration out of stations. An additional advantage may lie in having the stations less deep than would otherwise be the case, but, conversely, greater depth of excavation in cut-and-cover between stations might be impracticable.

In dealing with shape, size and structure there are substantial differences according to whether construction is by bored tunnel or cut-and-cover. The choice of method is a complex matter, but cut-and-cover is likely to be preferred where it is possible to follow a shallow subsurface route without unacceptable disruption of streets and services, while deeper tunnelling is progressively more necessary in heavily congested city areas.

In cut-and-cover twin tracks are usually accommodated in a rectangular reinforced concrete box with a central wall or line of columns. Dimensions and clearances can follow normal railway principles adapted to the rolling stock provided, and any abnormal loads which may be specified.

In bored tunnels a circular tube with cast-iron or concrete lining is almost invariable with the diameter designed to be a minimum, because cost increases disproportionately as excavated area increases. Accuracy of alignment is then more vital than for any other tunnel, because clearances between rolling stock and structure are so small. For high speed operating, modern track is laid with precision, embodying transition curves and superelevation to which the tunnel lining must conform. Inevitable errors of shape and position may necessitate rebuilding of lining which is a troublesome and time-consuming operation, and sometimes hazardous. Within limits, redesign of the curves and realignment of the track is usually possible while maintaining basic standards.

Even in sound rock some lining is desirable to protect passenger trains against even trivial falls of rock which could be disproportionately troublesome.

In deep systems station tunnels are much larger than running tunnels; they may be bored, or may be built in deep pits from the surface. High standards of watertightness with aesthetically acceptable linings, finishings and lighting are called for. These requirements, together with those for track, signalling, fire precautions, passenger amenities generally, and ventilation and movement of air are discussed in greater detail in volume 2.

2.7 HIGHWAYS

Highway tunnels may be in any situation where tunnelling is possible.

For long tunnels with heavy traffic the proper furnishing and equipment become quite elaborate and are dealt with in more detail in Volume 2. Here the broad differences from other tunnels are outlined.

In comparison with main line railways a very important difference is the permissible gradient, which may be at least three times as steep. This is particularly relevant to subaqueous road tunnels where a descent of the order of 30 m is necessary in order to pass under a navigable waterway. A gradient of 3½% to 4½% is normal, and in special conditions, such as the Clyde tunnel, short approach ramps on a restricted site may justify as steep a gradient as 6½%. It may be noted that a 30 m descent at 3% requires a length of 1000 m, much of it in tunnel, whereas a 6% gradient halves this length.

In summit tunnels steep gradients do not usually present a problem as the climbing is done before the tunnel is entered and its profile is likely to be nearly level with sufficient slope only to ensure drainage.

Although steep gradients are negotiable by road vehicles there are three factors that make them progressively more undesirable. On a long ascent of 1 in 30 heavy vehicles fully loaded can usually maintain a moderate road speed, but if the climb is steeper many will have to change down to low gear and will slow down all traffic to perhaps 15 kph or less, thereby reducing the capacity of the tunnel. In climbing steep gradients the output of noxious fumes—CO from petrol engines and smoke from diesels—increases disproportionately and makes greater demands on ventilation. A third objection to a steep gradient is the danger from vehicles running down out of control.

There are, of course, mountain tunnels having an overall one-way gradient through spurs or ridges. This gradient must be governed by the ruling gradient of the road system as a whole, but with special consideration of the three factors described above.

Horizontal curves may be much sharper than for railways if necessary although both for construction and for use the driving of a tunnel on a sharp curve should preferably be avoided. A radius of 400 m is not difficult for construction or traffic, and even sharper curves are practicable. At the north end of the Tyne tunnel there is a semi-circle having a radius of 128 m, built in tunnel to allow the roadway to emerge in a level area and provide access to a road junction close to the river bank.

Vertical curves at change of gradient are subject to much the same considerations as on the open road, but with the additional factor, not generally of major importance, that in a dip the sight line is restricted by the tunnel ceiling. A radius of 1000 m gives a reasonable road profile, especially as very high speeds need not be usually allowed for.

Dimensions of highway tunnels have increased with the steady development of road vehicles. It has been usual to provide at least two traffic lanes rather than single lane one-way tunnels in which it would be very difficult to ensure access to accidents. Road widths for two lanes have grown from 4.9 m in the early Blackwall tunnel (1897) up to 7.3 m in many modern tunnels, and occasionally 7.5 m. Preferred headroom in British tunnels is currently 5.4 m but in Europe 4.5 m is widely accepted: sometimes, however, greater height is provided in port areas.

In all tunnelling, increase in width imposes increase of cost not only in direct proportion but additionally by the structural consequences of increased roof span. In circular tunnels the vertical diameter is similarly increased and the excavated area then increases as the square of the diameter. A further increase may be necessary in the strength of the lining and therefore in its thickness. For subaqueous tunnels whose level is determined by minimum cover under a channel or rock surface, extra height imposes a lower road level and in turn longer approach ramps. Constructional problems of increased size are discussed in a later chapter. Economy of dimensions is obviously most important, and any substantial increase can be very costly.

It is these factors that lead to the construction of two-lane units and discourage the provision of wider carriageways, which are sometimes suggested as advantageous in preventing complete stoppage when a single vehicle breaks down.

The Mersey Queensway tunnel (1934) is exceptional in having a four-lane carriageway in a single circular bore, although the road width is only 11.0 m. The tunnel was driven through rock in free air and the main midriver section is a 13.4 m diameter circle, cast-iron lined. There was much unoccupied space beneath the roadway, thought suitable for carrying a tramway but not so employed, and now used for ventilation and for cables and pipes.

Submerged unit tunnels of rectangular section are much less sensitive to increases of width; wider lanes and as many as are justified by traffic needs can be accommodated in a single structure. For this form of tunnel comparison with a multi-lane bridge, referred to above, is much more direct.

Highway traffic cannot be constrained to run on any precise track like rail traffic, and therefore it is essential to allow adequate tolerances in deciding on the dimensions of the tunnel. Construction must follow as precisely as possible the designed line and dimensions. As in the case of metro tunnels, but with rather more latitude, it is likely to be necessary to prepare a three-dimensional 'wriggle diagram' to fit the required road geometry into the tunnel space.

Ground encountered and methods of construction are so varied and cover such a range that no aspects peculiar to highways invite comment. Shield drives, drill and blast, TBM's, cut-and-cover, and submerged units all have their applications, and even the small tunnel methods, including timbered headings and pipe jacking may serve ancillary purposes.

Lighting and ventilation can both have important influence on construction. They are more fully discussed in Volume 2. Artificial lighting of a very high standard is essential in all but the shortest tunnels. The human eye adapts quickly, but not instantaneously to changes in luminosity and, particularly in driving from bright sunlight into a tunnel, the provision of graduated diminution of light is most valuable. Such devices as sun visors outside the portals and much extra lighting within the portal area are used. At night the lighting within the tunnel is likely to be as good as or better than that outside; the transition boost lighting is then switched off and perhaps the tunnel lighting is also reduced.

Ventilation is a large specialised subject in itself and can have an important influence on the construction of the tunnel. Working shafts can be planned for incorporation as permanent air shafts, and the air ducts and their connections into the tunnel space may provide complex problems of design, excavation and construction. The provision of power supplies and emergency arrangements for lighting and ventilation is again something to be planned from the beginning.

Drainage has two aspects: first, the drainage of ground water entering the tunnel through the lining and second, surface water entering at the portals, carried in on vehicles, perhaps as snow, and flooding from burst water pipes or from the use of fire hoses. The leakage of ground water is minimised by waterproofing techniques during construction and is not likely to be a major operating problem in a well-built tunnel even if water is deliberately drawn off from the lining in order to prevent pressure building up. Ground water in very small volume can however be a serious decorative problem, as major seepages quickly contribute to staining of any finishes and assist in the build-up of sooty deposits which may be corrosive. Seepage water will frequently carry lime or other salts in solution and deposit them where it emerges.

In summit tunnels and those on a steady gradient all water can be drained to the portals by gravity, but in a subaqueous tunnel or a city underpass all water drains to the lowest point where a sump and pumps must be provided of capacity adequate for all probable emergencies.

Other features special to highway tunnels are traffic control systems and fire precautions, and provision for breakdowns and accidents. All these systems have to be devised to operate in conjunction with a random sample of the public, many unfamiliar with the particular tunnel. One important feature should be an effective means of quickly stopping traffic from entering the tunnel if an incident gives rise to obstruction of the tunnel or any other danger.

Some highway tunnels permit the passage of pedestrians and cyclists. Such provision reduces the traffic capacity and adds disproportionately to the cost, unless separate space can economically be provided. Because overtaking in tunnels is not normally safe, cyclists will slow down the whole flow of traffic unless segregated. Also, cyclists and pedestrians will require a higher standard of dilution of carbon monoxide because they are exposed for a longer time and while taking exercise. If the obligation to allow pedestrians and cyclists the use of a tunnel is unavoidable, it may be met by provision of a separate tunnel, as described earlier.

2.8 WATER CONVEYANCE

Tunnels to convey water, or sewage, have various features in common. The principal characteristics required are smoothness, and watertightness, but the need for these differs in degree. There are important differences in tunnels required merely to allow water to flow through them at low pressure, often without the bore being full—free discharge tunnels—and those in which the water is under high pressure, possibly very high pressure, as in the supply to turbines or in a town supply aqueduct passing under a valley.

The importance of smooth lining depends on the velocity of the water and on the length of the tunnel. Smoothness is relatively unimportant for a canal where movement of water is minimal, but can be most valuable in maximising the flow through an aqueduct or in minimising loss of head in power development.

The requirement for watertightness is also dependent in part on pressure, internal or external. High pressure tunnels must be extremely well sealed against water losses which could quickly result in serious erosion. In sewers, also, water losses can cause dangerous contamination, and even very small leaks may be enlarged by chemical action. Drinking water supplies must be secured against contamination by inward leakage when running at low pressure. Canals are perhaps the least demanding in this respect, and indeed may benefit from inflow of water, but must be well sealed against outward leakage.

Another general requirement for water tunnels is a means of access for inspection, maintenance and repair. Any system of penstocks or dams must be planned with this in mind.

2.8.1 Canals

The absolute requirements of a canal tunnel are: a level tunnel, retention of water and space for passage of barges. These relate particularly to transport canals. In canals feeding hydro-electric schemes minimum loss of head may be the overriding consideration. The level requirement is, of course, fundamental, from which it follows that canal tunnels are through ridges only and cannot as such dip down under rivers. Sections of an irrigation or supply canal can, of course, be connected by inverted siphon pipes or tunnels under obstacles.

Retention of water can be very important, as canal tunnels are most common on a summit level, where adequate water supply is most difficult, but essential to the operation of locks on both sides of the summit. If water can be not merely conserved but collected within the tunnel it is an added advantage. Lining of the canal invert with clay puddle was a usual practice together with the use of timber boarding. Other linings such as brickwork may be required.

Historically, space for passage ranges from 2.2 m x 3.6 m for the British 'narrow boat' in the Harecastle tunnel up to 32 m x 15 m in the Rove tunnel. The Harecastle tunnel was one way only at a time and without a towpath, the boats being 'legged' through by the crew. The use of engines and of tugs in tunnels gave rise to ventilation problems which in some cases have been solved by use of an electrically operated tug.

2.8.2 City supplies

The supply of pure water by aqueducts to cities is an ancient need, often demanding tunnelling. The problems of collection and storage in mountain areas are discussed later; they are much the same whether the water is for town supplies, irrigation or hydroelectric power. The aqueducts from dam site to city may be pipes on or above the surface or in trench, but tunnelling becomes necessary where ridges have to be crossed, or economically advantageous, as in London's supply, where other obstructions are numerous and a convenient route is available at depth in the clay. For these tunnels generally the water is not at very high pressure and they may even flow only part full. The principal requirements are: an overall downward hydraulic gradient accordant with the required volume and velocity of the water, a smooth unobstructed bore, and a watertight lining.

Manning's formula, $v = k\ R^{2/3}S^{1/2}$ connects velocity of water, roughness of tunnel (k), hydraulic mean depth (R), and hydraulic gradient (S). k is a roughness factor ranging from 100 for a large tunnel with exceptionally smooth bore down to the very low value of 25 for an unlined rock tunnel. For a concrete lined tunnel an average value is about 80. From such figures as these the value of smooth lining can be assessed for particular conditions.

Although small losses of water by leakage from a short aqueduct may in themselves be unimportant unless the leaks are enlarged by erosion, the cumulative loss from a long aqueduct becomes excessive and therefore a high standard is essential. Contamination by inward leakage is less likely, except perhaps if the supply fails or the pipe is emptied, but must be guarded against.

Tunnel sizes are determined to carry the necessary volume, using the available gradient, and linings are specified to perform three principal functions: first, any required support of the excavated ground; second, provision of the appropriate smooth bore; third, watertightness in both directions. As there is a minimum practical diameter of about 2 m for tunnels other than the shortest headings, tunnel size will sometimes be more than is otherwise necessary. Control by penstocks, and facilities for inspection and maintenance will be necessary, and surge shafts may be provided. Rates of flow in city water supplies are likely to be much more uniform than in hydroelectric tunnels, or in irrigation schemes.

2.8.3 Irrigation

Tunnels may also be required at most stages of irrigation projects: for collection of water, bypass of dams, transfer of water, and distribution. The tunnel requirements and problems are very much the same as for city supplies. There are schemes combining the two, such as the very large Orange–Fish river scheme in South Africa. Combined irrigation and hydroelectric power schemes are also numerous.

2.8.4 Hydroelectric power

As described in Chapter 1, many modern hydroelectric generating stations are sited underground in mountainous areas, and the practice is growing of incorporating pumped storage systems in their design. The tunnelling becomes quite complex.

In terms of tunnel requirements there are, in addition to the access and cable tunnels, at least three separate functional tunnel types. These are:

a. Free discharge.
b. High pressure water tunnels ('penstocks') feeding into turbines.
c. Machine halls and transformer halls, excavated as caverns out of solid rock.

2.8.4.1 Free discharge tunnels

For collection of water from adjacent, or even remote, valleys, tunnels through mountain ridges may be required. Their function is simply to carry the available water as directly as possible to the main reservoir. The considerations as to size, gradient and lining are very much the same as for water supply; minor leakages are not likely to be important in themselves and smoothness of bore will only matter to the extent that it may make possible a smaller tunnel.

From reservoir to pressure shaft a tunnel will be continuously full, but at relatively low pressure. Minimum acceptable loss of head determines the appropriate diameter and smoothness, while a gentle gradient to permit the tunnel to be emptied for inspection and maintenance is usual. The hydrodynamic consequences of surges back from the pressure shaft must be studied in each particular system to ensure that no damage results and that no air is entrained.

Tail race discharge tunnels must be adequate to carry the maximum turbine discharge without any obstructive back pressure. As the turbines will be sited at a low level to utilise maximum head of water the gradient for the tail race will be as flat as possible. Therefore adequate area and smooth bore are likely to be important. There will also be a need to ensure against scour and erosion of the lining under any conditions of discharge. In some instances these discharge tunnels are designed to operate under pressure and require the provision of surge shafts or air shafts.

2.8.4.2 High pressure tunnels

The turbo generators being sited at low level as described above must be fed with high pressure water by the most direct line possible, minimising loss of head by friction and at bends and changes of section. A very steeply inclined or vertical shaft is usual. Hydrostatic pressures at turbine level are very high and additionally there are shock waves and surge pressures arising from the shutting or opening of turbine penstocks to vary output in accordance with demand. These pressures are kept within limits by the use of surge shafts or chambers. The hydrodynamic system comprising tunnels, shafts, pipes and turbines has to be designed as a whole to handle the very high energy of the moving mass of water. In the earlier stations it was usual for steel pressure pipes to run at the surface down the mountain side into the station, but pressure tunnels and steel-lined shafts now often take the place of the exposed pipes. In underground power stations this is necessarily the case as illustrated in the simplified section (Fig. 2.2).

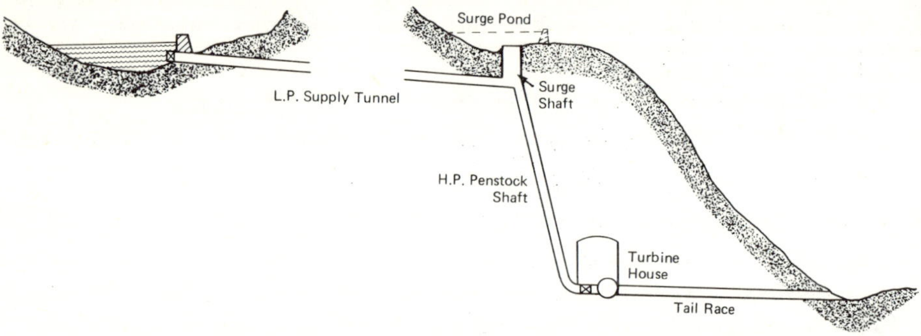

Fig. 2.2 – Diagram of hydraulic tunnels in underground power station.

The peculiar requirements of these pressure tunnels are therefore: a steep gradient requiring shaft sinking techniques rather than tunnelling and a lining capable of resisting high pressures and remaining watertight.

At great depths in rock it may be necessary to consider the natural undisturbed rock as being stressed hydrostatically under its own weight. After relaxation of this stress by virtue of the excavation the load may slowly build up again so that high stresses are ultimately imposed on the tunnel structure. To meet this condition the tunnel structure can be designed to comprise not only the lining but also such of the surrounding rock as can be integrated with it by sound construction methods.

The lining is likely to be a composite structure of concrete casing with a welded steel inner lining. The steel must not only be adequate in tension for the internal water pressure but must also be secured against the contingency of buckling when the tunnel is empty and external water and rock pressures develop. Against internal pressure substantial support to the steel plate is given by the concrete lining and also by the surrounding rock, but against external pressure the steel must be adequately anchored into the concrete.

High pressure grouting behind the concrete lining can be of great value, and indeed can be used to effect precompression of the concrete lining, which helps in avoiding tension from the subsequent high internal pressures. It was in the study of these varying stress conditions that much of the science of Rock Mechanics has been developed.

2.8.4.3 Machinery halls

Very large caverns are required for generating stations installed underground. Turbo-generators, transformers, switchgear and penstock controls must all be accommodated; and access for operation, maintenance and renewal will be needed, and cable runs and ventilation will also require tunnels or shafts.

The main chambers are usually much too large for full face tunnelling as a single operation. The chamber may have vertical walls and an arched roof with a span of 15 to 30 m. Sometimes the whole chamber is of horseshoe section. The typical procedure in its simplest form is for headings to be excavated along the springings of the arch followed by concreting of the skewback for the arch and then excavation for and construction of the arch, under the protection of which the bulk excavation can follow. The thickness of lining obviously depends on the quality of the rock, but unless the rock generally is reasonably sound it is unlikely that this type of station will be built. Much more complex sequences of construction may be necessary, depending on rock types and structures, and on the interaction of the various parts of the project (see Figs. 2.3(a) & 2.3(b)).

2.8.5 Dam bypass tunnels

In building a dam the problem of controlling the flow of water across the line of the dam, both during construction and in subsequent use, is frequently solved by tunnelling through solid ground at one side of the dam site. If this is through sound rock, lined as necessary, the river can be diverted from its normal channel while the dam foundations are laid in relatively dry conditions. If tunnels are of sufficient size to take the river in full flood work can proceed throughout the year without interruption. There is also the further advantage of allowing the dam to be built as a continuous structure without temporary apertures which have to be made good subsequently.

Whether or not diversion tunnels are used for construction, tunnels for draw off may be required. Water may be required from the same reservoir both for irrigation and for power generation, and there may also be a need for release of flood water. Separate tunnels in parallel can serve all these purposes. Except for power generation extreme smoothness is unlikely to be necessary but very high velocities and turbulence may be expected in dealing with floods, so that the tunnel must be properly lined unless the rock is absolutely sound.

The magnitude of such tunnels may be illustrated by those at the Mangla Dam on the River Indus where five tunnels of about 11 m diameter and each 500 m long were driven with a Robbins full face 'Mole'–later re-used for the Mersey Kingsway tunnel. The dangers of these great flows of water are apparent from the experience at the Tarbela Dam, also on the River Indus, where four diversion/draw off tunnels were used. When filling the reservoir in 1974 an enormous cavity developed over one of the concrete lined tunnels and some

Fig. 2.3(a) – Cruachan Pumped Storage – Machinery Hall: during excavation (Edmund Nuttall Limited).

Fig. 2.3(b) – Cruachan Pumped Storage – Machinery Hall: during installation of machinery (Edmund Nuttall Limited).

60 m of tunnel was washed away and a million cubic metres of ground, with loss of at least a year's use of the works.

2.8.6 Cooling water tunnels

Thermal and nuclear generating stations also require considerable supplies of water for efficient operation of the condensers, and tunnels are frequently necessary for the cooling water system. The usual hydraulic ideals of low velocity, large smooth bore, and absence of abrupt bends or changes of section apply to these tunnels so that pumping costs may be kept to the economic minimum.

The tunnels have to be driven in very varied and sometimes extremely difficult conditions where power stations are sited close to rivers and estuaries and varied alluvial deposits are encountered. Intakes and outfalls may have to be established in deep water, below the lowest tide level, involving difficult junctions between tunnel and shaft. Compressed air working is frequently necessary.

An unusual requirement for additional water at the oil fired Bankside power station in Central London, just across the river from St. Paul's Cathedral, was for gas washing to clean the sulphur oxides from the exhaust gas before discharging it to the atmosphere. Acid resistant lining in the discharge tunnel was necessary. As the river flow reverses with the tide, separation of the intake and outfall had to be effected by siting them on opposite sides of the river to avoid recirculation of warm water. The outfall tunnel was driven right across under the river bed into a shaft sunk from bed level. Another instance where difficult compressed air tunnelling was necessary through gravel and sand was for the nuclear generating station built on the Dungeness shingle promontory in Kent, where an old ship's timbers were encountered in the tunnel excavation. Submerged units sunk in the sea bed have been utilised at Kilroot in Co. Antrim.

2.9 SEWERS

All of the water that is supplied to a city has to be discharged again to a river system, and, because sewers start as pipes a little below ground level, the available head for discharge is normally very limited. In any case, there must be a continuous falling gradient along each sewer, no steeper than is adequate to carry the flow. Tunnelling becomes necessary when pipe depths become too great for trench excavation or when disturbance of the surface cannot be tolerated. Another tunnelling requirement arises when old sewer outfalls direct into a river or the sea are no longer permitted and new main collecting sewers must be built to carry the flows to treatment plants. Outfalls discharging limited quantities of untreated sewage into the sea are still considered satisfactory but must be carried out well beyond the tide line and may have to be tunnelled.

In comparing the mileage of tunnelling in Great Britain for various uses, other than in mine development, much the greatest is for sewers, and even although the tunnel diameter is not large, averaging perhaps about 3 m, the total excavated volume still exceeds that for any other purpose.

The primary needs for a sewer tunnel are precise gradient, smooth bore, watertightness and resistance to corrosion and erosion.

The flow in a sewer will vary greatly from dry weather flow to storm flow and the gradient must be such that adequate velocity is maintained at all times to carry solids and to prevent sand and grit from settling out. The capacity must be such as to carry the maximum storm flow. In order to maintain velocity when flow is minimal a large bore tunnelled sewer may be designed with a small half round channel in the invert, or as an egg-shaped section. Ventilation to avoid accumulation of noxious gases must always be provided.

The importance of watertightness is obvious. Brickwork of very high standard was widely used in the past; concrete is now commoner. If cast-iron is used for the structural lining it is likely to have an inner lining of brick or concrete. For the invert, blue engineering bricks provide a hard smooth surface resistant to corrosion. Very corrosive effluents from factories have sometimes to be accepted, or may be discharged accidentally. Scouring by sand and grit can also cause serious erosion unless a hard invert is provided. In hot climates, particularly where scarcity of water results in sluggish flow of concentrated sewage, linings above water level may be severely attacked by bacterial and acid action.

The smallest practicable size for driving a tunnel is about 0.7 m wide x 1.2 m high, excavated as a rectangular timbered heading, within which a pipe can be laid up to about 0.5 m diameter. It will be appreciated that a man can only work with difficulty in such restricted space and that so small a tunnel can hardly be continued for any great length. A system of tunnelling with a shield of 1 m diameter followed by lining with concrete rings built of three circular segments has been introduced.

In urban conditions there is often very little choice of route for a sewer and practically no choice of gradient; it must therefore be driven through whatever it meets, and the strata immediately beneath a city are likely to be very variable. Pipe jacking and tunnel jacking are of growing importance for sewer construction. The technique is more fully described in Chapter 4. Limitations on length are not serious if access manhole shafts can be utilised as thrust pits. Control of gradients may be found difficult.

The difficulties of bad ground increase disproportionately in large tunnels so that the relatively small size of sewer tunnels in some degree counterbalances the ground difficulties. The size also restricts the use of mechanised equipment, but tunnelling machines can be employed for larger trunk sewers where the diameter exceeds 3 m.

Sea outfalls can with advantage be tunnelled to beyond the low water mark, where pipe lines whether on the surface or in trench, would be vulnerable to storm damage and subject to the very corrosive conditions of intertide exposure.

An interesting development is the Chicago scheme to use deep level tunnels of large size to store temporarily storm water flows from the shallow combined sewerage system, so that pollution of Lake Michigan by uncontrolled and untreated overflow is avoided. A pumped storage scheme for generation of electricity is combined with the scheme.

For sewer tunnels additional requirements are the provision of means of access for inspection, cleaning and maintenance.

2.10 SERVICE TUNNELS

Increasing use is being made of tunnels to carry cables and piped services either under rivers or in cities. Where electricity cables carry much power they necessarily produce heat which must be dispersed. In some instances air cooling may be adequate but cooling by circulating water may be necessary. The Severn cable tunnel, 3600 m long, had to provide for heat removal amounting to 5000 kW by means of water circulated in pipes carrying the cables which are at 400 kV. Heating effects from trunk mains carried through highway tunnels may require special provisions.

Consideration has been given to the operation of cables at temperatures close to the absolute zero (−373°C) to take advantage of superconductivity, which eliminates all electrical resistance and the consequent heat losses. Such cables requiring very special arrangements for refrigeration and a very high standard of heat insulation would almost inevitably have to be carried in tunnels.

Telephone networks also make use of tunnels on a substantial scale in cities, allowing access for laying and inspection and room for expansion, and providing protection from the hazards of ducts laid in the ground.

Gas pipes are easily accommodated in tunnels specially provided for the individual service. Precautions are necessary against leakage and to monitor any escapes which might result in the build-up of a toxic or explosive atmospheric mixture. Proposals have been made for tunnels to be used for combined high pressure and low temperature storage and supply. A single kilometre of 3 m tunnel could provide a large storage capacity and reduce greatly the need for surface storage and also for big mains to handle peak flows.

It may be asked why so few examples exist in cities of tunnels designed to carry in a single subway the multiplicity of services required which are instead buried separately in trenches under the highways. The answer lies partly in the *ad hoc* manner in which the towns and cities–and the services themselves–have developed, partly in the numerous companies and authorities involved and partly in quite genuine difficulties of mutual occupation and safety.

Fig. 2.4 – Water mains in service tunnel under Manchester Ship Canal (Liverpool Corporation).

Among the special problems of mutual interference of services are: flooding from leaking or burst water mains, risks of explosion from leaking gas, induced electric currents interfering with telephone circuits or causing electrolytic corrosion, heat effects from electric cables carrying heavy currents. All those problems, as well as details of points of entry of each service to the tunnel and arrangements for repair and maintenance, need careful study and planning, not only for a common services tunnel but similarly if any such services are to be run through a highway tunnel. In a long highway tunnel the explosion risks from a very high pressure gas main might be quite unacceptable.

Nevertheless, there must be great potential savings to the community in this area. The cost of a sewer tunnel of minimum economic diameter is, in advantageous circumstances, no greater than the basic cost of digging a trench and providing and installing a 0.5 m diameter piped sewer. That comparison is valid without making any allowance for the disruption caused to surface traffic by the roadworks involved, and emphasises yet another important reason why service tunnels are so rare. The undertakings which dig up our highways to install and maintain services do not have to pay for the disruption they cause (indeed would be exceeding their terms of reference to do so), and the resulting distortion of the overall economics of the operation militates strongly against the construction of tunnels.

2.11 STORAGE AND OTHER TUNNELS

2.11.1 Storage

Storage underground is an ancient practice. Its most extensive modern use is probably for oil in bulk, having advantages of security against damage and fire, and causing minimal visual intrusion in areas where landscape is considered important. In suitable strata unlined cavern storage is possible at depth, oil drawn off being replaced temporarily by water to maintain a pressure balance until the oil supply is replenished. Tunnelled storage has also been used extensively to provide protection for explosives and other defence supplies and is a possible method for disposal of radioactive waste.

Underground car parks might be considered either as storage or as a facet of transportation. They are almost always constructed in cities, usually as basements in buildings, or under park areas.

2.11.2 Protection

Provision of a protected environment has been an important preoccupation of governments, particularly as a defence against nuclear weapons. Tunnelled shelters and control centres have obvious merits, particularly if massive rock is accessible. The special problems are perhaps less those of tunnelling than of structural resistance to explosives, and preservation of life and activity for a long period in a sealed environment.

2.11.3 Military

For the sake of completeness the sappers' art should be included. It has an ancient history in the undermining and breaching of fortifications, castles and city walls. Whether or not espionage tunnels fall into this category is a debatable question. The actual tunnelling methods are substantially those of driving of headings, with necessary timbering. Extensive tunnelling operations largely in chalk took place in France in World War I. More recently military tunnelling activity has been reported from Korea.

2.11.4 Mining

Apart from actual mining operations, many access tunnels and shafts are required in opening up coal and ore mines. The methods and machinery developed are very relevant to tunnelling for other purposes.

BIBLIOGRAPHY

The bibliography appended lists important books of general application to tunnel construction. A comprehensive bibliography will be included in Volume 2.

Virtually the whole range of tunnelling might be seen as covered by this chapter, but because highway, metro and railway tunnels are treated in more detail in separate chapters in Volume 2 the relevant brief bibliographies appear there. Tunnels conveying water and sewage are not further described as such in the book and some relevant publications on these subjects are therefore listed below.

General Tunnelling

Copperthwaite, W. C., *Tunnel shields and the use of compressed air in subaqueous works*, Constable, 1906.

Hewett, B. H. M. and Johannesson, S., *Shield and compressed air tunneling*, McGraw-Hill, 1922.

Richardson, H. W. and Mayo, R. S., *Practical tunnel driving*, McGraw-Hill, 1941 (Revised edn., R. S. Mayo, 1975).

Hammond, R., *Tunnel Engineering*, Heywood, 1959.

Pequignot, C. A. (ed.), *Tunnels and Tunnelling*, Hutchinson, 1963.

Szechy, K., *The art of tunnelling*, 2nd English edn., Akademiai Kiado, 1973.

Muir Wood, A. M., 'Tunnelling', chp. 30 *in* L. S. Blake (ed.), *Civil Engineer's Reference Book*, 3rd edn., Newnes-Butterworth, 1975.

Golder Associates and James F. MacLaren Ltd., *Tunnelling Technology*, Ontario Ministry of Transport and Communications, 1976.

Water, Sewage etc.

Humphreys, G. W., 'The main drainage system of London', *Min. Proc. Instn. Civ. Engrs.*, 1917, **204**.

Kent, C. G. and Morris, J. C., *The construction of main sewers including the timbering of excavations,* Longmans, 1937.

Watson, D. M., 'West Middlesex main drainage, *J. Instn. Civ. Engrs.*, 1937, **5** (Apr.).

Scott, P. A., 'A 75 inch-diameter water main in tunnel: a new method of tunnelling in London clay', *Proc. Instn. Civ. Engrs.*, 1952, **1** (May).

Fulton, A. A., 'Civil engineering aspects of hydro-electric development in Scotland', *ibid*, 1952, **1** (July).

Tattersall F. and others, 'Investigations into the design of pressure tunnels in London Clay', *ibid*, 1955, **4** (July).

Jaeger, C., 'Present trends in the design of pressure tunnels and shafts for underground hydro-electric power stations', *ibid*, 1955, **4** (July).

Cuthbert, E. W. and Wood F, 'The Thames to Lee tunnel water main,' *ibid*, 1962, **21** (Feb). *See also* discussion 1962, **23** (Dec.).

Guthrie Brown, J., *'Hydro-electric engineering practice',* **1**, *Civil Engineering*, Blackie, 1964.

Young, W. and Falkiner, R. H., 'Some design and construction features of the Cruachan pumped storage project', *Proc. Instn. Civ. Engrs.*, 1966, **35** (Nov.). *See also* discussion, 1967, **37** (Aug.).

Haswell, C. K., 'Thames Cable Tunnel', *ibid*, 1969, **44** (Dec.). *See also* discussion. 1970, **47** (Oct.).

Various authors, 'The Chicago Sewer System (Tunnel and Reservoir Plan – TARP)', *Proceedings Rapid Excavation and Tunnelling Conferences*:

1st, Chicago, 1972, Session 3, Chapters 57–61.

2nd, San Francisco, 1974, Session 4, Chapter 18.

3rd, Las Vegas, 1976, Session 11, Chapter 44.

4th, Atlanta, 1979, Session 19, Chapters 93–96.

Household, H., *The Thames and Severn Canal*, David & Charles, 1969.

Haswell, C. K., 'Tunnel under the Severn and Wye estuaries', *Proc. Instn. Civ. Engrs.,* 1973, **54** (Aug.). *See also* discussion, 1974, **56** (Aug.).

Crockett, A. S. and Dugdale, J., 'Development of Edinburgh's sewage disposal scheme', *ibid*, 1976, **60** (Feb.). *See also* discussion, 1976, **60** (Nov.).

Lander, J. H. and others, 'Foyers pumped storage project: planning and design'.

Land, D. D. and Hitchings, D. C., 'ditto: construction', *ibid*, 1978, **64** (Feb.). *See also* discussion, 1978, **64** (Nov.).

Hadfield, C., *British Canals*, 6th edn., David & Charles, 1979.

Muir Wood, A. M. and others, 'Dams and their tunnels', *Int. Wat. Pwr. Dam Constr.*, 1980, **32** (Feb., Mar., Apr. and May).

3

Investigations

3.1 PLANNING

Before a tunnel can be planned in outline and designed in detail much information on the physical aspects of the project must be sought in addition to, and having a direct bearing on, economic studies. The need for detailed and extensive investigation is probably greater than for most other types of construction. Tunnelling is necessarily costly, but it is false economy to save on the information required to make the best choice of line, level and methods.

This chapter deals with the physical aspects which must be investigated before and in the course of tunnelling. There are also the problems of economic utilisation, including traffic surveys and forecasts for transport tunnels, and general cost/benefit studies taking account of environmental aspects and, particularly in urban conditions, problems of disruption during construction.

The topography of the area concerned must be studied to the fullest practicable extent, together with the history of any relevant ground disturbance, as must its geology and geotechnology. Of course, the extent and scope of the investigation will be dependent on the size of the project and on existing knowledge and experience of the characteristics of the ground from previous excavations and works. However good the past records, each new project will benefit from additional borings and other site exploration.

Investigations for a tunnel should be a continuing activity throughout its planning, design and construction. As each item of information is utilised new and more detailed problems appear and further investigations become necessary.

The general planning of possible routes and levels based on the broad topography is followed by more detailed examination of possible alternatives to the point where the most favourable alignment apparent can be selected. Even at quite a late stage a substantial change of level or line may be found advantageous, where more competent rock may be found, or ground more suited to particular equipment. In any case further and fuller information on the ground structure through which the tunnel is to be driven will be a continuing requirement, and it is likely that there will be a need for probing ahead through the tunnel face where there are any doubts about variations in the ground.

3.2 TOPOGRAPHY

The first approach is the study of existing maps to the largest available scale, including contours or other information on levels. A site inspection will be advisable at an early stage and as often as possible thereafter in order to appreciate the significance of the features mapped and the general characteristics of the area. From the first tentative tunnel routes drawn on the map the extent of more detailed topographic survey required becomes apparent. Figure 3.1 shows a contoured topographic survey, used to enable the best route for a tunnelled railway to be selected from several possibilities.

Chapter 9 describes surveys more fully. They will usually be tied in to the national grid or other reference system and will comprise a triangulated network or a closed traverse establishing a coordinate system for the tunnel, together with infilling of details as the layout of the project is determined. Accurate levelling related to national bench marks and establishing additional reference bench marks for site control is a part of the basic survey. The choice of permanent reference points in positions which will remain undisturbed through the construction is fundamental.

The requirements will vary considerably according to the purpose of the tunnel as described in the previous chapter, and the character of the terrain.

The work may be in open country, mountainous or alluvial, or in a city centre with large and important buildings, a suburban area or an industrial area. The survey may be required to include accurate location of existing tunnels whether to avoid interference or to join up. Metro station work is particularly demanding in this respect, while in sewer work levels are of the utmost significance.

Aerial photographic survey can be most valuable both in speed and in accuracy of recording detail. It is well known that in the countryside it may sometimes reveal the unexpected presence of ancient workings.

The framework of the survey should be used as a basis of a full and accurate record of the construction including more particularly such ultimately concealed parts of the structures as temporary shafts and adits, or any areas of extensive overbreak in tunnel, or temporary excavations refilled.

3.3 SITE HISTORY

It is often important, particularly in mining and industrial areas, that the history of the site should be known. Mine workings have often been abandoned without particulars being incorporated in maps, but sometimes records exist in various

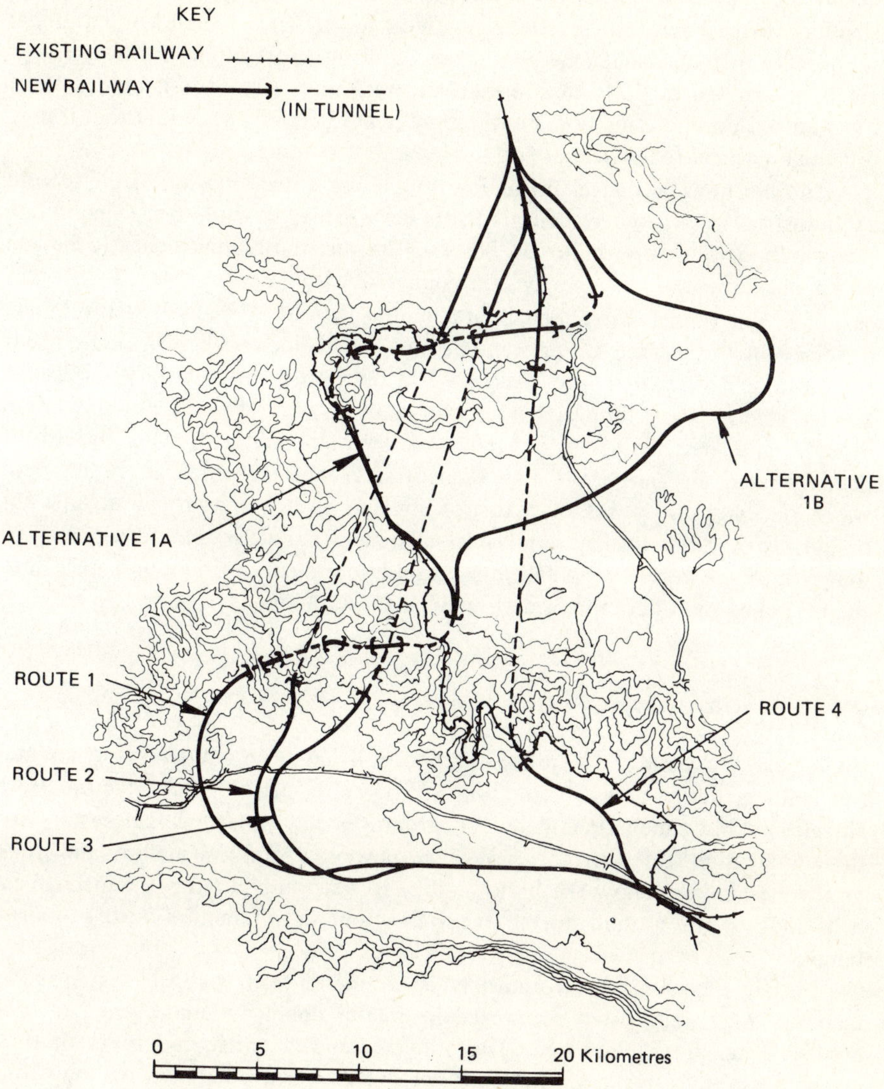

Fig. 3.1 – Contoured topograhic survey.

archives, or in deeds relating to the purchase of land from which the presence of abandoned shafts and adits can be ascertained or at least suspected. Quarries also have often been refilled after being worked out and at such a site a tunnel supposedly in rock might emerge unexpectedly into soft filling. Wells, likewise, are a hazard, particularly in some urban areas underlain by reliable aquifers. Even an old borehole may be troublesome if not properly sealed, in that it may provide a passage for water, or for the escape of compressed air.

Another unusual danger in shallow tunnelling is the pollution of the ground by industrial wastes, where contaminants have seeped down into the underlying strata over a long period. In the Blackwall tunnel the ground beneath the site of an old gasworks was heavily contaminated with tarry residues. In another case of a cable tunnel through chalk at a depth of about 20 m, hydrogen sulphide dissolved in the ground water was encountered. Under a steelworks site spent sulphuric acid discarded into soakaways in the past made necessary the use of special acid-resisting linings. Under old colliery waste tips excess of carbon dioxide and deficiency in oxygen has been encountered. This hazard of deoxygenated air can occur also in natural ground in which air rather than water occupies the pores of a soil or stratum with a content of organic silt or pyrites or other readily oxidisable substance. The deoxygenated air may be drawn into the tunnel by a fall in atmospheric pressure or may be forced into the tunnel by use of compressed air in adjacent workings.

3.4 GEOLOGY

Geological survey and geotechnical studies of surrounding ground are fundamental to planning, design and construction. Stratigraphy, petrology and tectonics are all relevant but much finer detail of structures and variations is necessary for tunnelling than for broader geological purposes. Soil mechanics and rock mechanics studies follow the basic geology to assist in predicting the mechanical behaviour of the ground during excavation and its interaction with supports, temporary and permanent.

Initially, geological information is available from published maps and memoirs which are often supported in greater detail by unpublished records accessible to those interested. The alluvial deposits, drift and types of rock occurring in the area and their interrelationship will be shown on maps but precise interface positions and changes in texture and strength are likely to be of great significance in tunnelling and must then be the subject of fuller exploration.

Any records of previous borings may be of great value. Inspection of exposures in quarries, pits, shafts and mines is almost always worthwhile and also in cliffs and stream beds and any excavations. Records of earlier tunnels should be sought and also of wells and deep foundations.

The major items of geological and geotechnical information required include:

1. Geological description with details of lithology and variability.
2. Location and orientation of discontinuities and planes of weakness relative to tunnel excavation–bedding planes, joints, faults, shear zones.
3. In situ stresses.
4. Geomechanical properties.
5. Ground water.

These are discussed in more detail in Volume 2.

3.5 SITE INVESTIGATION

The principal means employed to obtain additional knowledge of the tunnel are borings, trial pit shafts and exploratory headings. The objective is to identify and locate precisely each stratum relevant to the proposed tunnelling and to assess its characteristics and variations, and its behaviour during excavation.

The strata are identified and their characteristics ascertained from samples taken and from inspection and tests both in situ and in a laboratory. The whole programme needs to be very carefully planned so that the right questions are asked and the necessary answers obtained as economically as possible and without overloading a particular project with a mass of superfluous or irrelevant tests and studies. This again emphasises the advantages of a progressive investigation programme in which the areas of uncertainty are narrowed down.

As an example, a roadway may be required to form the descent from the surface through an open cut approach into a tunnel which passes under an obstacle. If the tunnel section can be sited so as to be wholly or mainly within a thick stratum of uniform character then borings for that section need be only sufficient to check that uniformity and to ascertain the relevant mechanical properties and behaviour, always subject to the proviso that thicknesses may be less than expected and characteristics may vary locally. Any place where an anomaly, such as a buried channel or a fault is suspected will require much more detailed study.

The approach ramps which may be through much more variable strata including soft water-bearing ground will also call for a detailed study with careful consideration of the methods of construction. These might include open cut construction supported on piles followed by cut-and-cover tunnelling involving the use of ground water lowering techniques and ground treatment by injection grouting.

Variations from typical characteristics may occur locally due to weathering, particularly in the vicinity of a fault or fissure, or an ancient land surface. As an example, the London Clay of Eocene age has a thickness of up to 100 m or so, but its surface has been eroded in the past and it is cut into by old channels or swallow holes, so that any tunnel driven near its upper surface may encounter such an anomaly unexpectedly.

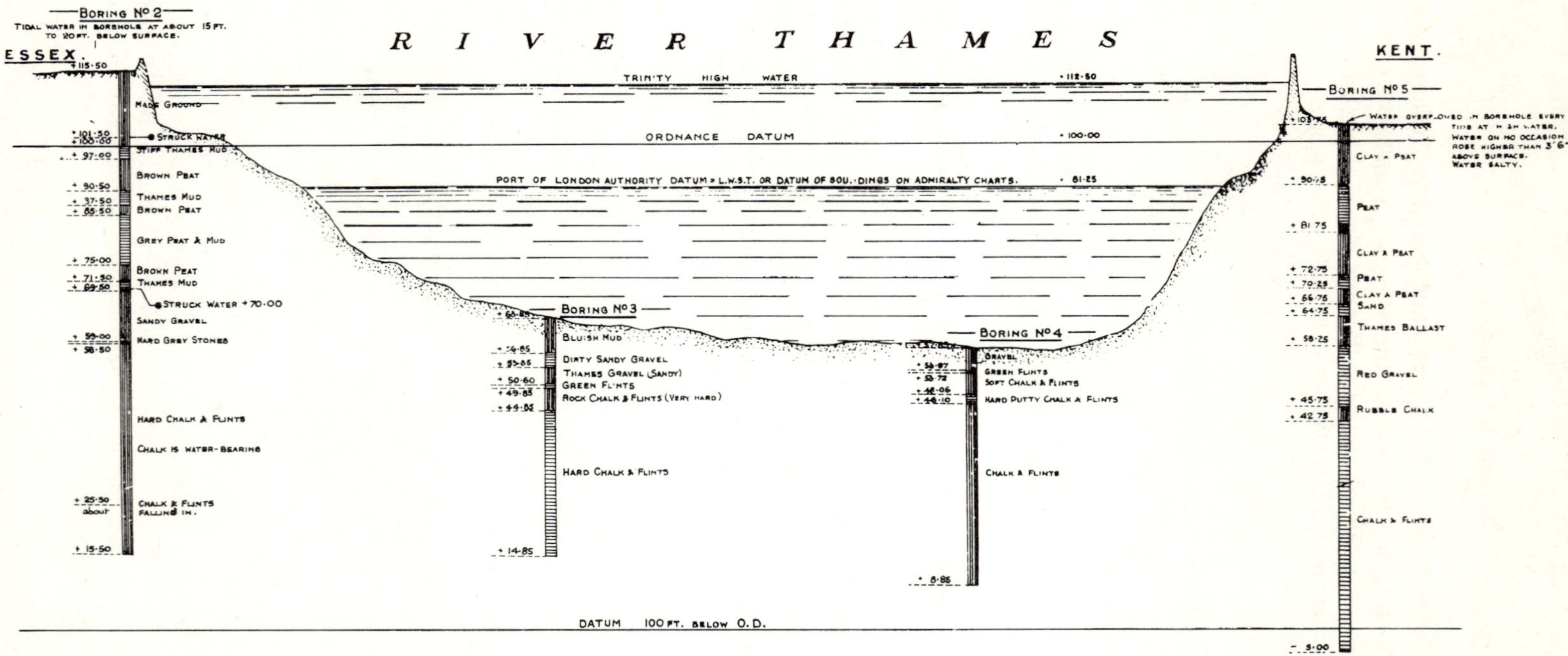

Fig. 3.2 – Section showing preliminary boreholes for Dartford Tunnel. Note that vertical scale is 10 times horizontal scale.

No pattern of borings can be sufficiently closely spaced to identify all such anomalies but the number of borings will in general be influenced by the variability expected or found.

In the first stage of site investigation by borings there should at least be sufficient to check the presence and levels of the expected strata and their principal characteristics. Some of these boreholes should be taken well below the intended invert of the tunnel to ascertain what lies below and to provide information on which a judgement can be formed as to the best level for the tunnel. Such deep borings may also identify the presence of aquifers under pressure (artesian water) or other significant features.

Following this first stage, tentative decisions on the tunnel route and suitable methods of construction are possible, and a second series of borings will furnish reasonably complete sections along and across the tunnel, but always in the knowledge that conditions between boreholes cannot be interpolated with absolute confidence. The closer and the more numerous the boreholes the more reliable is the cross section if similar strata are encountered consistently. Particularly in alluvial ground changes may be very rapid; in built-up areas and areas worked for minerals former excavations may have disturbed the uniformity of pattern; in rock unexpected faults, fissures and zones of weathering may occur.

The third stage of site investigation is to check at critical points, as design and geometry of the tunnel are finalised, to fill in gaps of information to resolve uncertainties. A pilot tunnel or exploratory heading may assist in these functions. Even after all these investigations excavation in any tunnel is exploratory and in many cases probing ahead should be adopted as a standard procedure.

The sinking of boreholes is by far the most widely applicable method of investigation for reasons of flexibility, speed and economy. They may range from hand augering to depths of a few metres in holes of about 40 mm diameter up to 1 m diameter pits, but for most tunnel purposes boreholes will be in the range of 100 mm to 400 mm diameter with depths normally less than 100 m, but much more where necessary.

3.6 METHODS OF BORING

The various methods of boring may be listed as:

1. Hand auger boring, for shallow holes in soils.
2. Light cable percussion boring adaptable for almost all ground other than deep holes in hard rock.
3. Mechanical augering, for cohesive soils.
4. Rotary open hole or core drilling, for rock.
5. Wash boring, for sands, silts or clays where sampling is unimportant.

Hand auger boring is very limited in its application to tunnelling. Depth cannot exceed about 5 m and the hole is uncased.

Light cable percussion boring is the most adaptable system for general use. The equipment comprises tripod, winch, light cable, drill rods, excavating tools and sampling tools. Relatively light and simple equipment can produce most of the essential information and samples. In its most portable form, the equipment can be dismantled into small units and transported to and set up on any part of

Fig. 3.3(a) – Drilling rig and tools – Mobile drilling rig, with percussion and auger capability (Engineering Laboratory Equipment Ltd.) . . . .

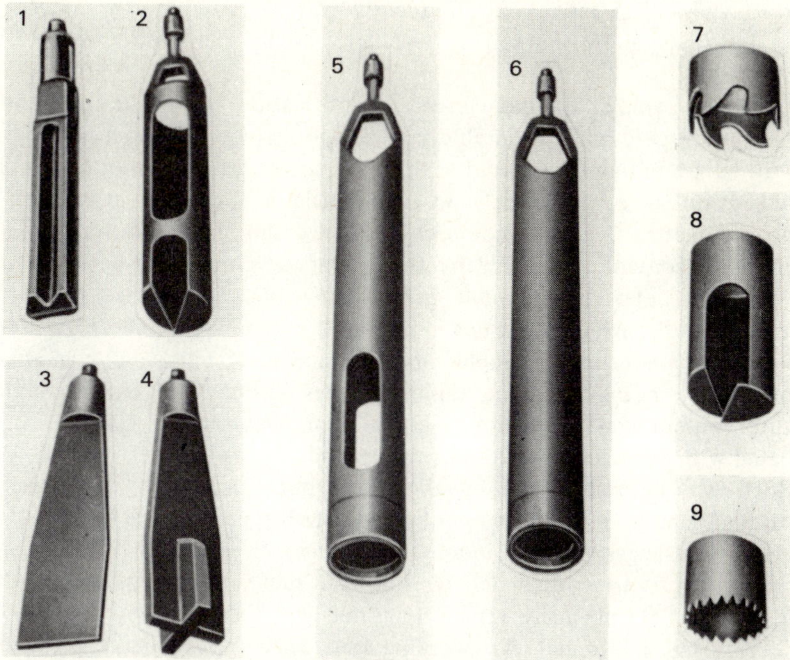

Fig. 3.3(b) – Tools for use with drilling rig. 1 Drilling bit for rock: 2 Auger bit for soft ground: 3 & 4 Chisels for breaking out rock: 5 Clay corer: 6 Shell for cleaning out hole after drilling: 7, 8 & 9 Gravelling, auger and serrated shoes for use with shell (Engineering Laboratory Equipment Ltd.).

any site, without requiring the construction of special access roads or elaborate levelling of a working site. it can be set up and operated in a limited space with a minimum of preparation and disturbance. To ensure the best use of resources, and reliable information with properly taken and prepared samples for tests, highly expert supervision is essential, but relatively unskilled labour can be used for much of the site work.

A hole is made by percussion methods, lifting and dropping a shell bit at the bottom of the hole for granular strata, or turning by means of rods an auger bit or clay cutter for cohesive soils, or lifting and dropping a chisel bit for rock or boulders. As the hole is sunk it is normally lined with steel tubes driven down. Unless water is present in the ground, a piped supply is needed for use of the shell in granular material to assist in collecting and drawing up the material.

In all these methods the excavated debris is extracted and examined at the surface, samples being preserved as appropriate. The samples are inevitably much disturbed and unless taken very frequently will be mixed by the process of drilling but they allow the soil at each level to be identified and classified broadly.

To obtain samples for more precise identification and for testing, boring must be interrupted while a suitable sampling tool is mounted and used in the undisturbed ground at the bottom of the hole. Granular soils cannot be sampled without disturbance, particularly where particle size is large in relation to borehole diameter. Degree of compaction is inevitably disturbed, fines are lost and moisture content is changed by the operation; sampling is better in very large diameter holes. In situ cone penetration tests can be used to provide information on behaviour of the ground.

Relatively undisturbed samples of clays—and marginally silts—can usually be obtained by means of suitable sampling tubes skilfully used. Grain structure, moisture content and strength properties from tests on the samples can be correlated with expected behaviour of the ground.

Rock will be fragmented by the chisel but can usually be identified. Undisturbed samples require the use of an annular bit, producing a core.

Mechanical augers are continuous flight augers usually with a hollow stem and having an outside diameter of 150–250 mm and a stem of half that diameter. They are normally vehicle mounted and therefore require adequate road access and a firm and level working area. They are used in cohesive soils, and have a hollow stem, which is plugged in general use, but can be used for sampling or for core drilling into rock below the soil. Larger augers drilling out a one metre diameter inspection shaft can be very useful in suitable conditions.

Rotary drilling is used mainly in rock but also in hard cohesive soils and the bit may either cut out all the material leaving an open hole or be of annular form preserving a relatively undisturbed core. Cores are usually 70 mm diameter but may be only 55 mm, or may be larger.

Diamond bits are usual and hollow rods to carry down fluid to lubricate and flush the operation.

3.7 SAMPLING

If open hole drilling is employed any required samples have to be taken separately, interrupting progress for the purpose. In any strata affecting tunnel construction, a continuous core is very desirable to help in determining the presence of discontinuities and the degree of fragmentation of the rock. Horizontal bedding and other joints will be reasonably well shown in a vertical bore but vertical and steeply inclined joints are unlikely to be adequately represented. Inclined boreholes designed to cut across such joints can, in special circumstances, provide very valuable additional information.

Wash boring is a relatively cheap, quick and easy way of probing through sands, silts or clays but information of the character of the soil at any level is minimal. It can provide a useful check on rock level where cover is shallow and variable.

Sampling may be adaptable to many purposes but not all may be relevant to a particular project. In granular soils particle size and grading, density, moisture content, and permeability are the principal items on which information is sought. Particle size and grading in particular will govern the behaviour of the soil in tunnelling and the passage through it of water and other fluids such as grouts or compressed air (Fig. 3.4).

In cohesive soils the same items are repeated with the addition of strength characteristics, namely: shear strength and compressive strength and also Atterberg limits which provide a guide to changes in plasticity with varying water content.

To allow most of these to be determined in a laboratory it is vital that the samples should be as nearly undisturbed as possible and that they should be carefully sealed off against change of moisture content immediately after securing the samples.

In rock the samples are used to determine corresponding characteristics; more particularly: density, permeability, crushing strength, mineral composition, and sometimes elastic moduli. Natural water content may be significant in some rocks and under some conditions. Of major importance is the structure of jointing and fragmentation in the rock and any changes caused by deep weathering. These properties are not readily mapped from test samples and may have to be explored by site tests and geophysical methods.

In all sampling the determination of the properties of the soil or rock sample specimens is subordinate to assessment of properties in the mass as actually encountered *in situ* when tunnelling. The hardest rock if sufficiently fragmented by faulting and crushing may behave in the mass like a granular soil. Under high pressures and stresses and in the presence of water shaly rock may resemble clay in plasticity, flow and swelling. Weathering from a surface exposed in the past or from a fault may disintegrate an otherwise sound rock at depth. These features may not be disclosed by borings and samples and may only appear during construction, but some indications can be obtained from site testing or geophysical exploration, or from exploratory shafts or headings.

3.8 IN SITU TESTS

As part of a scheme of exploratory borings *in situ* tests of ground behaviour can be valuable. The Standard Penetration Test measures the local resistance to the driving of a cone or sampler at the bottom of a borehole by use of a falling weight. In cohesive soils the Vane Test for resistance to rotation of a vane evaluates local shear resistance. Permeability may be assessed in a borehole, or an isolated length of a borehole by measuring acceptance of water under constant head or under falling head, or outflow of water. In a trial pit or shaft

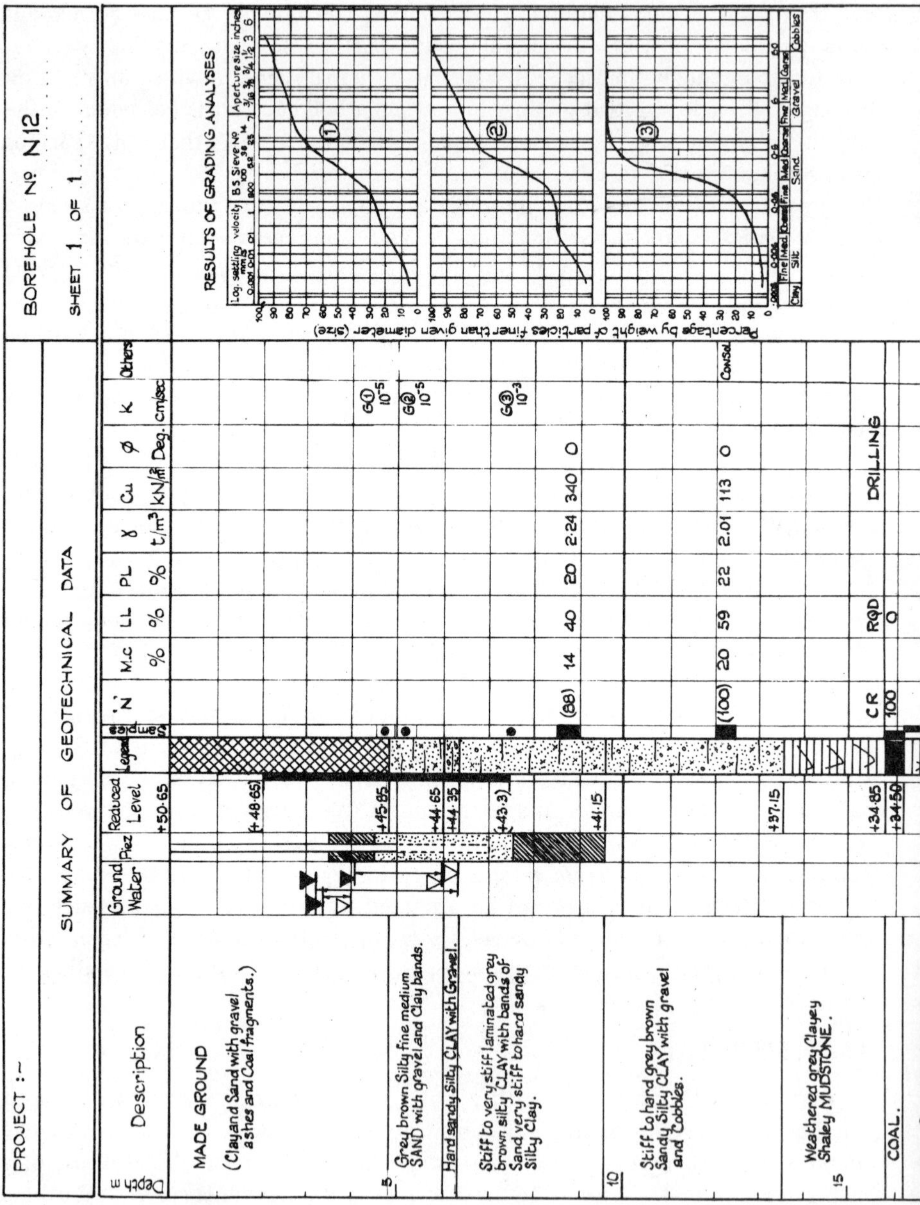
PROJECT :–
SUMMARY OF GEOTECHNICAL DATA
BOREHOLE Nº N12
SHEET 1 OF 1
Depth m
Description
Ground Water
Piez
Reduced Level
Legend
Samples
'N'
M.c %
LL %
PL %
γ t/m³
Cu kN/m²
φ Deg.
K cm/sec
Others
+50·65
MADE GROUND
(Clay and Sand with gravel ashes and Coal fragments.)
(+48·65)
+45·85
5
Grey brown Silty fine medium SAND with gravel and Clay bands.
+44·65
Hard sandy Silty CLAY with Gravel.
+44·35
Stiff to very stiff laminated grey brown silty CLAY with bands of Sand very stiff to hard sandy Silty Clay.
(+43·3)
+41·15
10
Stiff to hard grey brown Sandy Silty CLAY with gravel and Cobbles.
+37·15
Weathered grey Clayey Shaley MUDSTONE.
15
+34·85
COAL.
+34·50
(88) 14 40 20 2·24 340 0
(100) 20 59 22 2.01 113 0
ConsoL
G① 10⁻⁵
G② 10⁻⁵
G③ 10⁻³
CR 100
RQD 0
DRILLING
RESULTS OF GRADING ANALYSES
Log. settling velocity mm/s
BS Sieve Nº
Aperture size inches
Percentage by weight of particles finer than given diameter (size)
Clay
Silt
Sand
Gravel
Cobbles
Fine Med. Coarse
①
②
③

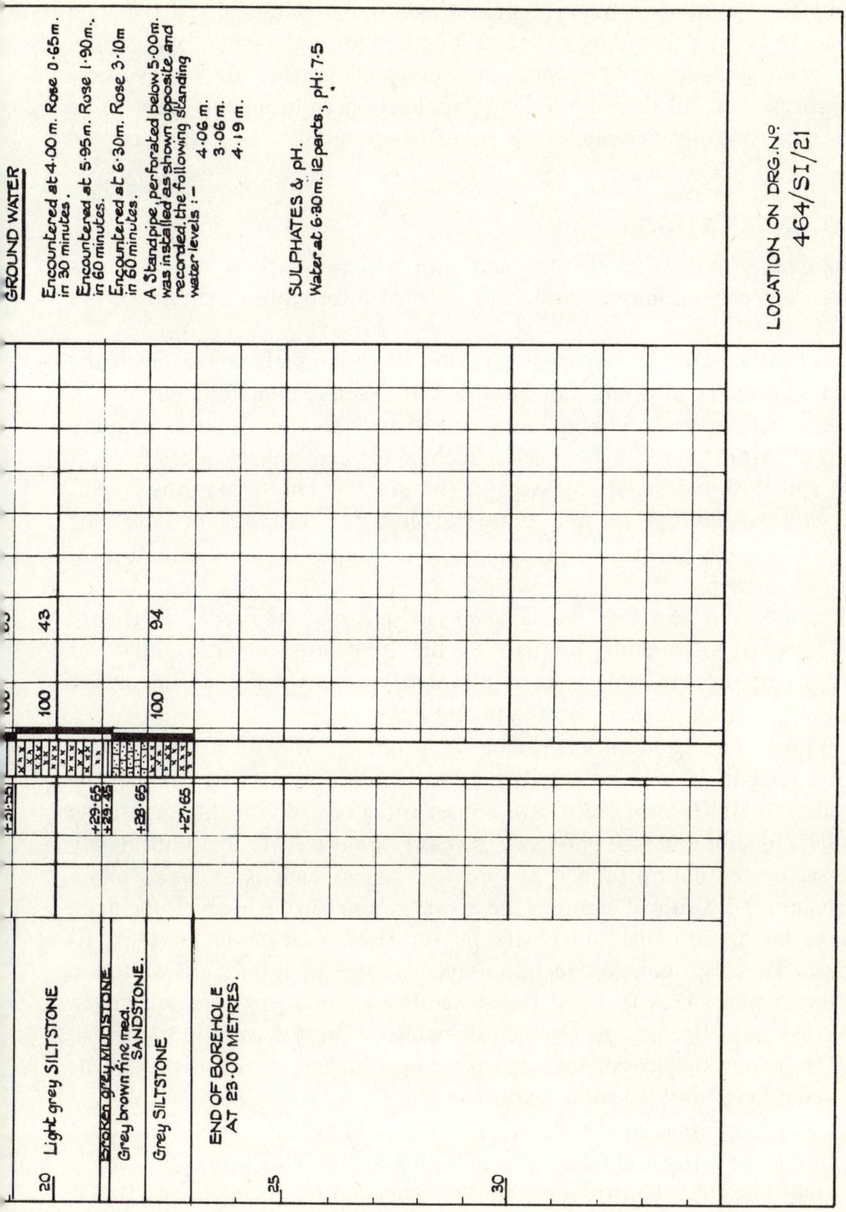

Fig. 3.4 – Typical borehole log, with grading curves.

of sufficient size plate loading tests may be made to determine the bearing strength of the ground at that level.

Geophysical methods include seismic refraction and electrical resistivity on land surfaces and seismic profiling and echo sounding for underwater exploration. For large scale geological phenomena magnetic and gravimetric surveys may provide information. All these methods in application to tunnelling need to be based on and carefully referenced to a sufficient number of fully sampled boreholes.

3.9 TRIAL EXCAVATIONS

Within their limitations of availability and cost, trial excavations in pits, shafts or headings will provide more useful and detailed information than any other method.

In pits, excavated under expert supervision from the surface and lined, the nature and succession of strata can be observed exactly, numerous samples of any size can be obtained with minimal disturbance of the samples, and any appropriate *in situ* tests can be made. Such a pit can remain accessible for inspection and further tests at any stage of the project. The limitations on this are: first, site availability; second, time and cost of construction; third, the presence of water may necessitate pumping with consequent disturbance; fourth, depth may be restricted.

Lined shafts have most of the same merits and may be carried to greater depths. They may also comprise part of the tunnelling scheme, either for temporary access or with a permanent function, but this will limit the choice of site and may restrict scope for tests and samples.

For tunnel construction a heading is probably the most informative exploratory system. It may be driven into the hillside as an adit, or may be driven from a shaft. It may constitute a pilot tunnel driven on the line of the main tunnel and subsequently enlarged, or it may be driven to serve ultimately as a drainage or ventilation tunnel, or to carry services such as cables or pipes. Even although a pilot tunnel explores the actual ground through which the main tunnel is to be driven, there may still be surprises in store in effecting its enlargement. The disproportionate importance of size in tunnel excavation is discussed in Chapter 4. A fault, of trivial significance in a small tunnel, driven and supported quickly, may give serious difficulty when the unsupported span of ground is greater and it remains open for longer and when the driving of the pilot may itself have initiated some disturbance.

This was exemplified in the Mersey Kingsway tunnel (1971) where a fault required only some extra timbering and propping in the pilot tunnel, but made necessary multi-heading construction in that area during enlargement. In the Eisenhower Memorial tunnel, also mentioned in Chapter 4, a pilot tunnel was driven to study the ground thoroughly, but over a lengthy section enlargement proved unexpectedly difficult because of squeezing rock.

Ground exploration is, of course, only one function of a pilot tunnel: others are access for ground treatment, reduced face loads and extra support in main tunnel, improved access, ventilation and drainage.

3.10 HYDROLOGY

The importance of water to tunnelling can hardly be over-estimated and information on its various manifestations is essential. It appears at the surface as springs, streams, rivers, lakes, estuaries and the sea. Below ground it is present as pore water almost everywhere below the water table, but is usually present also above the water table where held by capillary action or by adsorption; it may be static or may flow through permeable ground or fissures, or as underground streams, particularly in limestones. Its relevance to tunnelling is fourfold:

1. It may flood a tunnel, breaking in at an exposed face, or as surface water entering from a shaft or adit.
2. The presence of water in the ground adds hydraulic pressure to the loading on a watertight lining.
3. Pore water modifies significantly the plastic properties and strength of soils and water in fissures may reduce the strength of rock by lubricating joints.
4. Water intruding into beds of anhydrite or other minerals may act chemically on them causing swelling and disruption.

These aspects are further discussed in Volume 2.

3.10.1 Surface Water

Records of river levels and flows and lake levels are frequently available and are particularly useful if correlated with rainfall. In the absence of records systematic observations may be advisable.

Where tunnelling is in the vicinity of tidal water, tide tables recording as fully as possible times and heights of tides are indispensable together with any available information on tidal surges which may be superimposed on normal tides and, in estuarine conditions, any records of river floods.

In addition to surface flooding problems, changes in these surface levels will as a rule eventually affect water table levels in the ground and therefore pressures and rates of inflow in tunnelling. If compressed air is being used adjustment of working pressure will be required. The surface openings of all shafts or adits, particularly if sloping downwards, should of course be sited above maximum flood levels or be provided with appropriate protective walls or bunds or with flood gates. In urban work, the possibility of surcharges in adjacent sewers should not be overlooked and also possible backflow of water along pipes which have been intercepted but not adequately sealed off.

Another aspect of water standing in lakes or marshes or in peat is that methane gas may be generated where vegetation has decayed in the sediments; its explosive characteristics call for special precautions.

3.10.2 Ground Water

The water table is the significant level of ground water. In any stratum it is the level at which water will stand in an open hole connecting only to that stratum. It must be appreciated that the water tables in successive strata in a borehole may differ. For example, the water table in the shallower soils overlying an impermeable clay may be a little below ground surface, but penetration of a permeable stratum below the clay may tap artesian water having a level of water table above ground surface.

In variable ground with lenses of granular soils alternating with silts and clays there may be 'perched water tables' over limited areas.

The determination of water table levels is an important function of site investigation by borings and requires careful observation of water levels and flows in boreholes. As the operation of boring may involve the addition and extraction of water, natural undisturbed levels can usually only be observed during interruptions of boring long enough for equilibrium to be established. Every opportunity to make such records should be utilised. To ascertain longer term variations in water level, standpipes or piezometers carefully set into boreholes to respond to a single stratum may be established and observed over any suitable period.

Permeability of water-bearing strata is an important factor in determining the inflow of water to be expected and its influence on ground behaviour. It is also fundamental to any injection grouting processes which may be contemplated (Volume 2).

3.11 GASES

Gases in tunnelling are significant because of explosive or toxic properties. The most common as a tunnelling hazard is probably methane (Chapter 4) which may originate in strata having an organic content and may emerge in adjoining areas. It is particularly common in the vicinity of coal seams but also arises from decaying vegetation in lake beds or from peaty deposits and sometimes from rubbish tips. It is considerably lighter than air (rel. density 0.56) and is easily ignited in mixture with air at concentrations of 5% or more with a resulting explosion. If strata of this nature are present tests for methane should be made in any investigation.

Hydrogen sulphide (H_2S) may also appear as a product of organic decay and can be highly toxic. It may be present in existing or abandoned sewers.

Carbon monoxide (CO) is most familiar as a constituent of petrol engine exhaust gas but may be generated in smouldering coal seams or any fires where combustion is incomplete. It is highly toxic and is very slightly lighter than air (rel. density 0.97).

Carbon dioxide (CO_2) results from complete combustion in coal or other fires or where acid acts on limestone. It is not itself toxic, but can be dangerous by displacing air or reducing the proportion of oxygen present in the atmosphere. It is denser than air (rel. density 1.53) and therefore tends to accumulate in pits and sumps.

Sulphur dioxide (SO_2) is toxic. It results from combustion of sulphur and its compounds and may appear naturally in volcanic areas.

3.12 STRUCTURES

Investigations into existing structures below ground should be comprehensive, not merely to ensure clearances, but also to allow the risk of settlement to be minimised.

Most existing tunnels are recorded but may need to be surveyed precisely and in detail unless very adequate clearances from the projected tunnel and its zone of disturbance are assured. Deep foundations and piles are not always as well documented as is necessary and may have to be the subject of special surveys, trial pits, or borings.

Early mines, whether for coal or other minerals, are rarely recorded in detail and may be altogether forgotten. In any such areas the preliminary investigations need to be extensive and the methods of working adopted should take into account the contingency.

BIBLIOGRAPHY

The requirements outlined in this chapter are described in considerable detail in a series of publications of the TRRL and reference should also be made to British Standards and Codes of Practice, and other publications listed below.

Glossop, R. and Skempton, A. W., 'Particle-size in silts and sands', *J. Instn. Civ. Engrs.*, 1945, **25** (Dec.).

Harding, H. J. B., 'Site investigations, including boring and other methods of subsurface exploration', *ibid*, 1949, **32** (Apr.).

In situ investigations in soils and rocks, British Geotechnical Society, London, 1969.

Safety in wells and boreholes, Institution of Civil Engineers, London, 1972.

Subsurface exploration for underground and heavy construction, American Society of Civil Engineers, New York, 1974.

B.S. 5573, Code of Practice for safety precautions in the construction of large diameter boreholes for piling and other purposes (revision of CP 2011), British Standards Insitution, London, 1978.

B.S. 5930, Code of practice for Site Investigations (revision of CP 2001), British Standards Institution, London, 1981.

Laboratory and Supplementary Reports, Transport and Road Research Laboratory, Crowthorne. There are also special site investigation studies, listed in the main bibliography (Volume 2):–

Dumbleton, M. J. and West, G., *Preliminary sources of information for site investigations in Britain, LR 403,* revised edn., 1976.

Dumbleton, M. J. and West, G., *Guidance on planning, directing and reporting site investigations, LR 625,* 1974.

Dumbleton, M. J. and West G., *A guide to site investigation procedures for tunnels, LR 740,* 1976.

BRE/TRRL Working Party, *Probing ahead for tunnels: a review of present methods and recommendations for research, SR 171,* 1975.

4

Tunnelling in Soft Ground

4.1 SCOPE

Tunnel construction is a complex operation and is only efficient to the extent that the various elements are satisfactorily integrated in the construction process. What is under discussion here, and in Chapter 6, is work at the face in bored tunnelling, the essence of which is excavation from inside and underneath. Three items from the list of fundamentals in Chapter 1, namely: excavation, immediate support, and management of water, interact so closely in soft ground that they cannot be isolated. Permanent support is more fully considered in Chapter 8, dealing with lining of tunnels. Survey also receives separate treatment, although it is vital at every stage of excavation in giving accurate guidance, for which purpose it is essential that clear sight lines must be available, unobstructed by plant and equipment or any other obstacle.

All recent development has been towards integration of design and construction as a unified industrial process, labelled 'systems engineering'. Many of the problems are logistical, concerned with the best methods of removing the excavated spoil and bringing in materials for lining, and the provision of essential services–light and power, pumping and water supply, ventilation, etc.– and the continual extension forwards of those services as the tunnel advances. Spoil handling is of particular importance to the economics of rapid tunnel excavation and the choices are briefly outlined below. It remains of the utmost importance that any system adopted is sufficiently flexible to cope with the unforeseen hazards almost inevitably encountered in course of excavation.

In these chapters the aim is to concentrate on the basic work at the tunnel face, that is to say the excavation and support of the ground and related management of ground water. Although soft ground and rock tunnelling differ obviously in their practices, they also have much in common, and the same principles, and many similar techniques, are applicable, more especially in difficult ground in which there is a need for continuous and detailed support of exposed ground. Much of what is said here is therefore relevant to the chapter on tunnelling in rock.

Fig. 4.1 – Spoil disposal by skips on rail track.

The fundamental difference lies in the elastic properties and strength of intact rock which are not greatly modified by the presence of water, and which allow a cavity to remain self-supporting, in contrast to the lower strength of soft ground and the dependence of its properties on the presence and movement of pore water and its pressure pattern.

The geologist uses the term 'rock' comprehensively for all deposits with which he is concerned, but his 'unconsolidated sediments' correspond to the tunneller's soft ground, typically such deposits as clays, silts, sands, gravels. These are characteristic of valleys and plains, which are the usual sites of cities, and therefore much urban tunnelling, which is generally shallow, is in soft ground. Mountain tunnels, in contrast, are cut through ridges, which have resisted erosion by reason of the presence of hard rock. Subaqueous tunnels may be of either type according to whether the river crossed is currently running in and eroding a rocky bed, or has, above rock level, sedimentary deposits of adequate substance and thickness for the tunnelling process adopted.

The association of soft ground tunnelling with urban needs is reflected in metro construction, sewers, water supply and other services. In some cities rock is encountered widely at shallow depths and most tunnelling must be in rock. In metro construction there is often a choice between cut-and-cover and bored tunnelling, discussed in Volume 2. Sewer tunnels, of small to medium size, probably involve a greater mileage of tunnelling than all other uses added together. They are likely to be shallow tunnels driven through very varied strata, often in built-over areas. Water supply tunnels are less characteristically urban where they transport remote supplies citywards, but they are also driven within metropolitan boundaries for trunk mains in preference to pipes laid in trench.

4.1.1 Spoil Handling

The economic importance of efficient spoil handling and disposal is particularly relevant in a long tunnel, whether in soft ground or rock. The problems are logistical rather than structural. For every tunnel the choice of appropriate methods must be made in the light of the expected volume, the rate of output fixed by the excavation cycle, the working space underground and at the surface, the machinery and equipment available and the ultimate disposal facilities or requirements.

So many different types of handling equipment, frequently designed for use in mines, are available, and such varied combinations are possible, that detailed consideration in the present context is not practicable.

In general, where work proceeds from a shaft, there are four phases to consider, namely: loading at the face, transport to the shaft, hoisting and ultimate disposal. At all stages the nature of the material, its bulking, fragmentation and particle size, and water content affect significantly the choice of plant and necessary capacity. The choice of transport method, broadly rail skip, conveyor, road vehicle or slurry pumping, is the key to all the other phases. The commonest method is probably by rail skip hauled by battery locos or, in large well ventilated tunnels, by diesel locos. In small tunnels man haulage may be appropriate. This system involves track laying and maintenance, with frequent extension as the face advances and provision of passing facilities to allow empty skips to be brought up as the loaded skips are moved back. Loading at the face presents problems of access to the skips, depending on their capacity in relation to volume of output.

Material from tunnelling machines is likely to be passed back by conveyors to a loading point. In small hand excavated tunnels direct shovelling into skips may be necessary, or there may be room for a mechanical shovel, or a loading conveyor.

At the shaft a hoist may be installed with a cage into which skips may be run to be lifted to the surface, or a crane may be used. Discharge can be either direct into a road vehicle, or to a spoil heap or a storage hopper, providing a reservoir to smooth out operations. Other shaft hoisting operations are, of course, possible such as a bucket elevator, or a conveyor in an inclined shaft.

For the main transport from face to shaft, or to portal, or for part of the way a conveyor may be preferred to rail skips, but, of course, other transport is necessary to carry lining materials to the face. Conveyor systems may require transfer points which can cause difficulty if there is excessive spillage.

In some tunnels, where there is direct road access through the portal, road vehicles are used for the whole operation, with mechanical loading at the face.

Another system is conveyance by slurry pumping. This is an integral feature of the Bentonite Shield method. It has also been used in other tunnels such as the first Dartford Tunnel, driven in compressed air through chalk. The spoil was crushed and mixed to a slurry with water, flints being either screened out if large, or crushed. The slurry was pumped out through a 150 mm pipe to settling lagoons. One great advantage was that no 'locking out' operation was necessary.

Ultimate disposal may be by tipping to waste locally, or for reclamation of old quarry sites, or for building road or other embankments.

4.2 PATTERN OF GROUND STRESS

In undisturbed ground, whether rock or soft, there exists a state of stress increasing with depth. The largest component is normally vertical compression proportional to the weight of overburden, usually accompanied by horizontal compression. The magnitude of the horizontal stress is commonly about a quarter of that of the vertical stress, although conditions may be very different where geological forces, recent or ancient, have operated, and where rock structure is not uniform.

The presence of a cavity within the ground mass inevitably changes the pattern of stress. The simplest illustration of the nature and order of magnitude of the stress changes is, perhaps, the two-dimensional case of a circular hole

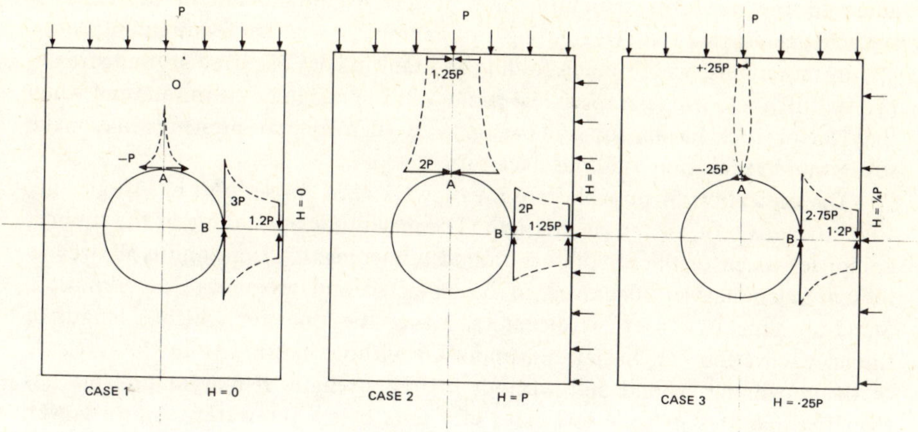

Fig. 4.2 – Patterns of ground stress.

representing a tunnel cut out of a uniform elastic solid in which there is a vertical stress, P, and a horizontal stress H (Fig. 4.2).

Three cases are considered:

$$H = O, \quad H = P, \quad H = \tfrac{1}{4}P$$

The second case represents a hydraulic pressure pattern, and the third case is more typical of natural conditions.

The stresses of particular interest are the circumferential stresses at the crown of the tunnel (A) and at the axis level (B). In the first case, with vertical loading only, tensile stress of P develops at A and compressive stress of 3P at B, but if followed outwards for a distance of one radius along the vertical and horizontal radii respectively both revert nearly to the general stresses which are respectively zero and P.

In the second case, compressive stress of 2P occurs at both A and B, again diminishing outwards to the general stress P at a distance of one radius.

In the third case, a small tensile stress is experienced at A, and a compressive stress of about 2.7P at B.

In all cases, radial stresses are necessarily zero at the cut surface. They increase outwards to their general values. Rock bolting may be used to increase them.

The implicit arching action is apparent, with the need for horizontal thrust or tensile strength at the crown, and the increased loading at springing. What is also of importance is the rapid diminution outwards of stress changes, showing such a tunnel to be very localised in its effects.

The assumptions of uniformity and elastic behaviour are rarely borne out in practice, but the indication of ground stresses in the range −P to +3P is helpful, when correlated with the compressive strength of the ground, as a guide to the need for structural support. Local containment of the exposed surfaces against ravelling, weathering, or swelling is a separate requirement.

In mountains, where much folding and faulting has occurred ground stresses may be high and irregular. Residual geological forces can be enormous and when they act on rock having some plasticity, very high squeeze pressures may make soft ground excavation methods necessary.

The indication of possible tension shows a need for support in soft ground, which has little or no tensile strength. The magnitude and timing of the support comprises much of the art of soft ground tunnelling. If the ground is allowed to loosen much heavier loads are to be expected and progressive deterioration. Stand-up time is a useful concept, as being the time for which a length of tunnel excavation can be left unsupported without detriment to the integrity of the remaining ground and without loss of strength. It may range from less than five minutes in very soft clay, silty sand below the water table or gravel, up to 24 hours or more for dense sand or gravel with a binder, or stiff continuous clay.

4.3 EXCAVATION IN SOFT GROUND

In the tunnelling sense soft ground broadly comprises gravels, sands, silts, clays and alluvial deposits generally. From the aspect of behaviour Terzaghi has classified such ground into the categories: Firm, Ravelling, Running, Flowing, Squeezing, and Swelling. To a very great extent properties and behaviour will depend on water content and movement of water, the management of which is a vital aspect of safe and efficient working.

In soft ground, of any category other than firm, some form of immediate support is always necessary, by timbering, shield, permanent lining or a combination of these. Water management may employ variously: pumping, ground water lowering, freezing, grouting, and compressed air.

4.3.1 Granular soils

Gravel or sand, unless cemented naturally or by grouting, will tend to slip at any exposed face until its angle of repose is attained; any inflow of water is liable to wash out the fine sand. This sort of movement is quite unacceptable on any but the smallest scale and timbering at the face becomes necessary to prevent its initiation. The crown of the tunnel is, of course, particularly vulnerable to gravel flow, and timber poling is necessary to hold the top except where a hooded shield is effective or suitable ground treatment. Support at the face in compact gravel or sand is rather to prevent initiation of movement than to carry any great thrust, because in compact granular material, in the absence of disturbance, arching action will develop quickly ahead of the face. There is, however, little or no cohesion and no tensile strength in such ground and any small cavity developing must be quickly secured to prevent loosening of a large volume. Close boarding of a face, using narrow horizontal timbers advanced one at a time from the top downwards, keeps to a minimum the exposed ground and the risk of a slip.

Water flowing in at the face may wash out the fines and loosen the ground. This emphasises the importance of really close-fitting timbers, but also of lowering by underdrainage the water level at the face as close to invert as practicable.

To effect this a filter well in the invert, or sunk from the surface outside the tunnel may be used, or relief drainage in the lined section of the tunnel where the ground is already secured. In special cases compressed air working or grout injection may be employed. If water-bearing gravel or sand is likely to be encountered for a long distance the Bentonite Shield may be the best solution.

4.3.2 Silt

Silt is much finer grained and calls for even more care, particularly in the presence of water. Its cohesive properties are very sensitive to water content. It is characteristic of silts that there is a very small moisture range between the plastic limit and the liquid limit. If allowed to dry out silts become brittle and crumble away readily; if wet they quickly become fluid. Their permeability is lower than that of sands, but high enough for water to be absorbed or discharged quite readily. Compressed air is a most valuable aid in tunnelling silt. Close timbering is essential unless the conditions are particularly favourable —a high clay content in the silt, no excess water, rapid progress of excavation. Grouting is usually unsatisfactory in silt because of failure to permeate the fine pore structure unless very expensive chemical grouts of low viscosity can be used. The presence of thin layers of silt interleaved with sand and gravel can inhibit effective grouting. Freezing has proved satisfactory in such conditions.

4.3.2.1 Organic soils

Mention should also be made of organic silts and clays, in which decaying vegetable matter is a typical constituent. Their physical properties may differ appreciably from inorganic soils, their permeability being low and their compressibility high. The organic content frequently gives rise to discharge of noxious gases, typically: carbon dioxide, sulphuretted hydrogen, or methane. If encountered in tunnelling very adequate ventilation becomes necessary and monitoring of incidence. The explosion hazard from methane gas is more fully described in Chapter 6.

4.3.3 Clay

Clay has very valuable cohesive and plastic properties which can provide excellent tunnelling ground except where too soft or badly fissured.

Its permeability is low and except to the extent of fissuring provides a seal against entry of water. It may, however, swell slowly when exposed in an excavation and exert increasing active pressure on any support or lining. This may make the time occupied in a cycle of operations critical.

The plasticity of clay allows it to deform under change of loading and to creep in at the face, slowly in stiff clay but more rapidly in soft clay; the deformation must be restrained within small limits if disturbance of the ground and its stress structure is to be limited. Heavy timbering becomes necessary for a big face but not necessarily close boarding.

Water entering along fissures may wash out material and enlarge the water passage; it may also soften the clay. All wet spots in a clay face must therefore be regarded with suspicion. This is particularly so where the tunnel is near the upper surface of the clay stratum and is overlain by water-bearing sand or gravel, as is often the case in London.

The swelling of clays is usually attributable to pore water moving into the space where previously existing stress has been relaxed by the excavation. The pore water moves slowly inwards from the surrounding ground, not outwards, as sometimes supposed, from the exposed surface. The phenomenon is particularly significant in overconsolidated clays, which previously have over a long period been subjected to heavy loadings squeezing out much of the pore water.

In the extreme case, clay or silt may be so soft and plastic that a shield can be advanced with a closed face, or with small shuttered openings through which as much ground as necessary is extruded into the tunnel. At the other extreme, clay may be so stiff and sound that, in a small tunnel, no immediate support is required.

4.3.4 Tools

The basic tools of excavation where explosives are not used are the pick and shovel, but with mechanical aids and elaborations. In clay tunnelling the pneumatically-powered clay spade and pick are the most used digging tools, to break out and trim the face. Where size of tunnel and other conditions permit, the operation may be more fully mechanised by use of the roadheader type of machine which excavates by rotation of a multi-pick head on a hydraulically operated arm. Continuous excavation may be by a drum digger or similar full face rotating machine, or by oscillating cutting heads. Tunnelling machines are more fully dealt with in Chapter 7. The Greathead shield also may function as an excavating, or more precisely, trimming tool.

4.4 TIMBERED HEADINGS

The fundamentals of soft ground excavation are best dealt with by describing examples: first, a timbered heading in sound clay, followed by the more elaborate procedures necessary in a piled heading in running ground.

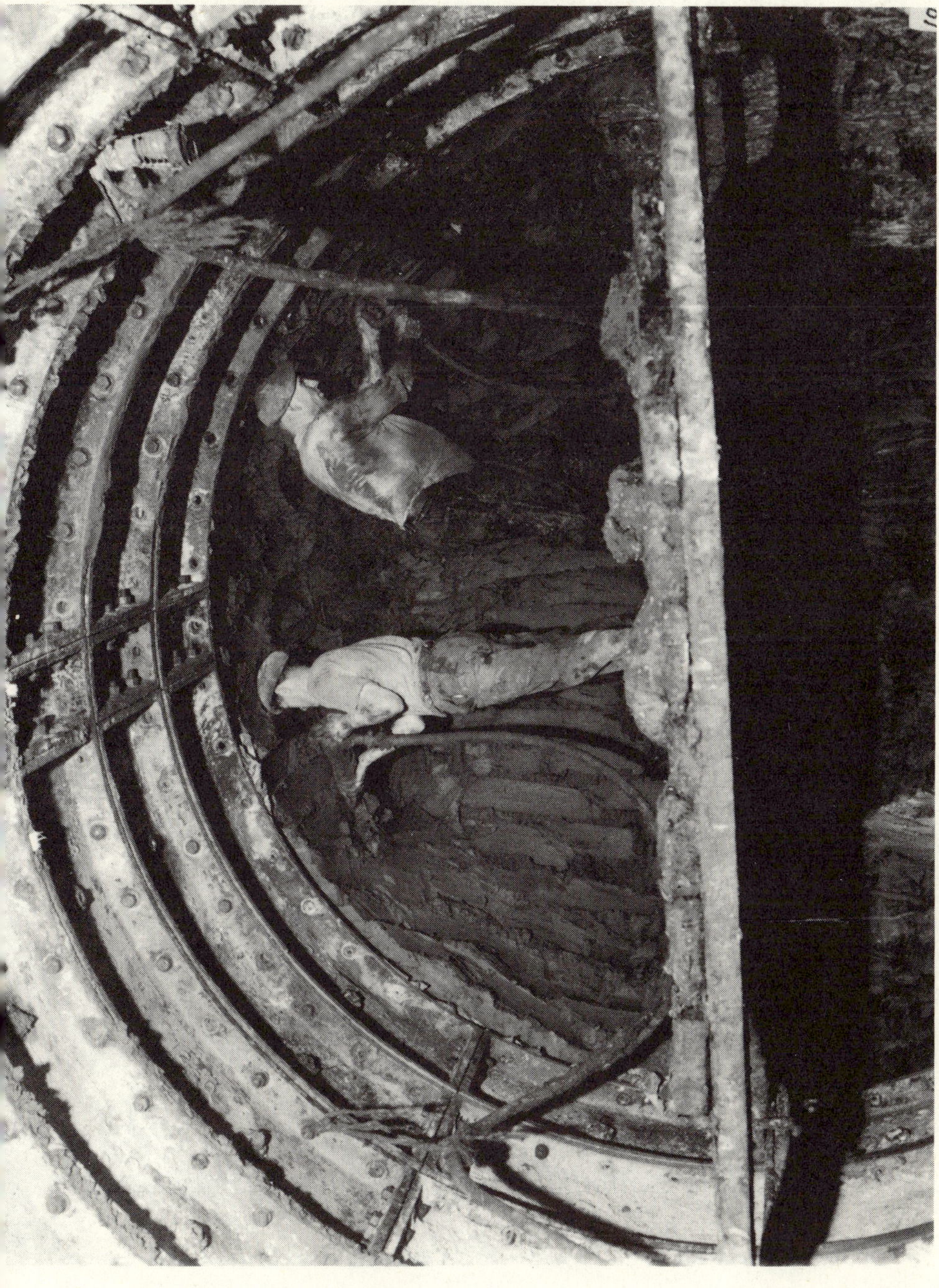

Fig. 4.3 – Excavation in stiff clay, without timbering, for C.I. tunnel lining.

4.4.1 Headings in clay

Headings like this may be used for such purposes as access to form a break-up for a larger tunnel, or for laying a sewer or other pipe, or for construction of walls preliminary to excavation and roofing of a passage or chamber. They may also function as pilots ahead of a large tunnel for advancing and supporting the crown, or for construction of arch skewbacks or wall footings or for ground treatment by grout injection.

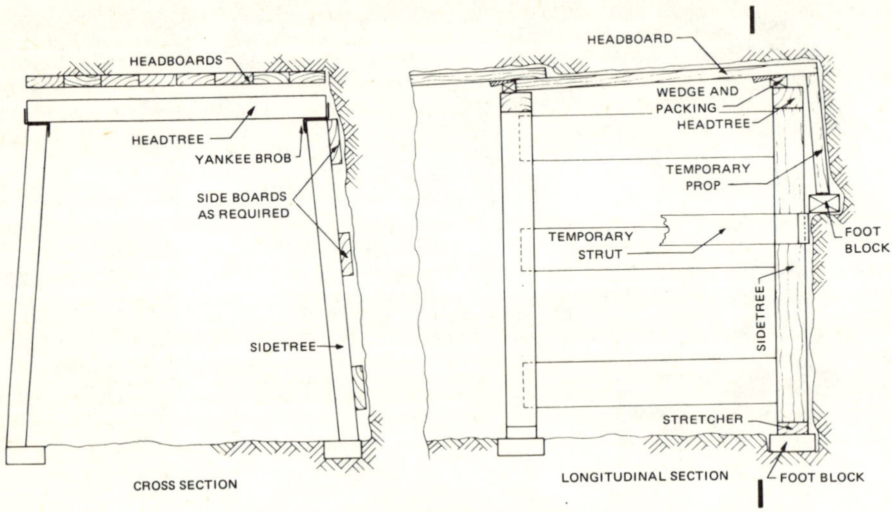

Fig. 4.4 – Timbered heading in clay.

Figures 4.4(a) and (b) show a heading about 1.5 m x 2 m in section being advanced in 1.2 m settings in sound clay. To advance the heading by one frame:

1. The top is excavated for a length over the width of two or more boards.
2. The headboards are tucked back over the previous headtree and propped temporarily at or near the face. These props must be secured at the toe, resting on footblocks with the clay face below strutted against shear failure.
3. The whole width of the top is excavated progressively, and the remainder of the excavation is completed.
4. A headtree is fixed in place to take the load from the headboards.
5. Sidetrees, wedged up to the end of the headtree, are caught at the top by 'yankee brobs' of bent steel plate and rest at the base on footblocks. They are set with 'sprag' of 1 in 12, or more, so that any lateral loading from the ground tightens them.
6. Side boards may, or may not, be required.
7. The temporary props at the face are removed, everything is wedged up tight, and all is ready for another setting.

4.4.2 Forepoling

In running ground, such as water-bearing sand or silt, or in fragmented or squeezing rock, wherever it is dangerous to expose even small areas for a short time—where, in fact, the stand-up time is so brief that support must be fixed in advance of excavation—forepoling may be found necessary. This is the use of poling boards or steel spiles as horizontal piles driven ahead of the tunnel face before it is opened up. The hooded shield embodies this principle, but in some conditions grout injection or freezing may offer alternatives.

A piled heading of similar dimensions to that already described exemplifies the principles, but details are likely to be much varied to suit the ground. Support of the face in such ground by close boarding adequately strutted at all stages is essential. Figure 4.5 illustrates the description.

The narrow head boards are cut to a chisel edge at the leading end and are driven into the ground ahead through the last completed setting. They have to be inclined upwards at about 1 in 6 so that they pass above the position of the next headtree with sufficient clearance for the next setting of boards to be driven under them and above the headtree. The upward slope may be greater than this clearance requires because it may be determined by the board having to be started with its tail beneath the previous headtree.

In the simplest case, the head boards, starting at the centre are first driven to their full length, and the side boards are similarly driven, inclining outwards. The top face board can now be removed, allowing the ground behind it to be scratched out to about half the length of the head boards. The top face board is advanced and refixed in this new position and is strutted back. The whole

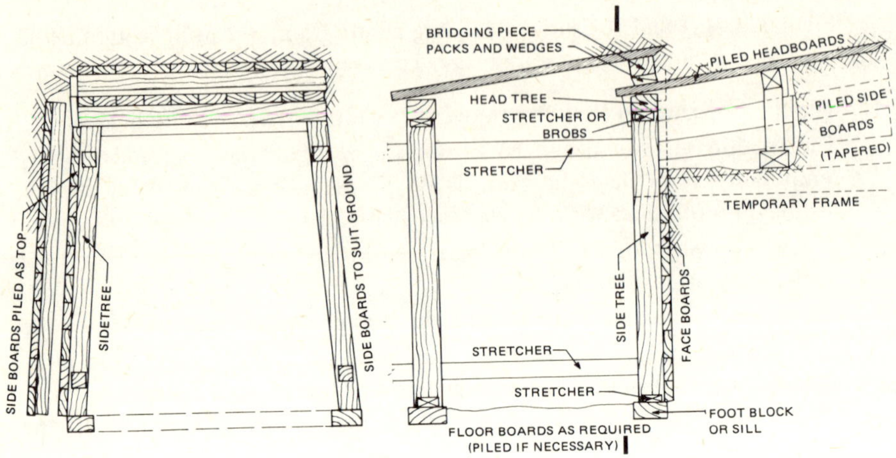

Fig. 4.5 – Piled heading.

face is advanced to the half length in this manner, one board at a time, working downwards until a headtree, sidetrees and a sill, constituting a midway frame can be completed to support the head, side and face boards. These are securely wedged at the frame, which is also strutted back to the frame behind. The excavation is again advanced one board at a time to the full length of the setting, where a permanent frame is similarly fixed but with bridging pieces and packings carrying the head and side boards which are packed out to allow for insertion of the next setting of boards.

The midway frame may be removed if the boards are sufficiently strong, or may be left in place if necessary.

As always in tunnelling, the exact procedure and details will be governed by the actual behaviour of the ground, the dimensions of the heading, the timber sizes being used and the skill and speed of the tunnellers. The necessary upward slope of the head boards and the outward inclination of the side boards leave the angle between top and sides exposed, unless tapering boards are used and very accurately driven. Any exposed area of ground must be spanned by short timbers, while small gaps may be stuffed with bags, straw, etc.

In some ground it may not be practicable to drive the piles their whole length at one time and it may be necessary to drive a little beyond half way, excavate, provide temporary support and wedge up and then redrive ready to excavate the second half.

The essence of the whole operation is carefully planned workmanship to ensure that no large area of ground is ever left unsupported, and that every piece of timber is accurately placed and tightly wedged at all times except when actually being moved.

4.5 CIRCULAR TUNNEL SEGMENTAL LINING

For a tunnel with segmental lining driven in soft ground without a shield, timbering of the top and face is likely to be required.

The structural strength and spacings of the timbers will be determined by the nature of the ground, the time during which the support is required, and the importance of safeguarding the overlying ground against any settlement. Figure 4.6 shows a typical case of a small tunnel of about 3.5 m diameter where close boarding is appropriate. In stiff clay such close timbering is unlikely to be necessary except during periods when work is interrupted, but the principles of supporting top and face should be followed at every ring by provision of the main structural elements of the support system.

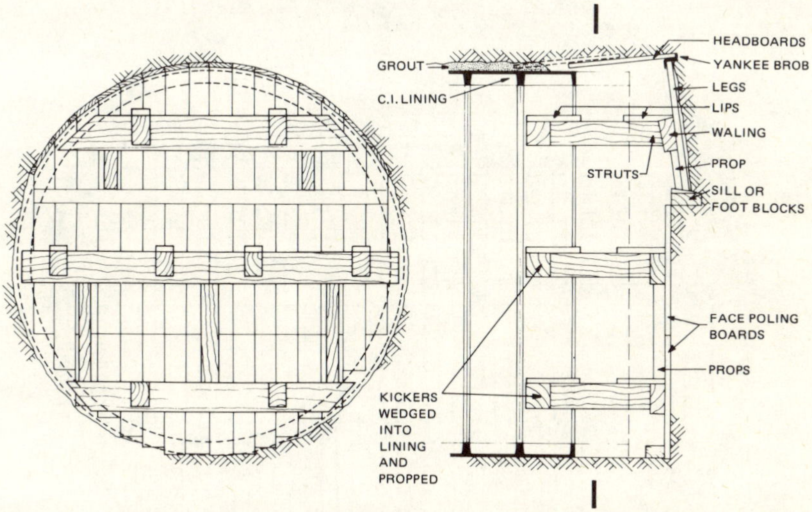

Fig. 4.6 – Close boarding in a small tunnel.

In the case illustrated the tunnel is advanced one ring at a time (exceptionally a two-ring step may be feasible). The cycle of operations, starting after erection and grouting of a ring is:

1. Excavate top and provide support by head boards which rest at the back on the completed ring and are supported by a front 'leg' set on a footblock on a bench in the clay face. The excavation proceeds from the centre working outwards to both sides until the top bench (about 1 m to 1½ m deep) has been advanced the distance of one ring.
2. Excavate a second bench. For this some head boards at the perimeter may or may not be required. Face boards and a waling are assumed necessary. The footblocks of the top legs must not be undercut and the face of the ground on which they rest must be kept strutted to secure it against shear failure.

3. Complete excavation to invert, shaping invert carefully so that segments can be set in place accurately.
4. Erect segments from invert upwards to key, checking accuracy of positioning of each segment. Unless the ground is unsafe, head boards can usually be recovered as segments are erected.
5. Pack at front of ring and grout up with cement grout.
6. The timbers are moved forward progressively into the next ring.

When the full face in a big tunnel has to be fully timbered, experience and craftsmanship of a high order are called for, Fig. 4.7 shows such a face.

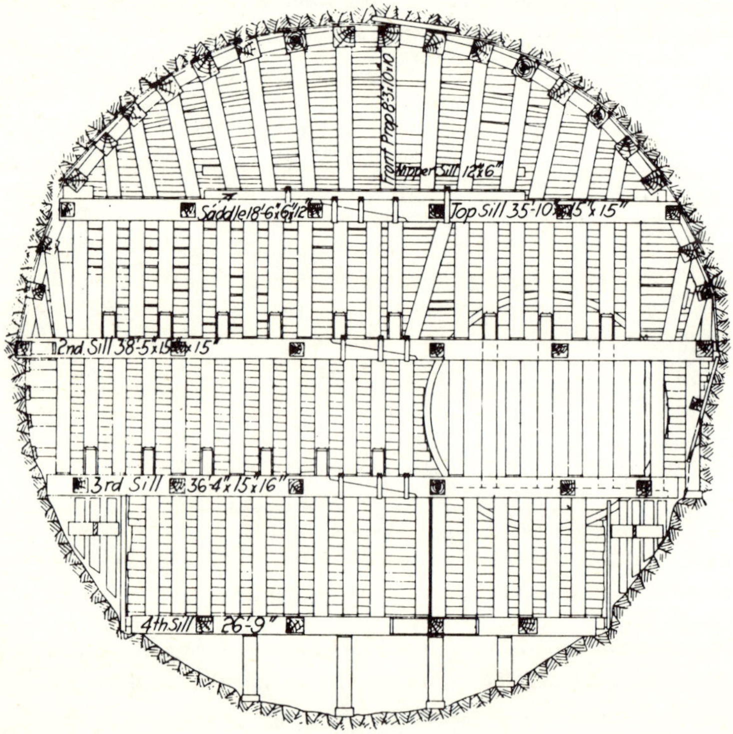

Fig. 4.7 – Timbering of 11 m face on the extension of the City and South London Railway.

4.6 GREATHEAD SHIELD

The Greathead shield, devised originally for tunnelling in clay under the River Thames, is primarily a device to provide continuous ground support. It can also in minor degree function as a cutting and trimming device. The shield comprises a cylindrical steel shell stiffened by ribs and diaphragms and provided with hydraulic jacks to propel it forwards. In addition to these basic elements many ancillary devices may be incorporated in or attached to a shield.

The fundamental operations in using such a shield are:

1. Excavate ahead, timbering face as necessary.
2. Shove shield forward one ring length by jacking from completed lining. The shield may be used to trim the excavation neatly provided excessive thrust is not required.
3. Withdraw shove rams and build segmental lining in clear space behind shield.
4. Grout up all voids between segments and ground.

Repeat cycle of operations.

4.6.1 Ancillary equipment

Among the ancillary devices which may be used are:

Face diaphragm
Vertical and horizontal stiffening girders
Hood
Bead
Tail
Segment erectors

The face diaphragm helps to stiffen the shell but primarily enables the face to be shut off in the event of a run of ground. It must, of course, have apertures for access to the face so that men may get in and out to excavate and timber the face. In a small shield a sliding door, or grooves to take stop logs, may be suitable. Complete closure of the face may be considered unnecessary but a muck trap may be used. In this the diaphragm is in two parts, the front one covering the upper half of the face with a margin below axis, while the rear diaphragm covers the bottom half and a margin above axis. In a big shield (9 m diameter) such as that used at the first Dartford tunnel, there may be three such muck traps at different platform levels. A single trap of this kind can function as a water trap in emergency if the tunnel is being driven in compressed air.

In general, face timbering will be fixed in front of the diaphragm as excavation proceeds downwards from the top but some special devices have been used. In the Blackwall tunnel (1897) the shield was provided with two complete diaphragms equipped with air locks intended to make possible the use of higher air pressure ahead of the diaphragm than in the tunnel. It had also a system of steel shutters at the face, which replaced timbering, and could slide back into the shield under control of guide screws during a shove by the main rams. The use of differential air pressure was never in fact practicable.

The extreme in soft silt is tunnelling by displacement where the shield is shoved ahead without any opening or with a small opening only. In the river bed of the Hudson river this technique was employed in the original rail tunnel and also in the construction of the Holland and Lincoln road tunnels. In the latter, with apertures in the shield of only 0.5% of the face area, it was reckoned

Fig. 4.8 – Rear view of Greathead Shield in Dartford Tunnel, showing main rams, thrusting against completed C.I. lining, segment erector arm, muck shoots and skips for disposal.

that 30% only of the ground was excavated and 70% went in 'heave'. The soft silt was extruded through the shield in a plastic condition and sliced off. A similar technique was adopted in the Chicago Subway construction, described in detail by Terzaghi in which the aperture in a 7.6 m shield with 7 m cover was varied from 4% to 20% in an attempt to control the heave. In fact, the heave ahead of the shield ranged from 0.03 to 0.10 m and was followed by a subsidence of about 0.15 m.

Vertical and horizontal girders may be necessary to stiffen the shield and the diaphragm. They can also serve to provide working platforms at different levels and to divide the face into working compartments.

Such girders can carry a system of face jacks which support the face timbering and which can be designed to maintain pressure on the face during a shove and yet to allow the shield as a whole to move forward into the face without crushing the timber. Sometimes 'gun struts', where the inward movement is restrained by friction from screwed clamps, are preferred to the hydraulic complexity of jacks.

Hood: A hood is a forward extension of the cutting edge in the upper half of the shield which acts like the piles in forepoling and is kept buried in the ground ahead to secure the roof at all times. In gravel where the necessary force to drive in such a cutting edge is excessive and might cause dangerous disturbance, clay pocketing can be employed, but it will only be unwillingly resorted to. In this hand operation a handhole is formed in the ground ahead of the cutting edge and is refilled with puddled clay, and the operation is repeated all round the hood until there is a prepared clay part ring into which the shield can be shoved, the clay serving both to provide a water seal and to lubricate the passage of the shield.

Bead: The radius of the cutting edge is often enlarged to be a centimetre or two greater than that of the shield skin by attaching a bead. The objective is to reduce friction between the ground and the skin when the shield is being driven forward, and to reduce stresses in the shield (and in the ground) when minor directional changes are required. Beads do not always extend around the invert of the shield. Clearly beads can contribute to the settlement of ground and structures above a tunnel drive.

Tail: An extension of the shield skin forming the 'tail' is required unless the ground is so reliable that it will stand up unsupported during the erection and grouting of the lining. Normally the lining is erected within the protection of the tail which still overlaps it by a small amount after the shove. There is a tendency for grout deposits to build up on the tail skin, and it may sometimes be necessary to stop and break them off.

Segment Erectors: In any but the smallest tunnels, erection of precast segments makes necessary mechanical lifting equipment. The design and dimensioning of segments may be governed by the problems of handling and erection. A large shield will usually be equipped with one or two radial arms in the tail which can lift and place segments. The arm, operated hydraulically, is pivoted near the tunnel axis and is extensible, with a head devised to pick up the segments from the delivery point in the invert and lift and place them in position for bolting up.

4.6.2 Developments

The drum digger and other digger shields have been developed from the Greathead shield by devising rotating heads to cut into the face as the shield is advanced. This is only suitable for ground with adequate stand-up time to allow timbering to be dispensed with. The linings also have been developed as part of the same system, using wedging out of the lining directly against the excavated ground in place of bolting and grouting. These machines and methods are more fully described later.

The Bentonite shield may be the ultimate development of the closed shield for use in granular water-bearing soils. In it there is a complete closed diaphragm ahead of which any space is charged with Bentonite mud. There is a cutter head rotating full circle and the spoil, immersed in Bentonite mud, is passed back through a special device set in the diaphragm functioning as a valve. The whole face is thus pressurised and the excavation and disposal is fully mechanised. This machine also is more fully described in Chapter 7.

4.6.3 Steering and guidance

Control of a shield to keep it on line and level is not an easy problem. Precise survey and setting out are necessary and at the most frequent intervals practicable the shield's position must be accurately checked and recorded and also its attitude in terms of 'lead', 'look-up' or 'overhang' and 'roll'. Errors must be corrected as quickly as practicable, normally by selective and controlled use of the shove rams. The accuracy of the last completed ring must also be ensured. These problems are more fully discussed in Chapter 9.

In controlling a shield the ratio of length to diameter is of great importance. If the ratio is much more than about 1 : 1 it becomes difficult to correct errors of direction; the shield has cut for itself a passage through the ground which leaves little room for turning.

4.7 TUNNEL JACKING

4.7.1 Pipe jacking

In recent years pipe jacking methods have been developed and elaborated to tunnel jacking. No precise borderline is definable, but when a man can operate in the pipe, somewhere above 1 m diameter, it may be considered as a form of tunnel construction. Even at smaller diameters it may be a useful ancillary to tunnel construction as, for example, when pipes are jacked ahead at close centres to form a roof beneath which a tunnel can be mined, or to retain roof and sides of a small passage. The technique is most suitable for relatively short straight lengths of small diameter in soft ground, but the limitations imposed by all these factors can be overcome.

Pipe thrusting, in which the ground is displaced and not excavated, is essentially a small bore pipe method. Horizontal auger boring is feasible for pipe lines up to approximately 1½ m diameter but for limited lengths. In pipe jacking, a line of pipes is jacked into the ground, being furnished with a cutting edge or shield at the leading end and with means for removing the displaced ground through the bore of the pipe. Diameters up to 3 m are not unusual and lengths of the order of 100 m, while longer drives are possible with intermediate jacking stations. Tunnel jacking may be the more appropriate term as the size increases and as rectangular or other special purpose structures rather than pipes are required. Pipe jacking is first outlined here.

Pipe jacking has been widely used as an alternative to open trench work, or lengths of timbered heading, where surface access is difficult and where settlement at the surface must be minimised. Sewers and culverts or other pipe crossings under a railway or road embankment, or under a canal or other obstacle can be so constructed, as can pedestrian subways at ground level through an embankment. The method also lends itself to sewer construction, where manhole shafts divide up the work into relatively short straight lengths and can be used as successive thrust pits. The technique has been adopted and adapted for more complex structures through embankments, further referred to later.

Its direct cost is usually greater than cut-and-cover, but surface disturbance and disruption are much less in congested areas, and working areas need not be so large.

The essential features of a pipe jacking installation are:

1. Thrust pit
2. Jacks and thrust ring
3. Pipes
4. Cutting shield

1. The thrust pit must be at tunnel depth and of sufficient length to accommodate the jacks, the thrust ring, and one length of pipe, with space for insertion and any jointing detail. The rear wall of the pit must be adequate to take the maximum thrust from the jacks and to spread that load into a sufficient area of ground. For large pipes and long drives the load is likely to be several hundred tons at its maximum as the drive nears completion. The whole success of the operation may depend on having ample jacking capacity and proper support for it. In a shallow pit or soft ground the back wall of the pit may not offer adequate support and it may be necessary to drive piles for the purpose, or install ties through the embankment or ground anchors.

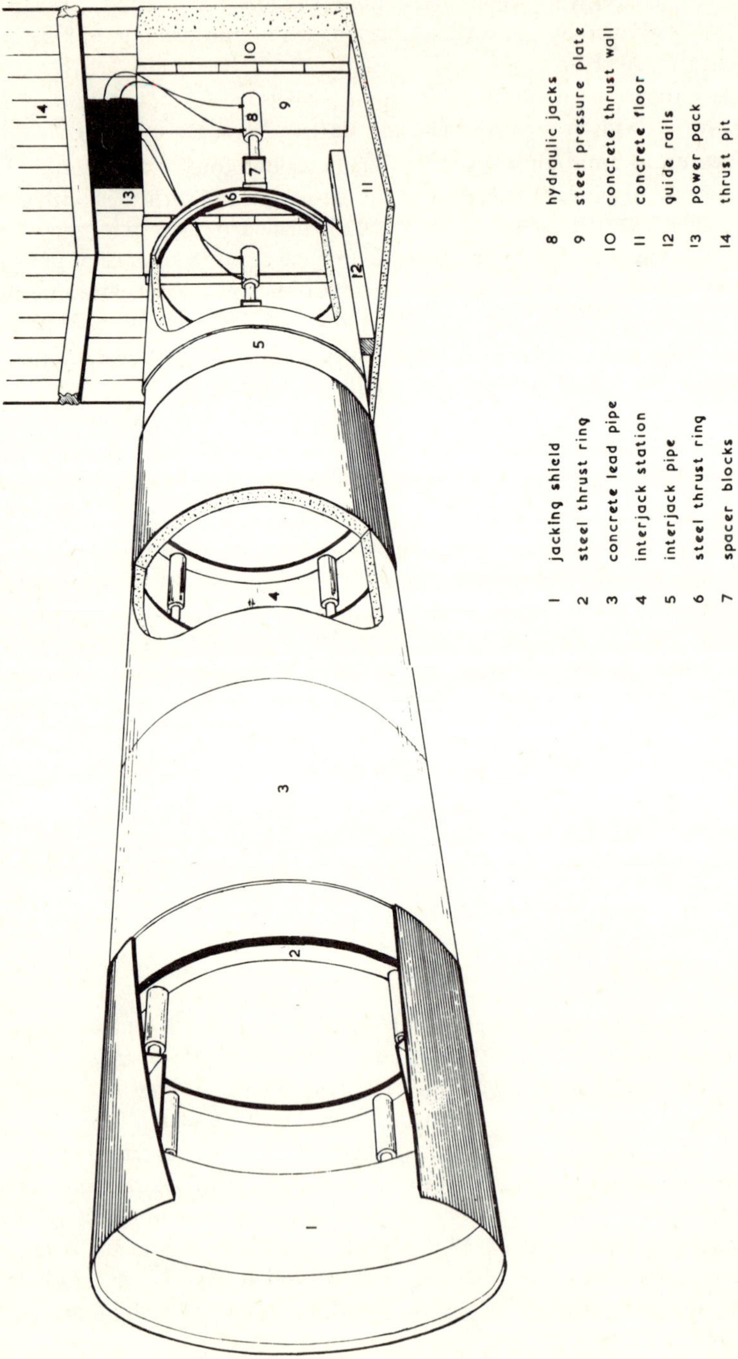

Fig. 4.9 – General arrangement of pipe jacking equipment (Pipe Jacking Association).

2. Hydraulic jacks, operating at high pressures to achieve compactness, are installed on the floor of the pit. There may be one, two or more jacks, depending on the diameter of the pipe. They are symmetrically disposed and act on the rear of the pipe line through a thrust ring designed to bear safely and evenly on the pipe end. The stroke of the jacks is likely to be substantially less that the pipe lengths and a series of spacer blocks of increasing length will be necessary for successive insertion after each stroke until an advance of a full pipe length has been effected.

   As the whole length of pipe is in contact with the ground the frictional resistance to advance increases progressively. It is possible to limit the resistance by installation of intermediate jacking rings whereby the leading end of the pipe only is first thrust forward and the rear part is then pushed up to it by use of the pit jacks. A further elaboration is to use a separate set of jacks to advance the shield only at the leading end and to follow up with intermediate jacks and pit jacks. In such cases it may be advantageous to be able to control the shield jacks individually so as to guide the shield and correct any misalignment.

3. Pipes. For many purposes reinforced spun concrete rings, rather thicker than ordinary sewer pipes are suitable, with joints substantially watertight but allowing some flexibility. Provision for grouting may be advisable not only to seal off any residual leakage and fill voids, but also to permit lubrication of the sliding surface during jacking, by Bentonite slurry or other suitable fluid.

   Steel pipes of normal diameter may be employed in long lengths, but satisfactory external protection against corrosion is difficult to ensure because of scratching during thrusting. They may advantageously be used as sleeves through which a smaller service pipe is threaded. A pit long enough to enter each pipe length and to accommodate the jacks is, of course, necessary and there must be room for them to be threaded down any shaft and turned into position.

4. Cutting shield. The minimum necessary is the provision of a sharp but robust circular cutting edge secured to the leading pipe. For a steel pipe, no more than the pipe itself may be necessary. Beyond this, the shield may be elaborated up to the dimensions, strength and complexity of a soft ground tunnel shield, with hood, thrust jacks, face jacks, platforms and working compartments, stiffening rings and diaphragms, as pipe jacking develops to full scale tunnel jacking.

   The essential difference from orthodox soft ground tunnelling then is that erection behind the face of segmental rings is dispensed with, the complete rings comprising the lining being built into the tunnel structure at the pit and thrust forward.

4.7.2 Tunnel jacking

As a tunnelling operation in its own right, jacking has been used for drives up to 4 m diameter and a record drive was completed in Chicago (1977) for a sewer with outside diameter 2.9 m for a length of 490 m.

A number of underbridges to carry roads through railway embankments have been constructed in the U.K. with the aid of jacking techniques, usually to construct the abutments and foundations in advance of the deck, which may be slid across subsequently in open cut. In most cases multiple prefabricated units have been jacked through in stages at one, two or three levels. In some cases half of the complete bridge up to 9 x 4 x 21 m has been built at each side and jacked forward in open cut, not really constituting a tunnel operation, except possibly for guide paths first built in timbered headings.

As compared with shield tunnelling, the limiting factors are:

1. Thrust
2. Size and weight of units
3. Control of alignment

1. The thrust necessary for large bores may be 1000–1500 tons. It can be minimised by ensuring a smooth outer surface without lipping at the joints and by lubrication with Bentonite slurry or otherwise at the sliding surfaces. The thrust can be further reduced by separating that for the shield or cutting edge from the friction on the lining. A separate set of shield jacks are provided for the shield and boring machine which may slide forward from within a protective 'can' as it advances, the 'can' being anchored in the ground to provide the thrust reaction. After advance of the shield the 'can' is in turn moved forward leaving a rear shield, or tail, into which the lining can then be shoved. If necessary additional intermediate jacking stations are possible. An adequately supported thrust block at the rear is, of course, essential.

   Another thrust problem, particularly with a rectangular section with shallow cover, is the drag exercised on the block of soil immediately above the structure. Patented devices, known as 'drag sheets', have been used where strong flexible sheets, of thin steel or fabric, are fed in and anchored back as the units are advanced in order to separate the force of the sliding friction from the ground above and reduce disturbance.
2. Problems of transport and handling down the shaft will normally impose a weight limit for each unit, possibly up to 25 T.
3. Alignment within 1% or so is very difficult to maintain in simple pipe jacking, even when the initial pipe is set most precisely. Jacking independently at the cutting head as referred to above can give control comparable to that of an orthodox TBM, and intermediate jacking with individually controlled jacks can help.

   Within the scope of these limitations, the method may be competitive, for sewer work in particular, with shield tunnelling in which a segmental lining has to be erected and grouted as part of the sequence of operations at the working face. The limitations on length are not serious in sewer work, where manholes at moderate intervals are normally required. Control of level is important.

The face support, and management of water present very much the same problems as in orthodox shield tunnelling with similar solutions, such as ground treatment and ground water lowering. The tunnel jacking method is one of growing importance but not for major tunnels. It may be noted that although most usually applicable to soft ground it is readily adaptable to soft rock, and hard rock can be dealt with if encountered.

4.8 SIZE OF TUNNEL

It will already be clear that the size of a tunnel may be fundamental in governing methods of excavation. In opening up a face, disturbance of the existing equilibrium of forces and stresses inevitably results in movement and in redistribution of forces. Among the difficulties which increase with size of face, mostly in disproportionate degree, are:

1. Volume of ground disturbed and extent of movement.
2. Access to all parts of face.
3. Strutting of exposed face.
4. Volume of excavation per ring and time of cycle of operations.
5. Water pressure differential between top and bottom of face.
6. Mixed face—more than one type of ground exposed.
7. Falling materials.

1. In a small face the volume of ground moving and relaxing will not be great and can often be tolerated or kept within acceptable limits by relatively simple timbering or other temporary support.

In a large face, the effects of relaxation of stress penetrate further ahead of the face and a larger volume of ground is involved over a longer period. Much stronger support becomes necessary to prevent weakening of the whole ground structure. Also, the risk of intersecting some fissure or other anomaly is much greater. The time factor is further disadvantageous in that there is a longer interval for trouble to develop between release of support of the ground ahead and its removal by excavation followed by new support. With increasing size full face excavation may cease to be practicable and the use of a pilot tunnel or a series of headings may be required. These limitations of size will apply much earlier in soft ground than in rock or in stiff clay.

In the Dartford tunnel (1963) one length of 9 m diameter tunnel through soft silty clay was driven in compressed air without a shield. A low level pilot of 3.5 m diameter with C.I. lining was first driven, followed by a pair of 3 m diameter similar top pilots. The full sized tunnel, also with C.I. segmental lining, was then excavated and built round this trefoil pattern of pilots. At no time was there a large area of face exposed and requiring support and at the same time the cycle of exposure and lining was greatly speeded up by the reduction in volume of clay to be removed per ring. There was also the advantage of anchoring back the face walings to the pilot tunnel iron which was bolted throughout.

This principle of using a sequence of small manageable headings to overcome conditions too difficult for full face working is repeatedly found to be the best expedient, even in rock. A small heading may be driven to get in the crown of a tunnel, or a pair of headings for the springing of an arched roof, or the whole of an arch may be built up progressively in this manner.

An outstanding example of this is the Eisenhower Memorial (originally Straight Creek) road tunnel through the Rockies in Colorado. Very severe rock pressures and movements made it necessary to build the main arch over a length of some 50 m by constructing a concrete arch in a sequence of thirteen headings, or drifts. Over adjacent lengths five or more headings were used to establish footings, springings, and crown.

2. The second aspect of size is that of ready access to all parts of the face, which is necessary for access and support. 'Man is the measure of all things', and accordingly, when a man's reach is exceeded stagings or other platforms must be provided, or with mechanised working gantry, jumbo shield or excavating machine. Brunel's shield was built up from 36 unit cells each about 2 m high x 1 m wide. Over a century later the Dartford tunnel shields were designed with 16 compartments separated by platforms and vertical diaphragms, the largest compartment being about 2½ m square.

3. The total thrust to be provided in strutting of the exposed face will increase with size not only in proportion to the increased area but also because of loss of arching action and the increased vertical depth on which pressure acts. Additionally, supporting walings and beams carry the loads across longer spans and must be proportionately stiffer.

The point may be illustrated simply by some rough approximations. Doubling the diameter of a circular tunnel quadruples the face area, the ground pressure may be supposed to be one and a half times as great, so that six times the load must be carried on beams twice as long, producing bending moments multiplied 12 times in a full length beam.

4. If the length of advance in each cycle of operation remains the same, the volume of material to be excavated and transported will increase in direct proportion to the area of face, or to the square on the diameter. The times for excavation and disposal and for the other elements in the total cycle tend to increase in larger tunnels. Therefore, the time interval between opening of the ground and provision of support will be substantially increased and heavier ground loadings are likely to develop.

5. In permeable saturated ground static water pressure at any level will be proportional to depth below the free surface. In fact, because of inflow, static pressure cannot develop unless the face is fully closed, but water problems will in any case increase disproportionately. Where compressed air is necessary the problems arising from difference of pressure between top and bottom of a big face can be formidable.

6. In a large tunnel the likelihood increases that the strata encountered will not be uniform but that different types of ground will be encountered at the top and bottom of the face.

Particular difficulties will be apparent if such strata as water-bearing gravel, sand or silt are met either at the top or at the bottom of a face, or if rock requiring the use of explosives occurs in a tunnel designed to be driven through soft ground. This was exemplified in both the Hudson river tunnel (1906) and the original Blackwall tunnel (1897). In the second Blackwall tunnel (1963), water at artesian pressure occurred in the invert so that the preparatory injection grouting scheme had to seal off both crown and invert.

7. The hazards from falling materials are obviously much greater in a large tunnel both because of the height of fall and the need for stagings, from which materials can be dropped.

It will be obvious that tunnels cannot just be scaled up, but that, even where a pilot tunnel has been successfully driven, the full scale enlargement presents further problems and difficulties.

There is also a minimum limit on dimensions as a man must have working room for himself, his tools and his timber within the tunnel, unless it is so short a heading that it can be constructed by reaching in from the ends. Apart from this, economic considerations of access and use of equipment may well make it cheaper to construct a 3 m tunnel than one of 2½ m diameter, but the actual limiting size will depend on the many construction factors of the particular project. Anything less than 2½ m is not likely to be economic for long tunnels. In the long London MWB water tunnel through the clay 2.54 m was adopted as a standard, and in the Scottish hydroelectric developments in tunnelling through hard rock the lowest cost per unit length was found to be for a tunnel of about the same diameter. For very long tunnels without intermediate working or ventilation shafts, the minimum economic size tends to increase with length, especially if the additional space has permanent value.

4.9 MANAGEMENT OF WATER

In most tunnels water is present in the ground to be excavated; a dry tunnel is unusual. Management of the water is a vital feature of excavation. It is probable that water more than any other factor has caused failures or unforeseen difficulties in tunnel construction. It is not only the spectacular troubles of inundation, as in the Thames tunnel, the Severn tunnel, the Lötschberg tunnel that emphasise the dangers from water and the importance of its management, but the multitude of small tunnels where costly delays have been caused by water, and subsidence and damage have resulted. The management of water is much more than a mere problem of oilskins and pumps.

Water, as a problem in tunnel construction, requires to be studied from three aspects:

1. Hydrology of the ground, both as existing before disturbance and as modified by the tunnel excavation and ancillary operations.
2. Soil mechanics, more particularly the changes in the properties and behaviour of the ground arising from changes in the pore water content and seepage pressures, and also any chemical reactions.
3. Logistics of provision and siting of pumps and pumping mains.

4.9.1 Hydrology

The inflow of water into the tunnel will be dependent on the water content and permeability of the various strata exposed and adjacent. The volume of water available to recharge the strata is important as is the rate of recharge.

In a subaqueous tunnel in which the source of the ground water is sea, river, or lake, the quantity is virtually unlimited. Much may depend on the existence or not of impervious strata protecting the tunnel excavation and on their thickness and characteristics. Although inundation from overhead is obviously the major hazard, artesian water in the invert may also be of importance.

The protection may be a bed of clay which virtually provides a water seal, but its continuity in adequate thickness above the tunnel must be constantly checked as the bed may thin out unexpectedly or its surface may be penetrated by a buried channel, or other hollow eroded in the past and refilled with alluvium. Such anomalies are not uncommon at the upper surface of the London clay which is usually overlain by water-bearing gravel.

Thin clay cover cannot be expected to provide much mechanical strength, therefore immediate support of the excavation followed as quickly as practicable by permanent lining and grouting becomes essential. In such ground any appearance of moisture at the face has to be regarded with extreme suspicion.

Such a protective stratum may be altogether absent as in the Blackwall tunnel (1897) where shield, compressed air, and an almost closed face supplemented by a clay blanket in the river bed were necessary and barely adequate. In the second Blackwall tunnel (1963) slightly more favourable ground was chosen, but with the added assistance of chemical injection grouting to reduce the permeability of the sand and gravel and to give it chemically some cohesive strength. Similar grouting treatment for a length of the invert also was necessary where artesian water was present.

A subaqueous tunnel may be in rock not completely impermeable but structurally adequate. Again there is a danger from buried channels and also fissures and faults.

Under the river Mersey such a buried channel of glacial origin eroded below the general bed level into the Triassic sandstone was encountered in the driving of the Mersey Railway tunnel (1886). There was generally at least 5 m of rock cover but locally the crown of the tunnel was in the clay and sand filling 1 m or so above the sandstone. The Mersey Queensway road tunnel (1934) encountered the same channel but the crown of the tunnel was kept low enough to provide a metre or so of rock cover, the pilot tunnel being lined locally with cast iron as a precaution. At the site of the Mersey Kingsway road tunnel (1971) this buried channel did not appear to be present, but a fault was encountered filled with soft material and communicating directly with the river. Protection of the critical length was provided by a reinforced concrete hood constructed in a series of small headings.

Where sources of water supply are not unlimited or where the recharging rate of the ground is limited by low permeability and adequate thickness, various methods of ground water lowering may be possible, as by well points, drainage tunnels, or well shafts with pumps. An invert heading kept in front of the face may be sufficient or pumping from a sump at the face.

Reduction of permeability by injection grouting to fill the pores can be used to check inflow, as can freezing. Compressed air also serves this purpose.

In mountain tunnels the water hazard is usually different. Fissures filled with water and recharged from the surface may be encountered, the water being at high pressure and possibly very hot. Such flows are likely to diminish with time both in volume and temperature but can be a very severe hazard. There is the added risk of such flows loosening and carrying into the tunnel large volumes of sand or other soft ground. A further hazard, particularly in limestone, is that of solution caverns filled with water and fed by underground rivers.

Examples of mountain tunnels in which water was a major problem include: the Lötschberg railway tunnel (1913), the Moffat railway tunnel (1928) through the Rockies in Colorado, the Tanna railway tunnel (1934) in Japan, the Tecolote water tunnel (1955) in California, the Litani water tunnel (1965) in Lebanon.

The level of the water table is not a simple concept in the context of tunnelling. In relatively shallow tunnels the strata are likely to be very variable with different permeabilities and rates of recharge. The significant questions are whether the ground is saturated and at what pressure, and what balance will be achieved between drainage into the tunnel, or ancillary drainage system, and recharge. This is a problem of dynamic rather than static equilibrium. Pressure gradients and seepage flows develop depending on permeability, source of water, and rate of pumping, and the excavation methods must take full account of this pattern.

4.9.2 Soil Mechanics

The interaction of water and solids in soils is a fundamental part of soil mechanics. The presence of water and its movement affects significantly the structural behaviour of soils exposed in tunnel excavation.

In the simplest case a damp sand will have some cohesive properties and will not run freely whereas the same sand either dry or saturated loses much of its cohesion and runs readily if unsupported. Water flowing through granular soil will tend to wash out the finer particles and will thus enlarge the passages, increase the flow and also cause settlement. A suitably graded gravel will, however, function as a filter retaining solids while allowing the water to pass.

Fine sands at the bottom of an excavation may boil up under the seepage pressure of rising water, creating an artificial quicksand, which can however be controlled by a proper gravel filter or by reducing the volume of seepage. The true quicksand results from a loose particle structure.

In silts water content is all important as described above. The original builders of the Hudson river tunnel in 1879 thought that regulation of compressed air pressure had provided the key to this problem, but found that stability could not be achieved.

In clays also water content and pore pressure control behaviour and properties, but changes are relatively slow so that time is usually available for the tunnelling cycle.

Following exposure by excavation there is a release of stress at the surface and increase of stress at depth as a consequence of which the clay acting as a sponge tends to draw in pore water from behind, which in many clays results in swelling. The time elapsing between excavation and final structural lining will govern the extent of this swelling and the resultant loading on the lining. In the case of a grouted precast lining some swelling usually occurs before grouting and some continues to develop and to impose a pressure on the completed lining.

4.9.3 Logistics

It should be clear that control of water before and as it enters the tunnel must be closely integrated with the method of excavation and support. The handling and disposal of the water is equally important.

In subaqueous tunnels it is usually necessary to drive much of the length on a falling gradient of about 3% to 5% for a highway tunnel. Water then inevitably runs forward and accumulates in the invert at the face where construction is in progress. All water must be pumped back along the tunnel and up to the surface where it can be discharged. Pumps at the face have to be moved continuously forward and therefore must be relatively light and mobile, transferring the water back to a fixed sump where a more permanent and mechanically efficient installation is possible. Clearly as much as possible of the water entering behind the face in the completed tunnel should be intercepted by a dam and sump before reaching the face.

If work is being carried forward beneath a river from a shaft it is likely that construction of a main sump and installation of permanent pumps there will be adopted. The pumping capacity must be adequate for all probable requirements with also ample standby capacity both against breakdown of plant and against unexpected influx of water.

In a long tunnel there are likely to be a series of pumps and sumps between working face and main sump at the shaft. Capacity of pumps and sumps, pipe sizes, and power supplies all need to be carefully planned to meet the maximum demand, bearing in mind that failure or inadequacy of the system may result in flooding of the tunnel and loss of the pumps.

4.10 ART OF TUNNELLING

In tunnelling the problems of excavation are more fully three dimensional than in other types of excavation. They are even four dimensional in the sense that time is a major factor because the stresses in ground that has been opened up are continuously changing and timing of support can become critical. The special characteristic of tunnelling is that ground is excavated from inside and from underneath. Therefore, it is, to a greater degree than in other excavation, exploration of the unknown. The weight of ground is always poised above, and its support at all stages must be ensured, as well as its protection from loosening by water.

The development of the various mechanised tunnelling systems should not lead to the conclusion that the tunnel miner's traditional skills in excavation, ground support and management of water are no longer required. Where ground conditions are suitable it is now possible to construct long lengths of tunnel with 'moles' or tunnel boring machines, erecting whatever permanent lining may be necessary in comparative comfort and safety. Even in ideal conditions it remains necessary to carry out much critical work by hand mining for such items as access headings, shield chambers, openings, junctions, and ventilation connections. The construction in tunnel of a station for an underground railway provides an example, where the station tunnel drive is not long enough to justify the installation of a fully mechanised system and where a whole network of tunnels has to be constructed at various levels, often beneath valuable surface property using great care and all the traditional skills.

In excavating tunnels in soft ground there is sometimes the likelihood and always the possibility that unexpected conditions, sometimes dangerous, will be disclosed as the face advances, however detailed and elaborate the preparatory investigations and studies.

Traditionally, skilled and dedicated tunnel miners with experienced supervisors lived with and cared for the working face, augering ahead to prove the ground, and placing support timbers where necessary. Work in difficult ground often proceeded for twenty-four hours a day for seven days a week, for reasons of safety as well as to increase progress, and it must be appreciated that continuous progress itself contributes to safety because of the limited 'stand-up' time of some ground.

With the mechanisation of tunnelling and the increased difficulty of arranging for work to be continuous over holidays, there is a danger that the traditional virtues may be lost. However modern the method, the need for caution remains and precautions such as regular inspection of the workings and the availability of emergency face supports and such items as standby pumps are as important as ever.

BIBLIOGRAPHY

The subject is dealt with extensively in the classic books, in particular those listed below, and incidentally in many papers and articles about particular projects which are included in the general bibliography in Volume 2.

Hewett, B. H. M. and Johannesson, S., *Shield and compressed air tunneling*, McGraw-Hill, 1922.

Richardson, H. W. and Mayo, R. S., *Practical Tunnel Driving,* McGraw-Hill, 1941 (revised edn., R. S. Mayo, 1975).

Terzaghi, K., 'Shield tunnels of the Chicago subway,' *J. Boston Soc. Civ. Engrs.*, 1942, **29** (July).

Terzaghi, K. 'Liner-plate tunnels on the Chicago subway', *Trans. Amer. Soc. Civ. Engrs.,* 1943, **108**.

Terzaghi, K. and Richart, F. E., 'Stresses in rock about cavities', *Geotechnique*, 1952, **3** (June).

Peck, R. B., 'Deep excavation and tunneling in soft ground', *Proc. 7th Int. Conf. Soil Mech. Fndtn. Engng.,* Mexico, 1969, State of the Art Volume.

Muir Wood, A. M., 'Soft ground tunnelling', *in The Technology and Potential of Tunnelling,* South African Tunnelling Conference, Johannesburg, 1970.

Peck, R. B. and others, 'State of the art of soft-ground tunneling', *1st North American Rapid Excavation and Tunneling Conference,* Chicago, 1972, **1**.

Szechy, K., *The Art of Tunnelling,* Akademiai Kiado, 2nd edn., 1973.

Bartlett, J. V. and King, J. R. J., 'Soft ground tunnelling', *Proc. Instn. Civ. Engrs.*, 1975, **58** (Nov.). *See also* discussion, 1976, **60** (Aug.).

Proctor, R. V. and White, T. L., *Earth tunneling with steel supports,* Commercial Shearing and Stamping Co., Ohio, 1977.

5

Compressed Air Working

5.1 USES AND LIMITATIONS

The use of compressed air in tunnel construction is to make practicable subaqueous tunnelling in conditions which would otherwise be unduly hazardous or impossible. It acts by providing a counterbalancing force against the inward pressure of water, reducing the inflow and supporting the ground. It is entirely practicable for men to work in compressed air, but there are physiological limitations and risks which make necessary strict observance of precautions, more particularly at higher pressures.

5.1.1 Decompression

The difficult physiological problems are those of decompression. During work under pressure air dissolves in the blood and is carried to the various tissues. The excess oxygen creates no serious problems except at very high pressures, nor does the nitrogen until pressure is reduced. Under reduction of pressure the nitrogen tends to form bubbles in the tissues and blood stream, and if these accumulate they may exert painful pressure on nerves and block the circulation with possible serious consequences. Decompression procedures therefore must be devised to minimise the risks. The principal means is controlled slow decompression providing enough time for harmless desaturation. The time necessary for decompression may be reduced by limiting the time of exposure to pressure to avoid saturation.

5.1.1.1 Diving

Diving has the same problem both in the ancient traditional diving to seek pearls or for salvage or other purposes and in helmet diving and modern scuba diving. Important differences from tunnel working are the individual self regulation of the diver and the higher peak pressures, the relatively short time of exposure to pressure, the problem of regulating decompression by slow ascent by stages, the temperature of the water and strong currents, and the extra difficulty of providing for any necessary recompression. The tendency has therefore been for diving saturation to be restricted by limiting the time spent at high pressure so that a quicker ascent is possible, whereas in tunnelling as long a shift as practicable is worked by a whole gang at moderate pressure, with the necessary sequel that decompression times, in somewhat more comfortable conditions, may be very lengthy. Techniques of saturation diving are being developed, and systems in which men remain under high pressure for days continuously. In very deep diving, gas mixtures such as oxygen and helium can be used to replace compressed air.

5.1.1.2 Flying

Decompression is also a problem of high flying in unpressurised aircraft. At about 5500 m atmospheric pressure is half that at sea level and at about 8500 m a third. Valuable research into the physiology of decompression in flying was therefore undertaken during and after World War II.

Airline services operate with pressurised aircraft. The pressure is not normally allowed to fall below that at 2500 m, namely 0.75 bar abs, but even so those who have been working in compressed air at 1 bar(g) or more are wiser not to travel by air for at least 24 hours after decompression.

5.1.2 Application

Its historical development as an aid to tunnelling towards the end of the 19th century is described in Chapter 1. Its earliest use was in dock work, but the main development was for river crossings by rail and road such as the Hudson river tunnel, the Blackwall tunnel and the Sarnia tunnel, and also for water and gas mains crossing under rivers and for metro tunnels and sewers at limited depths in water-bearing strata under cities.

These uses all continue to be important with the addition of structures associated with the development of electric power such as cooling water intakes and outfalls and cable tunnels for river crossings.

For major submarine tunnels such as the Channel tunnel and the Seikan tunnel the depths are too great for compressed air working. Their construction depends on finding rock strata competent, after suitable treatment and lining as necessary, to resist the stresses and water pressures.

5.1.2.1 Caissons

Historically use the of compressed air working for caisson sinking preceded that for tunnelling. It was a direct development from the diving bell and was first used in shaft sinking for mining operations by Triger in 1841 at Challonnes on the River Seine. Its earliest use for bridge foundations was for Rochester bridge on the River Medway in 1851. In shaft sinking as a preliminary to tunnelling caissons are frequently used, and the method has also been employed for sinking prefabricated lengths of tunnel at relatively shallow depths where horizontal tunnelling was difficult.

In caisson work the water pressure is balanced at a single level, that of the cutting edge, and working pressure increases directly in proportion to depth. Problems of access differ considerably from those in tunnelling. Air locks are usually small, and decanting, referred to below, becomes the standard method of decompression for all but the lowest pressures.

5.1.2.2 Tunnelling

Compressed air tunnelling therefore is a valuable technique but calls for very careful planning and extensive provision of plant and equipment. Men must be trained in the proper use of air and must accept the disciplines of regulation and of medical supervision. In tunnelling about 3½ bar is the upper limit in standard rules. Higher pressures are practicable and are experienced in deep diving, but permissible exposure times at such pressures are so short and decompression times so long as to be uneconomic for tunnel work unless the circumstances are quite exceptional.

5.2 PHYSICAL ASPECTS

5.2.1 Units

In defining pressure a wide range of different units are employed. The most familiar is probably pounds (force) per square inch (lb/sq in) often written 'psi'. Other widely used units are kilograms (force) per square centimetre (kg/cm^2) or feet head of water or atmospheres. With the increasing use of the S.I. system, based on the metre, kilogram and second, it appears likely that for tunnelling the 'bar' will prove useful and convenient, as it is sufficiently close to 1 atmosphere to be interchangeable in ordinary use. The kilogram per square centimetre is also nearly the same and is sometimes described as a 'technical atmosphere'.

Air pressures are almost invariably stated as the excess over outside atmospheric pressure, sometimes written as 'pounds per square inch gauge' (psig). Absolute pressure is calculated by the addition of atmospheric pressure to the gauge pressure, and is significant mainly in the study of physiological effects. It is sometimes denoted by the symbol 'ATA' for atmospheres absolute. In what follows pressure will generally be stated in bars (gauge pressure), except where other units are necessary for clarity or in quoting descriptions or official regulations. The tables which follow give (1) conversion factors between the different units and (2) equivalents in the other units for pressures in bars up to 3.6 bar.

5.2.1.1 Conversion Tables

Table 1 Conversion factors for principal units

		bar = N/m^2 x 10^5	atm	psi = lbf/in^2	kg/cm^2	Fresh Water ft	Fresh Water m	Sea Water ft	Sea Water m
bar		1	0.987	14.504	1.0197	33.46	10.20	32.64	9.95
atm		1.01324	1	14.696	1.0332	33.90	10.33	33.07	10.08
psi		0.06895	0.06805	1	.0703	2.307	0.709	2.250	0.686
kg/cm^2		0.981	0.968	14.224	1	32.81	10.00	32.00	9.76
Water— fresh	ft	0.0299	.0295	.433	.0305	1	0.305	0.975	0.297
	m	0.0981	.0968	1.422	0.100	3.28	1	3.20	0.975
sea	ft	0.0306	.0302	0.444	0.312	1.025	0.313	1	.305
	m	0.1005	.0992	1.458	0.102	3.36	1.025	3.28	1

Table 2 Conversion table 0.1–3.6 bar

bar	atm	psi	kg/cm^2	Fresh Water ft	Fresh Water m	Sea Water ft	Sea Water m
.1	0.10	1.5	0.10	3.3	1.0	3.3	1.0
.2	0.20	2.9	0.20	6.7	2.0	6.5	2.0
.3	0.30	4.4	0.31	10.0	3.1	9.8	3.0
.4	0.39	5.8	0.41	13.4	4.1	13.1	4.0
.5	0.49	7.3	0.51	16.7	5.1	16.3	5.0
.6	0.59	8.7	0.61	20.1	6.1	19.6	6.0
.7	0.69	10.2	0.71	23.4	7.1	22.8	7.0
.8	0.79	11.6	0.82	26.8	8.2	26.1	8.0
.9	0.89	13.1	0.92	30.1	9.2	29.4	9.0
1.0	0.99	14.5	1.02	33.5	10.2	32.6	10.0
1.2	1.18	17.4	1.22	40.2	12.2	39.2	11.9
1.4	1.38	20.3	1.43	46.8	14.3	45.7	13.9
1.6	1.58	23.2	1.63	53.5	16.3	52.2	15.9
1.8	1.78	26.1	1.84	60.2	18.4	58.8	17.9
2.0	1.97	29.0	2.04	66.9	20.4	65.3	19.9
2.2	2.17	31.9	2.24	73.6	22.4	71.8	21.9
2.4	2.37	34.8	2.45	80.3	24.5	78.3	23.9
2.6	2.57	37.7	2.65	87.0	26.5	84.9	25.9
2.8	2.76	40.6	2.86	93.7	28.6	91.4	27.9
3.0	2.96	43.5	3.06	100.4	30.6	97.9	29.9
3.2	3.16	46.4	3.26	107.1	32.6	104.4	31.8
3.4	3.36	49.3	3.47	113.8	34.7	111.0	33.8
3.6	3.55	52.2	3.67	120.5	36.7	117.5	35.8

5.2.2 Supply of Air

The first questions to be resolved in making use of compressed air for tunnelling are:

1. What is the maximum air pressure foreseeable?
2. What is the appropriate working pressure?
3. What volume of air must be provided?

From the answers follow the planning of bulkheads, air locks, compressor installations, and ancillary plant and equipment, and also appropriate medical precautions and procedures. Extra fire precautions are always necessary because of enhanced rates of combustion and difficulties of access.

The physical interactions of groundwater, compressed air and soil structure in the process of tunnelling are complex and must be examined in some detail. Pressure and volume of air have to be estimated initially with an ample allowance for possible contingencies, in order to equip the site with appropriate plant, but precise requirements can only be determined by practical experience in the progress of the work. Security of air supply in adequate volume is vital for safety wherever loss of air pressure can result in inundation or collapse of the tunnel. Not only must compressor capacity be ample for all foreseeable needs but independent standby power should be ensured.

It is usually relatively simple to fix a maximum for pressure, namely, the hydrostatic head resulting from a water column from lowest tunnel invert level to highest water level including allowance for tidal surges or flood waters where appropriate. There may however be an additional requirement if work must be done ahead of a shield in liquid mud whose density is of course greater than that of water.

The appropriate working pressure is discussed below, but as a starting point a hydrostatic balance level about two thirds of the height up the exposed tunnel face may usually be taken.

The quantity of air to be supplied is much more difficult to determine. The first requirement is to supply at least a minimum for each man working in the chamber. The figure in current British practice is 10 ft^3/min (5 l/s) measured at the working pressure. Compressor capacity is usually measured as the volume of free air per minute and therefore in defining the compressor capacity the above figure will be increased in proportion to the absolute pressure. This air volume should of course be delivered at a point sufficiently close to the face to ensure that men working there have an adequate supply.

In clay, or other ground of low permeability, there may be very little loss of air from the working chamber except through use of the air locks, and an exhaust pipe may be necessary, but if the ground is at all open there will be substantial leakages through the face and through the joints of the completed lining. A figure of 60 to 100 l/s per square metre of the tunnel cross section is suggested for moderately open ground. More may be required if the face is very open, and if the lining of the completed tunnel is not adequately sealed by grouting and caulking. This requirement, plus air consumed in the use of the locks will be well in excess of the minimum man requirements.

At the first Dartford tunnel, low pressure compressor capacity provided on each side of the river was 7000 l/s for the main supply in three electrically driven compressors and 2800 l/s in six diesel-driven standby compressors. The total was thus 9800 l/s for a face area of 86.5 m^2 which is about 80 $l/s/m^2$ main supply plus 30 $l/s/m^2$ standby. This proved more than adequate in use.

5.2.2.1 Deoxygenated Air

In the context of air supply it may be appropriate to refer to the hazard of deoxygenated air. If air displaces the pore water from a stratum containing oxidisable material the oxygen may be wholly or largely absorbed leaving mainly nitrogen. If this gas is then drawn into the tunnel because of a drop in pressure, men may be in danger of asphyxiation. Several instances have been recorded, most commonly due to oxidation of an organic stratum but also of iron pyrites. In one instance of Metro station work adjacent chambers were being worked at different air pressures and deoxygenated air in an intervening stratum was displaced by the higher pressure air into the chamber operating at the lower pressure. The remedy is of course provision of adequate fresh air where work is being done, and particularly after a shut down.

5.3 EXCAVATION AND SUPPORT

5.3.1 Gravel and Sand

The case of tunnelling in gravel or sand beneath the water table may first be examined. It is assumed that the gravel is connected to an extensive source of water and cannot be dewatered satisfactorily by pumping.

A tunnel face completely sealed by a watertight diaphragm would be subjected to the full water pressure of the hydrostatic head from the level of the water table, which, of course, increases from top to bottom of the face. The application of air pressure in the tunnel equal to the water pressure at mid level would produce an overall equality of thrust, but with excess pressure of water externally in the bottom half and excess air pressure internally in the upper half. Opening up of the face then produces a tendency for air to escape in the upper half and water to flow in below as shown in Fig. 5.1.

The air pressure effectively reduces the water pressure acting at the face but does not reduce the pressure of the gravel itself or its tendency to slip in, or to be washed in where the water is still flowing.

The air will escape all over the upper half of the face but more particularly so at the top where the pressure difference is greatest. In escaping it expands and displaces water and fine material and may well develop and enlarge a channel or

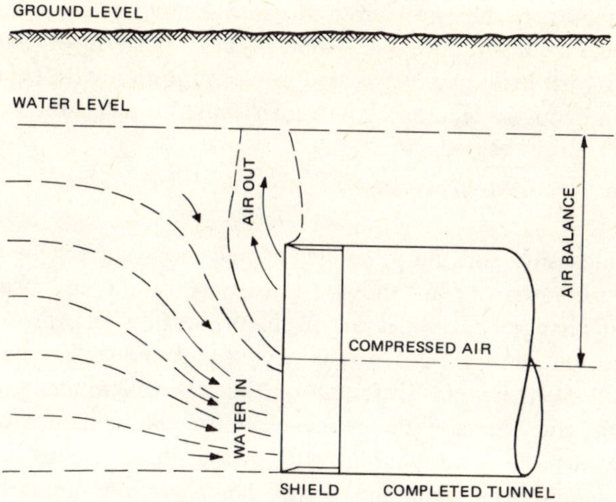

Fig. 5.1 – Air balance in tunnel.

pipe up to the surface, through which large volumes can escape, perhaps catastrophically. Such a sudden catastrophic escape constitutes a 'blow'; air pressure is lost and the tunnel is inundated until the channel chokes and pressure can be restored. On the other hand, the displacement of the pore water changes the soil from a saturated to a moist condition in which capillary forces provide a little cohesion and the stability of the mass is improved. The shield is designed to minimise the risk of catastrophic failure, and may incorporate devices such as a 'hood' at the cutting edge kept buried in the ground ahead, and shutters at the face or safety doors through a diaphragm.

The choice of air pressure for practical use is a matter of judgement. The level in the face at which the selected air pressure balances the hydrostatic head is a convenient measure to define pressure at any time. If the balance level is low in the face, the tunnel is kept drier but the risk of a blow is greater, the air consumption is higher and men are subjected to a higher pressure.

If the balance level is high in the face, the tunnel will be wetter with more danger of ground washing in and it may become more difficult to excavate and line the invert.

As already suggested, a balance level about two thirds of the way up the face is usually suitable as a starting point but will almost inevitably need to be adjusted to suit the particular tunnel. On general grounds of reducing risk of a blow and exposure of men, the lowest practicable pressure is to be preferred, but increased pressure may be needed for limited periods as when high tidal levels are significant, or perhaps for the invert construction.

It is not unknown for tunnel miners to ask for higher air pressure to get drier working conditions, but such increases should be carefully controlled. The drying up effect may be short lived because although the extra air pressure initially pushes back the ground water, a new pattern of flow may soon establish itself with little gain. Increased air consumption with enhanced ground disturbance and longer decompression times must be taken into account if the working pressure is increased.

5.3.2 Silt

For finer grained silty soil the principles of balance are the same but the lower permeability to water and air and the sensitivity of the soil structure to pore water content change the immediate problems. Inflow of water and escape of air will be less, but close support to the exposed ground surface will be required. Any loosening of the grain structure of the silt will immediately reduce its shear strength and increase its permeability so that collapse of the ground becomes possible, the water-bearing silt flowing like a viscous liquid. In major inundations water-borne sand and gravel can also flow similarly and a long length of tunnel can be filled.

The first use of compressed air without any shield in the Hudson river tunnel was originally thought to have solved the problem of tunnelling in river silt by regulating air pressure to a nicety so that the silt was maintained in a surface dry condition retaining its cohesion. The equilibrium was much too unstable for safe working until the Greathead shield in combination with cast-iron segmental lining was introduced, whereby exposed surfaces of ground were kept to a minimum and were provided with immediate support.

Silt continues to be the most unpredictable type of soil and its presence in a tunnel face always calls for caution. Its cohesive properties are unreliable and it cannot be readily strengthened by injection grouting because of its fine grain and low permeability. Very fluid resin grouts can be used but injection is necessarily slow and costly and it may be difficult to confine the material to the required stratum.

5.3.3 Clay

A characteristic of clay is that it is nearly impermeable and, except through fissures, there will neither be inflow of water nor escape of air through a homogeneous stratum. Nevertheless, use of compressed air can contribute greatly to safety in subaqueous tunnelling.

Even if the clay is stiff and unfissured and of sufficient thickness to provide adequate cover above the tunnel, compressed air may properly be employed both to provide continuous support to the roof of the tunnel and to be immediately available if the protecting layer above the excavation is unexpectedly lost, being breached by a slip or a fissure, or by a sudden dip locally in the clay surface because of erosion of an ancient surface by a stream or by weathering. Where the provision of compressed air is precautionary the tunnel may be worked at a nominal pressure of 0.3–0.5 bar unless and until an actual or anticipated emergency requires an increase.

If the clay is soft, silty or fissured or the stratum is of uncertain extent and thickness the use of compressed air may be the only practicable means of tunnelling safely. It cannot, however, be relied upon to support blocks of clay exposed in the top or face because air pressure will permeate fissures and may even assist in dislodging such blocks of clay. The air pressure will, of course, continue to act in relieving the thrust from deeper in the ground, but proper timbering of the face and top is essential to avoid this hazard.

In clay, escape of air at the face is normally very small and ventilation for the miners must be provided as discussed earlier.

5.3.4 Chalk

Chalk is a relatively soft limestone of Cretaceous age. Its properties range widely and its variations along with length of a tunnel are not always predictable. It is of low permeability but is frequently much jointed and fissured, particularly at shallow depths. It may contain cavities formed by solution action and 'pipes' leading down from its surface filled up with gravel or clay from an old land surface. These features were the cause of many difficulties in Stephenson's Watford railway tunnel in 1834–37, when, of course, compressed air working had not been developed.

The water which is to be controlled by compressed air enters through fissures and joints. It most usually comes from above but may sometimes be artesian. The usefulness of compressed air depends very much on the particular circumstances of the tunnel as may be exemplified by three instances.

The first Dartford tunnel was driven under the River Thames using shields and compressed air, The under-river drive was mainly through chalk, but with limited overhead cover; the under-land approaches from the portals descending towards the river banks exposed gravel, silty clay and peat above the chalk. The character of the chalk varied from very soft 'putty' chalk, probably caused by surface weathering in glacial times, to soft but coherent rock at greater depths but with numerous joints and fissures and bands of flint nodules. Conditions were wet, but controllable by moderate air pressure; generally speaking the hydrostatic balance level in the main tunnel was near or above the crown (maximum about 2 bar). Injection of grouts was used to minimise air losses through the gravel and reduce permeability.

In contrast, the Channel tunnel (further described in Chapter 10) was not to use compressed air although also in chalk, because the depth was much too great for appropriate pressures to be acceptable for men to work in, and because a suitable stratum was located in which the chalk was almost impervious and largely free from open joints and fissures, having a large clay content and in fact resembling in character an extremely stiff homogeneous clay.

Another contrasting tunnel in chalk was the cable tunnel under the River Thames at Tilbury which was sited at a depth of over 50 m in the hope of driving through better chalk with greater freedom from water-bearing fissures. The work was executed in free air, but a few large fissures were encountered requiring extensive grouting.

5.4 PHYSIOLOGICAL EFFECTS

Men are quite capable of hard physical work in compressed air at moderate pressures which in tunnelling are normally limited to about 3½ bar. In deep diving below 35 m, pressures are higher and different problems develop, both by reason of pressure and the marine environment. In tunnelling men enter through an air lock in which air pressure is gradually raised to that of the working chamber and they leave through the same lock. The principal physiological effect observed by a man entering is the equalisation of pressure in the ears. The inner ear is brought up to pressure through the narrow Eustachian tubes and if these passages are clear and unobstructed, no discomfort will be felt. If the passage is restricted, or is obstructed as when suffering from catarrh, it may be necessary to raise the pressure very slowly or it may prove impossible for the individual to enter without severe pain in the ears or ultimately a damaged ear drum. Various aids to opening the passages and permitting equalisation of pressure are useful: the act of swallowing can help, or nose blowing with the nostrils held. It must be recognised, however, that personal idiosyncrasies have to be respected and that for some individuals entry may not be possible.

Once the working pressure is attained without damage to ears a man can operate quite normally under pressure. It is sometimes suggested that the extra oxygen supply enhances his physical capacity.

The major physiological problems arise during and after decompression, and may not be immediately obvious. Under pressure additional atmospheric nitrogen, which constitutes four fifths of the atmospheric air, dissolves in the blood stream and progressively in the various tissues to which blood is fed. Nitrogen is particularly soluble in fatty tissue. Some tissues quickly approach saturation but others continue to take up further nitrogen for several hours.

On decompression the excess nitrogen must be discharged again. If decompression is sufficiently slow the emerging nitrogen will be carried to the lungs by the blood and there will be discharged without ill effects. Decompression may result in the formation of bubbles in the blood or tissues and if these bubbles are not harmlessly carried to the lungs by the blood stream, they may lodge in the blood vessels or tissues with painful and possibly serious consequences. There is still much debate on the precise details of the physics and chemistry and physiology of compressed air sickness. Decompression rules and timetables necessarily continue to be formulated empirically, guided by theories but evaluated by practical experience.

The most common phenomena are 'the bends', when a bubble lodges near a joint and exerts painful pressure on a nerve, and 'the niggles', minor irritations of a similar nature. These may occur during decompression, but more often may not be apparent for an hour or more afterwards. They can usually be relieved quite readily by recompression. The original working pressure, or a little more, may be necessary but often a lower pressure gives relief. Slow controlled decompression again, under medical supervision, usually resolves the problem.

More serious cases of decompression sickness, the so-called 'Type 2', occur when a bubble forms in the viscera, lungs, spinal chord or brain. Unless treated, paralysis or death may follow, and frequently did in the past, Again, recompression almost invariably relieves the symptoms, but decompression may have to be extremely slow, possibly extending over several days with reversals if symptoms reappear, to ensure that there is no recurrence.

There is a third form of illness which only develops over a period of months, namely aseptic bone necrosis, in which a bone, most often in hip or shoulder, is attacked and its structure damaged. If the attack is near a joint surface it may eventually interfere with the use of the joint and become painful and may require major surgery. If remote from a joint no ill effects may be noticeable. Diagnosis is by radiography but is difficult in the earlier stages.

Correlation with exposure to compressed air working was slowly established from about 1942 onwards. There was a medical study in New York State, but despite care in formulating and operating decompression rules a substantial number of cases were diagnosed in the construction of the third Lincoln tunnels (1953–57), in most of which the men affected had long records of past exposure to compressed air. Heavy compensation costs were incurred.

Medical records initiated at Dartford tunnel (1957–59) and elaborated at Clyde tunnel (1958–62) and subsequently were studied under the control of the Decompression Sickness Panel of the Medical Research Council and established clearly that aseptic bone necrosis is a form of compressed air sickness. The detailed study is particularly difficult because of the lengthy period of observation necessary and the problems of obtaining, maintaining and diagnosing X-ray records of the changing work force, whose employment in compressed air may be occasional and intermittent.

It is thought that extended decompression times should reduce the incidence. New tables have been devised accordingly and show promising improvement.

5.5 DECOMPRESSION TABLES

Pioneer work on the subject of decompression sickness was done by J. S. Haldane before 1908, for an Admiralty Committee concerned primarily with diving. As a basis for decompression tables he used a pattern of saturation times for different tissues, with corresponding rates of release of dissolved gas. He found that rapid decompression from a level of 2.3 atm absolute appeared safe, that is to say from

1.3 bar (19 psig). He adopted a ratio of 2 : 1 for the absolute pressure drop but expressed doubts as to the applicability of this ratio at very high pressures. Thus from a pressure of 1 atm rapid decompression to atmospheric pressure was permitted, from 2 atm the immediate drop was to ½ atm, and from 3 atm to 1 atm. Thereafter pressure had to be released very slowly, the rate depending on pressure and length of exposure. This formulation of the problem has been used ever since in the preparation of many tables, but details have been modified and assumptions as to tissue saturation times have been varied. There are other theoretical approaches. It is now clear that the 2 : 1 ratio does not ensure freedom from bubble formation in all conditions and that the initial rapid drop should not be so great.

In 1936 when the construction of the Dartford tunnel was under consideration, a committee of the Institution of Civil Engineers recommended rules for decompression based on Haldane's studies and on practical experience. Pressures up to 1.5 bar were considered not to require slow decompression, but for higher pressures tables were provided defining the amount of the first stage rapid decompression and a uniform rate for the succeeding slow decompression. The time of exposure to compressed air is relevant because it takes 4 hours or more for saturation of 'slow' tissues, which are likewise slow in releasing nitrogen. Accordingly, the tables permitted speedier decompression for short exposures. The main table set out, for pressures of 22 psi (1.5 bar) and over, the initial drop permissible in two minutes and the times for further uniform slow reduction to zero for times of exposures of: 6 hours or more, 4 hours, 3 hours, 2 hours, 1 hour, ½ hour. A revision of these tables was formulated in 1951 and incorporated in new Statutory Regulations in 1958; the decompression times were extended somewhat and the threshold for timed decompression was reduced to 1.24 bar (18 psi). It was not expected that these tables would eliminate 'bends' but it was considered that an incidence rate of 2% of compressions would be acceptable, because, except for unusual Type 2 cases which form a very small proportion of the whole, there appeared to be no permanent damage.

5.5.1 Blackpool Tables

When the statutory tables were worked to over a period cases of illness were encountered even below the 1.24 bar threshold and the incidence of bone necrosis was found to be widespread. Following Decompression Sickness Panel Studies a committee of the Construction Industry Research and Information Association re-examined the subject and published a report in 1973 recommending a new Code of Practice. It incorporated new tables with longer decompression times and with the threshold level reduced from 1.24 bar to 1 bar. There were also proposals for various other medical precautions and organisational arrangements, including X-ray examinations to be reported to a central register. The new tables are known as the 'Blackpool Tables' from the site of their first experimental use and have been widely adopted in Great Britain. At the maximum pressure of 3.4 bar, exposure exceeding 4 hours requires to be followed by 6 hr. 35 min. decompression time to follow. As a 12 hours limit for shift working is imposed on exposure time plus decompression time, this allows a maximum working shift time of 5 hrs. 25 min. at this pressure. Apart from this limitation, which takes effect above 2.8 bar, the Code permits the working of a full 8 hour shift, the assumption being that saturation is reached after 4 hours and that further exposure adds nothing to hazards.

5.5.2 New York split shifts

In New York very extensive tunnelling experience in the East River railway tunnels and elsewhere in the period 1904–21 and subsequently, led to the adoption of the split shift system in which working time at higher pressures was severely limited, a substantial intermediate period in free air being required.

Prior to 1909 shifts of 3 hrs. in, 3 hrs. out, 3 hrs. in for pressures exceeding 2.2 bar (32 psi) were worked and this pattern was progressively developed so that in the Lincoln tunnel (1953–57) the permissible shifts were typically:

Up to 1 bar (14 psi):	3 hrs. in,	2½ hrs. out,	3 hrs. in
At 3.4 bar (50 psi):	½	5	½

Decompression times were: 1 bar: 6 min.
3.4 bar: 37 min.

In 1960 these decompression times were increased for the second period to 9 mins. and 62 mins. respectively.

The limitations make impracticable substantial work in compressed air above 2.4 bar.

5.5.3 Washington State tables

In 1963 after further studies, Washington State adopted new standards which have been adopted with modifications in other states and countries. They are based on a single working period restricted progressively as pressure is increased but can be adapted to split shifts. Eight hours is allowed below 1.25 bar (18 psi) reducing to 3 hours above 3.31 bar (48 psi). The decompression is operated in 3 or 4 stages, the total time ranging from 16 mins. after 8 hours exposure at 1.0 bar (14 psi) to 249 mins. after 3 hours at 3.4 bar (50 psi). For comparison the Blackpool tables require 20 mins. and 200 mins. respectively.

The split shift system is now generally considered undesirable as actually increasing the hazards by doubling the number of decompressions. Complete desaturation is not achieved in the rest interval.

5.5.4 Comparison

A comparison of decompression sickness rates at various pressures using the Blackpool and Washington tables showed significantly lower 'bends' rates for the former despite restrictions on shift length in the latter.

There is a possibility that the limitation of exposure time at the higher pressures provided for in the Washington tables may be effective in minimising bone lesions in comparison with the longer exposures permitted in the Blackpool tables.

An interesting aspect of the Blackpool tables is the adoption of staged decompression, the pressure being held at each step of 0.2 bar (or 4 psi) for the number of minutes tabulated. An advantage is that it is easier to be precise in holding at a level than in operating a prescribed rate of drop between levels. The last step from 0.2 bar to atmosphere eliminates the uncertainty arising from leakages as the pressure helping to retain the lock drops to zero and the tightness of its seal is lost.

5.6 INCIDENCE OF ILLNESS

The physiological hazards arising from working in compressed air have not been eliminated but have been brought under control. Type 1 illness is accepted by those choosing to work in compressed air as an occupational risk with no long-term consequences. Individuals appear to differ considerably in their susceptibility and on any job there is likely to be a progressive elimination of the more susceptible. Some acclimatisation of men is observed over a few days if work continues at a steady pressure, but this is lost after absence of a week or ten days, and in any case, does not appear to provide any protection against an increased pressure.

Type 2 illness is rare but potentially dangerous; immediate and properly controlled treatment by recompression appears to be almost invariably effective. The recorded incidence is about 0.05% of decompressions.

The third hazard, that of bone necrosis, has been found by X-ray examination in a substantial proportion of compressed air tunnellers and in a smaller proportion of divers. Only about a quarter of the cases detected are in joint surfaces and produce symptoms. Because of its slow development and the difficulty of diagnosis a centralised system of records has been developed at Newcastle-on-Tyne to assist research into elimination of causes. Strict observance of improved decompression tables with accurate records is the most hopeful action.

5.7 MEDICAL CONTROL

In many countries regulations provide for the health and safety of men employed in compressed air. The precise official regulations applicable must of course be ascertained and worked to as a minimum, but additional safeguards may be specified or adopted for a particular project.

The rules become progressively more restrictive above a pressure of 1 bar or thereabouts, and may include the following requirements:

1. Medical examinations of workers before employment and periodically during employment including radiographic records if practicable.
2. Issue of identification discs to ensure proper treatment (by recompression) in the event of collapse away from site.
3. Strict control of entry and all decompressions by reliable lock keepers and proper detailed documentation of all exposures and decompressions.
4. Provision of changing and working facilities and rest rooms to enable men to remain on site for an hour or more after exposures over 1 bar.
5. Provision of a medical lock solely for therapeutic use and in charge of a trained medical attendant working under the direction of a doctor.

5.8 PLANT AND EQUIPMENT

The physical effects of compressed air working have been described above and the principal items of plant and equipment have been listed but are now described in more detail.

5.8.1 Bulkheads

The first essential is to seal off the working space from the outside air, closing off all openings. If the tunnel is worked from a single shaft the seal may be effected by an air deck across the shaft or by a bulkhead within the tunnel. If other shafts or adits lead into the working space they must likewise be sealed. Air locks provide passage through the bulkhead or air deck for men and materials.

The placing of the main bulkhead will be chosen to suit the particular project. It may be advantageous to provide initially an air deck and airlock in or at the top of an access shaft and to replace or supplement these later, when the tunnelling is sufficiently far advanced, by a bulkhead across the tunnel and two or more air locks through it. Unless the tunnel is small and pressures low it is essential to have separate man locks and muck locks to avoid mutual interference. Very careful planning of the lock arrangements is necessary to ensure that progress is not unnecessarily restricted by inadequate locking facilities.

A tunnel bulkhead must be a substantial and well built structure because very large forces are involved in a tunnel of large cross sectional area. For example, a 9 m diameter tunnel has a cross sectional area of 63.6 m^2; with a pressure as low as 1 bar the thrust on this area will be about 650 tonnes force.

Mass concrete is the most usual material for construction and a thickness of 0.4 to 0.6 times diameter is usual. Calculations of strength appropriate to the maximum possible air pressure should take account of the large holes required for air locks, but may usually assume arching action within the thickness. The whole thrust has to be transmitted by shear or otherwise into the tunnel lining behind the bulkhead and from the lining into the ground. A sufficient length of completed lining or other structural arrangement must therefore be ensured to distribute this thrust.

In addition to the air lock passages all services into the tunnel have to pass through the bulkhead including low pressure and high pressure air pipes, water supply, electric power, light and control cables, telephone cables, pumping mains and possibly such special services as hydraulic power, slurry discharge pumping or bentonite slurry supply. These have to be planned in advance and built in during construction of the bulkhead. As there is no reasonable possibility of making additional provision once the bulkhead is in use, ample spare capacity is advisable ensured by including a number of spare pipes blanked off until required. See Fig. 5.2.

Special care is necessary in concreting such a bulkhead to ram the concrete tightly round each service pipe, leaving no voids for escape of air and also to ensure careful sealing and grouting into the tunnel lining all round with particular attention at the soffit where the concrete may shrink away from the lining.

The bulkhead is ultimately demolished to clear the finished tunnel and this must be done without damage to the permanent lining into which it is anchored. Steel structures have been used for air bulkheads but are less easily adaptable to site requirements, and present the same problems of shear anchorage into the lining, and of providing for the passage of services.

An alternative to use of a bulkhead is the construction of an air deck in an access shaft if there are no other openings into the tunnel or if all are sealed off. The problems of making the deck airtight, and providing for the passage of services are much the same as for a bulkhead. Special care is necessary in dealing with the thrust upwards from the air pressure acting on the underside of the deck. Part will be taken safely by the dead weight of the deck itself, but any part transferred to the shaft lining needs to be examined critically against possible failure from breakage of the shaft lining in tension or shear, or loss of ground friction by reason of excavation or penetration of water or escaping air.

5.8.2 Air Locks

An air lock is an intermediate chamber between the open air and the working chamber through which passage may be effected between the spaces maintained at differing pressures. In tunnelling a lock functions to provide controlled entry to compressed air workings and exit therefrom. The lock is provided with two doors hinged to open respectively from atmosphere into lock chamber and from lock chamber to working chamber. It must be airtight and capable of withstanding the maximum pressure to which it may be subjected. The doors should be of steel accurately fitted and furnished with rubber or other sealing strips to ensure

Fig. 5.2 – Services into tunnel through bulkhead at air lock.

airtight closure and should be hinged as shown in the figures to allow even bearing all round, assisted by differential air pressure.

The principle of operation of an air lock is simple. In order to pass into the working chamber, which is at a pressure higher than atmospheric, the inner door of the lock is closed and the lock chamber is at normal atmospheric pressure while the outer door is open to the free air. A man enters the lock chamber, closes the outer door behind him, and compressed air is fed into the lock chamber through a control valve until the pressure is the same as that in the workings. The inner door is then free to open and entry to the tunnel is possible. In order to come out of the tunnel into free air the operation is reversed, compressed air being released from the air lock chamber to the atmosphere until the outer door can be opened.

Locks are required for the passage of men and materials, and although the principles are the same, there are important differences in detail. Locks are almost always cylindrical and the internal diameter is preferably 2 m or more. For men, decompression rates must be precisely regulated and from high pressures the time required may be a matter of hours, during which the lock is not available for other use. Dimensions of the lock must be such as to accommodate comfortably on seats the full numbers of shift workers and others who may be emerging at one time. The door dimensions (say 0.8 m x 1.2 m) need only be such as to allow a man to pass, or in an emergency a stretcher but should preferably be larger. The air feed and discharge pipes are usually limited to 25 mm diameter and may be fitted with automatic valve control to accord with the appropriate decompression tables. The rapid stage of decompression produces sharp cooling of the contained air and formation of fog which should be counteracted by heating during that stage. The lock is normally under the control of an external lock keeper who must have a window opening to see into the lock and telephone communication with the interior of the lock and the working chamber. Fig. 5.3.

For materials, not accompanied by men, the lock dimensions must accommodate the required number of waggons or skips being used, and must be long enough to take the longest timber, pipe or steel member likely to be required. In addition adequate clear space must be allowed for the door at the outer end to swing into the lock. A separate 'timber lock' comprising a long pipe with air doors is possible, but not often found advantageous. The door dimensions of a materials or 'muck' lock will be governed by the cross section of the loaded skips, but are of course limited by the diameter of the lock: 1.2 m wide x 1.4 m high is practicable in a lock of 2 m diameter.

For a materials lock, restricted solely to that use and never used for men, the air supply and exhaust pipes may be much larger: 100 mm diameter is common, to give the quickest practicable compression and decompression. Visibility is not important and heating is unnecessary.

In practice most materials locks, even if primarily used for their single

Fig. 5.3 – Two air locks for men and/or materials. Note: bulkhead in background, doors opening inwards, pressure gauges.

purpose, are equipped so that they can be utilised in emergency as man locks. They must then have small diameter air supply and discharge pipes, windows, gauges, telephones, seats, etc.

In addition to separate man and materials locks, consideration must be given to provision of an emergency man escape lock if the tunnel is liable to the hazard of complete inundation from the face. Such a lock will be sited as high up as possible in the tunnel to be above flood level. It should at all times be at working chamber pressure with its inner door open so that it can be entered without delay. It should be capable of accommodating the full number of men who may be in the working chamber at any time.

There is sometimes also a need for an engineer's survey lock for the work of carrying survey lines forward and for repeated checking as the tunnel face advances. These operations require lengthy occupation of a lock during which time neither men nor materials can pass through. It may be possible to utilise an existing lock if there is a weekend or other shutdown or if duplicate locks have been installed but the survey work is essential and time consuming, however costly interruption of passage may be. An engineer's lock has sometimes been installed in tandem with an emergency lock, extending the lock outwards so that the refuge of the emergency lock chamber remains open to the working shift even when survey operations are in progress. This can also serve for emergency access of a rescue gang into the tunnel.

The most common type of lock is the 'boiler lock' which comprises a cylindrical steel chamber riveted or welded of about 2 m or more in diameter and 5 m or more in length with steel end doors. It is normally installed so that the body of the lock is in free air while the end having the door opening outwards into the working chamber is built into a bulkhead if sited in the tunnel, or secured by a flanged connection if used at the top of a shaft (see figures). The cylinder is then in tension when the lock is under pressure. If the lock is in the open it is desirable to protect it against extremes of temperature by screening or insulation.

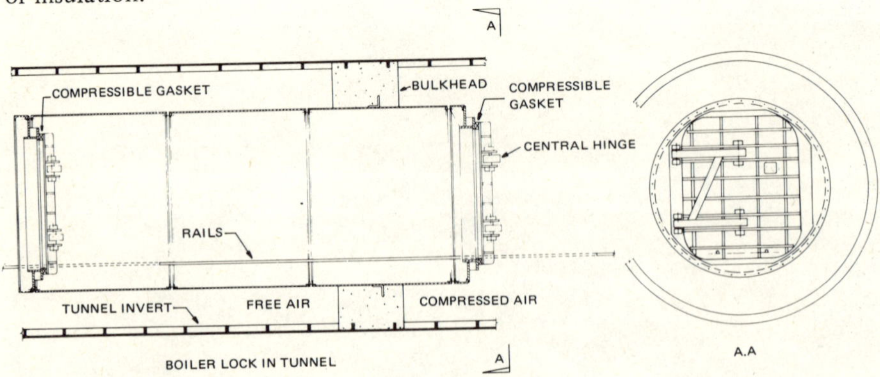

Fig. 5.4(a) – Boiler lock in tunnel.

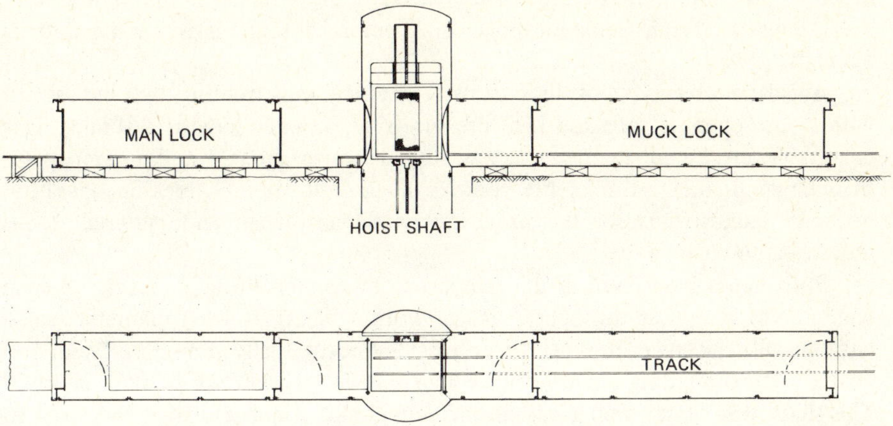

Fig. 5.4(b) – Section plan and elevation of T-locks at shaft head.

A lock used for materials is usually furnished with a rail track for passage of skips but there must be gaps in the track to allow the air doors to be closed, short lengths of rail being dropped into prepared seatings when the door is opened.

An alternative type of air lock is formed by constructing two bulkheads, with doors opening into the high pressure side, across a length of tunnel. This is suitable in a small diameter access tunnel, up to about 4 m diameter or so. In a larger diameter tunnel the volume of air used at each locking operation becomes excessive. Also, the provision of separate man and materials locks requires a duplicate adit.

Both types were used in the first Dartford tunnel. Initially, access on each bank of the river was through a vertical working shaft of 18 ft diameter sealed with an air deck. At each shaft boiler locks led horizontally through a T junction into a vertical winding shaft within which a hoist transported men or materials. The shaft was offset well clear of the tunnel and connected to it by two 4 m access headings. In each of these adits an air lock was constructed as described. The surface locks were used initially for sinking the shaft and constructing the pilot tunnel but the bulkhead locks were used for constructing the main tunnel, allowing the shaft to be used in free air. The tunnel was driven from both sides of the river simultaneously and these were connected by the completed pilot tunnel but had to operate at different pressures; therefore a mid-river air lock utilising two bulkheads across the pilot tunnel was necessary, fitted with four doors because the difference of pressure might operate in either direction.

Another type of air lock is the 'caisson lock' which is of limited capacity and not suitable for any extensive tunnel work. It comprises a vertical cylinder forming the materials lock and designed to accommodate a cylindrical bucket handled by a crane. The lower door of the lock is hinged to open downwards and is mechanically operated from outside, but the upper door is a separate circular lid which has a central rope gland and which lifts off with the bucket for discharge. When the bucket is returned and lowered into the shaft the lid is clamped down onto the top of the chamber. The man lock is a 'blister lock' on the side of the shaft immediately below the materials lock. It is necessarily of very limited capacity and quite unsuitable for use by large numbers or for lengthy decompressions.

5.8.2.1 Decanting

Where lock accommodation is so restricted but long decompressions are nevertheless necessary, a process known as 'decanting' can be used, subject to observance of supplementary regulations. The men are rapidly decompressed to atmospheric pressure in the small lock and go without delay into another more commodious lock in which they are recompressed as quickly as practicable to the original working pressure. Their decompression then follows the normal rules.

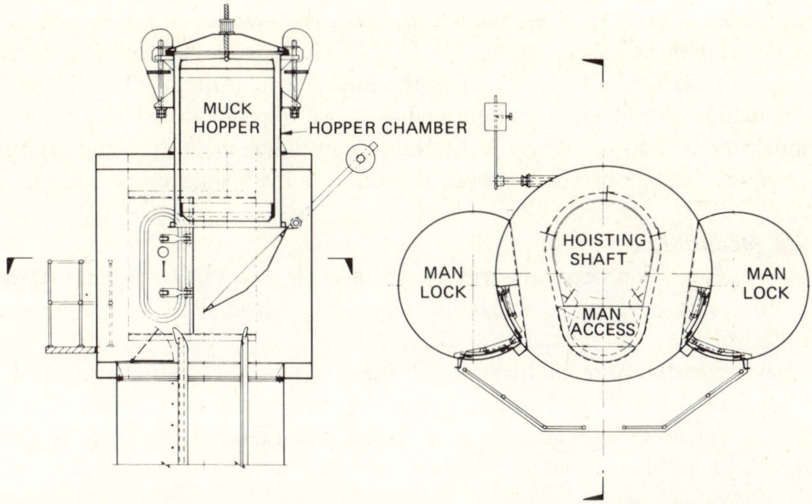

Fig. 5.4(c) – Sectional plan and elevation of Caisson locks.

Other types of lock may be adopted for particular circumstances but the principles are the same. It has sometimes been the practice in a long tunnel to construct a bulkhead part of the way along it and instal a second lock maintaining a lower pressure in the first section of the tunnel than at the face. This was considered helpful in providing an intermediate stage of decompression but it is not currently thought to be of any real advantage in that respect. It would certainly make it more difficult to ensure proper control of individual decompression times.

5.8.2.2 Medical Lock

Recompression is the necessary treatment for all Type 2 cases and many Type 1 cases of compressed air sickness. A separate medical lock at site becomes essential when such cases are to be expected.

It is normally of cylindrical form like a boiler lock, but the treatment chamber has its door at one end only, the other end being closed off. For entry and exit a small man lock is provided (see Figure 5.5). There is also a small lock for passing in meals or medical requirements quickly and easily.

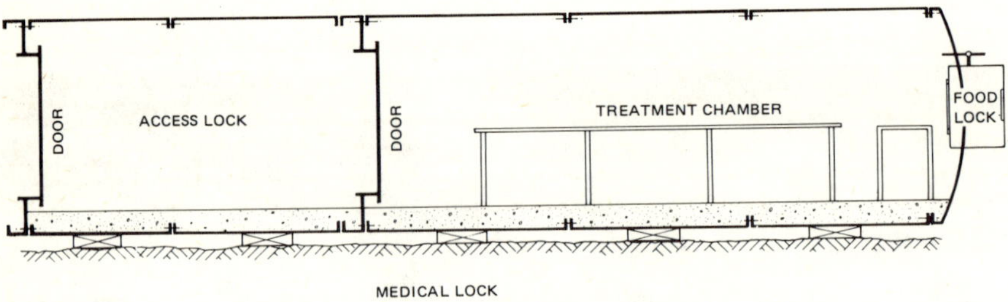

Fig. 5.5 – Medical lock.

5.8.2.3 Equipment of Locks

The principal requirements have already been mentioned in describing the functioning of air locks. Little need be added in respect of materials locks unless they are required for occasional use by men, but for man locks a fuller description of operation, equipment and lock management is appropriate. In what follows a large tunnel using moderate air pressures is taken as the basis. Some simplification is possible in small tunnels at low pressures, while extra precautions are of course necessary if pressures are high.

A man lock must be operated under the control of an experienced lock keeper, who remains outside in free air and who is responsible for ensuring the proper regulation of all compressions and decompressions. Compression during entry may be delegated to a responsible and experienced man passing through the lock because the acceptable rate of pressure rise is a matter for each individual and does not involve hidden after effects. Decompression, however, must be controlled exactly in accordance with the prescribed tables and, except in emergency, should not be effected by anyone other than the responsible lock keeper.

There must be fixed in a place readily visible to the lock keeper and also to a man inside the lock, gauges showing accurately the pressure in the working chamber, and the current pressure in the lock and also a clock by which decompression can be timed. A bullseye window, in each door, is required so that the lock keeper can see those in the lock and take any necessary emergency action, and so that those in the lock can see gauges and clock when required. Telephones between lock keeper, lock chamber and working chamber are a most valuable provision.

The arrangement of air valves provides for normal operation by the lock keeper but also for emergency action by those in the lock. The piping is shown diagrammatically in Figure 5.6. Automatic control of the exhaust valve to the predetermined programme of decompression is sometimes employed. For a materials-only lock larger pipes can be used to operate more quickly and controls can be simplified.

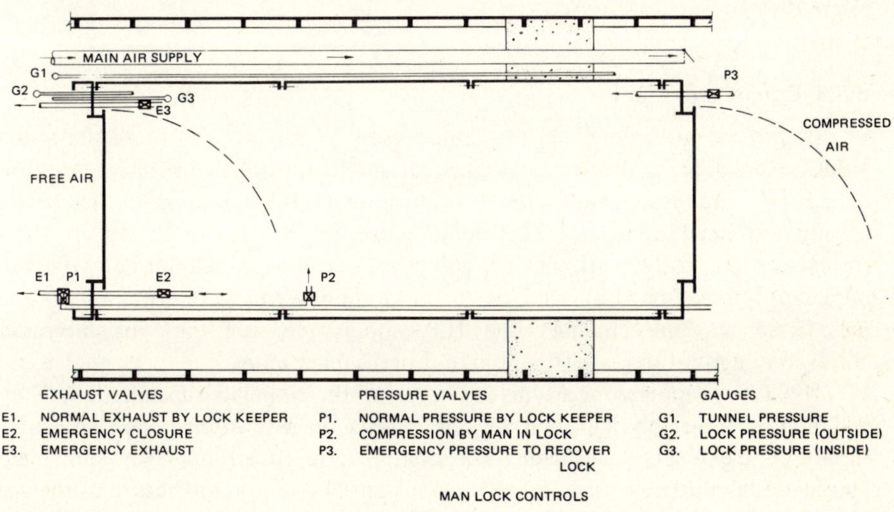

Fig. 5.6 – Controls in man lock.

Where men may have to spend an hour or longer in a lock, provision of relatively comfortable conditions becomes increasingly important. Adequate seating is essential with proper back support; at least 0.5 m and preferably 0.6 m width per man should be allowed and 1.3 m^3 of air space. This implies a diameter of at least 1.82 m. In a small lock with many men CO_2 levels may rise unacceptably during a long decompression and additional ventilating air will be required. This must be fed in at or above the lock pressure and preferably from a fresh air supply, and not from the working chamber, particularly in the event of a fire in the working chamber; the exhaust volume must of course be increased correspondingly.

Other requirements are a boarded floor, good lighting, heating for the rapid decompression stage, sanitary facilities and careful fireproofing against the enhanced hazard of fire in compressed air.

For a medical lock the requirements are more extensive. The compressed air supply should be fresh and always available at a pressure at least 0.7 bar above working pressure. An accurate recording gauge should show the pressure in the chamber to the medical attendant who should be responsible for controlling the pressure. Ventilation should be provided and properly regulated. Windows, telephone, lighting, thermostatically controlled heating, sanitary facilities and strict fire precautions are required. The chamber should be furnished with a full length bed and blankets.

As the medical lock will usually be sited on the surface it needs to be screened or adequately insulated against overheating from the sun and against cold. Most of these matters are more fully dealt with in the CIRIA Code referred to above.

5.8.3 Compressor Plant

In compressed air tunnelling the compressors supplying air to maintain the requisite pressure in the working chamber up to a normal maximum pressure of 3.5 bar, are described as the low pressure (L.P.) compressors. There are usually also high pressure (H.P.) compressors for power supply to air tools for excavation, drilling, air winches, and other purposes, which normally operate on a supply pressure of about 7 bar or more; in order to work efficiently in the pressurised working chamber the H.P. supply pressure must be increased correspondingly so that a differential of 7 bar is maintained.

The L.P. compressors are at the heart of the tunnelling installation. They will almost invariably be sited on the surface in a position where an ample supply of clean air is available. The compression of air has two immediate physical consequences: first, there is a substantial rise in temperature of the air; and secondly, there is an increase in humidity, because the compression increases the mass per unit volume of the water vapour present. Cooling of the air and removal of excess water are therefore required.

Accurate control of pressure is necessary and a quick response to any sudden demand for more air.

Figure 5.7 shows a plan of a compressor house on a large job.

Clean air supply is most important and filtration may be necessary in some situations. Siting of the intake should be such as to avoid fumes from any plant or traffic.

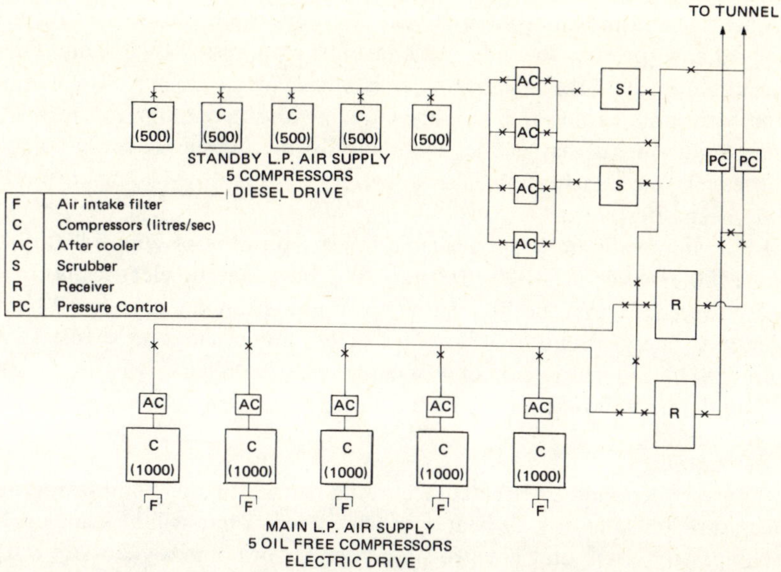

Fig. 5.7 – Compressor house.

5.8.3.1 Compressors

The quantity and pressure required have been discussed above. Except for small tunnels at low pressure the volume of air to be compressed is large and the plant is correspondingly large and heavy. It is advantageous to install a number of compressor units rather than a single one of the maximum capacity so that varying requirements may be met economically and the supply is not completely disrupted if one machine breaks down. Extra standby for emergencies has already been referred to.

L.P. compressor units are available for compression of about 150 l/s up to about 1500 l/s (say 300–3000 ft^3/min) and they may be electrically driven or diesel powered. It may be noted that to supply air at the rate of 60 $l/s/m^2$ in moderately open ground (see 5.2.2) a small tunnel of 2 m diameter would require 190 l/s and a large tunnel of 9 m diameter 3800 l/s.

(Note: 1 l/s = .001 m^3/s = 2.12 ft^3/min)

5.8.3.2 Cooling

The necessary cooling is effected according to the type of compressor and the temperature and quantity of heat to be extracted. There will be some cooling in the compressor itself, and if a two-stage compressor is employed there will be an intercooler incorporated, but in most tunnel installations there will also be an aftercooler through which the compressed air passes on its way to the tunnel supply main. Condensation of water vapour is to be expected in the aftercooler which must be drained effectively. Even after cooling the compressed air remains extremely humid as fed into the tunnel and as subaqueous tunnels are intrinsically wet the atmospheric humidity in the working chamber is usually 95–100%. Any lowering of pressure is likely to result in fog.

5.8.3.3 Oil separator

In most compressors some oil vapour or mist will be discharged with the compressed air, and should be removed by means of an oil separator, which may be a chamber packed with wood wool or other filtering medium, or a chamber utilising centrifugal action in the air stream, or may incorporate both devices. Oil free compressors are manufactured.

5.8.3.4 Receiver

The system may incorporate a receiver to act as a reservoir and regulator of compressed air and to smooth out fluctuations of pressure. Compressor output may be controlled automatically in accordance with the defined limits of receiver pressure. Silencers and pressure reducing valves may be useful additions to the circuit.

5.8.3.5 Air mains

The pipe lines to the tunnel are preferably in duplicate so that damage to one does not shut down the supply. They should be fitted with non-return valves as near as possible to the point of entry to the working chamber. Large diameter mains of 250 mm or more are to be preferred to ensure that air velocity at full capacity does not exceed about 10 m/s and thereby minimise loss of head.

5.8.3.6 Gauges

Gauges to indicate the pressure in the tunnel should be fitted in the compressor house and there should be telephonic communication between compressor house and working chamber.

5.8.3.7 H.P. Compressors

As already stated, higher output pressure is necessary for efficient operation of tools, that is up to about 10 bar. As with the main L.P. compressors, air as delivered into the tunnel should be clean and free from oil. Coolers with drainage and oil separators are therefore required in the delivery line.

5.8.4 Electrical Supply System

The principal aspects in which the electrical supply requirements for a compressed air tunnel differ from those for a wet tunnel in free air are the absolute dependence on compressor availability and the enhanced fire risk in the working chamber and air locks.

Supply of compressed air must be absolutely assured where there is any possibility of collapse or inundation. If compressors are electrically driven a completely independent duplicate supply is essential, or alternatively, diesel powered standby compressors of full capacity which can be quickly brought into use.

An independent emergency lighting circuit is even more important in compressed air workings than in free air and a telephone system independent of other circuits.

The greater fire risk in compressed air arises from the increased oxygen per unit volume of air. Oil filled electrical switchgear must therefore be fully safeguarded, or avoided altogether.

5.8.5 Pumping out water

The normal arrangements for removing water from a tunnel are dealt with elsewhere. In compressed air working it is common to utilise the air pressure to 'snore' out water. At the end of an armoured hose a filter to exclude stones is attached and the hose is connected through a cock to a pipe leading back to the free air side of a bulkhead. The filter is lowered into the water in the tunnel and, with the valve open, the compressed air, partly entrained with the water, drives and carries it through the pipe to discharge into a sump in free air. The 'snorer' is also sometimes a useful means of drawing fresh air forwards to improve ventilation at the face.

5.9 FIRE RISKS

Adequate supplies of water for fire fighting are important. The fact that outlets must discharge against the increased air pressure is not important if the supply is from the surface because the extra hydrostatic head will compensate.

Apart from the special risk to oil filled electrical plant, there is a greatly increased hazard from all combustible material. A major hazard can arise if heavy timbers supporting the face or elsewhere become ignited as combustion will be much more rapid than in free air and the site of the fire may be inaccessible. Lowering of air pressure to allow water to enter at the face may sometimes be a useful expedient, but with the disadvantage that the drop in pressure will generate fog and thereby add to the problems of fire fighting.

Fires in compressed air should therefore be fought by men familiar with the tunnel workings and with compression and decompression procedures in air locks, which may constitute a serious obstacle to ordinary fire brigades.

BIBLIOGRAPHY

The bibliography of compressed air working has two distinct aspects, the physical and the physiological. While the general pattern of a brief chapter bibliography followed (in Volume 2) by a more extensive and comprehensive main bibliography, is appropriate to the engineering aspects, much of the physiological material is so special to this chapter that the relevent section which follows will not be fully repeated.

Engineering aspects of work in compressed air are described in the old standard texts and others, supplemented extensively by Instn. Civ. Engrs. and other papers, a few of which are included below.

Greathead, J. H., 'The City and South London Railway; with some remarks upon subaqueous tunnelling by shield and compressed air', *Min. Proc. Inst. Civ. Engrs.,* 1895, **123**.

Hay, D. and Fitzmaurice, M., 'The Blackwall Tunnel', *ibid*, 1897, **130**.

Copperthwaite, *Tunnel Shields and the use of compressed air in subaqueous works,* Constable, 1906.

Hewett, B. H. M. and Johannesson, S., *Shield and compressed air tunneling*, McGraw-Hill, 1922.

Norrie, C. M., 'The River Hooghly Tunnel', *Min. Proc. Instn. Civ. Engrs.,* 1933, **235**.

Richardson, H. W. and Mayo, R. S., *Practical tunnel driving*, McGraw-Hill, 1941.

Groves, G. L., 'The Ilford Tube', *J. Instn. Civ. Engrs.,* 1946, **26** (Mar.).

Singstad, O., 'Historical development of subaqueous tunneling', *J. Boston Soc. Civ. Engrs.,* 1949, **36**.

Pequignot, C. A. (ed.), *Tunnels and tunnelling*, Hutchinson, 1963.

Kell, J., 'The Dartford Tunnel', *Proc. Instn. Civ. Engrs.*, 1963, **24** (Mar.). *See also* discussion, 1963, **26** (Dec.).

Haxton, A. F. and Whyte, H. E., 'Clyde Tunnel: constructional problems', *ibid*, 1965, **30** (Feb.). *See also* discussion, 1967, **37** (July).

Falkiner, R. H. and Tough, S. G., 'The Tyne Tunnel: construction of the main tunnel', *ibid*, 1968, **39** (Feb.). *See also* discussion, 1968, **41** (Nov.).

Glossop, R., 'The invention and early use of compressed air to exclude water from shafts and tunnels during construction', *Geotechnique*, 1976, **26** (June).

Physiological effects

Books and papers above are not repeated here:

Hill, L., *Caisson sickness and the physiology of work in compressed air*, E. Arnold, 1912.

Report of the Committee on regulations for the guidance of engineers and contractors for work carried out under compressed air, Institution of Civil Engineers, 1936.

Paton, W. D. M. and Walder, D. N., 'Compressed air illness: an investigation during the construction of the Tyne (Pedestrian and Cyclist) Tunnel, 1948-50', *Medical Research Council Special Report 281,* HMSO, 1954.

The Work in Compressed Air Special Regulations, S.I. 1958 No. 61, HMSO, 1958.

Golding F. C. and others, 'Decompression sickness during construction of the Dartford Tunnel', *Brit. J. Industr. Med.,* 1960, **17**, p. 167.

Duffner, G. J., *Decompression sickness and its prevention among compressed air workers*, Metropolitan Engineers Report, Seattle, 1962.

Safety standard for compressed air work, State of Washington, Department of Labour and Industries, 1963, Ch. 20, part 2.

McCallum, R. I., Walder, D. N. and others, 'Bone lesions in compressed air workers, with special reference to men who worked on the Clyde Tunnels 1958-1963', *J. Bone and Joint Surgery,* 1966, **48B** (May).

Compressed Air Safety Orders, State of California, Divison of Industrial Safety, 1966.

The problem of decompression sickness in diving and civil engineering, Annual Report 1965-66, Medical Research Council, HMSO.

McCallum, R. I. (ed.), *Decompression of compressed air workers in civil engineering,* Oriel Press, 1967.

Bennett, F. B. and Elliott, D. H. (eds.), *The physiology and medicine of diving and compressed air work,* Bailliere, 1969.

Medical Research Council Decompression Sickness Panel, 'Decompression sickness and aseptic necrosis of bone: investigations carried out during and after the construction of the Tyne Road Tunnel (1962-66), *Brit. J. industr. Med.*, 1971, **28**.

Walder, D. N. and McCallum, R. I., 'An objective appraisal of the Blackpool (U.K.) and Washington State (U.S.A.) Decompression Tables', *Proc. 5th Int. Hyberbaric Cong.*, Vancouver, 1973.

A Medical Code of Practice for work in compressed air, Report 44, Construction Industry Research and Information Association, London, 1973. 2nd edn. 1975, 3rd edn. in preparation.

Medical Research Council Decompression Sickness Panel, *Experience with a new decompression table for work in compressed air, Technical Note 59,* Construction Industry Research and Information Association, London, 1974.

Construction Industry Research and Information Association and British Tunnelling Society, discussion on 'The recommendations for work in compressed air and experience with the new decompression table', *Tunnels and Tunnelling,* 1974, **6** (Dec.) and 1975, **7** (Jan.).

Aseptic Bone Necrosis, (Proceedings of a symposium, Newcastle upon Tyne, 1976), *Technical Note 12*, CIRIA Underwater Engineering Group, London, 1977.

6

Rock Tunnelling

6.1 GENERAL REQUIREMENTS

In comparing rock with soft ground as a tunnelling medium the obvious contrasts are that rock is so much more difficult to break out that it usually requires drilling and blasting, and that the requirements for ground support call for a different approach. Of course, soft rock may be tunnelled using soft ground techniques, which indeed have also to be adopted with modifications for very heavy loadings encountered in squeezing rock, but in typical rock tunnelling long lengths can be excavated with such immediate ground support as is judged necessary, leaving the permanent lining as a separate and largely independent operation. If the excavation and ground support can be so executed as to develop effectively the arching action of stresses within the rock mass, the effect of the lining may be merely to seal in and secure that stress pattern, while providing additional strength to absorb long term changes resulting from relaxation in the rock, and also, of course, to provide a suitable interior finish.

Tunnelling in rock by drilling and blasting is unavoidably a cyclic operation and not continuous. A careful balance in the manning and equipment for the different phases is essential for efficient progress. The elements of the cycle are:

Drill face to appropriate pattern and depth
Withdraw drilling equipment
Charge with explosives and withdraw all men
Fire charges
Ventilate to remove fumes
Bar down loose rock and make safe
Remove spoil
Install immediate support

Where the tunnel is not driven full face but with top heading and bench, there will be a more complex dual cycle.

Machine tunnelling with a full face rotary boring machine is more nearly a continuous operation, as described in Chapter 7, but fixing of support may

Fig. 6.1 – A 7 m dia. tunnel excavated by drilling and blasting in hard rock. Note that rock is self-supporting, and provision of blowing

Fig. 6.2 – Trimming gantry in hard rock tunnel (Edmund Nuttall Limited).

impose a cyclic pattern in some degree. Methods of spoil disposal are discussed in general terms in Chapter 4. A factor of particular importance in rock tunnelling is the fragmentation of the rock which has considerable bearing on the suitability of handling machinery.

6.2 STRESSES

The stress patterns described in Chapter 4 for homogeneous ground uniformly stressed are equally valid for idealised rock, but in folded and faulted strata conditions can be very different. Residual geological forces may be heavy, and discontinuities in the rock may make the stress pattern far from uniform.

In tunnelling through rock the objective is so to construct the tunnel that the stresses and deformations in the ground and in the temporary and permanent supports are at all times kept within safe and acceptable limits. As operations proceed stresses are progressively redistributed, with relaxation in some areas and intensification in others. In particular the tunnel roof near the face is usually supported by transverse arching action bearing on the sides and by longitudinal arching bearing on the strata at the face and on the completed and supported roof behind. Advance of the face withdraws the front support and lengthens the effective span of the longitudinal arch, imposing added load on the rear supports as shown diagrammatically in Figure. 6.3. In this situation timing can be of vital importance. Where the work is advancing quickly and supports are provided close to the face all may be well, but any delay may allow serious increase of load and disturbance of rock structure to initiate overstressing or collapse.

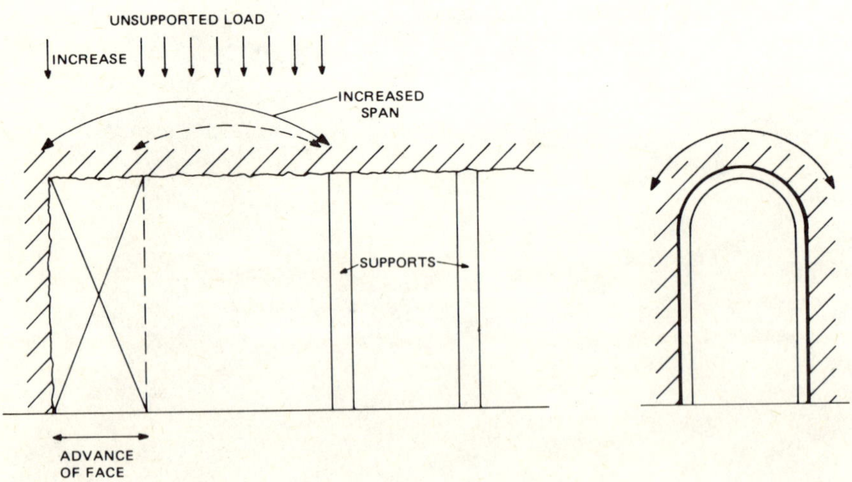

Fig. 6.3 – Increase of load as face is advanced.

The magnitude and pattern of the redistribution of stresses depend on various factors, including:

Rock type
Strength and other properties of the rock
Jointing and other discontinuities
Stresses existing in the rock
Size and shape of tunnel section
Type and strength of immediate support
Permanent lining
Construction methods and procedures

Stresses may be assessed empirically, or by mathematical and laboratory analyses, including the use of the finite element method, practicable with computers. The accuracy of results cannot be better than the accuracy with which the factors above can be ascertained, and some of those factors will vary unpredictably within the rock mass. However careful and elaborate the calculations, experienced judgement at site is absolutely essential.

In the actual construction progressive adjustment of stress occurs as some of the rock around the tunnel fails locally or moves along joint planes relieving stress concentrations. A final equilibrium is attained when peak stresses, no greater than confined rock strength, redistribute further back into the surrounding rock mass. Zones of failed rock may then remain in such areas as the springline and the crown, but needing only minor local support or containment.

6.3 ROCK CHARACTERISTICS

Rock structure is discussed in Volume 2, but the following notes outline rock characteristics in the context of drilling and blasting. Rock type is a very general term embracing a wide variety of geological factors ranging from basic formation–igneous, sedimentary, metamorphic–to specific properties such as texture, mineralogical and chemical composition, age and origin, anisotropy, degree of alteration, hardness. The term can usefully indicate the approximate grouping of such properties as compressive strength and modulus of elasticity, and can further be qualified by description as 'hard', 'medium', 'soft', used as a relative term appropriate to the rock type. Thus a 'hard' quartzite may have a compressive strength of over 210 N/sq mm, whereas a 'hard' sandstone may be about 120 N/sq mm.

Rock quality has to be assessed in the context of drilling and blasting. Very few rocks are too hard to be drilled with ordinary tungsten carbide tipped drills; some weak clay-like rocks tend to choke the drills, and also may shatter locally not pulling the round clearly. Fragmentation to a suitable size for handling may require study. Rock hardness controls the drilling rate but is not a well defined quantity and is not to be equated to compressive strength, or abrasiveness. Uniformity, texture, toughness to impact are all relevant to drill performance.

An important characteristic of excavation by drilling and blasting is the profile produced. Although the drilling pattern is chosen with this in mind, the geological features are dominant. In a strongly jointed blocky mass overbreak must be expected. Preferential failure on the weakest plane cannot be avoided.

6.4 DRILLING EQUIPMENT

Pneumatic rock drills are the most widely used type in mining, quarrying and civil engineering generally. They are operated by compressed air at pressures of about 7 bar (100 psi) and may be percussion or rotary or a combination of both actions. They range from hand-held units to heavy boom mounted 'drifters' fitted on 'Jumbos' which are mobile gantries of robust construction capable of being advanced to the face for drilling and withdrawn clear of the firing and spoil removal until the next cycle. Screw or chain feed mechanisms may be used to advance the drill in the hole. This equipment is particularly suited to large and long tunnels through reasonably uniform rock, where technique need not be varied greatly.

Hydraulic drill systems are coming into favour again in some tunnelling applications. Electric power drives an internal hydraulic motor. The advantages claimed are higher power output, lighter weight, reduced noise, reduced fogging, absence of compressed air lines. Higher production rates and lower operating costs are also claimed, but initial costs are higher and specialised maintenance is required.

For smaller tunnels the preferred tool is the lightweight drill which can be lifted, handled and operated by one man, and which is usually operated with the support of an 'air leg'. Since the invention and development of the tungsten carbide tipped bit in recent years, drilling speeds have advanced greatly.

6.4.1 Drilling Pattern

The number, arrangement and depth of holes and the charge weights and firing sequence are decided to suit the size, shape, and particular conditions of the tunnel, the available equipment and the blasting characteristics of the rock.

The basic drill pattern consists of:

1. Cut holes
2. Cut spreader holes
3. Outer rows and perimeter holes
4. Lifters or floor holes
5. Easer holes

Fig. 6.5 illustrates by way of example burn cut and angle cut patterns.

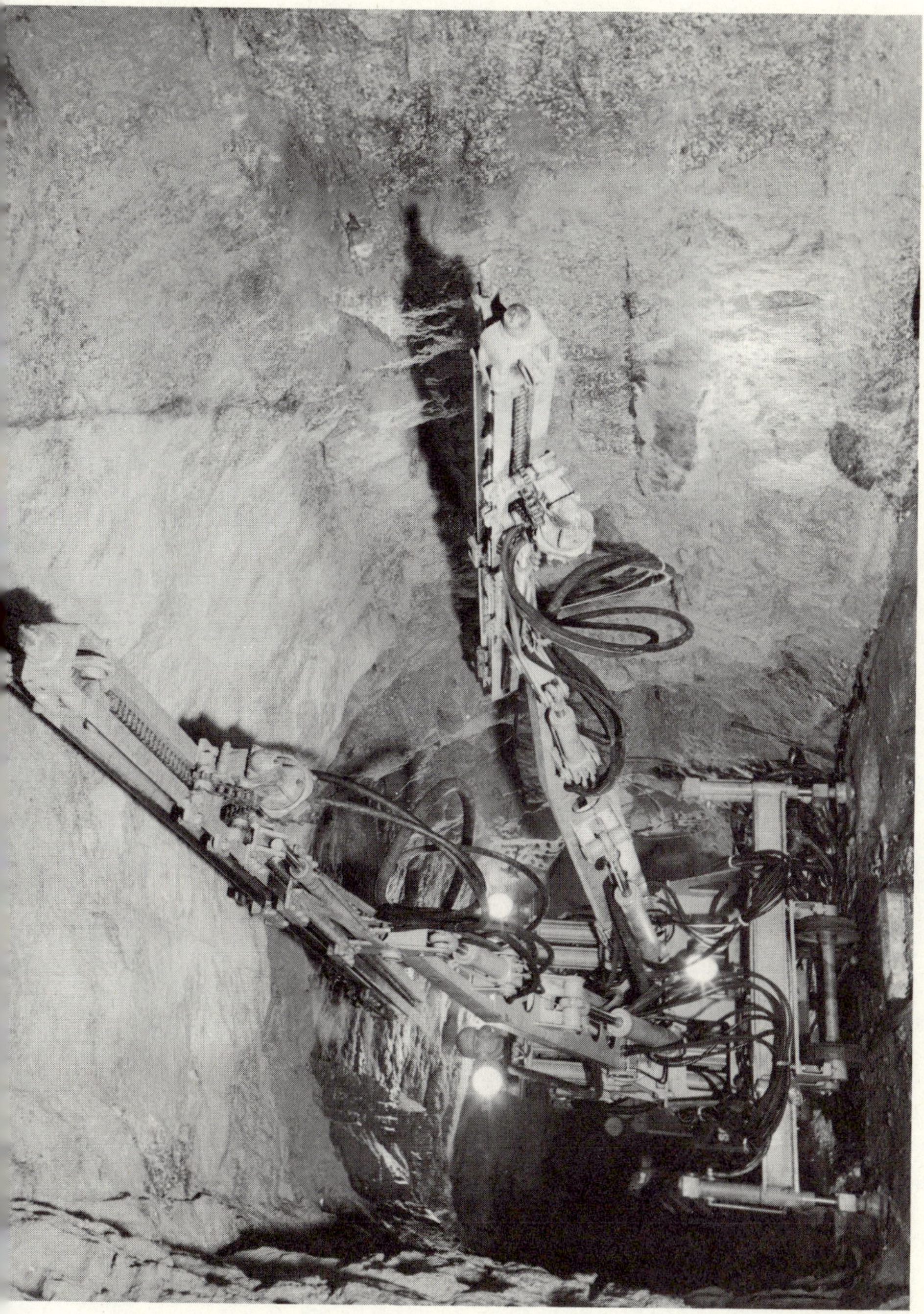

Fig. 6.4(a) – Twin boom drilling rig (Compair Ltd., formerly Holman Bros. Ltd.).

Fig. 6.4(b) – Lightweight hydraulic drill with telescopic leg (Torque Tension Ltd.)

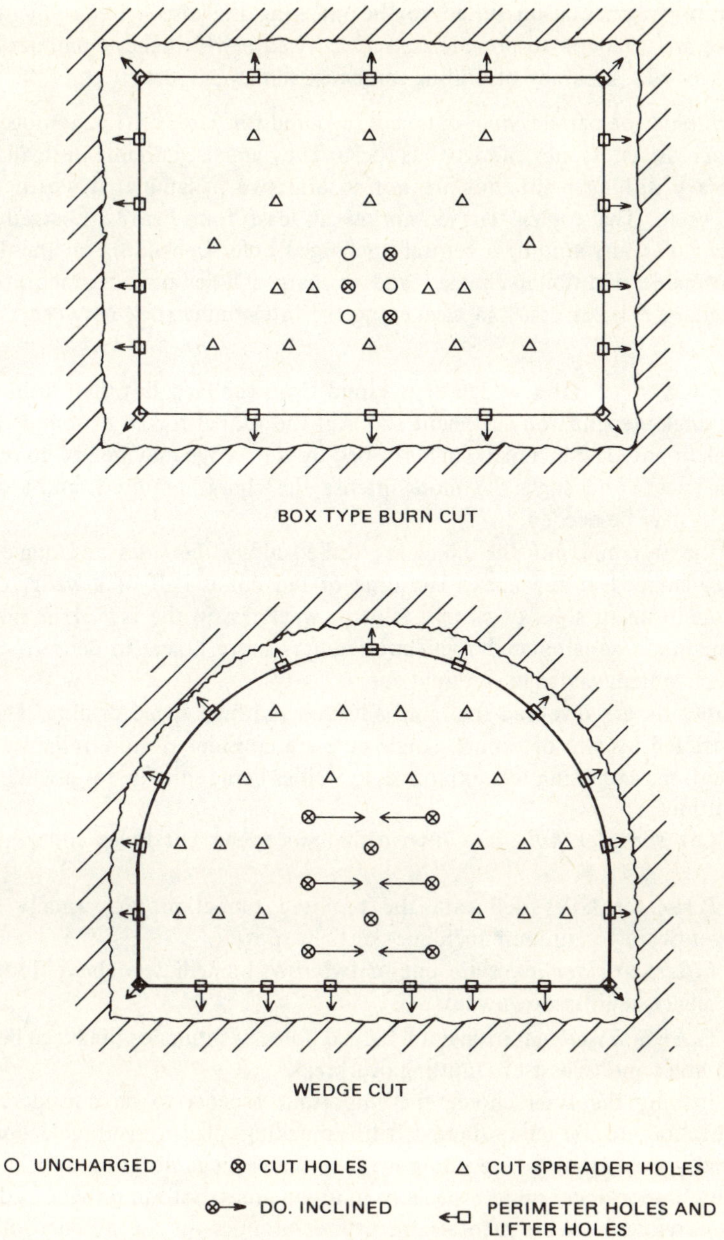

Fig. 6.5 – Drilling patterns.

1. *Cut holes*. The pattern is developed around the chosen cut with the aim of ensuring fragmentation suited to the mucking and disposal equipment. The cut holes are intended to provide first a cavity into which the remainder of the face may break. Accuracy of drilling can be extremely important.

(a) *Burn cuts* (or parallel hole cuts) are favoured for hard, brittle, homogeneous rocks such as sandstones and igneous rocks. They are also suitable in small drives using heavy drilling plant, but are not so effective in laminated weathered or faulted rock. The typical cut comprises at least four heavily charged small diameter holes surrounding a central uncharged hole. Depending on the drilling plant available additional charged and uncharged holes may be incorporated. The uncharged holes assist in weakening the intervening rock between charged holes.

(b) *Angle cuts* cut out a wedge or pyramid from the face, but their drilling at a suitable angle depends on sufficient room in the tunnel for an economic length to be taken out. The included angle formed by the wedge may be 90° to 60° but the more acute the angle the more intense the charge required, and a double wedge cut may be needed.

In the pyramid cut the holes are drilled along the sides and angles of a pyramid, charged at the apex. The drag or fan cut is half of a wedge cut, to drag from floor or sides or from a fault or weakness in the face. It is favoured in sedimentary or laminated rock but requires more holes. In dense rock it is less effective than wedge or pyramid cut.

Burn cuts are favoured for rapid advance and high speed drilling. They are not restricted by size of tunnel. Angle cuts are considered more reliable, giving less vibration, and using less explosive, as well as being suitable for a wider range of conditions.

2. *Cut spreader holes* are intermediate between cut holes and perimeter holes.

3. *Perimeter holes* delineate the required tunnel profile, usually angled slightly outward to minimise incidence of tight spots.

4. *Lifters or floor holes* are one or two rows immediately above floor level, usually angled slightly downwards.

5. *Easer holes* are additional uncharged holes to influence breakage between charged holes and to assist in limiting overbreak.

In making the final choice the important aspects to be considered are: fragmentation—to be co-ordinated with mucking plant; overbreak control—particularly important where lining involves backfilling; vibration and noise—of increasing importance, more especially in urban areas, both in respect of damage to structures and environmental amenity; economics—in the absence of other restrictions the fastest sustained advance is likely to result from the longest practicable rounds, and fewest holes per round, but the drilling and mucking cycles must be compatible within the shift time.

6.5 EXPLOSIVES

There are two main categories of authorised explosives: 'permitted explosives' approved for use in gassy conditions, more particularly where the presence of methane is possible, as in coal measures; and those for general use in non-gassy conditions. They are further classified as gelatinous, semi-gelatinous and non-gelatinous, and slurried explosives. For tunnelling, gelatinous and semi-gelatinous explosives are almost always employed, having high water resistance and high density (in the sense of concentration of explosive energy). They have to be used in long small diameter holes and must therefore be sufficiently sensitive for the detonation to travel without fail the whole length of the long narrow column. The choice of the particular grade of explosive will depend in part on the hardness of the rock and on the wetness at the face.

The handling, storage and use of explosives in a tunnel requires particular care. Every stage of the process from delivery at site to ultimate use should be most strictly regulated and should at all times be under the control of named responsible persons, who will ensure compliance with statutory requirements, safety codes, and other rules and procedures special to the project. When a tunnel is advancing steadily a very regular routine will usually be established for each shift and it becomes very important that the hand-over from one shift to the next is properly organised and that there is no laxity in observing safety rules.

6.5.1 Detonators

Electrically activated delay detonators are normally used. There is debate over the relative merits of half-second or milli-second detonators (.025 to .075 sec). The half-second delays have advantages in allowing the successive 'pulling' on the face to develop more fully, and in lessened shattering at the perimeter, but they increase the possibility of sympathetic detonation. A combination of milli-second and half-second detonators may offer advantages on large faces with complex patterns.

6.5.2 Charging procedures

Column loading, following thorough clearing out of the hole, is the feeding of the primer charge and then the other charges down the hole end to end until the required total charge is placed, leaving uncharged perhaps 300 mm at the top. In spaced loading where a maximum charge was not required it was the practice to space out the charges with short pieces of timber, but it is now better practice to employ low density semi-gelatinous charges. Stemming with clay cartridges at the top to confine the explosive charges adds to the efficiency of the blast but because of time and cost is not always practised. Gelatinous explosives are always tamped into the hole with a wooden rod. ANFO, which is a mixture of ammonium nitrate and hydrocarbon oil, is sometimes used, but not in wet areas, as are explosives of lower density. It can be charged into the hole by means of special pneumatic equipment comprising a pipe and pressure vessel, safeguarded against build-up of static electricity.

6.5.3 Firing arrangements

Standard precautions prior to charging at the face include the disconnection of all electric power from the jumbo or other plant and the removal of all power leads to a distance of 70 m or so from the face.

Firing circuits may be coupled to the mains or may be operated from magnetic-type exploders having a capacity of 100 detonators.

Series or parallel circuits are used, the latter being the more usual in complex tunnel work. The series circuit can be more simply tested; with a parallel circuit additional care with detonators is necessary. The firing line consists of twin insulated cables connecting to a firing switch up to 300 m from the face. They should be located on the opposite side of the tunnel from lighting, power and other services, and should be inspected and tested to ensure that they are undamaged and are properly insulated prior to firing. Detailed safety rules should be made and observed.

Having fired a round the first essential is to ensure that the air is cleared of poisonous nitrogen oxide fumes and that no misfires remain unexploded. Watering down of the muck pile is likely to be the first operation, followed by clearing up of the floor and scaling down of the sides and roof to remove all loose rock which might fall. Mucking plant for disposal to the conveying or transporting plant should have output compatible with the rest of the cycle of operations. It should have a good command of the face area so that it can scale the sides and face and clear the corners all without excessive manoeuvring. It may be rubber tyred, crawler mounted, or rail track mounted. Although the independent machines are more versatile they may be liable to heavy repair and maintenance costs, so that the rail mounted equipment may be the more robust and economical if chosen to suit the job.

The main types of plant are: face shovel, with built-in chain conveyor; throw over shovel loaders emptying direct into trucks or skips; side tipping loaders. The latter two have the disadvantage of requiring forward and reverse movements for each bucket load.

6.6 VENTILATION

6.6.1 Methane

The properties of methane gas have been described in Chapter 3. Its presence in a tunnel constitutes a hazard because it forms an explosive mixture with air at a dilution of about 20 times by volume. The use of 'permitted explosives' has been mentioned above, but the continuing presence of methane may impose the need for extensive precautions. Mining regulations on the subject are detailed and strict. The gas commonly originates in carbonaceous rocks, more particularly in coal seams and oil-bearing strata, but also in peat and other organic deposits. It can travel laterally for considerable distances through joints and fissures, and can be transported in solution by water. It may appear in a tunnel as a steady infiltration or suddenly when a pocket is penetrated by a drill.

Sampling and testing to detect its presence where there is reason to suspect it, and routine monitoring when identified may be vital to safety. A supply of air adequate to dilute and remove the methane is required. Concentration exceeding 1% is considered highly dangerous, and at one quarter of that level precautionary measures are likely to be essential, particularly against possible ignition by any flame or spark or hot spot. The installation of flameproof electrical equipment in such conditions may be obligatory. Obviously, even the use of 'permitted' explosives must be regulated accordingly.

6.6.2 Removal of fumes

Forced ventilation is necessary to remove fumes, heat, and dust from the working areas. There is a choice between systems feeding fresh air to the vicinity of the face, from which the contaminated air flows back along the tunnel, and systems extracting contaminated air at the face, thereby drawing fresh air in along the tunnel. Combined or alternating systems are also possible. The most suitable system depends on the project as a whole. If the vicinity of the face is the only work area it may be advantageous to make it habitable as soon as possible after blasting by direct supply of fresh air, but if follow-up work on supports and lining occupies the excavated tunnel, the blowing back after firing at regular intervals of a cloud of fumes and dust could be unacceptable, and might make an extract system preferable. Another factor is that firing might be timed at a change of shift when time for clearance of the air could be made available. The problem can be even more complex if a network of interconnecting tunnels is under construction. The whole problem of ventilation is one of great importance to be studied carefully in advance. In addition to methane if present and fume clearance, ventilation must deal with vehicle and plant exhaust fumes, drilling dust, blasting dust, vehicle and plant heat generation, geothermal heat, and body heat. Dust suppression is usually managed satisfactorily by water spraying, but there are usually regulations relating to the danger of silicosis which must be strictly observed.

6.7 DRAINAGE AND PUMPING

The volume of water entering a rock tunnel may vary from no more than that introduced for hosing down and dust suppression to the volumes encountered in subaqueous tunnels short of catastrophic inundation. In mountain tunnels springs of very hot water may be experienced, and jets of water at high pressure.

It is not proposed to generalise about these exceptional major inflows at this stage except to suggest that where the possibility exists contingency planning in advance is prudent, with particular attention to the safety of men underground. Inundation is not instantaneous and practical escape ways and methods can be devised.

Where inflows are normal the risks of washout in rock are much less than in soft ground tunnels, although soft filling and pulverised rock may be washed out of a fault zone, or a bed of weakly cemented or uncemented sandstone may be encountered, for which soft ground techniques are appropriate.

If the tunnel is driven on a rising gradient water entering can usually be drained under gravity back to and out of the portal, presenting few problems. On a falling gradient, as necessarily in a subaqueous tunnel, gravity will drain all water towards the face and collecting and pumping sumps are necessary. Fixed pumping mains will be installed to carry the water back to a main collecting sump near the portal, or at a working shaft, from which the water must be lifted to the surface and discharged.

The selection of pumps will be governed as to size and number by the expected inflow, preferably with a large contingency margin in addition to maintenance and repair requirements, to cover unexpected flows. The type will depend in part on the solids content of the water and the working head. Small readily portable pumps are required at the face and other working areas, with larger fixed installations to the rear. Abrasive and corrosive action from the solids and from saline content may impose very severe wear on pump impellers and bearings.

Waterproofing problems are discussed in Chapter 8.

6.8 IMMEDIATE SUPPORT

The general problem of stress in the rock during tunnelling has already been discussed. The need for immediate support, that is, support prior to the permanent lining, will depend on the geological detail, namely: rock type, joint pattern, dip and strike, discontinuities, ground water, rock material properties, state of weathering or alteration of rock mass. The need for support will further depend on construction factors, namely: size and shape of tunnel, direction of drive relative to dip, excavation method and procedure. Immediate ground support has two objectives: first, to give local support to rock which might fall out or be seriously displaced, and second (and perhaps most important), to guide the development of the natural arching action and to preserve it.

The degree and quality of this arch depends on the state and nature of the rock and particularly its competence. The simple picture, greatly modified by rock structure, is that an arch forms above the excavated roof leaving a roughly triangular zone which may need added support. If it is allowed to relax the weakness may spread upwards to a greater height and heavier support becomes necessary. The choice of method of arch preservation depends on the factors listed above and also on other factors such as permissible ground subsidence, permanence of opening, economics of lining.

6.8.1 Arch ribs

The traditional timbering has been largely replaced by steel supports, but does have the advantage of being more easily cut to length and shape, and perhaps of giving more warning of subsequent movement. It can also be adapted more readily to varied cross sections and special requirements. Where timber is plentiful and steel scarce and costly it may be the right and obvious choice. Timber which has to be left in, however, is to be avoided because of its eventual decay.

The most generally useful support is the arched steel rib shaped to conform to the intended cross section of the tunnel. A rolled steel channel or H-section is usual. It is designed to carry in the first place the loosened rock beneath the natural arch by means of blocks and wedges which should be spaced as closely as practicable to get even bearing and minimise bending movements in the rib. At the same time adequate footblocks or sills are provided and wedged to take the vertical loading, and strutted apart or otherwise secured against horizontal displacement.

Where there is loose incompetent rock between ribs, laggings spanning between ribs may be added, open spaced or close-boarded as circumstances dictate, and packed up to the rock as necessary. The laggings are preferably of steel for permanent work but may be of timber in temporary pilot headings. See Fig. 6.6.

The system is adaptable to a wide range of excavation procedures. A tunnel boring machine should produce a smooth circular cavity and, if ribs are employed, they will be fully circular with suitable facilities in the machine for their handling and erection. It may well be, however, that the immediate erection of a segmental precast concrete lining is to be preferred in such a case.

A circular invert is not favoured for other methods of rock excavation because it is time-consuming to trim to shape and makes vehicle access difficult. The preferred shapes are a semi-circular arch in the top and a straight or horse-shoe shaped lower half.

Such a tunnel may be excavated full face, or in two stages with a top heading and bench initially, followed by bench excavation to invert. In the latter case semi-circular ribs are fixed in the top half, supported at springing on sill beams, which in turn are underpinned by the ribs for the lower half. During the transition period relaxation of the top rib support must be prevented, possibly by use of hydraulic jacks and folding steel wedges.

Close monitoring of progressive increases in loading of ribs as the face advances may allow support requirements to be more accurately assessed. Early

Fig. 6.6 – Arch ribs and laggings in tunnel in chalk.

and efficient installation of ribs and better natural arching may suggest a wider spacing between ribs, or conversely unexpectedly heavy loading may require added strength in support. Steel ribs may ultimately be incorporated in the final permanent lining, being taken account of in the design, or in some cases may be dismantled for re-use.

6.8.2 Liner plates

These are steel plates generally flanged for bolting together, which can be used to cover and support the excavation, in combination with, or independent of, steel ribs. They are generally similar to a cast-iron lining, but normally lack the strength and corrosion resistance requisite for a permanent lining, and are ultimately superseded by a concrete lining cast *in situ*.

6.8.3 Rock bolting

This is a system for strengthening and supporting the rock surrounding the excavation by means of bolts anchored deep in the rock and stressed to compress it from the face. The term is also used to cover untensioned dowels grouted in as reinforcement, further described below.

The primary function of rock bolting is to maintain the integrity of the stressed rock so that it acts more effectively as an arch or beam spanning across the excavation, or so that it secures any loose rock or thin stratum at the surface of the cavity by anchoring it at depth. The relative importance of these aspects depends on the particular circumstances in the tunnel.

The object of rock bolting in the case of circular or arched openings is to create a zone of radial compression in the rock and to reinforce the zone of highly stressed rock round the cavity. In the roof of a rectangular heading the bolts should bind together the rock layers to form a fixed beam stronger than the sum of the individual layers, and to suspend weaker strata from a stronger zone above.

For these purposes the bolts should be installed as soon as possible to secure the ground before bed separation or joint movement takes place. Secure anchorage of the bolts must be effected.

The life of the bolt system must be adequate for its function. If it is merely to give temporary support until a permanent lining is installed, or until a pilot heading is enlarged, corrosion is unlikely to be important, but loss of tension due to slip at the anchorage or yield at the face must be minimised. Where the bolts have a permanent or a long-term function protection against corrosion becomes important. Corrosion resistant alloys may be used and protection by grouting can be specified. Resin grouts with the dual functions of anchoring the bolts and protecting against corrosion are used.

Anchors are of three main types: the expansion shell, the slot and wedge, and the chemical grout socket. The expansion shell consists of a wedge-shaped plug, screwing on to the end of the bolt and thereby pulled out against outer wedges, or leaves, which are forced outwards against the sides of the hole. Slot-and-wedge bolts are anchored by driving the slotted end of the bolt against a wedge at the end of the hole, thereby expanding the end of the bolt against the sides of the hole. The hole in this case requires to be drilled to depth precisely. The rock must be sufficiently strong to stand the local stresses imposed by the driving and wedging actions. Chemical anchorage is by filling of the space between bolt and hole with polyester resin grout, either locally at the anchorage or throughout its length. Resin and catalyst accelerator are prepared in cartridge form and after insertion in the hole are mixed by insertion and rotation of the bolt. Devices to retain the grout in an upward inclined hole are necessary until the grout sets and develops its strength. Fig. 6.7(a), (b), (c).

Point anchor bolts may be converted to column grouted bolts to improve anchorage in soft ground and to provide protection against corrosion. Cement grout is usual but resin grouts may be preferred. An injection pipe through the washer is used and an air release pipe from the back of the cavity, which must be sealed off at the rock face. Resin grout cartridge systems are also available.

Untensioned column grouted dowels placed in a regular pattern in a rock mass may provide suitable reinforcement to maintain stability or absorb peak stresses. They may be deformed steel bars encased in cement or resin grout or timber dowels in resin grout. The latter are useful where the rock is to be mechanically excavated later. They were used in the form of hollow bamboo rods in conjunction with advance grouting from the Mersey Kingsway tunnel pilot, before its enlargement by machine.

Bolts are usually of medium tensile steel bar up to about 30 mm diameter with one end threaded for nut and washer and the other adapted to the chosen type of anchor. The washers are flat plates to bear against the rock surface, drilled for grouting if required. Washers profiled to deform at the specified tension can be obtained. A swivel hemispherical washer to assist in getting uniform bearing can be provided. Fig. 6.7(d).

Accuracy in the diameter of the hole is important, wide enough to give clear passage to the bolt assembly, but tight enough to fit the anchor without excessive clearance. Precise depth is important for wedge anchors and for resin capsules.

Tensioning by torque wrench is usual, but direct hydraulic tensioners are better. Rock bolting machines have been developed for use in mines.

The bolting system has to be designed to suit the tunnel, the rock and the method of tunnelling. Bolts should normally cut across the rock stratification and always should be anchored in competent rock, but variation of length so that all anchors are not in one horizon is advantageous.

Tunnel dimensions and drilling equipment may impose limitations on length, but there is usually little to be gained beyond a depth of twice the radius of the tunnel. Another guide for length is that in jointed rock it should be not less than three times the width of joint blocks, to ensure anchorage in layers at least two

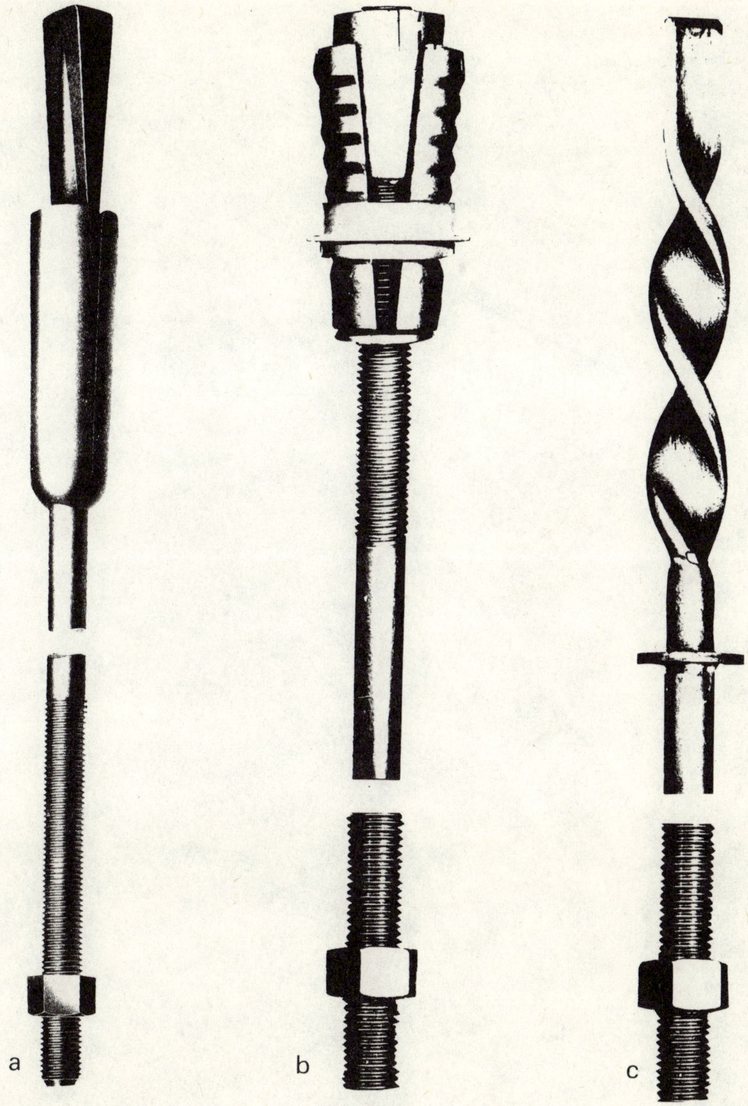

Fig. 6.7 – Rock bolts: (a) Slot and wedge; (b) Expansion shell; (c) Chemical grout (b and c: Torque Tension Ltd.).

Fig. 6.7(d) – Examples of various types of rock bolts and washer plates (Torque Tension Ltd.)

blocks behind the surface. Spacing should be regular and preferably about half the bolt length or less. Practice in Australia based on work by the Snowy Mountain Authority aims in arched openings at creating a compression of 70 kN/sq m (10 psi) in the zone of uniform compression.

In large cavern excavations long rope anchors may be used.

6.8.4 Sprayed concrete

Sprayed concrete or Shotcrete, can be used to assist in providing effective immediate support. It is more fully discussed in Chapter 8, as contributing to the permanent lining. If used near the face in conjunction with drill and blast excavation high early strength is essential to resist blast damage and allow build-up of the desired thickness. Accelerating agents are commonly added, aiming at an initial set after about 30 s and a final set after about 120 s. Mesh reinforcement may be fixed close to the rock in advance. Steel fibre reinforcement improves flexural strength.

6.8.5 Segmental linings

In an accurately bored circular tunnel precast concrete segmental linings (described in Chapter 8) can be used in conjunction with grouting to provide immediate support behind the cutter head and shield. Expanded linings are not likely to be used for any but soft ground because of the danger of severe point loadings from high points and trapped fragments.

6.9 GROUTING

In rock tunnelling the function of grouting in the context of excavation and immediate support is primarily in sealing fissures in the rock and displacing water from sheared zones and generally improving the competence of the rock mass. Backfill grouting between an immediate support system and the rock can be utilised to increase contact area and preload the rock. Its subsequent use in conjunction with segmental linings has been mentioned. It also may be applied in sealing off leaks at a later stage.

Rock tunnels sometimes require extensive grouting operations to seal faults and fissures which would otherwise allow great quantities of water into the tunnel. There are obvious advantages in discovering the requirement for grouting by drilling ahead of the tunnel face and monitoring the flow of water encountered during drilling. The grouting operation can then be effected before the advancing tunnel encounters the fault and the flow of water.

BIBLIOGRAPHY

General books, see bibliography to Chapter 4; particular projects, see general bibliography in Volume 2.

Rankin, W. W. and Haslam, R., 'Modern blasting practice in tunnelling operations', *Civ. Engng. Publ. Wks. Rev.*, 1957, **52** (Feb., Mar. and Apr.).

Fish, B. G. and Westwater, R., 'Blasting practice in shaft sinking and tunnelling', in *Symposium on shaft sinking and tunnelling,* London, 1959.

Lang, T. A., 'Theory and practice of rock bolting', *Trans. Soc. Min. Engrs. AIME,* 1961, **220**, 333.

Deere, D. U., 'Technical description of rock cores for engineering purposes', *Rock. Mech. Engng. Geol.,* 1963, **1**, 18.

Fogden, C. A. and Garrod, A. D., 'Hazards from blasting fumes in normal rock tunnelling with particular reference to nitrogen dioxide', *Proc. Instn. Civ. Engrs.,* 1965, **31** (July). *See also* discussion, 1966, **34** (Aug.).

McGregor, K., *The drilling of rock*, C. R. Books, 1967.

Proctor, R. V. and White, T. L., *Rock tunneling with steel supports* (including K. Terzaghi, *Rock defects and loads on tunnel support, an introduction to tunnel geology*), Commercial Shearing & Stamping Co., Ohio, 2nd edn., 1968

Stagg, K. G. and Zienkiewicz, O. C., *Rock mechanics in engineering practice*, J. Wiley, 1968.

Deere, D. U. and others, *Design of tunnel liners and support systems*, U.S. Department of Transportation (NTIS, PB 183799), 1969.

I.C.I. Blasting Practice, Nobel's Explosives Co. Ltd., Scotland, 4th edn., 1972.

Barton, N. and others, 'Engineering classification of rock masses for the design of tunnel support, *Rock Mech.,* 1974, **6**, 189.

Jacobs Associates, *Ground support prediction model (R.S.R. concept),* U.S. Bureau of Mines (NTIS, AD 773018), 1974.

Bieniawski, Z. T., 'Rock mass classifications in rock engineering', in *Proc. Symposium on exploration for rock engineering*, Johannesburg, 1976.

Langefors, U. and Kihlstrom, B., *The modern technique of rock blasting*, J. Wiley, 3rd edn. 1978.

BS 5607, Code of Practice for the safe use of explosives in the construction industry, British Standards Institution, 1978.

Schach, R. and others, *Rock bolting, a practical handbook*, Pergamon, 1979.

Hoek, E. and Brown, E. T., *Underground excavations in rock,* Institution of Mining and Metallurgy, London, 1980.

7

Tunnelling Machines for Soft Ground and Rock

Twenty five years ago a separate chapter would not have been necessary to cover the use of excavating machinery in tunnelling work. In recent years, however, progress has been so rapid that a survey of current plant is likely to be out of date before it is published. Better then to emphasise the important principles rather than catalogue the current solutions. It is convenient to deal separately with soft ground and rock machines, but they have much in common.

7.1 HISTORICAL DEVELOPMENT

7.1.1 Soft Ground

The circular Greathead shield of 1869, in which excavation was manual, forms the basis for all subsequent tunnelling machines for soft ground. They are in essence shields with mechanised excavation devices built in, or excavators adapted for use within the protection of a shield.

The tube railway developments in London clay disclosed conditions which encouraged the use of mechanical excavation in the 1890's. On the Central London Railway two mechanical methods of excavating the clay were tried in conjunction with the Greathead shield in 4 m diameter tunnels.

First was the Thompson ladder excavator which was a track mounted pick and bucket ladder, as its name implies, based on the same principles as the ladder dredger. This machine proved indifferent; apart from its predictable tendency to leave its rails, it could not excavate the whole face, and the mounting of the tools on an endless chain was vulnerable. The second method, the 'Price digger' of 1897, introduced the full face rotating cutting head, with picks (including gauge cutters) mounted on the head. The first models were electrically driven through an axle which proved to be a weakness, and on the subsequent Hampstead and Piccadilly tubes the cutting head, still electrically powered, was driven through gears and a pinion and large diameter internally toothed ring well forward in the shield, greatly reducing the bearing and torsional problems encountered in the first layout.

Fig. 7.1 – Thomson's Digger, January 11th 1898.

Under favourable conditions the 'Price diggers' progressed much faster than the Greathead shields, especially when efficient belt conveyors were introduced, but they were initially difficult to steer. With care and experience in driving quality of work improved, but the Greathead shield was considered more satisfactory in respect of accurate alignment and freedom from subsidence of the lining.

Fig. 7.2 – Markham Digger Shield as developed from Price Rotary Excavator in 1936. Average speeds of over 50 m per week were sustained.

In 1903 Price digger shields were driving up to 80 rings (41 m) of 4 m diameter cast-iron segments per week with a record of 56 m.

Between the wars 'digger shields' (or soft ground tunnel borers as they were termed in the U.S.A.) were developed gradually, always working on the rotating cutting head principle and always within a Greathead shield with the exception of an occasional machine for hard dry self-supporting clay.

7.1.2 Rock

In rock, drill and blast excavation was commonplace in mines in the 18th century, but a major advance took place with invention and development of pneumatic and steam powered percussion rock drills in the first half of the 19th century, and the invention and application of dynamite in the 1860's (used in the St. Gotthard tunnel 1870 onwards).

The development of drill and blast methods has been continuous since then, and accounts for the vast preponderance of rock mining and rock tunnelling excavation to this day.

The removal of the debris after blasting, whether by rail, truck or belt conveyor, has also developed over the years, with considerable ingenuity in design to enable powerful plant to operate in confined spaces.

The drill and blast methods and the associated spoil removal are covered elsewhere in this book and more comprehensively by mining literature and are not considered further in this chapter.

The one early and splendid example of the rock tunnel borer is Colonel Beaumont's machine built by John Fowler & Co. in 1881, used 1881–2 on a pilot bore for the Channel tunnel at Folkestone working in chalk, and in 1883–4 to drive pilot and ventilation headings (some of them still visible) through Bunter sandstone for the Mersey railway tunnel.

Whilst coal-cutting machinery, and other mining plant not discussed here, were occasionally adapted for use in tunnels, no remarkable developments in tunnel boring machines took place until the 1950's. By that time the advances in metallurgy, hydraulics, the technology of lubrication, bearings and other branches of mechanical engineering were sufficient to allow machine excavation to be considered seriously, and the rising cost of labour gave an extra incentive to development.

Until the 1950's, no tunnelling machine had been produced capable of excavating hard rock. Soon thereafter the race was on to produce tunnel boring machines (often referred to as 'TBMs') which would cope with an ever wider range of rocks.

7.2 SOFT GROUND TUNNELLING MACHINE TYPES

7.2.1 Full face machines

A rotating cutting head is mounted at the front of a Greathead shield. It can be carried either on a central shaft or on peripheral roller bearings. Occasionally it is driven by powering the central shaft, but more usually by an annular gear; some of the configurations have been developed at least partly to avoid the patents of others. Radial hydraulic motors have become a clear favourite for powering the head, and a number of compact motors (depending on the diameter of the tunnel) are mounted around the head. Typical speeds of rotation are in the range 2–6 revs. per minute.

The cutting head has usually four to six spokes, radiating from the centre of the head at its front, and on it is mounted an array of cutting tools, normally picks, which scrape material from the working face as the head rotates. Behind or between the spokes are buckets or scoops which carry the spoil upwards so that gravity deposits it on a conveyor for removal to the rear of the machine.

The spokes are normally disposed so that the centre of the cut face is slightly in advance of the periphery. Thus the cut face forms a shallow cone with its apex at the centre. This gives some directional stability as the tunnelling machine is jacked forward, and also allows the spoil conveyor to be well forward.

The cutting tools at the centre naturally have very limited travel and an interesting series of designs has resulted. Some have favoured an embryo auger at this point, known as a 'stinger', which must give some additional marginal stability although it is vulnerable to damage when directional changes are made,

Fig. 7.3 – 'Open Wheel' Machine. Typical specification for 3 m dia. machine with reversible cutting action: Power 130 kW; Wheel torque 15500 kg m; Thrust 1200 T; Weight of shield and wheel 25 T (Zokor).

either to negotiate intentional curves in the tunnel or more particularly when sharp corrections have to be made to keep the tunnel on the desired line and level.

The diameter of the excavation is determined by the outside, or 'gauge', cutters. In very soft ground it has occasionally been the practice for the cut diameter to be slightly less than the diameter of the cutting edge of the Greathead shield, leaving a thin slice to be trimmed off by the cutting edge. Normally, however, the cut diameter is marginally greater than the shield diameter to reduce the friction on the skin of the shield during driving. The gauge cutters, of course, travel further than any of the other tools and are subject to greater wear. The wear on the cutters and the consequent reduction of the cut diameter may cause increasing difficulty in jacking the tunnel machine forward.

The direction of rotation of the cutting head is generally reversible, so that any tendency for the shield itself to contra-rotate can be corrected. Sometimes retractable fins are fitted which, when extended out into the ground, act to prevent or correct rotation of the shield. Fins leave a disturbed path in the ground which is often undesirable.

7.2.1.1 Oscillating cutting head

This tendency of the shield to rotate may first have led to the idea of having an oscillating as opposed to a rotating cutting head. Several soft ground machines have been used in which the cutting head oscillates through 45° or more, driven by hydraulic jacks, so that each spoke of the cutting head with its tools scans the face like a windscreen wiper.

The drive jack mechanism for an oscillating head is mechanically simpler and therefore should be cheaper and more reliable than a number of hydraulic motors. In practice, the motors, which have a multitude of other applications in mining and industry, are competitively priced, very reliable, and easily replaced as a standard unit.

One disadvantage of an oscillating head is that each 'spoke' has to carry a full array of cutting tools and if any fail the gap is more serious than on a rotating head. Nevertheless, oscillating heads have given very satisfactory results.

7.2.1.2 Variations to suit ground conditions

The basic machines, just described, operate within a Greathead shield having a circular cutting edge, a steel tail (within which segmental linings are erected) and hydraulic jacks for pushing the machine forward off the completed tunnel linings. Such machines can operate in stiff clay, cemented sands, or dry silts with adequate 'stand-up' ability.

If ground conditions are sufficiently favourable, it is possible to dispense with the tail, and to build flexible jointed segmental linings as described in Chapter 8.

If, on the other hand, the soil has a tendency to move in at the cutting face—either a very soft clay squeezing in or a less cohesive soil running in—then with

Fig. 7.4 – Kinnear Moddie Drum Digger as used on Victoria Line. 4 m dia.

such a machine the spaces between the spokes can be plated over at the face, leaving slots (possibly of adjustable size) to reduce the ingress of soil. It is hoped thus to ensure that just the right amount of spoil is removed per unit of tunnel advance so that subsidence above the tunnel is negligible, but, in practice, that can only be achieved in homogeneous ground, and the method is unsatisfactory when, for instance, most of the face is in stiff clay, but includes a stratum of dry sand or soft mud which flows freely through the slots however narrow they may be. The thrust applied in supporting the face adds to the frictional load on the cutter motors.

Various machines have been designed to allow the cutting head to move forward independent of the shield itself. This is advocated partly to allow excavation to proceed constantly, even while a ring of lining is being erected, and partly to improve the steering characteristics of the machine. If the cutting head is adjustable relative to the axis of the shield, then it can excavate ahead of the shield in the most desirable direction irrespective of the actual attitude of the shield, and when the shield is jacked forward it will tend to move into the excavated hole rather than drive perhaps further off line or level. This arrangement has not been widely favoured perhaps because it makes for more mechanical complications in the already congested shield, and if the ground is good enough to allow such unsupported excavation in front of the shield, then steering and other problems tend not to be important. On the other hand, the ability to retract a stalled cutting head a few centimetres has obvious advantages.

Another variation is the articulated machine. The front shield unit, with cutting head, is as previously described but in place of the integral tail is another short length of shield comprising a thrust ring, which is separate from the shield proper and which can be expanded to grip the surrounding ground. The machine moves in pulses by successively pushing off the expanded thrust ring whilst excavating, then relaxing the grip of the thrust ring and pulling it forward, to be re-expanded against the ground. This modification is believed to have originated in the United States where it is common practice, even in soft ground, to use steel ribs, or rings, with laggings for immediate ground support. Such lining cannot normally accept the thrust of the shield rams, so the thrust ring or some other device is required. The articulated machine is able to negotiate tighter curves than a rigid shield, which can be an important advantage for sewer and water tunnels.

7.2.1.3 Subsidiary equipment

Full face machines, whether in soft ground or rock, require considerable service facilities behind the shield. These are either towed by the shield on sledges running on the invert of the completed tunnel, or rail mounted. Typically, a soft ground machine would require power cables and transformers, arrangements for unloading, storing and erection of tunnel segments, a control console, electro-hydraulic power packs for shield rams and drive motors, grouting equipment, and an integral spoil removal system to transport spoil far enough to the rear for loading either into large shuttle cars or trains of skips or on to a main conveyor.

7.2.1.4 Economic utilisation

The full face tunnelling machines just described entail a major investment. They are unlikely to be efficient in ground conditions other than those for which they are designed, and can be a major source of delay and expense if they encounter unfavourable conditions.

The initial outlay includes not only the machine and subsidiary plant but also all the provisions and facilities at site, the layout of shield chambers, access tunnels, railways, shafts, winding and surface spoil disposal arrangements, all geared to deal with the special characteristics and high output of the machine. Such a machine, organised for a major drive, will have a long train of support plant behind it and cannot achieve its full potential until sufficient length of tunnel has been driven and lined to allow space and time for all the following plant to be installed and working.

For a full face tunnelling machine to justify itself economically there must be time and capital available for obtaining the machine and all its subsidiary requirements and setting it to work. It is unlikely to be advantageous unless the drive exceeds 1 km and is in reasonably uniform conditions.

7.2.2 Independent excavators

In contrast, simple Greathead shields can be launched from shield chambers close to the working shaft or even from the shaft bottom itself and can soon be working at full output. This advantage and others have led to the development of a family of excavating machines some of which can be brought in to work in conjunction with a basic Greathead shield and others of which are actually mounted on a modified shield structure, although normally removable. These machines increase the progress and reduce the labour of excavation without involving the delay and expense of setting up for and using a full face tunnelling machine.

7.2.2.1 Boom cutters

The boom cutter is one such development. The basic mechanism was originally developed as a tractor mounted road header machine for mining work. The boom of such a machine can be mounted within, or upon, the shield structure and operated so that a rotating cutting head on the front of the boom scans the working face. Spoil falls and is carried back by conveyors for disposal.

The boom may be telescopic, and incorporates the motor and drive shaft. The cutting head, which rotates about the boom axis, may be conical or flat faced, but carries cutting tools or picks to excavate the ground.

Other configurations of cutting head are available which may most easily be described as operating on the lawn-mower principle.

The type of cutting tool or pick mounted on the cutting head can be selected to suit the type of ground to be encountered, but large cutting heads have been found to be less susceptible to clogging in softer ground.

This type of excavator allows easy access to the face if support of the face or ground treatment is found necessary.

Another development for self-supporting ground, but giving at any time the option of reverting to hand excavation, is the claw-armed digger shield. Semi-circular cutter arms are mounted on pivots close to either side of the vertical (or the horizontal) axis at the front of a Greathead shield. When retracted they lie close to the cutting edges of the cylinder and from this position can be thrust forward and round so that they slice out a hemispherical cavity ahead, whose shape contributes to the stability of the excavation. The cutter profile can be designed to cut out other shapes of tunnel cross section.

7.2.2.2 Bucket excavators

In the same way that the boom cutters were introduced into the traditional Greathead shield, so were bucket excavators. Initially they consisted of a telescopic arm with fixed bucket which cut down the tunnel face and raked spoil back to a conveyor. Later developments utilise a jointed arm and variable angle bucket based on the tractor mounted back-actor principle. This method of excavation is often proposed for a mixed or broken face of gravel, clay or soft rock as the bucket can bring considerable force to bear to break out hard material, but full consideration must be given to methods for support of the face.

7.3 THE BENTONITE SHIELD

The difficulties of driving tunnels through cohesionless ground such as sands and gravels, especially below the water table, have been rehearsed elsewhere in this book (Chapters 4 and 5) together with the special measures such as compressed air, chemical treatment or freezing. These difficulties have had the dual effect of making such tunnels expensive and also of discouraging the driving of tunnels in such strata in the first place.

Since 1964 the authors have been concerned with the invention and development of methods of tunnelling which to some extent overcome the difficulties. The initial concept was very similar to an orthodox full-face TBM but with a pressure bulkhead so that the cutting head operated in a chamber filled with a thixotropic bentonite slurry kept at a pressure sufficient to support the face and keep back ground water.

The slurry from the chamber, carrying excavated spoil, is circulated through a control gate and pipe line to a separation plant whence the spoil is carried away to disposal. The slurry is reconditioned and returned to the chamber.

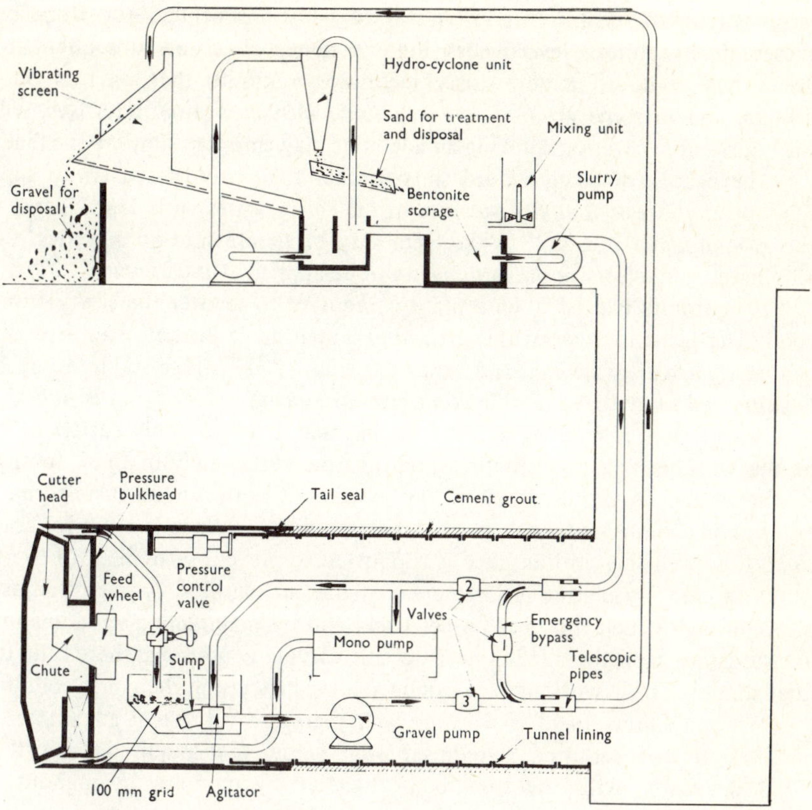

Fig. 7.5 – The Bentonite Shield – diagram of slurry circuit.

The first tunnel built by this method was successfully accomplished in 1971–3 on the proposed line of a new London tube railway near New Cross in south-east London.

7.3.1 Variations

Meanwhile in continental Europe and more especially in Japan, where a huge programme of public works was in progress in the 1970's, the bentonite shield or slurry shield method was developed on a large scale and in various forms. In 1979 it was reported that out of over one hundred tunnel machines at work in Japan, no fewer than thirty-five used the slurry shield principle.

The original machine in London used bentonite slurry in the belief (subsequently confirmed) that it would act as it does in 'slurry trench' or 'diaphragm wall' methods of construction. One important effect was that the settlement of overlying ground and buildings was greatly reduced to acceptable levels. The use of a potent thixotropic slurry is still to be recommended for work below vulnerable structures. Nevertheless, the experience gained on subsequent tunnel drives has produced a new slurry technology akin to that used in oil well drilling, and in some grounds the soil itself, with or without additives, will, in the course of excavation, provide an adequate clay slurry to support the face.

The beauty of using a fluid slurry rather than compressed air to support the working face is that the pressure in the slurry varies with depth just as does the ground water pressure, so the instability of air pressure on a vertical face is eliminated and excessive pressure need not be applied at the crown.

If bentonite is used, then a plant is required to recover the slurry from the spoil. The plant is sited on the surface except in large diameter tunnels. Stones can be removed by sieving, and other particles by centrifuge and hydrocyclone. Settling tanks have been used in some circumstances.

A variation, also using a full face machine, is to add only sufficient water at the working face to turn the spoil into the consistency of thick toothpaste so that it can be extruded behind the machine. Clearly such a system can only be used on cohesive soils such as clay or clayey silt, and there is cause for concern as to how well the working face is supported to prevent settlement of overlying buildings, etc. By plating in large areas of the cutting head or having adjustable slots through which the spoil comes back, and by maintaining but adjusting the forward thrust on the cutting head to suit varying conditions, there is no doubt that support of the working face can be achieved except in the worst conditions.

When a slurry shield is working well there are very great advantages. The number of men required is reduced even below that required for orthodox TBM operation, with the spoil being handled by pumping throughout. Conditions in the tunnel are greatly improved by the reduction in materials handling, and the absence of noise, dirt and dust. But things can go wrong, and if so, the presence of the full face machine and the pressure bulkhead complicate corrective measures. It is difficult to get at the working face (even to replace cutting tools). It is often necessary or prudent to make provision for pressurising the tunnel with compressed air when required to carry out various operations, such as repairing the seals around the tail of the machine. These seals gave much trouble in the early tunnel drives, and without them ground water, grout and slurry can escape into the tunnel.

Bentonite shield tunnelling is still a specialised and expensive business but the system can now offer substantial savings of time and money compared with earlier expedients.

7.4 OPERATION OF MACHINES

7.4.1 Hydraulic system

Soft ground machines, like Greathead shields, are normally advanced by thrusting off the tunnel lining built immediately behind the machine. The pistons of an array of hydraulic rams around the perimeter of the machine extend backwards, transmitting the force to the lining through shoes designed to spread the load. Small machines operate on as few as six rams whereas the largest machines (over 10 m diameter) may have as many as forty. The maximum force per ram varies between 10 T and 160 T. The power is transmitted through an hydraulic system incorporating safety devices, pressure limiting switches and control valves from one electrically driven pump which can develop pressures of up to 35 or 40 N/sq mm but generally kept around the 15 N/sq mm level.

Industrial power packs of electric motors driving hydraulic pumps and hydraulic fluid reservoirs are installed to operate each major portion of the hydraulic system, the cutting head, the shove rams, and the erector mechanism for the tunnel lining, although in the smaller diameter tunnels a combined power pack system is usual. Power generation will obviously vary to suit the type of machine used, but typical of the 4 m diameter digger machines used on construction of the Victoria Line under London through clay was a 200 kW McAlpine machine, although power requirement can be as high as 450 kW.

7.4.2 Steering

Steering of the shield is effected by selective use of the thrust rams. Position and attitude of the shield must be ascertained relative to the specified alignment and to the actual position of the last completed rings. As discussed in Chapter 9, the required centre line may be defined by plumb lines and levelling rods, or a laser beam projected on to a target on the shield. Calculation of the appropriate corrective measures must not be delayed; a computer may be helpful, but experience of the responses of the shield is essential.

Modification of the attitude of the rings, whether for a specified curve or correction of errors, may be by packings in the circumferential joints or by introduction of special segments forming a tapered ring.

When a soft ground machine is excavating the rate of advance of the cutting head may be limited by the resistance at which it stalls or by clogging of the spoil handling system whether by excessive volume or stickiness or spillage at transfer points. Too slow a rate can waste time and power and on occasions can result in large coagulated masses of small debris, or on other occasions may exacerbate a dust problem.

Comparison of the steering characteristics of many shields and tunnelling machines has shown that steerability is dependent on the ratio of diameter : length of shield. If this exceeds 1.4 alignment control is usually satisfactory; below 1.1 it is almost always difficult.

7.4.3 Support at the face

The subject is discussed in Chapter 4 but certain aspects are particularly relevant to machine excavation. The normal action of taking a cut from the face precludes support for that area at that moment, whether the cut is by hand or machine, except so far as fluid pressure contributes to support as in the Bentonite Shield. Stand-up time is therefore very critical. Hand excavation allows support to be maintained over most of the face while a small part only is advanced, but with mechanical excavation on a larger scale the problem is rather different. The choice of machine obviously should take into account the expected behaviour of the ground, assuming that normally it will stand up throughout the cycle of operations. Abnormal, even if not unexpected, conditions may demand face support. The immediate problem is then that of access to secure the face, possibly followed by hand excavation over the critical length. Access may be very difficult with some types of full face rotary cutting heads, but those plated between spokes can give partial support even when cutting, and even if brought to a standstill may be manageable.

With boom cutters and other independent excavators the conditions more nearly resemble hand excavation on an enlarged scale, and the machine can usually be drawn back, or moved to one side to allow timbering or other free support. With such machines special supporting devices can be incorporated in

Fig. 7.6 – Bucket Excavator. The bucket is capable of 360° rotation and considerable break out force can be applied anywhere in the tunnel face. Note the 'poling plates' around the top of the shield which can be jacked forward independently, and the jack operated breasting plates which can be hinged back, or which can help to support the upper half of the tunnel face. This 9 m dia. machine was built for the Tehran Metro (Zokor).

the shield, such as the hydraulic face jacks of hand shields, or plates hinging out from the cutting edge as were used in the 6 m shields in the Washington, D.C., and Baltimore metro construction. Hydraulically operated platforms, or 'tables', at different levels in the face can provide a series of shelves limiting inflow of dry sandy material to its angle of repose. In the Hamburg metro such a system was adopted for the top half of a mixed face, the lower half being in free standing clay.

In another support system used in Frankfurt and Essen metros, the need for direct face support was minimised by using a 'blade shield', in which the periphery of the cutting edge is built as an array of blades which can be advanced individually, much as in 'piling' ahead of a face.

These devices, of increasing elaboration, are seen by some as a substitute for the traditional skills (now regrettably rare) of the soft ground tunnel miner. They are indeed an adaptation of his techniques, but there is less flexibility in their use and it is more difficult for those operating them to form a judgement of the response of the ground.

7.4.4 Special protection of plant

Mechanical and electrical plant working in a tunnel under construction is usually subjected to very severe conditions.

Water leaking or flowing into the excavation is usually unavoidable, but it may carry in solution corrosive salts. Fine dust particles of highly abrasive character may also arise from the breaking up of the rock or from deposits of sand and may be carried in the water, or as airborne dust, settling on damp surfaces and forming a grinding paste which can destroy bearings.

The need for careful waterproofing of all electrical equipment and connections is obvious, but additional screening should be provided wherever practicable.

Mechanical bearings, especially those near the working face, need very special protection; complex seals may be used, supplemented by automatic continuous grease injection. The surfaces on which the seals bear also need protection against wear and corrosion, as do moving surfaces in the hydraulic system, such as thrust rams and control valves where heavy chrome plating or use of special alloys may be thought necessary.

7.4.5 Grouting

The structural function of grouting is discussed in Chapter 4 and elsewhere, but its possible leakage from the back of the lining into the tail of the shield or outside the tailskin is particularly a hazard to proper progress of a machine. The tail may become 'grouted in' if it stands too long or grout may build up thickly on the outside of the tail and hamper movement by displacing excessive ground.

In the Bentonite Shield the tail seal is particularly important to avoid contamination of the grout by leaking bentonite slurry or of the bentonite slurry in the face by cement.

7.5 ROCK TUNNEL MACHINES

The development of modern rock tunnelling machines has been a feature of the last three decades, and followed swiftly in the wake of the production of reliable machinery for use in excavation of soft rock in the mining industry. By 1956 tunnelling machines were being used in medium rock conditions such as sandstone and limestone. Although they were not then considered to be economically successful, they provided the basis for future development.

Many of the problems encountered with the mechanics of these machines were similar to those met with on the soft ground machine development: but the main area of concern was with the cutting tools, the constant renewal of which, almost on a daily basis, was a considerable item of cost and prevented worthwhile rates of progress being achieved. The first successful hard rock borer was produced in 1957 by the Robbins Co. in America. The machine, about 3 m diameter, was more robust and powerful than previous models and, after various improvements and adaptations, was able to bore for days before disc-cutter replacement was necessary, the rock encountered (in a Toronto sewer tunnel) being inter-bedded siliceous limestone and shale with compressive strengths ranging up to 140 MN/sq m.

Whilst the success of this first machine led to some others being over ambitious and committing TBMs to drives beyond their powers, improvements and developments have been continuous in recent years.

The designers of TBMs now understand much of the nature and magnitude of the forces which the frameworks of their machines must endure; the mechanism for cutting rocks of an ever-widening range of types is understood reasonably well, and a range of cutting tools, especially disc cutters, of lengthening working life, is available. A major problem remains in the intrinsic variability of rock in texture, hardness, jointing, and fissuring, angle of dip, water content; a machine must be adaptable.

7.5.1 Full face cutters

The use of a full face TBM assumes the ability to cut in rock a cylindrical hole, self-supporting for a sufficient time. The machine typically comprises a fixed head, supported by jacks on shoes in the invert and at the sides, containing the drive motors and a rotating cutting head carried in a large and heavy thrust bearing. Behind, and attached to the fixed head, is a chassis on which are mounted electro hydraulic motors, conveyor, rear jack, and an expanding ring or series of pads which can grip the tunnel walls and against which forward thrust can be made. The whole mechanical system must transmit the torque and thrust necessary for boring, while remaining accurately on the centre line, and steerable.

One of the shortcomings of drill and blast tunnel construction is the heavy incidence of overbreak. Despite improved techniques, a very considerable (and unpredictable) quantity of rock is excavated outside the desired tunnel profile, and has to be replaced, generally with concrete, in the course of supporting and lining the tunnel. The operation also disturbs and weakens a further area of rock.

One of the great advantages of the TBM is that, unless the rock is so cracked and fissured that blocks fall into the tunnel, it cuts a smooth bore of the correct profile which not only gives a useful reduction in the amount of excavation but also makes the best use of the rock strength and facilitates the construction of any tunnel lining whether it be *in situ* concrete, or precast segmental lining.

It has been shown that a TBM can bore a smooth tunnel which will stand without temporary supports through rock which, if blasted, would require steel ribs and laggings or other support measures. It is interesting to observe that, with rotary boring, rock tunnel technique approaches that in soft ground, by adopting a circular form and that it also lends itself to use of precast segmental linings. Figs. 7.7(a) and 7.7(b).

If there is a danger of rock falling behind the cutting head, it is common practice to provide a steel canopy or hood on the fixed head to protect men and machinery in advance of tunnel support being installed, and a drilling platform for rock bolting overhead may also be provided.

If the rock is heavily fractured or unstable the machine may need to be mounted in a shield in which even the method of forward thrust may be similar to that for a soft ground machine.

7.5.1.1 Cutter head

The design of the cutting head varies for different types of rock. Drag bits on radial arms may be adequate for soft rock such as chalk. Disc cutters, multiple discs or roller cutters are necessary for harder rocks, so mounted on a strong cutting head that as the head rotates successive tools following circular paths cut and split the rock in the most economical manner, taking advantage of the configuration of the rock face left by the previous tool. Fig. 7.8.

7.5.1.2 Rotation

Rotation of the tunnel machine contrary to the direction of the head is resisted either by skin friction of the shield to the ground or by the side and vertical thrust rams pushing anchor shoes against the excavated sides of the tunnel. Unlike the soft ground machines, the head cannot be reversed to counteract this roll tendency without resetting the tools, as the cutting tools fixed to the head generally are only capable of effectively cutting the rock in one direction.

7.5.2 Boom cutters

An alternative excavating method is the use of boom cutters, which, as for the full-faced machines, may or may not be mounted within a protective shield.

Fig. 7.7(a) Shielded full face machine 3.9 m dia. used for water supply tunnel in Italy. Machine provided with rear hood

Fig. 7.7(b) – Rear view showing probe drill mounted on hydraulic arm within the machine body. One of the difficulties of machine tunnelling is to reconcile rapid progress with any necessary probing ahead to locate hazards (Robert Priestley Ltd.).

Fig. 7.8 Full face hard rock machine. Note the massive construction of the cutter head and its roller discs (Thyssen (G.B.) Ltd.).

Fig. 7.9 – Track mounted boom tunnelling machine with spiral cutting head. Debris is swept back automatically through the machine on to the belt conveyor behind (Thyssen (G.B.) Ltd.).

The cutting head can take various forms, the design taking account of the strength of the rock and the loads imparted through the cutting tools. In self-supporting rock conditions the boom may be mounted on a crawler chassis, which may have a cover for protection of the operator from falling debris. Forward thrust is usually taken to the ground *via* the crawler tracks but some strutting to the side walls may be necessary. Fig. 7.9.

7.5.3 Excavators

Excavators working on the same principles as those for soft ground are also extensively used, being suitably more robust to deal with rock. Again they can be housed within a shield if ground conditions so dictate. The largest example of this type of machine is the 'Big-John' used on the 12 m diameter Huttegg Central Section of the Seelisberg Highway tunnel between Italy and Switzerland. This 450 T excavator rested on two runways mounted within a shield and was moved fore and aft by two 400 T hydraulic jacks. The telescopic boom was radially positioned by a 100 T jack whilst two 400 T jacks manipulated the digging action of the bucket upper teeth (weighing 0.6 T each) in removing the jointed and fractured shale, and drawing the excavated material back on to a conveyor. Final trimming of the tunnel shape was by the cutting edge action of the shield. This arrangement averaged 8 m per day advance.

Many modifications have been devised in an effort to improve the cutting efficiency of the machines and cutting tools. One such device to reduce the forward thrust required is by undercutting of the rock face. Separately driven cutter heads mounted at an angle to the main axis on a rotating head attack the face in a helical path following the perimeter like a screw thread.

Direct thrust of tools against the face is not required. The cutter tips act at some depth below the face which results in a groove being formed from which the major part of the rock breaks away towards the free face created during the previous pass of the head.

This radial undercutting action has been found to require as little as 30% of the thrust requirements of more conventional full face rotating machines. This obviously has the advantage of considerably reducing the load on the head's main bearings and the anchorage faces for forward movement of the machine.

Steering of the machine is claimed to be far easier, and tighter curves are possible, and by suitable arrangement and gearing of the individual cutter heads tunnels of other than circular cross-section are possible. The machine is obviously complex to manufacture and maintain and would appear likely to operate best in homogeneous rock.

The system has been further developed to produce the Mini full-facer. A single rotating cutter head is mounted on the end of an arm which can be manoeuvred vertically up and down. Cutting starts at the invert and the cutter head is swung upwards in an arc towards the crown. On completion the cutter head is lowered, the machine advanced a pre-determined distance governed by the strength of the rock and the process repeated. This configuration enables tunnels as narrow as 1.3 m and of 2.2 m height to be formed. Pressure pads grip the sides of the tunnel to resist any thrust and excavated material is moved backwards on to a conveyor system by the downward swinging cutter head.

7.6 OPERATION IN ROCK

7.6.1 Power requirements

Power requirements tend to be much greater for rock machines than for soft ground due both to the larger thrust forces required and to the greater expenditure of energy on cutting per unit volume. The power is transmitted by electro-hydraulic means and can be as high as 5000 kW as provided for the 'Big-John' excavator and supporting equipment, or as low as 400 kW on a Robbins full face machine of 3.5 m diameter used on the Kielder project in north-east England.

7.6.2 Protection

In rock tunnels immediate overhead protection in bad ground is mainly afforded by the provision of a shield above the tunnelling machinery. Some support of the excavated face, and shielding against falls, can be given by the cutter head of the machine, but any requirement for continuous general face support suggests the need for some change of techniques.

Access to the face suffers the same limitations as for soft ground machines and is easier when the cutting mechanism can be manoeuvred out of the way, as with boom cutters and excavators.

7.6.3 Linings

In situ concrete lining can be used where the cut rock stands without immediate support, or where temporary ground support is feasible by arch ribs or rock bolting and possibly sprayed concrete. The machine must be designed from the start with the necessary capability to accommodate such support.

Precast segments are a valuable alternative. They can readily be installed in a smooth cylindrical excavation, in contrast to the difficulties of erection in an irregular blasted profile.

7.6.4 Disposal

The excavated material has to be collected at the face and transported through the various types of machine on to a conveyor which may be of rubber belt, chain or plate type. Any size of material can be handled, depending on power drive available and adequate clearance, but problems arise if some of the fragments are too large, blocking passage, or perhaps worse, too fine. Dust, or fine, gritty slurry can form a very abrasive mixture and can choke transfer points or accumulate obstructions. Some full face machines are designed with a diaphragm close behind the face to trap dust and provided with water spray damping.

7.7 MERSEY KINGSWAY TUNNEL MOLE

Many of these problems in rock TBMs can be exemplified by the experiences on the Mersey Kingsway Tunnels (1971, 1973). A highway tunnel having a bore of 9.6 m, it was unique in being machine driven beneath the estuary of the river Mersey through Triassic Bunter Sandstone. The water-bearing sandstone is generally medium grained, weakly cemented, and with occasional silty parting layers. The general dip, but with superimposed current bedding, is at about 7° in the line of the tunnel, but with faults across the line. The bed of the river was believed to be rocky without any bed of clay, and wet conditions were anticipated, as was confirmed by a central pilot tunnel driven by drilling and blasting. A tunnel 'Mole', originally designed and used to drive diversion tunnels at the Mangla Dam, was acquired and was modified extensively for the new conditions. The diameter had to be reduced from 11.2 m, central access into the pilot tunnel was provided, and a system for handling and erecting a heavy concrete segmental lining as close as possible behind the face was installed. Disc cutters projected about 300 mm ahead of the face of the very substantial cutting head, so that space was clear for broken material to fall into the invert, from which it was picked up by peripheral buckets, and discharged *via* a shute near the top to a conveyor system and ultimately to a point 30 m behind the face. Up to 500 tons of forward thrust was available from rams acting on thrust pads jacked out against the completed concrete lining. Ten 75 kW electric motors spaced round the circle drove the cutting head which ran in a 5 m diameter roller thrust bearing. Renewal of cutters was effected at intervals of about 100 m by means of a short cross heading out of the side of the pilot tunnel. Figs. 7.10(a) and 7.10(b).

In operation considerable difficulties were experienced, attributable in large degree to: (a) the very wet and saline conditions under the river; (b) the way that some of the rock broke down to fine abrasive sand, which tended to become re-cemented in the invert; while (c) some layers broke off at the face as lumps, jamming rotation; and (d) fractured rock, more particularly in fault zones. The practical importance of such features, although known, was certainly not fully appreciated in the machine design.

The drive started downhill from the end of an open cut, as a consequence of which the water accumulated in the invert at the face, from which it had to be lifted at least up to the central pilot tunnel. The expected slow start with new systems and equipment was made worse by considerable overbreak

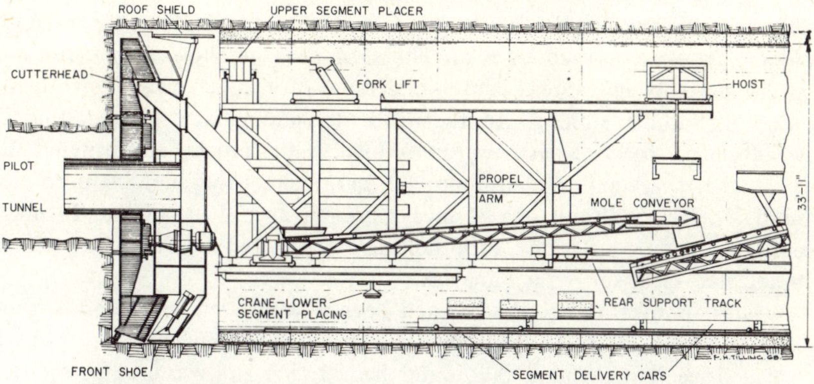

Fig. 7.10(a) – 'Mersey Mole' sectional elevation. Note the small cylindrical access into the pilot tunnel.

Fig. 7.10(b) – Plan view of Robbins machine preparing to drive into the Bunter Sandstone at the Wallasey portal in 1967.

in the crown, above which the depth to the top of the sandstone, overlain by boulder clay, was only about 6 m. The slow advance allowed time for overbreak to develop and spread and the shallow rock cover made less effective natural longitudinal arching. Another difficulty adding to delay was that the heavy machine, carrying also overbreak load in the crown, encountered silty strata in the invert and settled about 80 mm, making ring building even more difficult. Portal conditions nearly always present special hazards, and it might have been worth considering rock bolting from a top heading, but with all bolts and steel clear above the Mole cut.

A fault, known to be present, had been encountered in the pilot tunnel and secured with steel arches, timber and grouting, but when approached by the main drive it was found that water had eroded and was eroding seriously a filling of silty gouge in the fault gap which led up to the river bed. The rock on either side of the fault was more fissured than normal. The area was first extensively grouted from the pilot tunnel. As a safety measure the machine drive was stopped while a concrete hood 6 m x 2 m x 24 m reinforced with steel joists was constructed in three 2 m x 2 m headings driven successively side by side clear above the machine cut. After eight weeks' delay the drive was restarted and carried successfully through the fault zone. The emphasis here is again on the demand made by a machine for reasonable uniformity and on its lack of flexibility.

A further trouble, attributed in part to this delay in very wet conditions, was the failure of the 5 m diameter main bearing behind the cutting head. Water carrying grit had penetrated the bearing seals. The point at which the machine had stopped was fortunately in sound rock and was not abnormally wet. The machine could not be withdrawn through the completed tunnel because of the concrete lining, and the bold enterprise of replacement *in situ* was decided on. A spare bearing (in poor condition) was available, and was renovated and brought into the tunnel. The cutter head had to be drawn forwards from the fixed head so that the roller path might be pulled off and then the new bearing had to be fitted as a shrink fit in its place. All this is normally an operation requiring extremely clean dust-free conditions. Here, there was no danger of dry dust but water and grit were continuously and pervadingly present. A working chamber with an overhead runway beam was constructed by enlarging upwards with removal of two completed rings and some additional segments from the lined tunnel and the new bearing was suspended from the runway behind and above the cutting head. The cutting head was then anchored securely into the face of rock ahead and the Mole was withdrawn axially, for which an extraction force of 280 T was necessary. The new bearing was fitted to the Mole after heating to a temperature of 11°C above ambient to effect the necessary shrink fit. The Mole was then steered forward with great precision into the cutter head. After six weeks' delay the drive was successfully resumed.

Fig. 7.10(c) – 'Mersey Mole' breaking through on completion of its first drive to form a junction at the Liverpool side (March 1970).

The relevance of this to employment of such machines is to give emphasis to the need first for extreme precautions in protection and sealing machines against intrusion of abrasive grit and saline water, and second, for contingency planning for major repairs in any machine failure. Further severe damage to the Mole resulted a little later from a broken pinion tooth lodging in the ring gear, and extensive remedial work took about three weeks.

Duplication of the tunnel followed on completion of the first drive, with the machine modified in the light of experience. Handling and control of the spoil were improved. Slurry was drained and processed separately, as were blocks of rock. The overall tunnelling time was only half that for the first drive; 13 months against 27 months. It is of interest that a 3 m machine, designed, on the basis of experience in the first main drive, to cut a pilot tunnel for the duplication, had to be withdrawn because the volume and character of slurry generated could not be satisfactorily handled in the limited space available. The pilot was completed by drill and blast.

There is little doubt that machine tunnelling was the appropriate method for the project, and a recital of these difficulties simply points out the practical consequences of quite minor departures from the expected ground and water behaviour, and the necessity to have available engineering experience and resources to overcome them. The machine is not an automated answer to tunnelling but another tool to be adapted with skill and imagination to the conditions encountered.

7.8 EFFECTS OF MECHANIZATION

It is a truism that the design of tunnels must be closely integrated with the construction, and the advent of the TBM emphasises the need.

Unless the designer has a fully adequate understanding of the methods of construction and is competent to assess the suitability of the resources which can be made available, much time and money may be wasted. Contractors tendering for a tunnel in a period of a few weeks are in a very difficult situation if the promoter and designer have not taken every opportunity in the preceding months or years to collect and collate all possible information about geological and hydrological conditions, and have not so specified the project that the most economical method of construction, in time and cost, can be adopted.

Major schemes often require the design and construction of TBMs specially for the project. This process can take more than a year in itself and if the project is urgent and the other preparatory work is comparatively brief, then if machine tunnelling is considered to be the appropriate technique, earlier completion of the tunnels may be achieved if the promoter is prepared to order the design and manufacture of TBMs before signing a contract for the tunnel construction. Here is a situation which tests the confidence of the client in his engineers, and which tests his engineers' competence.

By close collaboration and carefully formulated contracts, advantages have been gained by ordering TBMs in advance, notably for London's Victoria Line tunnels, where the relative uniformity of the London clay and familiarity with machine performance and ultimate requirements simplified the complex relationships.

Another problem is introduced if a contractor already owns the only TBM apparently suitable for a proposed tunnel, and he is thus insulated against close competition from other contractors. This situation seldom, if ever, stifles competitive tendering, partly because it is generally necessary and quite costly to recondition and modify a machine for a new tunnel, partly because the fortunate owner of the TBM is usually acutely aware of its limitations, and partly because tunnel contractors are highly competitive and have an illogical desire to face the challenge of yet another tunnel.

7.9 CURRENT AND FUTURE DEVELOPMENT OF ROCK TUNNEL MACHINES

Some have dreamed of a machine capable of dealing with any stratum or combination of strata. Such a machine, if attainable, would be too expensive for normal use. It could only be done by fitting the ground to the device and not the device to any ground. Protective provision of a roof shield, a rock bolting platform, sprayed concrete, early erection of arch ribs, or of a segmental lining, all serve special requirements but a choice must be made. It is more realistic to aim to make machines which are more adaptable either in the course of a single drive, or for use on subsequent tunnels. Fig. 7.11.

For the full face rotating head TBM, better disc cutters and picks are being developed all the time: mechanical and electrical components are being improved in strength and reliability and better protected from dust and water. Working conditions for the operators are meeting higher standards in terms of safety, noise, temperature and quality of air. Fig. 7.12.

In any particular case much of the machine design is dependent on expected geological and water conditions. It may often be valuable to give maximum access, and essential to give some access to the working face to deal with special problems. It has been found desirable in some types of rock to design machines which allow sprayed concrete to be applied to the rock close up behind the cutting head. Such an arrangement is likely to conflict with the precaution of having a protective hood over the machine. Similarly, various ingenious methods have been designed to allow erection of steel ring beams or segmental lining to support the rock close to the face. On other occasions the machines have had to accommodate rock bolting procedures. All these special requirements have led to a proliferation of designs rather than a universal machine, and for good reasons.

The observation has often been made how wasteful of energy it must be

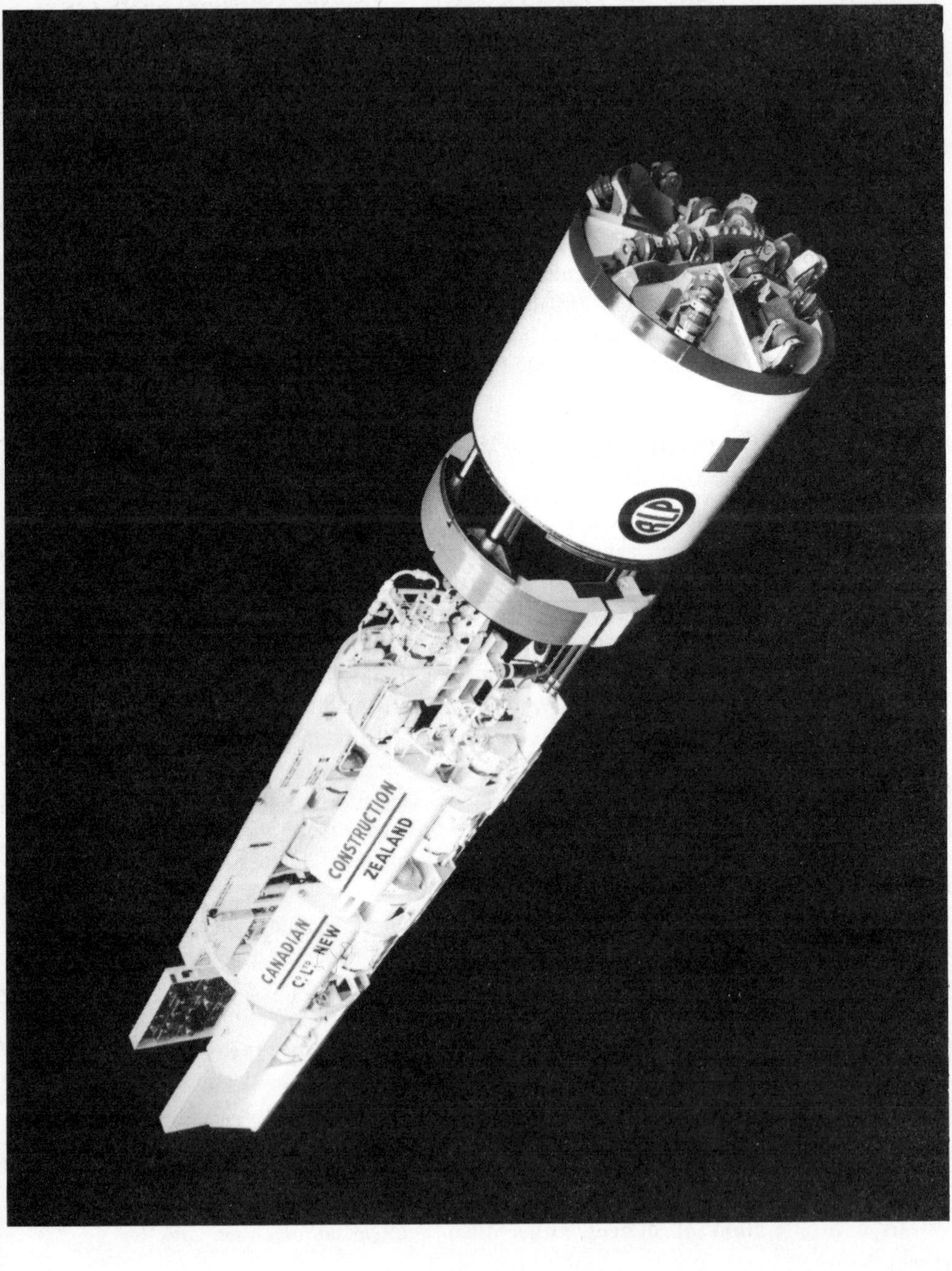

Fig. 7.11 – A versatile machine for use in New Zealand. Shielded full face machine 2.65 m dia. For use in hard and soft ground, utilising ... tion in hard ground or lining segments in soft. Cutting head can accept disc or pick cutters (Robert

Fig. 7.12 – A 10.7 m dia. machine which, with its fellows, has achieved unprecedented output in boring hard dolomitic limestone and shale for storm water reservoirs (the famous TARP project) under Chicago (The Robbins Co.).

to grind rock into small pieces for removal. Such comment does less than justice to the work which has been done to optimise the size distribution of TBM spoil by tool design and spacing but these cannot be changed to meet variations in the rock over the area of a large face or as the face advances. Even with today's high energy costs, the power used in excavation is a small proportion of tunnelling costs. But energy costs are rising, and in long tunnels, apart from direct cost, excess power used by the TBM generates much heat, which has to be removed at additional cost.

It is tempting to envisage a TBM designed to cut forwards ahead of the face a deep groove following the perimeter (thus giving the advantage of a smooth cut bore) and to follow by breaking the bulk of the rock core into large (but handleable) pieces by other means. For small tunnels the core splitting might be achieved by drilling a central hole from which mechanical devices would apply pressure to break the remaining annulus of rock. For larger tunnels a pattern of such holes would be necessary. It may be that hydraulic pressure or even small explosive charges could be used to break the rock. At present it would appear that the ever-improving performance of orthodox TBMs precludes the investment required to develop such a machine with no certainty of commercial advantage, but the time may come. An interesting method was recently used to construct a large tunnel for the Paris metro. In very soft rock, a machine akin to a chain saw was used to cut a thin slot ahead of the working face around the upper perimeter of the tunnel. The slot was then filled with concrete, forming a buried arch. The face was excavated within the pre-placed initial support and the process repeated. Such a method would not appear to lend itself to widespread use, and it is unlikely that the rate of progress would make it attractive for long tunnels. Nevertheless, it offers a new safeguard against settlement or collapse in, for instance, certain soft rocks which deteriorate rapidly on exposure to air. The 'chain saws' required are undoubtedly miniature rock tunnelling machines and it is interesting to note that they must incorporate probably unknowingly many of the principles of the original Thompson ladder excavator used in the 1890's, the first machine mentioned in this chapter.

BIBLIOGRAPHY

This is a new and growing subject involving machines of increasing diversity and complexity. The brief bibliography which follows indicates some sources of additional information on current developments.

Donovan, H. J., *Modern tunnelling methods*, Public Works and Municipal Services Congress, London, 1968.

Muirhead, I. R. and Glossop, L. G., 'Hard rock tunnelling machines', *Trans. Instn. Min. Metall.*, 1968, **77A**, 1. *See also* discussion, 118.

Megaw, T. M. and others, 'Mersey Kingsway Tunnel', *Proc. Instn. Civ. Engrs.*, 1972, **51** (Mar). *See also* discussion, 1973, **54** (Mar.).

Pirrie, N. D., 'The use of rock tunnel machines', *ibid*, 1973, **54** (Feb.). *See also* discussion 1973, **54** (Nov.).

Bartlett, J. V. and others, 'The Bentonite tunnelling machine, *ibid*, 1973, **54** (Nov.). *See also* discussion 1974, **56** (Aug.).

Robbins, R. J., 'Mechanised tunnelling, progress and expectations,' *Tunnelling 76, Proc. Int. Symp.*, London, 1976.

Johnson, A. D. and Parkes, D. B., 'The use of roadheading machines in the construction of the Liverpool Link Line', *2nd Australian Conference on Tunnelling,* Melbourne, 1976.

8

Permanent Linings

8.1 NEED FOR LININGS

Permanent linings are required in most tunnels, always in soft ground and frequently in rock. They are required for two purposes: structurally, to contain and support the exposed ground, and operationally, to provide an internal surface appropriate to the function of the tunnel. The sequence of lining operations is varied and often complex so that the terminology is sometimes ambiguous.

The term 'primary lining' in British practice is frequently used to refer to the function of permanent structural support, and 'secondary lining' to subsequent interior treatment. This is quite appropriate to much soft ground tunnelling procedure, but less so in rock tunnelling and would not be generally understood in the U.S.A. In American practice 'ground support' is the term used for the structural supports, which are classified as 'permanent' when specified and incorporated in the design calculations, and as 'temporary' when left to be provided by the contractor as found necessary for safety in construction. *In situ* concrete, or other material, is the 'lining', incorporating the earlier permanent supports. Further secondary finishings for operational or aesthetic reasons may be provided.

This chapter is principally concerned with permanent structural linings. Immediate temporary support requirements are discussed in Chapters 4 and 6. Secondary linings for operational purposes are referred to in Chapter 2, and discussed in some specialised contexts in Volume 2. Both temporary support and secondary lining may be essential aspects of the design and construction of permanent linings, as will appear in context.

8.2 PRIMARY REQUIREMENTS

The primary requirements for permanent structural linings are:

1. To provide necessary structural support
2. To control or eliminate inflow or escape of water
3. To accommodate the operational cross-section

In meeting these requirements the chosen lining system must be capable of safe and economic construction, and be adaptable to possible variations and unforeseen contingencies encountered in the progress of the work.

Adaptability of the lining system may be a very important factor, as the urgent need to change to a different lining even for a short length of difficult ground can cause much delay and disruption to the programme and result in heavy additional cost.

8.2.1 Structural support

When a tunnel is excavated the existing stable equilibrium of the ground is disturbed and a new stress pattern must be established with the aid of supporting structures. In sound homogeneous rock an arching action will develop, and added support may not be necessary, whereas in soft plastic ground something close to hydrostatic pressure is likely to develop, quickly or slowly, and make it necessary to provide support competent to carry the whole overburden pressure. In most tunnels the requirement lies somewhere between these extremes, and the art of tunnel design lies in assessing the real needs at each stage and providing accordingly, with proper margins for the unexpected. Timing of support is of vital importance, to restrain immediately any movement which might develop catastrophically, but to allow strain movements in the surrounding ground, with concomitant stress changes, to establish a new and stable equilibrium. It is a matter partly of calculation from ascertained properties of the ground, but largely of experienced judgement, how to achieve the best final result. The major question is that of foreseeing and controlling the stresses and movements in the surrounding ground as a whole, but there must also be provision against loss of ground by local frittering away, or ravelling, at the exposed surface, or collapse of loose blocks of rock, or erosion by water. These subjects have been discussed in the context of excavation in Chapters 4 and 6 and may be further referred to below when dealing with the various forms of lining.

Design of linings for strength has in practice been largely empirical, based on experience. In cast-iron segmental linings for soft ground tunnels there has almost always been such a substantial margin of strength to suit foundry practice and to stand up to handling and erection procedures that the close analysis of the finished structure was not pursued. Precast concrete segmental linings were developed largely by adaptation from cast-iron.

Much tunnelling in rock is carried out on the basis that the cavity formed will largely be self-supporting, but that as the excavation proceeds conditions will be continually assessed and supports provided where needed. Nevertheless, much valuable work done over the last quarter century has greatly increased detailed understanding of the nature of rock tunnel support, and now allows rational estimates of requirements to be made.

In soft ground, the basis of modern analysis and design is that ground and lining ultimately act together as a composite structure. The soil has both plastic and elastic properties, and the complex stress pattern which ultimately develops depends on the stress history of the components, determined by the sequence and timing of the operations of excavation and support. In addition to the direct hydrostatic pressures on an impermeable lining, possibly modified by leakage, ground water is an important factor in the stress-strain relationships in cohesive soils, involving soil mechanics studies of pore pressures, drainage and consolidation, and other aspects.

In rock, the properties of sample cores are by themselves an inadequate basis for prediction of behaviour. Of overriding importance is the joint structure of the mass, which is much more difficult to ascertain by any exploration in advance.

It is not intended here to present or further discuss the various mathematical formulations which are tools of tunnel design. It may be noted that the structural design process for a tunnel is not comparable with that for a bridge or other surface structure: not merely because the ground imposing the loadings is not fully known in advance, but because of the interactions between ground, construction procedures and structure.

An important aspect of lining design in soft ground is the change of shape of the excavated cavity. A circular hole will usually distort to an ellipse with its major axis horizontal, because of the action of vertical stress resulting from the overburden. A rigid circular lining within this will resist distortion and be subject to bending stresses, whereas a flexible lining will conform to the ellipse but develop passive earth pressure where it displaces the soil outwards. A flexible lining acting compositely with the soil offers a thinner and lighter design than a rigid cylinder which must resist heavy bending moments.

8.2.2 Control of water

In water-bearing ground substantial watertightness of the permanent lining is usually important, but absolute watertightness is very difficult, if not impossible, to achieve and is likely to be very costly. As in so many aspects of tunnel construction the choice of lining and waterproofing methods must be made in the light of costs, balancing initial expenditure against operation and maintenance.

In rail tunnels some leakage may be acceptable provided any water is deflected away from overhead wires and from other electrical equipment into a drainage system. In road tunnels a higher standard may be required, eliminating altogether 'droppers' in the traffic space and preserving clean and dry wall surfaces. In metros the running tunnels are comparable to main line railways but there may be more delicate electrical equipment to safeguard, whereas in the station tunnels a high decorative standard will be sought. In sewer tunnels watertightness is important to the degree that ground water must not carry in silt, or enlarge leaks, or significantly add to the volume of flow, and most particularly that outward leakage must not contaminate the ground and any aquifers.

Pressure tunnels in hydroelectric power systems are a special case. The lining, usually welded steel or *in situ* reinforced concrete, must be completely watertight against high internal pressure, and must also be watertight and mechanically stable against, possibly high, external pressure when emptied for maintenance or other reasons.

Deep subaqueous tunnels may also be subject to high external water pressure and, rather than design the lining to exclude it absolutely, controlled inflow may be accepted and dealt with by drainage and pumping. The pore water in the surrounding ground is thus subject to a pressure gradient and the ground develops stresses to carry the loading.

8.2.3 Cross section

In tunnels generally arching action is the preferred method of carrying ground loads. The tunnel roof is therefore nearly always in the form of an arch, except in most cut-and-cover and many submerged tunnels. The arch may be completed below springing as a horseshoe form or with vertical side walls or as a complete circle or oval, depending on the ground conditions and the function of the tunnel. The circular form offers major structural advantages in ground where loads from all directions are encountered. It is usually operationally best for carrying water or sewage, and it also is the preferred form excavated by a shield or full face machine. In railways, the structure gauge must be accommodated within the circle above rail level and the permanent way and drainage in the invert so that the circular area is less fully utilised than is a horseshoe section. In highway tunnels the roadway and the rectangular traffic space above it only occupy about 50% of the area, although the extra space may be utilised for ventilation and other ducts.

The horseshoe form is the obvious cross section in rock which is largely self-supporting, but where lateral pressures develop, a lining is necessary and the side walls must be securely anchored at the toe, either into sound rock, or by struts across the invert or by a complete invert lining. Particular care may be necessary to ensure that drainage water does not soften or erode clay or soft rock in the invert or at the bases of the side walls.

8.3 PRINCIPAL MATERIALS

The principal materials and construction methods for permanent lining of bored tunnels are:

1. *In situ* concrete
2. Sprayed concrete
3. Preformed segments bolted or flexible
4. Brickwork and masonry

8.4 IN SITU CONCRETE

Concrete placed *in situ* is the most widespread form of permanent lining in rock tunnels. It has the advantage that it can be designed to accommodate any desired shape of cross section. It is ordinarily cast with the use of a travelling shutter well behind the working face, both in time and distance, embodying any immediate permanent supports already in place, and possibly also temporary supports. As far as practicable the lining is designed to function in compression without heavy bending moments and therefore rod reinforcement is largely avoided although it may be needed where loading is uneven and variable.

It is often preferable to complete the concreting of roof and walls before the invert, but sometimes it is thought advantageous to cast the invert first, thereby providing a smooth continuous support for the main shutters. Conditions for concreting are not always favourable to obtaining a consistently high quality, and thicknesses are often such that only low strength need be specified. Mixing of concrete at the surface and placing by pump may be of limited advantage as the distance from the shaft increases. Transport by skip, with appropriate agitation, and transfer to the shutters by a pneumatic placer is of more general application. There is always difficulty in ensuring dense uniform concrete in the crown of the arch and particular care is necessary if voids are to be prevented. In any case special provision for grouting is likely to be required.

In a large tunnel there may be three sets of shutters, each say 10 m in length and so designed that they can be partly dismantled, or folded and moved forward through the two sets in use for re-use 30 m ahead. Fig. 8.1.

Where it is thought important to complete the concrete lining quickly after excavation shutters may be for a much shorter length, as little perhaps as the depth of each advance, possibly 1½ m, but much of the advantage of large scale *in situ* concreting is lost, and use of precast segmental lining might be more appropriate.

8.4.1 Waterproofing

Because of the conditions in which it is placed it is particularly difficult to ensure watertightness in *in situ* concrete tunnel linings. Construction joints can be spanned by waterstops, although they cannot readily be embedded accurately in the concrete and are liable to displacement when a pneumatic placer is used. The problem of the crown has already been mentioned.

A plastic membrane which can be hot air welded is found reliable where rock is largely free from running water. After excavation any substantial flows of water are sealed locally and piped down to drains. Fixing pegs with plastic covered heads are grouted in and the exposed rock generally is coated with sprayed mortar or concrete against which first a plastic 'fleece' may be secured providing a smooth surface for the membrane of plastic sheeting fixed by means of the pegs, edges and any defects being welded. The structural *in situ* concrete

Fig. 8.1 – Travelling shutter for *in situ* reinforced concrete lining.

lining is cast against this membrane. The piped leaks run freely and the 'fleece' remains permeable so that water pressure does not build up before the concrete is complete. They can then be left open, or back grouted and capped off as appropriate to the site and the use of the tunnel.

8.5 SPRAYED CONCRETE AND ROCK BOLTS

Sprayed concrete has become an increasingly widespread technique of ground support, by itself or in combination with such methods as rock bolting, steel arch ribs, and mesh reinforcement. It plays an important part in the so-called New Austrian Tunnelling Method (N.A.T.M.) whose basic principle is to ascertain and control the development of stresses and deflections and their interaction with supports and lining, and thereby to establish within the surrounding rock a load-bearing ring.

Sprayed concrete is applied in thicknesses up to about 75 mm to rock surfaces. Aggregate, up to about 12 mm, and cement are mixed in a gun and sprayed forcefully on to the surface, water, carefully gauged to give a low water/cement ratio, being added, either at the nozzle in the 'dry' process, or earlier at the mixing stage in the 'wet' process. In all cases there is a substantial volume of 'rebound' which can be kept as low as perhaps 10% with well designed equipment and skilled operators but may be as much as 50% in unfavourable conditions. Where desired, reinforcement can be embodied in the finished lining. It is desirable that the rock surface should be clean and free from running water. When sprayed forcefully against freshly exposed rock, at a velocity of the order of 100 m/s, the cement mortar in the mix is driven into fine cracks and fissures and also forms a skin on the surface, on which a layer thoroughly bonded to the rock is built up. This acts to assist in both the purposes of support referred to earlier, by bonding the rock mass into a monolithic whole and by preventingloose surface material from progressive deterioration.

It will be clear that the sprayed concrete as deposited is in no sense to be treated as an independent structural element but as an integral part of a thick rock cylinder. It is indeed considered important that the concrete should be thin enough to be flexible and to conform to changes of shape without starting cracks. Extra bending strength is claimed for sprayed concrete incorporating steel fibres, or possibly alkali-resistant glass fibres.

Rock bolts cannot in themselves constitute a lining in any sense, but are an ancillary means of support which can be particularly effective in conjunction with sprayed concrete. The support function has been discussed in Chapter 6. They contribute to the establishment of the stressed rock cylinder around the tunnel.

8.6 SEGMENTAL LININGS

Circular segmental rings of cast-iron or concrete provide an immediate permanent lining of great strength, provided that when erected they can be brought into close contact with the excavated ground by grouting or otherwise. Their use in tunnels dates from London's Tower Subway (1869) and may have derived from cast-iron 'tubbing' employed in shaft sinking for mines at the end of the 18th century. The Tower Subway pioneered the combination of circular form, Greathead shield, cast-iron segments and use of grout. When compressed air working was added, the Blackwall tunnel (1897) and the Hudson river tunnel (1906) and their successors in London and New York and elsewhere became feasible. Cast-iron lining became the standard for metro tunnels in London and was unchallenged in this field until the 1930's. Precast concrete was then introduced as an alternative on some sections of the Ilford extension of the Central Line, and since then the special merits of cast-iron have had to compete against the lower cost of concrete, and other particular advantages in special uses.

As permanent linings all the segmental types share the basic advantage of providing immediately on erection a strong supporting structure with useful flexibility, but in developing the interaction between lining and ground the timing of grouting operations is relevant. In potential for waterproofing the different types vary and their suitability must be assessed in relation to operating requirements and the nature of the ground. In shape the circular form is almost always adopted, except that egg-shaped sewers have been tunnelled with cast-iron segments. In some special cases junctions and transitions between circular cast-iron tunnels have been formed continuously with C.I. castings of complex geometry.

The principal variants in materials for segmental linings are in the use of grey cast-iron, S.G. Iron, welded steel, or precast concrete and in structure either bolted and grouted rings or flexibly jointed rings. The ordinary form of cast-iron lining with flanged segments bolted together will first be discussed in sufficient detail to allow its advantages and disadvantages to be appreciated; the merits of the variants will then be examined, providing a basis for evaluation for any particular application. Indeed, it is by such a process that the variants have been developed to meet constructional and operational requirements in different uses.

8.6.1 Bolted cast-iron rings

The tunnelling process for bolted C.I. rings is, in outline, that excavation is advanced one ring width at a time, and the new ring is immediately erected by bolting the segments to one another by the cross joint, and bolting them back to the circumferential flange of the previous completed ring. In a typical ring there are six or more flanged segments and a narrow 'key' (Figures 8.2(a) and 8.2(b). The segments are identical except for the two 'top' segments which are slightly curtailed at one end, and have there an oblique end flange to provide a tapered gap for the key; the erection of a ring can be completed from inside the excavation by final entry and bolting up of the key. Grouting up with cement grout to fill the working clearance between the skin of the iron and the ground should follow as quickly as practicable to allow the lining to carry the load from the ground without avoidable settlement.

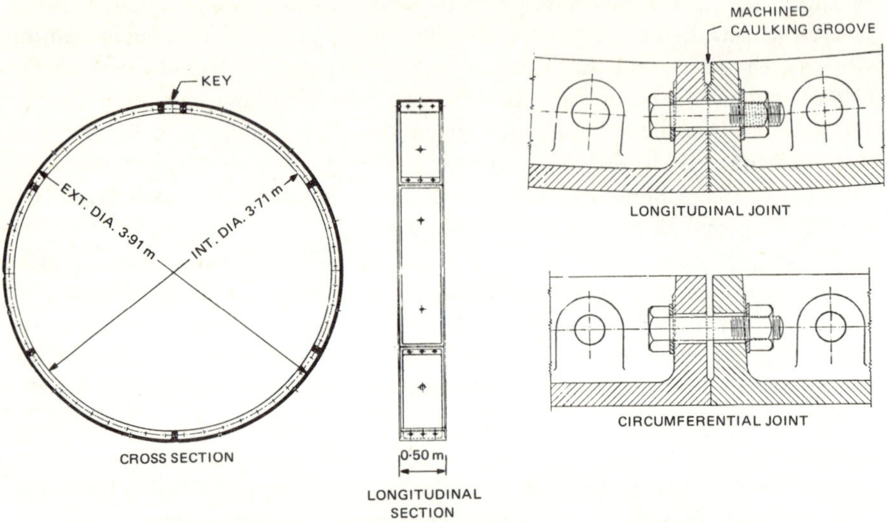

Fig. 8.2(a) – Typical ring of bolted C.I. segments.

Fig. 8.2(b) – Bolted segmental C.I. lining.

8.6.1.1 Cast-iron

Wide use has been made of grey cast-iron of a relatively low grade. It has very adequate compressive strength but relatively low tensile strength and is brittle. Its density is high. When used in circular tunnel rings most of the ground loading is carried as direct compressive stress. Bending moments, and the consequent tensile stresses, are usually minimal in these rings by reason of slight flexibility at the cross joints which allows the shape of ring to accommodate to a slightly elliptical form.

In transport, handling and erection the segments must be capable of standing up to rough handling in difficult conditions, including being dropped. When erected behind a shield the longitudinal thrust from the jacks is liable to be imposed unevenly on the skin of the segments. The proportioning of typical segments has developed accordingly. A factor limiting maximum weight and therefore sizes of segments for small and medium tunnels has been the need at times to be able to lift them into place finally by manpower. In larger tunnels mechanical lift devices can be accommodated, but conditions often remain cramped and access difficult.

Another limitation on design and dimensioning of individual segments lies in casting practice in the foundry. Experience dictates much of the detail in proportioning flange and skin thickness and connecting radii, with strengthening fillets and tapers to prevent shrinkage cracks and ensure dimensional accuracy and stability against distortion.

Dimensional precision, to ensure close fitting joints lending themselves to accurate erection and sealing against water, may be aided by accurate machining of the matching joint surfaces. The cross joints are nearly always machined in modern practice, and the circumferential joints are also frequently machined where the cost can be justified but may be left as cast and provided with packings. It may be noted that where a tunnel has to be built on a curve, whether by design or to correct deviations, tapered packings between successive rings are necessary, unless special tapered rings are specified for the purpose. Fig. 8.3. It further happens that, even with fully machined iron; errors may creep in and become cumulative where silt or other deposits intrude between rings, particularly where the invert has to be built in wet and dirty conditions, and that corrective packings to restore rings to a true plane face cannot be avoided.

8.6.1.2 Corrosion

Grey cast-iron has over a century proved itself to be surprisingly resistant in tunnels to rust and other corrosion. The castings after cleaning up from the moulds are usually hot dipped in a bituminous or similar solution, although any subsequent machining or drilling will obviously expose fresh iron surfaces which must be given appropriate additional protection. It seems that in the sand casting process a thin film of fused sand is embodied in the surface and offers valuable added protection. Alkaline conditions associated with cement grouting

Fig. 8.3 – Complex castings to provide for curves and tapers.

also are helpful for preservation, but cases of acid attack, probably from oxidation of sulphides in the ground, have been recorded.

Corrosive attack on the mild steel bolts in the flanges is also possible, but these can be painted or otherwise protected and are usually sealed by use of grummets against flow of water past the head and nut. Ultimately they are replaceable, although in a finished tunnel, in which ground stresses have stabilised, flange bolts serve little structural purpose.

8.6.1.3 Availability and cost

Availability of C.I. segments as such depends on the existence of suitably equipped iron foundries which can produce adequate supplies at the required dates. If sound and accurately dimensioned castings are to be ensured consistently considerable specialised experience is a great advantage. Conversely, large production runs over a lengthy period allow techniques to be improved and costs to be reduced.

Cost generally is now found to be substantially higher than that for concrete rings of similar size and type, but there may be valuable advantages, in many cases, in construction procedures, watertightness and operational convenience.

8.6.1.4 Waterproofing

As a material, sound cast-iron is itself completely impermeable except in occasional small zones of porosity where 'pinholes' may allow leakage of water at high pressure. These can usually be drilled out and plugged as thought necessary.

To back up the lining, the first defence against water is the grout layer injected through the lining to fill the voids left in the tunnelling process. Supplementary grouting at increased pressure is possible, but in water-bearing ground it is further necessary to waterproof, by caulking or otherwise, the cross joints, the circumferential joints, and the bolt holes and grout holes. In such cases it is usual for the cross joints to be manufactured with machined meeting faces, which leave caulking grooves about 6 mm wide and 25 mm deep in the exposed edges of the assembled joints. The circumferential flanges may be similarly machined, but they may be left unmachined in which case they usually have a much deeper groove, nearly the full flange depth, to accommodate wood or other packing, which can where necessary be cut out and replaced with caulking.

For the most difficult conditions strip lead or lead wool is favoured as the caulking material, one reason being that it can be recaulked if necessary subsequently. For most applications an asbestos-cement composition is found to meet requirements, at a substantially lower cost.

Bolt holes in flanges must also be sealed by fitting washers and grummets at each end. Plastic polythene grummets compressed by tightening of the bolts, squeeze into the neck of the hole and provide a satisfactory seal. Grout holes, through the web of segments, are preferably screw threaded so as to hold the grouting nozzle under pressure and to allow screwed plugs to seal them off.

Caulking and grummeting are costly and labour intensive operations. The substantial quantities in a major cast-iron tunnel may be illustrated by the 1.4 km long first Dartford tunnel (1963), in which joints caulked with lead strip amounted to 90 km and bolts grummeted to 260,000.

Other joint sealing techniques depend on the insertion of some form of plastic strip between segments during erection, but it is extremely difficult to ensure the necessary care and cleanliness in a necessarily wet and dirty tunnel.

8.6.1.5 Operational considerations

Where flanged C.I. rings are built into a tunnel the pattern of ribs may be advantageous in a few cases, as for fixing frequent cable and pipe brackets in a rail tunnel, but it may be necessary to provide a smooth and easily drained invert by means of concrete filling, or a complete smooth bore lining for a sewer. If a secondary lining is in any case necessary the ribbed iron may not be disadvantageous except in respect of the invert.

8.6.2 Spheroidal graphite iron

S.G. cast-iron has nearly the same chemical composition as grey C.I. but the graphite content is of nodular rather than flaky form. The effect is that its tensile strength is much greater and its compressive strength is also somewhat higher while it is less brittle and probably more resistant to corrosion. Its behaviour is more like that of mild steel, but it may be employed for cast segmental linings in exactly the same manner as grey iron. Somewhat thinner and lighter sections may be used, and with a structurally more economical distribution of material, provided that in the casting process distortion during cooling can be kept within acceptable limits.

The lower weight of material to transport and handle into place has obvious advantages in construction; in all the other tunnel lining aspects referred to it is virtually the same as grey C.I. Cost per ton is higher, which is not usually fully offset by the reduced weight, but valuable savings are possible in handling and time of erection.

8.6.3 Welded steel segments

Welded steel has not been widely used for segmental circular linings. It is sometimes employed at positions where abnormal loadings are imposed by adjacent structures or in junctions and other openings in a tunnel, for sake of the extra strength in bending and tension of steel compared with cast iron. Welded steel bolted segmental linings have been used for substantial lengths of tunnel in the U.S.A., and were adopted in preference to cast-iron for the circulating water tunnels at Dungeness nuclear power station, principally to sustain very high thrust loading from the shield during tunnelling.

Waterproofing presents the same problems as with other bolted linings unless welding is found practicable. Corrosion in unfavourable ground demands more attention than does cast-iron. There is no great difference in the corrosion resistance in the ground of the two materials but (1) steel lacks the surface film of fused sand on the castings and (2) if advantage is taken of the strength to use thinner sections the proportionate loss when attacked is greater. More elaborate protective coating methods may therefore be necessary.

8.6.4 Bolted precast concrete

The first major substitution of concrete for cast-iron was, as stated, for the Ilford extension of the London Central Line in 1939. The design and principal dimensions followed very closely the pattern of cast-iron segments. Flanges were substantially thicker, longitudinal stiffening ribs were introduced to carry shield jacking thrust, and the number of bolts in the circle was reduced. See Figs. 8.4(a) and 8.4(b).

As compared with cast-iron, high quality concrete requires greater thickness and must have steel reinforcement where tension may occur. The segments are more liable to suffer damage in handling and in jacking and also after erection because of greater rigidity; cracks may not appear until shown up by leakage later. Cross joints are usually erected with a bituminous packing. Waterproofing is more difficult than with machined cast-iron segments, which are likely to be specified where difficult conditions are expected. Dimensional accuracy and reliability demand a very high standard of manufacture, and this has been progressively improved with experience. Resistance to corrosion has not the long history in use of cast-iron, and in chemically aggressive ground the problems require careful study. Concrete quality and cover over reinforcement need to be controlled accurately, with use of sulphate resisting cement where appropriate.

A range of standardised sizes are available from specialist manufacturers and cost is of the order of one third less than for cast-iron rings. Fig. 8.4(c).

8.6.5 Expanded linings

The development of flexible joined expanding linings was an important departure from flanged bolted linings, whether of cast-iron or concrete, which formed a rigid continuous cylinder and had to be grouted immediately, as one phase of the cycle of operations, to secure the ground. The new technique is particularly suited to stiff clay. The principle is that the ring of segments is assembled directly behind the shield in close contact with accurately trimmed ground, and appropriate compressive hoop stress is applied by wedging or jacking. There is no gap to be grouted and the excavation can be advanced immediately wedging is complete. A further saving of time and cost is the complete elimination of bolts, both in cross joints and circumferential joints. These time savings are favourable to the use of digger shields because excavation is interrupted for a smaller proportion of the cycle.

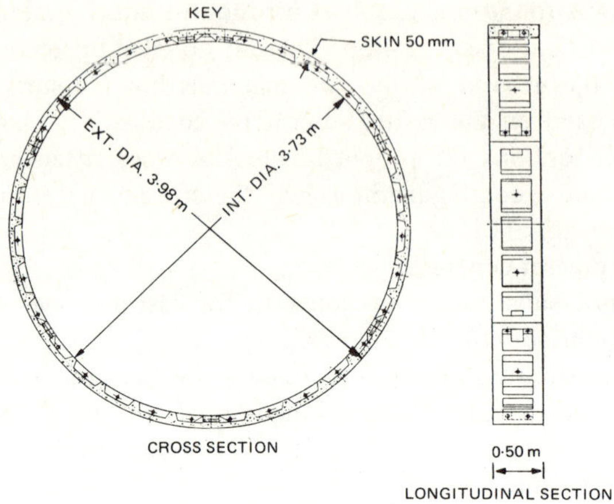

Fig. 8.4(a) – Precast concrete segmental lining. Original London Transport pattern.

Fig. 8.4(b) – Precast concrete segmental lining as used in Toronto subway.

Fig. 8.4(c) – Concrete mould for segmental lining (Charcon Ltd.).

Large scale applications of the principle were in the Potter's Bar railway tunnels (1959) and the Don Seg system used for a 27 km water tunnel across London (1959) and flexible cast-iron and concrete segmental linings designed for the Victoria Line (1968) and subsequent London metro tunnels. The Don Seg system uses 10 identical solid concrete segments in a ring, each tapered longitudinally but alternated in reverse directions. Very accurate casting is necessary. The rings are compressed against the clay by using the shield jacks to shove the segments into alignment against the taper. It provides a smooth bore lining. The Wedge Block system developed from this employs only a single longitudinally tapered key block, jacked in at the crown. In the Victoria Line concrete rings, jacks low on each side were used to stress the rings and the gaps were packed and wedged. Cast-iron rings embodying the same principles were used for the central section of the line, as providing a greater margin of strength and more flexibility in the thinner segment to adapt to uneven loading and distortion. Experiments in the use of S.G. iron made clear its advantages, but material costs for any form of cast-iron can only compete against concrete in special circumstances.

The difficulty of waterproofing any of these flexible systems is such that they are not normally suitable for water-bearing ground. Figs. 8.5(a), (b), (c), (d).

8.6.6 Grouted smooth bore lining

The requirement for a smooth bore for water or sewer tunnels and some others can be met directly by various systems of segmental concrete. Don Seg and Wedge Block have already been referred to, but they are not intended for grouting. Others are designed to be built on a former ring leaving an external space of about 25 mm to be grouted as with flanged and bolted segments. The Mini-tunnel is a special system for driving sewers of about 1 m diameter, a ring of three segments being erected behind a special shield, and grouted in. A concrete segmental lining was designed for Phase II of the Channel Tunnel (1974) which was erected immediately behind the shield and wedged back to the ground by a key segment. Each segment bore on the ground locally by four projecting pads, leaving space for grouting as found appropriate to the strata encountered (see Fig. 10.2(d)).

8.7 BRICKWORK AND MASONRY

Neither brickwork nor masonry is now a normal material for the structural lining of a tunnel. Most early British canal and railway tunnels were brick lined, the bricks often being manufactured on site. The whole 'English method' was based on alternating excavation and timbering with bricklaying and the necessary skills were highly developed; it would not be easy to revive them. Sewer tunnels were usually built in brickwork and hard blue brick was specified for the invert as the material best able to resist erosion by the scouring action of grit.

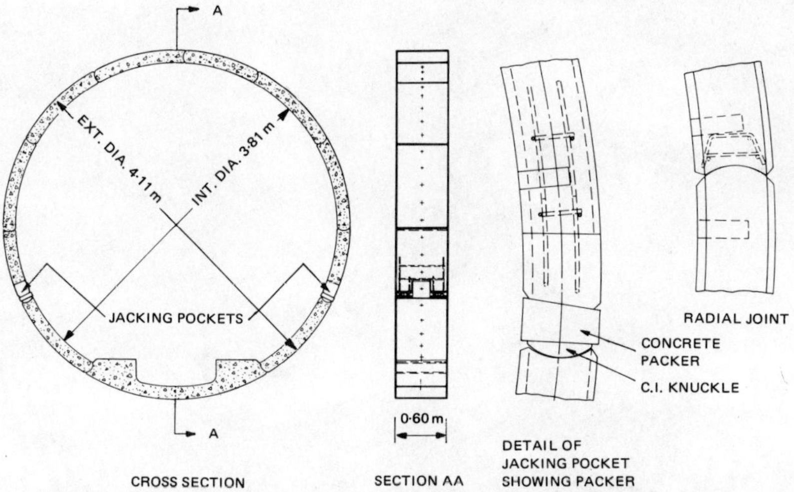

Fig. 8.5(a) – Expanded flexible concrete lining as used on Victoria Line.

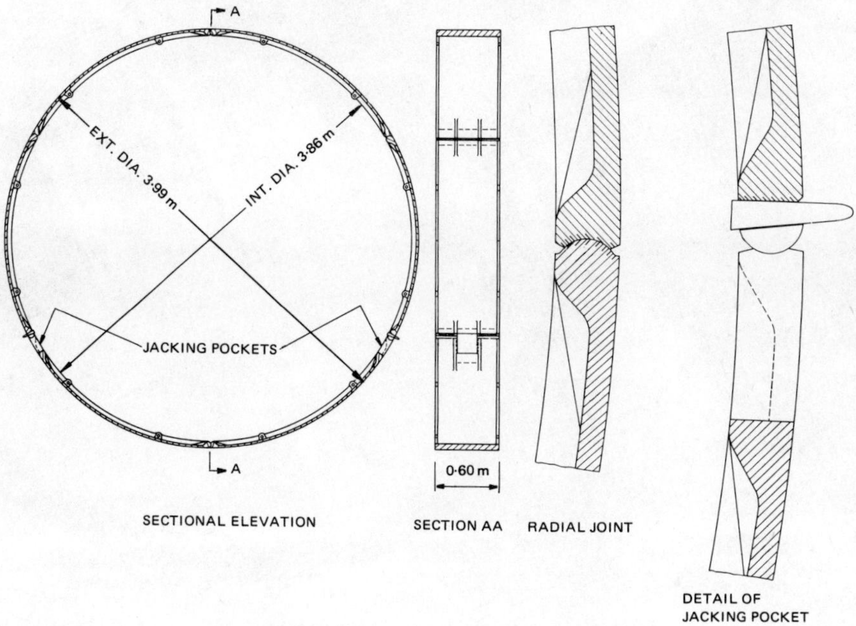

Fig. 8.5(b) – Expanded flexible cast iron lining as used on Victoria Line.

Fig. 8.5(c) – Completion of jacking in cast iron.

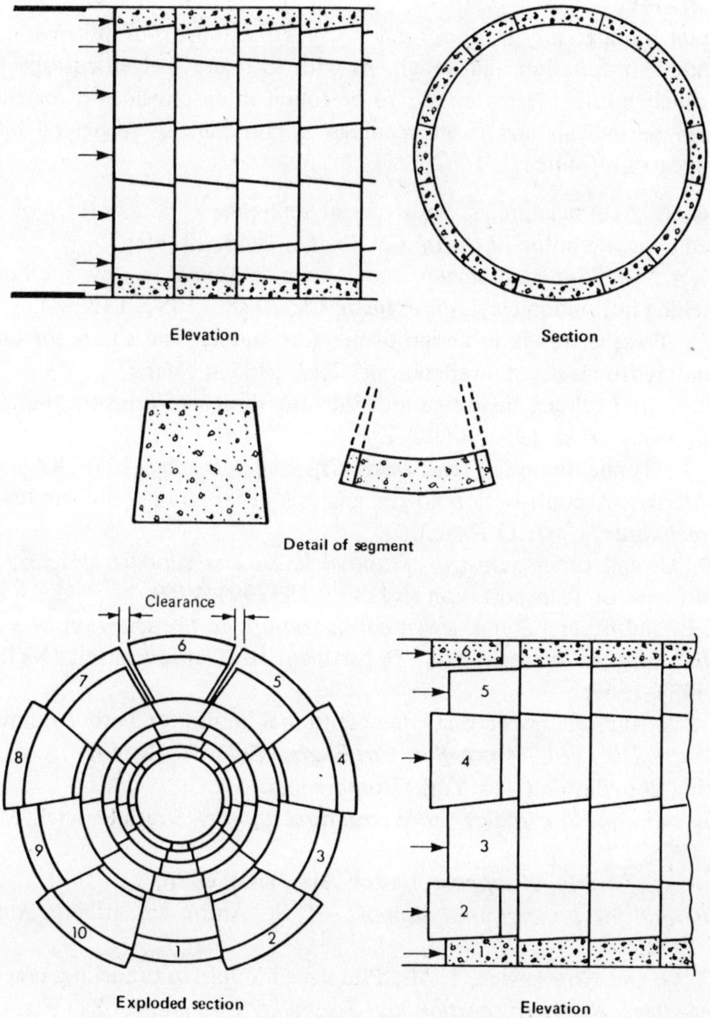

Fig. 8.5(d) – Don Seg lining.

Masonry was less commonly used where local stone thought to be suitable was available.

Problems of repair of these linings in old railway and sewer tunnels in particular may become increasingly important and may range from re-pointing and local replacement of bricks to much more extensive rebuilding, or possibly opening out completely.

BIBLIOGRAPHY

This chapter bibliography is a brief guide to further sources of information on linings and their function and design. As with so many specialised aspects of tunnels, much invaluable material is to be found in descriptions of particular projects, in periodicals and in proceedings of conferences. A general bibliography appears in Volume 2.

Groves, G. L., 'Tunnel linings, with special reference to a new form of reinforced concrete lining', *J. Instn. Civ. Engrs.*, 1943, **20** (Mar.).

Scott, P. A., 'A 75-inch diameter water main in tunnel: a new method of tunnelling in London clay', *Proc. Instn. Civ. Engrs.*, 1952, **1** (Nov.).

Jaeger, C., 'Present trends in design of pressure tunnels and shafts for underground hydro-electric power stations', *ibid*, 1955, **4** (Mar.).

Tattersall, F. and others, 'Investigations into the design of pressure tunnels in London clay', *ibid*, 1955, **4** (July).

Dawson, O., 'Tunnel linings', *Trans. Instn. Civ. Engrs. Ireland,* 1956, **83**.

Morgan, H. D., 'A contribution to the analysis of stress in a circular tunnel', *Geotechnique*, 1961, **11** (Mar.).

Deere, D. U. and others, *Design of tunnel liners and support systems,* U.S. Department of Transportation (NTIS PB 183799), 1969.

Peck, R. B. and others, *Some design considerations in the selection of underground support systems,* U.S. Department of Transportation (NTIS PB 190443), 1969.

Bartlett, J. V. and others, 'Precast concrete tunnel linings for Toronto subway', *J. Constr. Div., Proc. Amer. Soc. Civ. Engrs.,* 1971, **97** (Nov.).

Engineering Foundation (New York) Conferences:

Using shotcrete for underground structural support; South Berwick, Maine, 1973.

Shotcrete for ground support, Easton, Maryland, 1976.

Shotcrete for underground support III, St. Anton am Arlberg, Austria, 1978.

Tough, S. G. and Noskiewicz, T. M., 'Preformed linings in tunnelling practice', *Proceedings, Rapid Excavation and Tunneling Conference,* San Francisco, 1974.

Muir Wood, A. M., 'The circular tunnel in elastic ground', *Geotechnique*, 1975, **25** (Mar.). *See also* discussion, 1976, **26** (Mar.).

Lyons, A. C., 'Some developments in segmental tunnel linings designed in the United Kingdom', *Underground Space,* 1977, **1**, no. 3.

Craig, R. N. and Muir Wood, A. M., 'A review of tunnel lining practice in the United Kingdom', *Supplementary Report 335*, Transport and Road Research Laboratory, Crowthorne, 1978.

9

Survey and Control

9.1 SURVEY NETWORKS

Tunnel Survey, of which setting out is a part, comprises a number of phases and operations.

The initial stage is a topographic survey to provide a contoured map on which possible tunnel alignments can be laid out so that the best may be selected. This provides a network from which the tunnel alignment is set out on the ground and used for construction. The surface network must then be extended underground progressively as that tunnel advances, being used for detailed control of excavation at the face and lining. The underground network has to be repeatedly resurveyed and adjusted if accuracy in construction is to be maintained. There is usually a final operation of accurate survey of the completed tunnel.

The operations of survey and setting out are closely interdependent, particularly so in tunnelling, because lines cannot be carried ahead of excavation and survey stations in a tunnel are liable to be disturbed by ground movement as well as by construction operations.

There is inherently greater potential for precision in surveying than in setting out because the same stations can be surveyed separately and checked and resurveyed as often as necessary, whereas setting out is essentially the operation of fixing a station and marking it once and for all. Hence, apart from the danger of disturbance, there is an inherent limit to precision in setting out.

In tunnelling the whole surface network must be sufficiently established to include and relate both ends of the tunnel before work can commence, but the corresponding underground network can only be built up as excavation progresses and cannot be completed and finally checked until the tunnel breakthrough has been made. Essentially, the underground survey network depends in most cases on extension forwards from a relatively short base – indeed a very short base when access is by a vertical shaft – and extreme precision of angular measurement is therefore necessary.

The requirement of a high degree of accuracy is in conflict with the physical conditions in which the work must be done, particularly as regards the underground part of the survey in which conditions are often necessarily dirty and cramped and work has to be done under pressure of time, in addition to which the network also is usually constrained to a badly structured geometrical layout.

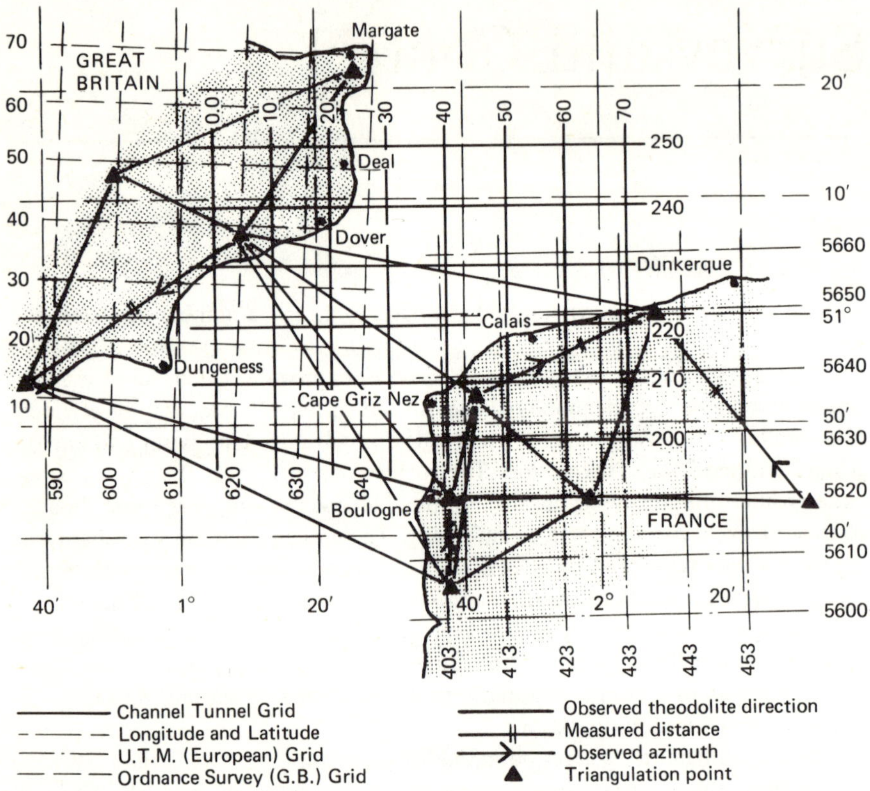

Fig. 9.1 – Channel Tunnel Survey. The figure shows the pattern of triangulation necessary in connecting the British and French survey grids, and in establishing a Channel Tunnel grid for use in the project.

9.1.1 Map Projections

A tunnel survey usually covers a sufficiently limited area for plane geometry to be applicable without reference to earth curvature, and is self contained. Correlation to a national or wider system may then be simply achieved by the incorporation of two or more common points in the network which may be utilised as necessary for transformation of coordinates.

For very long tunnels, such as the Channel tunnel, geodetic considerations arising from curvature of the earth's surface make the survey problems more complex. In that particular example further difficulties arise in the coordination of the differing national and international survey grids and also the long water crossing.

In long tunnels, such as those for water supply, aspects of curvature may be significant in control of levels and gradients, and also in convergence of meridians if true north is relevant to the survey. The essentially linear nature of such tunnels usually avoids the increase in complexity which would result from considerations of double curvature.

9.1.2 Accuracy

The tolerance acceptable in a completed tunnel will vary with its purpose but may be as little as ±40 mm for a metro tunnel in which a precast segmental lining is erected behind the advancing face. In water supply tunnels, obviously, local tolerances may be greater, but in sewer tunnels levels must be closely specified. In all cases, where a tunnel is driven from both ends to meet, accuracy of junction is a prime consideration and should normally be less than 75 mm. Where this cannot be attained with certainty, a long pilot heading may be required to effect a smooth junction.

The required angular precision of the survey lines in the tunnel will obviously depend on the length of drive from the nearest access shaft, which will be relatively short in metro tunnels, but may be several kilometres in water tunnels. For the Channel Tunnel 18 km drives from each side were contemplated. The desired survey accuracy may thus range from 1/20,000 to 1/200,000.

The skeleton of the main topographical survey requires to be of appreciably higher precision than the working lines. Accuracy of 1 in 20,000, which represents an angle of 10 sec. is acceptable for short drives, but for long drives higher accuracy is to be sought. 1 in 200,000 is an ideal figure with a one-second theodolite, but in practice 1 in 60,000 is unlikely to be bettered.

9.2 GROUND SURVEY

The surface survey may be based on triangualtion, traversing, trilateration or any combination. Triangulation is probably best in reasonably open country, but in city streets traversing is much more suitable. Trilateration has been very unusual except perhaps for local detail but modern EDM instruments may have advantages.

In urban areas roof top surveys may be helpful but tall buildings move with change of temperature and in strong winds, and additionally transfer down to ground level and below is difficult to effect accurately.

Detailing may be done manually from the survey lines, or may be provided by aerial photography where control points are identified, for which roof top stations have then considerable merits.

In busy city areas it is usually quite impracticable to survey at street level during normal working hours, and the work must be done in late evening, night and early morning, or at weekends. Night work has also advantages in instrumental accuracy when temperatures are steady, and changes from sunshine to shadow are avoided.

9.2.1 Layout of Survey

Although the prime function of the survey is to link the two ends of the tunnel with an accurate network, it should also be laid out to help in locating and mapping accurately relevant topographical detail.

In open country watercourses, points of reduced cover, and other relevant features will be required. In urban conditions, street frontages, cellars and indications of underground services are among the structures to be surveyed. The layout should be as simple and direct as is compatible with these requirements and with a sound geometrical network.

One important aspect of layout is that stations should be so sited that they can be preserved for re-use, either for checking or for extension of the network or further detailing.

9.2.2 Triangulation

In extensive surveys triangulation has been the most efficient method of providing an accurate network over large areas, because measurement of angles can be effected with precision over long lengths even where direct access between points is difficult, whereas distance measurement has hitherto been much more laborious and often impossible to effect with accuracy.

Electronic Distance Measurement (EDM) has changed the balance of advantage somewhat because direct access between stations is no longer required and network lengths can be measured wherever required.

The fundamental feature of triangulation is the measurement of all angles in a network of triangles and the calculation of all lengths, by methods of plane or spherical trigonometry from a principal base line laid out on the most favourable terrain available and measured with every precaution and refinement.

Base lines are usually laid out on ground as nearly level as possible and to be as long as practicable consistent with their ends being sited where they provide good sight lines to the other survey stations and 'well conditioned' triangles.

In the past they have usually been measured by use of 'Invar' steel bands and the methods remain useful.

The degree of precision necessary in any particular case will depend on the extent of the triangulation to be built up from it and the use to which it is to be put, but a brief outline of the precautions adopted may be helpful.

The base line is set out as an exact straight line in bays of about 30 m and level differences between the ends of each bay are ascertained precisely. The 'Invar' band is suspended as a caternary under known constant tension by use of weights and pulleys. Both ends are read using vernier or microscope as appropriate. Measurements are repeated as often as necessary to ensure the required precision. In the most important geodetic work further precautions and refinements will be necessary.

The alternative of EDM is now available and makes possible checks on length as often as necessary in the triangulation, avoiding the complete dependence on a single base length. Difference of level between the ends must again be ascertained. Offset stations may be established at either end of the base line to facilitate simultaneous reciprocal readings by use of two sets of instruments which can then be exchanged end for end. Air temperatures, wet and dry bulb, and air pressure should be recorded at the beginning and end of each set of measurements to provide appropriate allowances for refraction and frequency modulation.

In either method there may be a need to make corrections for earth's curvature and height above sea level.

9.2.3 Levelling

The survey network is in fact three-dimensional but is separated practically as far as possible into horizontal plane geometry and vertical height.

Levelling is preferably reintegrated with the triangulation or traverse so that all stations are observed. All levelling should be done in closed circuits, which are less laborious when automatic levels of acceptable accuracy are available.

Movement of level points is a particular problem in tunnel work as 'draw down' is likely to affect a considerable area. More especially where a tunnel follows a street line, it may be impossible to arrange for all points to be outside the area of potential settlement.

Subsidence must be considered probable, its magnitude depending on geological factors and construction methods. It may be important to observe its rate of development, for which levelling techniques of considerable precision will be required.

9.2.3.1 *Bench Marks*

For every project, however small, reference bench marks are essential, well outside the zone of influence of tunnel excavation, and founded in a stratum unlikely to be affected by the tunnelling or by its influence on the water table.

For larger projects, especially if of long duration, the provision of an adequate network of 'base' bench marks is appropriate. They may have to be founded at considerable depth in sound rock, being constructed of reinforced concrete, or a steel rod or pipe keyed into the rock and insulated from shear by a surrounding envelope of coarse sand. At the top will be a level pin of non-corrosive metal with a protective cover.

For a long tunnel a series of such 'base' bench marks will be of great value throughout construction.

They should be precisely levelled in relation to each other initially and at regular intervals, and any one can then be used locally for accurate measurement.

Circuit closures should be attainable within 3 mm over 1 km, the accuracy over longer circuits being proportional to the square root of the number of sights.

9.2.3.2 *Monitoring of Subsidence*

For structures, levelling points should be attached to columns, beams or foundations either as brackets securely bolted on or as pins or bolts firmly grouted in.

For ground movement a substantial block of concrete founded at a depth on a metre or more should carry a reference levelling point near the surface protected by a cover. Kerbs and similar ground level features are unsuitable for precise long term use as being liable to damage and to local disturbance.

9.2.3.3 *Precise Levelling Instruments*

The older types of precise level are capable of high accuracy but are more time consuming and more easily affected by temperature change and vibration from wind and traffic than are modern automatic instruments; in sunny conditions shading by an umbrella may be desirable but perhaps impracticable.

Most instrument manufacturers offer precise automatic levels. A nominal accuracy of 2 mm per km is appropriate for tunnel work.

Precise levelling staves having a body of metal or wood incorporate an Invar calibrated strip and have two reading scales with a fixed difference between them so that readings at difficult points on one scale are automatically checked. For 'change points' in a levelling run a solidly based point is essential. Suitable equipment is available.

All instruments should be checked regularly for damage or maladjustment: levels for collimation in particular, and staves for cleanliness of seating and for correctness of bubble level. An annual check by instrument technicians should be arranged.

9.2.3.4 *Procedure*

Normal good practice in levelling should be followed and for precision work particular attention should be given to a number of points.

Backsights and foresight should be equal and of moderate length, not generally exceeding about 35 m, to avoid substantial refraction effects and to allow scales to be read more quickly and accurately. Of course, when several points other than backsight and foresight are read from one station the sight lengths for these intermediate points may be unequal.

Both scales on the staff should be read and the difference checked within ±0.15 mm. Although two or more staves may be in use to expedite the taking of a series of levels, care should be taken to ensure that the same staff is used on all bench marks and in closing the run so as to eliminate errors of zero positions on the staves.

Where it is impractical to close the circuit in levelling a series of monitoring points the standard deviation of the scale differences (variance from staff constant) for all staff readings may be calculated as a check and should not exceed ±0.07 mm.

Precise levelling demands care and concentration and is time-consuming. Three assistants to the surveyor namely, an assistant to book the readings and two staff men, can contribute to speed and accuracy and security of equipment; if various levels additional to backsights and foresights are required, an extra staff man can prove helpful in speeding up the work.

The procedure for a record of readings should be: backsight on left scale, foresight on left scale, foresight on right scale, backsight on right scale.

Continuous sequence of observations should not exceed four to five hours even in favourable conditions if errors are to bc minimised. Ideal observing conditions are on a cool, overcast, calm day and at a time free from traffic and other vibration.

9.2.3.5 *Records of Subsidence*

Frequent monitoring of subsidence by precise levelling is advisable at intervals of time governed by the rate of subsidence. A graph of subsidence against time for each point is the best record, supplemented by longitudinal and cross sections. If an extensive area is involved a contoured plan of subsidence is likely to be useful. The initial pattern of points to be monitored should be planned with such records in view.

9.3 PHOTOGRAMMETRIC SURVEY AND MAPPING

Photogrammetry is the preparation of accurate maps from aerial or other photographs.

It is extremely useful for making the necessary detailed topographic surveys for tunnel works both in urban and rural areas, and, where the works are extensive, is likely to be more economical than equivalent ground survey, which cannot, however, be wholly eliminated as it is essential in the setting out and underground network.

Photogrammetry necessitates the use of elaborate and expensive equipment for photography and for transformation of the photographs to precise scales, and must be carried out by specialist organisations.

9.3.1 Photographic Equipment

The cameras must produce images of low distortion and uniform scale, for which stable focal length and a truly plane film are essential.

The intervals between exposures must be so controlled during flight as to provide ample overlap between successive pictures. 60% overlap in the direction of flight and 40% side lap are usual. An intervalometer is used to assist the navigation the timing being dependent on height, ground speed, and angle covered by the lens. Film sizes up to 300 × 300 mm are available. Stereoscopic pairs of photographs taken from the overlaps allow contour plans or three-dimensional models to be made.

9.3.2 Control

Although normally the camera is nominally vertical, errors are quite inevitable and a ground control network is necessary in order to orient and scale the photographs accurately. It is obviously desirable that the ground control points should be a part of the ground survey with which they must ultimately be integrated. Four or five points should be established so as to be clearly identifiable in the aerial photographs and of known co-ordinates and elevation.

Where modern digital design methods are to be used permanent control points rigorously corrected and precisely identifiable are essential.

Aerial triangulation is sometimes employed but rather for extensive topographic mapping than for tunnel survey. Photo co-ordinates are transferred progressively along overlapping pictures from one controlled base to another. Correction and adjustment of closing errors is likely to be necessary.

9.3.3 Mapping

The fundamental problem in mapping from photographs is the transformation of the image produced on the emulsion to an orthogonal projection on a horizontal plane. When working from a single photograph differences in ground levels introduce uncertainties into the mapping. Nevertheless a controlled mosaic using portions of photographs individually rectified and enlarged to the scale of known ground distances can produce a well scaled plan.

For most purposes requiring accuracy stereometric methods are used by employment of overlapping photography from which three-dimensional co-ordinates can be obtained. Fig. 9.2. There are two basic methods: optical or mechanical projection from glass based diapositives.

Optical projection is by use of lenses, similar to the camera lenses, projecting the image on to special tracing platens, whereas mechanical projection uses rods, pivoting at a point representing the projection centre of the camera to define the light rays. The latter system is more readily adaptable to varying focal length, different photographic formats, varied plotting scales and digitising of co-ordinates.

Where the requirement is simply to update accurate maps already existing, or to fill in topographic detail, simpler plotting instruments can be used.

9.3.4 Accuracy

The accuracy of the finished plan will obviously depend on the photographic equipment, the flying height and techniques and the methods of projection.

Rectified photographs alone are of very limited accuracy. Controlled mosaics are also limited, depending on control density and on elevation differences.

With more refined techniques and accurate plotters positional accuracy of points may be of the order of 0.012% of flying height for the range of 1000–2600 m. Height accuracy of points on the ground is of the order of 0.067%.

Fig. 9.2 – Air Survey for photogrammetry. The two photographs from a line of survey are not successive, an intermediate one being omitted, but even so there is a small overlap. Successive pictures overlap by 60% providing for stereometric plotting.

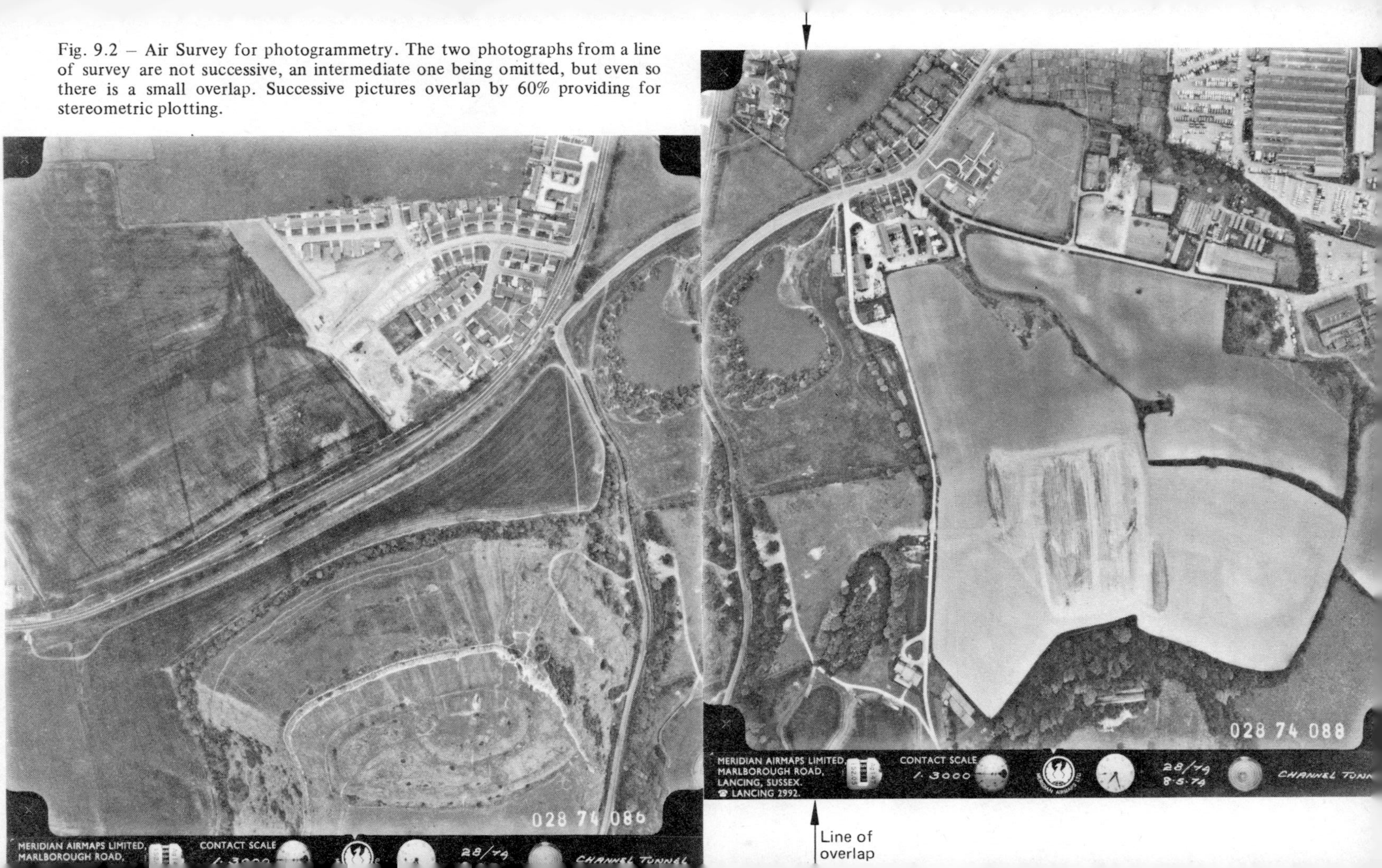

Terrestrial photogrammetry may be utilised for filling in detail on aerial mapping, for underground work, and for detail of near vertical features. Cameras, using glass plates, are oriented by theodolite. They are available in pairs with parallel focal axes for direct stereoscopic photography. Terrestrial plotters are normally of the mechanical projector type, adapted to the short focal lengths and precisely known orientation.

9.3.5 Topographical Detail

Much of the detail required for planning a project and for acquisition of necessary land can be obtained from aerial survey with a ground control network, but it is likely that ground survey will be necessary for precise delineation of critical features, particularly in urban work. Offsetting by tape from surveyed lines remains a very useful method for such features as building frontages, although direct reading tachometers may be found more expeditious where a substantial volume of work is involved. A small section of a 1/500 plan in an urban area is reproduced (Fig. 9.3) to show the mapping required for a metro system.

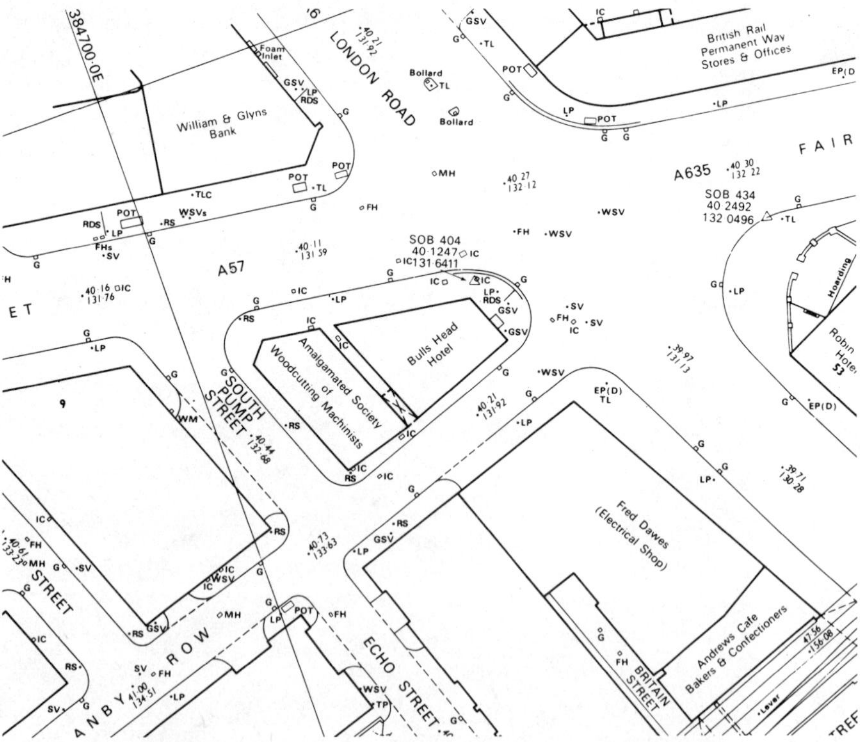

Fig. 9.3 – Mapping required for a metro system.

9.3.6 Automatic Plotting

Where three dimensional co-ordinates of any point are readily obtained computer directed plotters can be used for the automatic production of maps and associated sections.

9.4 BURIED SERVICES AND STRUCTURES

Records of the buried services of Public Utilities are rarely of the necessary precision; minor deviations during construction may be unrecorded, or positions may have been identified in relation to a kerb line and road level which have subsequently been altered. Manhole and valve positions help in location of services, and can usually be identified on aerial surveys, while detecting instruments may be helpful in tracing pipes and cable. Nevertheless, at critical sections, trial pits to expose the services and relate them precisely to the survey usually are necessary.

Few subsurface structures are connected by a reliable survey to an established grid system. Any tunnelling work in their vicinity usually requires a new survey, possibly with trial pits and borings. Even where the interior of a structure is accurately located, the actual thickness of a wall may be greater than supposed because of overbreak, and the lengths of piles may not be known. Other deviations from original plans may remain unrecorded.

Abandoned workings associated with mining or quarrying are rarely adequately recorded, but some help may be obtained from the local records of learned societies or other bodies (see Chapter 3).

It is worth repeating that all underground excavation whether temporary or permanent or overbreak should be recorded in a form which will be available for future reference. Mining regulations generally embody such a requirement.

9.5 SURVEY INSTRUMENTS

Theodolites and levels have undergone considerable design changes in recent years. For the same or higher accuracy they are more compact and simpler in use and are generally better protected against damage and intrusion of moisture and dirt.

Constrained centreing of theodolites, targets, electronic distance meters (E.D.M.'s) and optical plummets has been devised to permit accurate interchange, so as to minimise angular errors arising from errors of centreing over survey points.

9.5.1 Theodolites

For precise underground work on control surveys and deformation monitoring the 'one second' theodolite is most suitable. It reads to one second of arc and with suitable observing procedures the standard deviation of angles may be

±2 secs. for control surveys and less than one second from monitoring pillars at night.

Circle eccentricity is unlikely to be measurable in such modern instruments nor 'run' in the micrometer; they have 'double circle reading' so that opposite sides of the scale are read simultaneously.

For vertical angles most modern theodolites incorporate automatic vertical circle indexing, some with liquid compensators having no moving mechanical parts. This eliminates an annoying source of error where levelling of the vertical circle is omitted. Compensators are claimed to be accurate to one second. Adjustment errors are eliminated by observing in both telescope positions.

For underground survey a theodolite should be equipped with lighting of adjustable intensity for the circle and graticule. Right angle extensions to eyepiece and micrometer will be required for steep sights and also a striding level to ensure greater precision transversely. Other useful equipment includes optical micrometers, illuminated targets and special short range targets, and trivets for mounting the instrument on pillars or wall brackets. A top centreing mark for setting the instrument under a station fixed in the roof may also be requisite.

If setting out only, and not precision survey, is required simpler 10 second or 20 second theodolites may be adequate. Useful features are: additional anticlockwise circle divisions, self reducing tacheometer graticules, automatic vertical circle indexing and, again, a top centreing mark.

For very precise monitoring of horizontal movement there may be a use for 'geodetic' theodolites reading to 0.5 seconds but it is rarely that the one second instrument is not adequate.

9.5.2 Levels

Automatic levels have largely supplemented the tilting type for underground use. Speed, convenience and accuracy are markedly improved. There is no need to illuminate a level bubble and no danger of misreading it when poorly illuminated. Compensators are very reliable but occasional testing is advisable, and it is prudent to check against sticking by light tapping of the instrument.

A graticule without stadia hairs is to be preferred to eliminate misreading in poor light.

Staves can usually be of the reflective type for which lighting by reflection from a miner's cap lamp is adequate at usual ranges. A staff with 'even' ends, whose lengths is an exact number of metres, is of advantage when taking levels in crown and invert of a tunnel, because it need not be inverted for soffit readings.

The risk of misreading scale fractions because of reversal is eliminated, but of course the appropriate correction in whole metres must be made.

Observing techniques with precise levels have been referred to above. A recent automatic precise level has been designed to overcome the necessity for equality of sight lengths.

9.5.3 Lasers

There are two particular uses of lasers in tunnel survey.

That of most general application is to provide a visible guide line for control of excavation at the face. The older systems involved repeated fixing of sight lines using survey stations, in the crown of the tunnel from which plumb bobs were suspended so that survey lines might be projected forward to the face. Having established a point at the face it was necessary to make adjustments for any curves and to fix the axis level, then to determine the desired tunnel perimeter, the whole operation having to be fitted into the tunnel construction cycle.

The laser beam, set up and aligned by conventional theodolite techniques, provides a clearly identifiable line, and a recognisable spot continuously projected on to the tunnel face from an unmanned instrument, with obvious advantages in time saving, convenience and skilled manpower. The useful operational length is limited by natural dispersion in the tunnel atmosphere and of course by any requirement for driving on a curve. Necessary precautions include the use of intermediate screen to cut off the beam if the source is inadvertently deflected and regular checks by a surveyor and readjustment for curves.

A second application is to provide increased range as a carrier for electronic distance measurement.

The usual type in tunnels is the Helium-Neon gas laser. It is of low efficiency (0.1%), but has the advantages of running continuously without elaborate cooling devices, and of retaining spectral purity. The hazards to vision of direct observation are not great because of the low power of the beam; nevertheless in tunnel conditions, where general illumination is often poor and where consequently the pupil of the eye is dilated, laser beams should be sited to minimise the chance of direct impact on the eye of anyone in the tunnel.

9.5.4 Distance Measurement

The calibrated steel band, requiring calculation of corrections for level, temperature, tension and sag remains an indispensible tool for much tunnel survey. Banding is probably the most accurate nethod for lengths up to 100 m or so, and when surveying urban detail may also be the basis for offset measurements.

For large projects requiring precise work banding has given way to the use of electronic distance meters (EDM), which within limits of range measure lengths in a single operation.

There are savings of time and reduced opportunities of error compared with the multiple operations and calculations of corrections in banding. There is also the great convenience of measuring across difficult or inaccessible ground, as across water courses, railways, highways or precipitous ground or direct to elevated survey stations or within tunnels. Where all sides in a triangulation or traverse can be accurately measured precision in adjustment of the network is generally improved.

The instruments are of course much more expensive than steel bands but in an extensive survey there can be a very great saving in the cost and time of field-work. Experience in use of the instruments is necessary and knowledge of instrumental errors. There are in EDM work three principal sources of potential error, namely:

(1) levelling and plumbing of instrument and target over survey stations
(2) 'index', a correction particular to the instruments used
(3) 'cyclic error', which is inherent in the electronic resolution of the measuring frequency.

Regular calibration against an established base length is of great importance.

9.5.5 Recent Developments in Instruments

The recent advances have mainly been in distance measurement. The 'Mekometer' developed by the National Physical Laboratory has an accuracy comparable with that of an Invar band for ranges up to 3 km, the accuracy being; ±0.2 mm constant ± length × 10^{-6}.

Electronic tachemometers such as the Zeiss-Oberkochen Reg Elta 14 with an infra red carrier beam or the Aga Geodimeter 710 with a laser carrier beam dispense with scale circles and employ electronically coded angle measurement systems. They allow data to be observed and directly recorded on punched tape for computer processing. Observed or reduced measurements are also displayed. Accuracy in length measurement is about ±10 mm, and in angular measurement ±2 secs. A number of manufacturers are now in process of developing and marketing such instruments. They are costly, but on suitable projects are potentially very valuable. Figs. 9.4(a) and 9.4(b).

Conventional theodolites may be equipped with auxiliary EDM's such as the Wild D13 incorporating a slope distance reduction computer or the Kern DM 500, fitting round the telescope and powered by the instrument lighting circuit. Both systems are 'short range' (about 2 km) and use infra red carrier beams. Accuracy in distance is about ±6 mm.

Long range (up to 30 km) microwave EDM's such as the CA1000 Tellmometer have been greatly reduced in size, and in weight to about 3½ kg with an accuracy of ±15 mm ± length × 5 × 10^{-6}. They have the added advantage of providing radio communication. In tunnels, reflection from solids of the microwaves may preclude their use.

Gyro theodolites have been developed, which can be of particular assistance in carrying underground accurate orientation, (further referred to below). They comprise essentially a theodolite in conjunction with a suspended gyro. Experience in their use, and an understanding of the variables affecting accuracy are necessary if they are to be employed. Base line checks should be incorporated in all observations to minimise the variations that may occur in the spin-up of the gyro and in the so called 'tape constant' as well as possible diurnal variations. The instruments are sensitive to ground vibration which should be eliminated as far as possible during observations.

Fig. 9.4(a) – N.P.L. Mekometer in use in tunnel. Note plumb bob from survey station above, and support bracket secured to tunnel segments.

Fig. 9.4(b) – Kern Mekometer ME3000 (Kern, per Survey & General Instrument Co. Ltd.).

Canadian Geodetic Survey methods are claimed to achieve bearings with a standard deviation of ±2 secs by observation of oscillations when the instrument is clamped near North.

Even in local survey convergence of meridians may become significant if high precision is sought. The gyro-theodolite identifies true astronomical north for which the convergence at 45° latitude (λ) is about 30 secs per km of E–W distance, and is therefore significant in accurate survey work, and increasingly so, in proportion to tan λ at higher latitudes. At very high latitudes, exceeding about 80° the gyro-theodolite ceases to be useful as its axis is too nearly coincidental with that of the earth.

9.5.6 Tape Recorders

Recording of survey observations can be speeded up by use of miniature portable cassette recorders, which may save the services of a booker. Accuracy should be high because readings are spoken while viewing the scale. Relevant descriptive information is easily added at the same time, and the whole transcription time can be kept within usable limits by means of a microphone switch. The method is of particular value in reading repetitive data in such operations as precise levelling, stadia tacheometry and EDM tacheometry.

9.5.7 Calculators and Computers

The development of pocket and desk calculators has contributed greatly to survey work by providing flexibility in methods and allowing every sort of numerical calculation to be done quickly and accurately in the field and allowing immediate checks on complex observations.

Computers with their capacity for handling and analysing vast volumes of data make possible much more accurate adjustment of survey networks and can be programmed to provide without delay all necessary information for setting out and optimising curves within a tunnel.

9.6 SETTING OUT AND CONTROL

In short tunnels the centre line or an offset line may sometimes be established on the surface, but in general the starting points for tunnelling, whether through adits or from shafts, cannot be aligned directly but must be connected through a coordinate grid established in the survey. From permanent stations of the network a reference base line is established at each working site and this base line is then used to transfer bearings and distances into the tunnel workings.

The stations of the reference base line should be substantial and stable, sited where not liable to disturbance by excavation, or by traffic. It is not always possible to ensure absolute stability, in which case frequent checking from the survey network will be necessary.

If the tunnel is driven from an adit or from the portal survey lines can be extended into the tunnel and underground survey stations can be established either in the roof of the tunnel using plumb bobs to sight on and for centreing the theodolite or on brackets at the sides, or more rarely in the tunnel invert, where interference from traffic and from spillage is probable.

9.6.1 Shaft plumbing

If work is initiated from a vertical shaft the problem of transferring alignment accurately from surface to underground becomes difficult especially with a narrow and deep shaft. The simplest method is that of co-planing in which two wires carrying heavy suspended weights are hung down the shaft. They are set in exact alignment in the desired direction. Underground, a theodolite is set up and brought into exact alignment with the wires and used to fix permanent marks on the alignment. The positions of the wires are measured from the surface survey and their coordinates established, from which the whole coordinate system can be continued underground. The accuracy of the transfer, particularly as regards bearings, is restricted by the shortness of the base length between wires and by the difficulty of ensuring that the wires are at rest. It is essential for the surveyors to have unobstructed possession of the shaft for a considerable time. The method is obviously very direct, and can be used quite quickly for starting off a heading from the bottom of a shaft, but when the alignment must be extended for a long tunnel drive, the utmost care and refinement are necessary if bearings are to be maintained within a few seconds.

Another system of shaft plumbing is the 'Weisbach Method'. In this again two plumb wires are used, but, instead of seeking exact alignment between wires, surface station and underground station, closely approximate alignment only is required, with accurate measurement of the small angles and of the distances. Advantages of the system are that both surface and underground theodolite stations can be securely established in advance and that measurements can be repeated as often as thought appropriate and averaged. They can be checked at any time even with slightly different plumb wire positions.
Figure 9.5 shows the layout for the Weisbach method. A is the theodolite station and B & C the plumb wires. The distance, AB, should be as short as practicable and prefereably not greater than twice the distance between wires. The angle BAC should in no case exceed about 10 mins but it is better if it can be made as small as 30 secs.

Angles XAB, BAC are measured with the theodolite and distances AB, BC are taped. The angle BCA of the very flat triangle is very small and can be calculated by the sine rule, with the simplification that the angle and its sine are proportional, so that:

$$\text{Angle BCA} = \frac{AB}{BC} \times \text{Angle BAC}$$

The operations are repeated underground.
The points BC are then incorporated just as are any other stations.

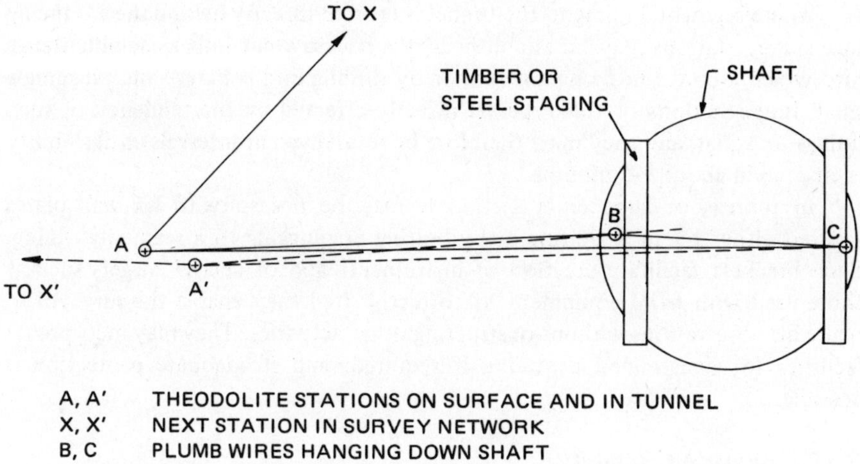

Fig. 9.5 – Shaft Plumbing, Weisbach method.

Of all the main survey operations it is probably shaft plumbing that is essentially least precise, and it should therefore be carried out with extreme care and repeated as an independent check.

9.6.1.1 *Gyro-theodolite*

It has already been observed that deep narrow shafts increase the difficulties of bearing transfer to underground work. The use of gyro-theodolites can therefore be advantageous, certainly at depth–particularly at depths exceeding 50 m. Accuracy of 20 secs. is readily achieved and can be improved by repeated observations, but it is essential that the instrument should be checked against a standard line before and after each set of observations.

9.6.2 Underground Survey

The essence of tunnel setting out is to provide accurate information at the working face of position and direction. The survey underground becomes an open-ended traverse in which disturbance of stations is always to be feared: it must therefore be the subject of repeated re-survey and correction as necessary.

The stations established should be simple and robust. They are exposed to particularly arduous service in drill and blast operations. For security and ease of use they are preferably fixed in the upper part of the tunnel.

In rock tunnels temporary stations may be secured to timber or steel supports, but these are likely to move under load progressively and it is preferable if possible to drill up into the rock soffit and insert a suitable plug from which a plumb line can be hung. The rock is not immune from movements and resurvey will be necessary. Where the station is being used directly to guide tunnelling it may be equipped with a graduated scale and movable cursor so that it can be adjusted exactly to give a predetermined line.

Where segmental lining to the tunnel is erected directly behind the advancing face survey stations may be established by a hacksaw cut into a metallic flange into which a cord can be hammered, or by drilling into concrete and plugging a small hole. Stations of this type are directly affected by the tendency of such linings to 'squat' and they must therefore be resurveyed at intervals until stability is ensured in about 3-6 months.

In tunnels of large cross section it may be necessary to fix wall plates secured either by rock bolting or by bolting or plugging to a segmental lining. Such brackets facilitate erection of instruments and of special targets such as those used with EDM equipment. If correctly sited they enable the surveyor to make his observations without obstructing other activities. They may incorporate facilities for constrained centreing if required, and if adequate protection is possible.

9.6.2.1 *Permanent Stations*

In some cases permanent survey stations are required within the finished tunnel as for example in metro systems with very small clearances. In an unlined tunnel, or one with segmental lining immediately erected, the principal construction stations may suffice, particularly if they are in the soffit and near the centre line. Where the finished lining is *in situ* concrete a brass plug and identifying plate can be cast in and surveyed after movements of the concrete have finished.

Secondary stations for such purposes as rail track alignment may also be required. They are readily established from the permanent stations.

9.6.3 Control of Tunnel Drive

There are, broadly, three aspects of survey work within the tunnel with considerable overlap between them as the face advances.

First, the initial survey for setting out, which must keep pace with the advancing excavation, is often executed in difficult conditions. Lengths of sight lines will be limited by the tunnel atmosphere and by the tunnel geometry and the needs of face control. Stations and level marks may be required at intervals as close as 15 m or may be possible at 150 m. Almost always there are risks of movement and damage which may be mitigated by careful referencing of points.

The second phase is that of the primary control survey, necessary to strengthen and check the initial survey. Initial stations can be used, or new ones established to provide a better layout of the principal lines. Work should be done under the most favourable conditions practicable with minimum air currents, clear air, and uniformity of temperature to minimise horizontal and vertical refraction.

This survey must keep pace with the tunnel advance and ensure that the initial survey is not extended for more than 3 or 4 stations without check from the primary control. Much depends on the stability of the stations, but regular resurvey is always necessary if accuracy of bearing is to be maintained.

The third aspect is that of face control to ensure that excavation is regulated within the required limits. For hand tunnelling or drill and blast work control of line at the face is still frequently exercised by visual sighting of pairs of string lines, with similar methods for level. Laser lines are now being adopted widely as they provide continuous visual indication of line and level and, although they too must be subject to regular checking and adjustment, they release the surveyor for less routine tasks.

Where a shield or full face machine is employed, additional survey checks are necessary to ensure not only that the shield is on line but also that its aspect is correct. There are three measurements to be verified:

(1) 'Lead' – the horizontal angle of the face of the shield, measured as the distance by which one side of the shield leads the other in relation to 'square marks' set out across the tunnel at right angles to the axis of advance, which may of course be on a curve.

(2) 'Look-up' or 'overhang' – the vertical angle of the face of the shield, measured as the distance by which the invert leads, or lags behind, the crown. It may be measured from a plumb line registering on a circular scale.

(3) 'Roll' – the rotation of the shield about the axis of the tunnel, measured by spirit level or plumb bob.

Where circular segmental linings are being built, whether or not with a shield, the leading ring needs to be surveyed for the same three measurements and in addition for shape as a true circle, or other specified cross section, and for the plane of the face. Unless these are controlled within close limits assembly of segments becomes progressively more difficult and the shape can be seriously distorted.

Increasing mechanisation of operations at the face makes access for survey lines and measurements more restricted and suitable sight lines hard to secure. This is more particularly the case with tunnel boring machines, yet their faster advance demands stricter control if large deviations are to be avoided. Laser beams projected to a screen directly visible to the machine operator can display to him visually the exact position of the machine. With added instrumentation it can also indicate lead, look-up and roll.

Errors of position cannot be corrected abruptly and the machine must be steered to bring it back on course when a deviation is observed. Predictors have been devised and fitted to indicate continuously a point a few metres ahead towards which the shield should be aimed so as to provide the optimum correction. In their simplest form these can be little more than two graticules through which the beam passes, fixed to the machine and separated as widely as practicable; the distance of the beam off centre on each graticule indicates errors of line and direction.

More elaborate devices may be necessary where machines are operating continuously and the laser beam can be split and carried round a more complicated light path to indicate the appropriate steering. It is worth noting that there are dangers of unwanted refraction in such light paths, caused by heating from motors and other equipment.

9.6.3.1 *Correction of Errors*

Careful and frequent checking at the face, or continuous control when a machine is in use, should prevent gross errors, but tunnels do go off line and corrective measures become necessary. Where the tunnel is initially unlined, with intermittent support only, local enlargement to clear tight spots is usually possible, but where the final lining is erected close behind the face special measures are necessary. With a continuous cast-iron or concrete segmental lining some adjustment is usually possible from one ring to the next by suitable packing or by use of special taper rings, (provided primarily for constructing curves), but it sometimes becomes necessary to make a fresh start with an independent ring and subsequently to dismantle and rebuild the badly aligned length.

9.6.4 Final Survey and Adjustment

Survey of the completed tunnel is usually necessary to check adherence to critical design requirements. This must be particularly exact in metro tunnels where clearances between rolling stock and lining must be maintained and where the track is laid to precise curves. The survey is to measure actual cross sections at appropriate intervals, which may be closely spaced. Each cross section is examined in relation to the specified centre line and clearances to identify any critical points requiring adjustment. It is usually possible to accommodate minor encroachments on the specified clearances by some adjustment of the track centre line and it is this modified centre line that is known as the 'wriggle diagram'. Specified limits of curvature and gradient must be maintained, but tangent points can be moved and small changes in curvature or gradient accepted. After development of the wriggle diagram there may remain tight points requiring cutting away or rebuilding, which can be a formidable problem in bad ground.

Various survey methods have been developed to speed up and simplify the survey as compared with the use of theodolite and level at every critical point.

They include:

Photogrammetry in which stereoscopic photography from precisely defined bases in the tunnel is used to locate in 3 dimensions any identifiable spot on the tunnel walls or lining. There are difficulties in ensuring suitable lighting and in uninterrupted access.

Graphical. Details of cross sections are prepared with the aid of a 'sunflower' device set up in precise relationship to the control survey. This is most suited to relatively small tunnels.

Geometric. In lined tunnels the specified geometry can be checked by a suitably designed template preferably in conjunction with EDM, the data being processed directly by computer.

9.6.4.1 *Adjustments*

Adjustment of Underground Surveys may be appropriate where there is redundancy within the network, or an external check such as a link to the surface survey through an intermediate shaft or adit, and ultimately when drives from opposite directions meet.

Internal redundancy of the underground network may be used most simply, and in many cases satisfactorily, to adjust coordinates by Bowditch's method, but the method of least squares is to be preferred if the correction is other than nominal and if there is access to a computer for the calculations.

Surface ties showing up an accumulated error in the underground survey may again be similarly used to recalculate and adjust its coordinates. The question then arises whether to adjust the tunnel excavation and lining to the corrected alignment, or to modify the design alignment to accommodate the tunnel as actually constructed. This is essentially the same problem as in the preparation of the 'wriggle diagram' and the choice can only be made after full consideration of the consequences for construction and future use.

BIBLIOGRAPHY

The list which follows refers first to the general field of tunnel survey, after which are noted references to some specialised aspects.

Peele, R. (ed.), *Mining Engineer's Handbook* (Section 18), J. Wiley, 3rd edn., 1941.

Jenkins, R. B. M., *Curve Surveying,* Cleaver-Hume, 1960.

Hurst, G., 'Survey and setting-out', Chapter 3 in C. A. Pequignot (ed.) *Tunnels and Tunnelling,* Hutchinson, 1963.

Wasserman, W., 'Underground survey procedures', *New Zealand Surveyor*, 1967 (Mar.).

Chrzanowski, A. and others, 'Underground survey measurements, research for progress', *Canadian Mining and Metallurgical Bulletin*, 1967 (June).

Clark, D., *Plane and Geodetic Surveying for Engineers,* Constable, 1972.

Peterson, E. W. and Frobenius, P., 'Tunnel Survey and tunneling machine control', *J. Surv. Map Divn., Proc. Amer. Soc. Civ. Engrs.,* 1973, **99** (Sept.).

Specialised references

Cooke, L. H., 'Underground orientation by exact and approximate alignments of plumb wires in one shaft', *Trans. Instn. Min. Metal.,* 1925, **34**, 391.

Shepherd, J. S., 'Precise azimuths from steep sights', *Bull. Instn. Min. Metal.*, No. 554, 1953 (Jan.).

Thomas, T. L., 'The correlation of local grid to national grid co-ordinates', *RICS Journal,* 1954 (Dec.).

Hallert, B., *Photogrammetry*, McGraw-Hill, 1960.

Thompson, M. M. (ed.), *Manual of Photogrammetry*, Amer. Soc. Photogrammetry, 3rd edn., 1966.

Davies, O. V., 'Lasers and distance measurement', *Surv. Rev.,* 1966 (Jan. and Apr.).

Thomas, T. L., 'The precision indicator of the Meridian, theory and application', *Proc. 3rd S. Afric. Nat. Surv. Conf.,* 1967.

Price, W. F., 'Lasers and laser safety', *Surv. Rev.,* 1974 (July).

Thomas T. L. and Asquith, D., 'The suspended gyroscope', *Chart. Surv., Land Hydrogr. Min. Q.*, 1975, **2**, no. 3 and 1976, **3**, no. 3.

10 The Channel Tunnel and Seikan Tunnel

The Channel Tunnel and Seikan Tunnel are subaqueous tunnels of so special a character as to deserve separate description and discussion.

Projects for the Channel Tunnel have a long history, stretching back to the beginning of the 19th century, but may at last be approaching realisation, although tunnelling has twice been started in earnest, only to be frustrated by political decisions, not by engineering obstacles.

The Seikan Tunnel is a comparable project to connect the main Japanese island of Honshu with the northern island of Hokkaido, passing under the Tsugaru Strait. It has been under construction now for over 10 years, but early completion of tunnelling is a reasonable hope.

The special feature of these tunnels is that they are very long crossings under the open sea, and they are therefore particularly vulnerable to the hazard of inundation at high pressure from a boundless reservoir, through undetected fissures or buried channels. Shorter undersea tunnels already built are those in the Chesapeake Bay bridge/tunnel crossing and the Japanese Kanmon tunnels. Other undersea tunnels contemplated are one between Denmark and Sweden, one under the Straits of Messina and one under the Straits of Gibraltar. The latter two are very ambitious major projects requiring novel technology; they have not been worked out in any such practical detail as have the tunnels here under discussion.

For the Channel Tunnel various options still remain open, but a railway tunnel has always seemed the most acceptable in terms of engineering and finance, and as causing no obstruction to shipping. For the Seikan Tunnel no alternatives to rail have appeared practicable, and the project under construction is for very high speed railway operation, combined with use by slower local services. The great advantage of rail as against highway is that ventilation requirements are manageable, and that the traffic is fully organised and controlled.

10.1 CHANNEL TUNNEL

Ideas for a fixed link across the 35 km wide channel between England and France, whether tunnel, bridge, causeway or some hybrid scheme, date back to the beginning of the 19th century. Many of the schemes were fantastic, others were too far ahead of current technology in their demands, and all called for heavy capital expenditure without absolute assurance of practicability or ultimate profitability. Railway development in both countries stimulated the promotion of schemes, and until the middle of the 20th century all but the most unrealistic proposals were for a rail tunnel.

Since then the growth of road transport and advances in technology have stimulated proposals for highway tunnels and bridges usually in combination with a rail crossing. The rail schemes have continued to be favoured as demanding smaller capital outlay and providing more certain returns.

10.1.1 Bridge and other projects

Bridge. Many bridge projects have been advanced but even if within proved engineering capacity they have generally been considered to present too great a hazard of collisions with shipping, both during construction and in service. The straits carry a greater volume of international shipping than any comparable waterway, with the added complexity of a great number of ferries cutting across the main channel passage. The establishment of a system of control, restricting shipping to defined safe lanes, is not likely to be quick or easy, particularly as the not infrequent conditions of storm or fog must be provided for. Some agreed rules are already in effect, and more precise control could no doubt be progressively effected. At best the dangers of faulty navigation and ships out of control would continue. The Chesapeake Bay bridge/tunnel crossing exemplifies some of the hazards to such bridges, even in enclosed waters under national control.

Additional disadvantages in service of a highway bridge are the exposure of vehicles to severe gales, and the cost and difficulty of satisfactory structural maintenance in such exposed and corrosive marine conditions.

Capital cost has usually appeared to be much higher than for a rail tunnel, but there would be advantages in independent traffic flow avoiding road/rail transfer delays. As compared with a tunnel the absence of any ventilation problem may be noted.

Highway tunnel. On considering the alternative of a highway tunnel as compared with a rail tunnel, a number of factors are very unfavourable to a roadway. Adequate ventilation of a highway tunnel to dilute and remove the noxious exhaust gases from internal combustion engines is a problem which increases unmanageably in a long tunnel carrying heavy traffic (see Volume 2). Artificial islands in the waterway might be an engineering possibility but would be very costly in construction and would be hazardous to shipping, perhaps also creating the risk of inundation of the tunnel. In such a long tunnel the hazards from vehicle breakdown and from inexperienced and largely untrained car drivers would be substantial. There would be no normal facilities for stopping ot relaxing over an enclosed distance exceeding 40 km, and the claustrophobic conditions might affect some drivers to a dangerous degree. The hazards arising from collision and fire are obviously much greater than with rail transport.

Bridge/tunnel. Schemes for composite bridge/tunnel construction relate primarily to highway use for which they are attractive in that:

1. they allow clear shipping lanes wider than any bridge span, and with no height limitation;
2. tunnel lengths are much shorter and offer some possibility of ventilation adequate for highway traffic.

They are unsuitable for railway operation because of the restriction of gradients to about 1 in 100: to emerge from a depth of as little as 20 m below water to a height of 50 m, plus a structure depth of 10 m, a length of 8000 m would be necessary, constituting a major barrier to shipping. Even for a highway gradient of 1 in 25 the corresponding length is 2000 m, but it could geometrically be laid out as a helix of 650 m diameter. Another obstacle to a railway bridge is the difficulty of accepting rail loads on a long, and necessarily flexible, suspension span.

Submerged tunnel. Schemes for prefabricated unit submerged tunnels have been proposed many times in the history of the project, but have never yet found favour in the ultimate comparison with a bored tunnel. The latest bridge/tunnel schemes incorporate such methods. Novel technology is needed, as briefly described in Volume 2.

10.1.2 Geology

The geology of the crossing at the favoured site for a bored tunnel is relatively straightforward. The presence of similar chalk cliffs on both sides of the straits suggested continuity beneath the sea, and this has been confirmed by numerous borings and surveys. The most favourable tunnelling stratum is a 65–80 m thick band of 'grey chalk', the Lower Chalk or Cenomanian, which is a chalk marl with substantial clay content contributing to plasticity and low permeability. It is easily excavated and self-supporting at a tunnel face for the necessary time. The pilot tunnels of 1882 remain sound. The Middle and Upper Chalk are whiter and harder but extensively fissured, and therefore water bearing. The Lower (grey) Chalk rests directly on the Gault clay, because the Upper Greensand is missing here. The Gault, again, is underlain by the Lower Greensand, typically water-bearing and highly permeable. In this area, all the strata dip fairly uniformly towards the north east, and therefore, in the absence of discontinuity, a tunnel can be planned to lie almost wholly within the very favourable stratum of the Lower Chalk, by aligning it towards the north east as it deepens and towards the south west as it rises to the surface.

Such a route has been adopted by all the most favoured proposals from the 1870's onwards, always with the proviso that more geological information was needed to determine precisely the best alignment and more especially to confirm the continuity of the stratum and the absence of fissures or buried valleys in the sea bed.

10.1.3 History to 1981

The first realistic proposal for a twin rail tunnel was advanced in 1856 by Thomé de Gamond and led to a series of projects and promotions.

Work on shafts and exploratory pilot tunnels was actually started on the French side at Sangatte in 1878 and on the English side near Dover in 1882, but military objections on the English side brought work to a standstill. Attempts to revive the project were rejected in England by Parliament in 1890, by the Committee of Imperial Defence in 1914, by Parliament again, by 7 votes, in 1930, always on fears of its potential for invasion. In 1957, there was an important joint initiative by the then Channel Tunnel Co., the Suez Canal Co., the French Société Concessionaire and the American Technical Studies Inc., who formed the Channel Tunnel Study Group. They organised and financed a comprehensive study covering traffic, geology, civil engineering and financial and legal aspects. They considered both rail and road crossings by bored tunnel, submerged tunnel, bridge and combined bridge/tunnel and reported strongly in favour of twin bored railway tunnels as appearing geologically feasible and being cheapest in construction cost and providing the quickest return. Following this, the two Governments concerned set up a working party of officials in November 1961 to report on this bored rail tunnel proposal, but also on a bridge proposal of a rival group. Their report in September 1963 again favoured strongly the tunnel proposal on grounds of feasibility, cost and economic return, with a proviso about further geological and other site investigations.

In 1964, the Governments announced their intention to proceed with the project. The Channel Tunnel Study Group were asked to organise further site investigations, which were carried out in 1964/1965 and included 73 marine boreholes, supplemented by geophysical profiling to provide fuller detail of the morphology and stratigraphy of the area, particularly on the proposed line of the bored tunnel. Data on winds, waves, tides and currents and transport of sediment were obtained, more particularly to assist in study of a submerged tunnel. A bored tunnel route within the Lower Chalk was proposed, conforming to a maximum gradient of 1%, a minimum gradient (for drainage) of 0.08% (1 in 1250), and a minimum radius of 4200 m.

In 1966, the Prime Ministers of the United Kingdom and France announced a decision to proceed 'subject to finding a solution for the construction work on mutually acceptable terms'. Three groups, in which financial interests had a major part, submitted proposals, but none was found acceptable. A new combined group was formed, which made revised proposals which were accepted early in 1972. This new Group organised two companies: yet another British Channel Tunnel Company, and the Société Francais du Tunnel sous la Manche.

Late in 1972 the British and French Governments concluded with the companies parallel agreements covering:

Phase I: for studies to be carried out immediately and completed by mid-1973.

Phase II: outline arrangements for shaft sinking and experimental tunnel boring up to 1975.

Phase III: for the major construction operations up to 1980, together with complex financial provisions in which the Governments undertook to guarantee up to 90% of the capital to be borrowed by the Group for the costs of construction.

In 1971–72 a further 17 marine boreholes and 8 land boreholes in the Folkestone area were sunk in coordination with another comprehensive geophysical survey of the route using much improved equipment. This allowed particular attention to be given to zones of weathering in the chalk. This geophysical work was very successful and gave greater confidence in the continuity between boreholes. No major structural fault zones were detected on the route and it was concluded that conditions were favourable to tunnelling; there was still no absolute certainty that no fissures or other discontinuities had been missed. The alignment of the tunnel was adjusted so as to pass the coast under the Shakespeare cliff rather than under Dover harbour, providing a working site well clear of the former heavily built up urban area. Fig. 10.1.

The Phase I programme was duly completed, with the tunnel design developed and brought up to date and with new and more detailed forecasts of traffic and revenue. A revised scheme was found necessary for the eventual sharing of receipts. A British Government White Paper, which sets out in considerable detail the case for the tunnel scheme, was endorsed by Parliament in November 1973, followed immediately by the initialling of a Treaty between the United Kingdom and France.

Phase II then started with the signature of Agreement No. 2. On the engineering side it included construction of access works on both sides and the driving of trial lengths of the permanent service tunnel, –2 km at the English coast and 1½ km at the French coast. It also included comprehensive and detailed studies for, and design of, the tunnels, the terminals, and the railway works and equipment and rolling stock, and also geodetic surveys to link the national grids. Thus, by May 1975 plans were to be brought up to the stage where contracts could be placed for the whole project, to comprise Phase III and to be covered by Agreement No. 3.

All was brought to a stop early in 1975 at a time when Phase II work was progressing satisfactorily and machine driving of the English service tunnel was just starting. The British Government decided not to proceed with ratification of the Treaty. Economies in spending were thought necessary, doubts were cast on the forecast rates of growth of traffic and particular emphasis was laid on the

Fig. 10.1 – Channel Tunnel – English working site. Shakespeare Cliff and Dover Harbour appear in the background. An access tunnel

greatly increased estimates of cost for the rail link to London. As had been the case nearly a century earlier, a very promising start with the engineering was literally submerged as a consequence of political and financial doubts and hesitations. The dependence of major projects on politics and finance may be unavoidable, but it certainly is very frustrating to engineers, particularly when work in progress is brought abruptly to a halt.

Not unexpectedly, renewed interest has been shown by those concerned, including an EEC Committee, in the subject of a cross Channel link. British and French railways have studied jointly a possible single rail track tunnel as being the least demanding on economic resources. In 1980, this and other schemes were submitted for consideration by the Transport Committee of the House of Commons. The proposals ranged from the single track bored tunnel to a high level road and rail bridge and included a comprehensive, and costly, scheme combining a railway in submerged tunnel with a highway part in tunnel and part on bridge, and with artificial islands.

The Parliamentary Committee in March 1981, reported in favour of a single track bored rail tunnel and service tunnel, as in the 1980 'economy' plan proposed by the railways and described below, but with a larger tunnel diameter which could at a later date be utilised to accommodate all ferry traffic, as in the 1975 plan.

10.1.4 Rail tunnel planning

It is worth examining the plans adopted for the 1973 scheme together with those for the minimal 1980 British Rail scheme, as exemplifying the principles discussed elsewhere in the book. Fig. 10.2(a).

In broad terms the 1973 plan was for a comprehensive rail ferry service combined with very high speed passenger trains between London and Paris and Brussels. The 1980 'economy' plan leaves the heavy vehicle ferry business to shipping and caters only for passengers, passenger cars, and normal railway freight including containers. Very high speed is no longer planned, and the initial proposal is that a single tunnel (plus a service tunnel) should take traffic alternately in each direction. The implications are explained below in more detail.

Alignment. The proposed alignments were determined by the geology as described above and by railway operational requirements. It was considered that the undersea tunnelling should be as near the base of the Lower Chalk as possible, avoiding by at least 10 m cover any weathered rock which had been identified in strata communicating with the sea bed.

The original scheme was intended to accommodate frequent long and heavy trains at speeds up to 200 km/hour, but in the revised scheme a normal working speed of 120 km/hour is contemplated with a possible increase to 160 km/hour.

Maximum gradient was to be about 1% and minimum for reasons of drainage 0.1%. Minimum horizontal radius is specified as 4200 m, originally determined

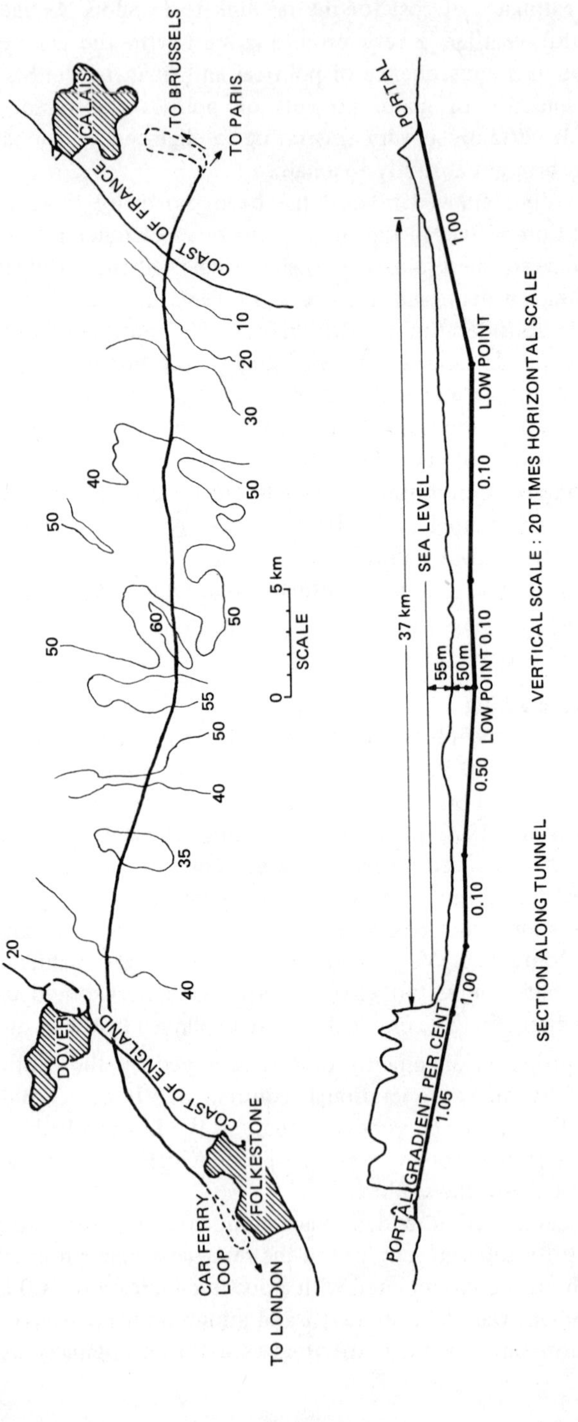

Fig. 10.2(a) – Channel Tunnel – plan and section showing alignment of 1973 scheme.

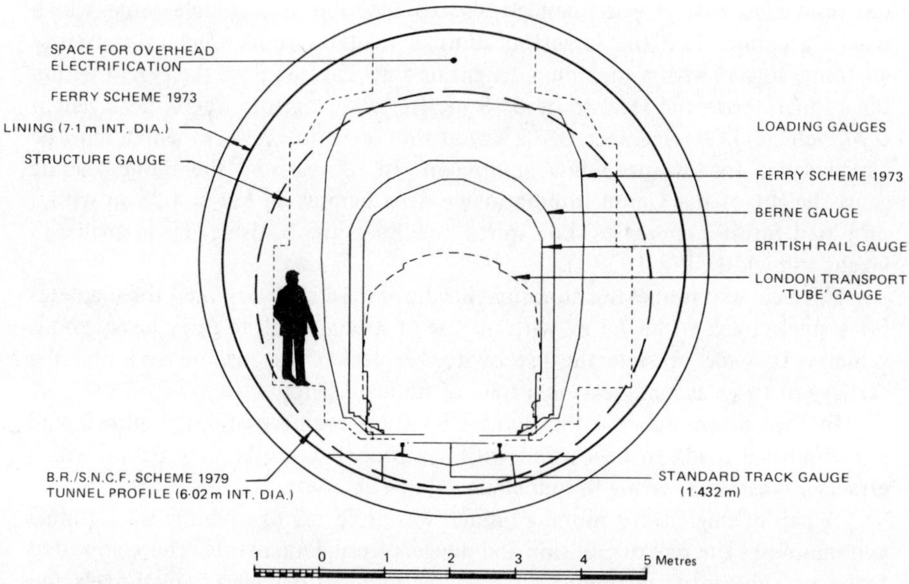

Fig. 10.2(b) – 7.1 m dia. circular lining accommodating standard loading gauges and proposed (1973) gauge together with 6.02 m circle proposed for 1980 scheme.

in relation to a design speed of 200 km/hour and with rail cant of 50 mm. The radius is retained but maximum cant is reduced to 40 mm.

The major problem of gradient is on the English side where the topography, with high coastal cliffs, fixed the rail level at portal as about +42 m, while the requirement for rock cover where the tunnel passes under the coast line imposes a rail level of −40 m. At a gradient of 1% the necessary length of descent is 8 km, which emphasises the importance to capital cost of adopting the steepest acceptable gradient. In fact, after careful study of train loads in relation to available locomotives, it was found possible originally to steepen the gradient to 1.03% and thus shorten the tunnel length by about ¼ km, and in the current proposals to steepen the gradient further to 1.11% with further substantial economy in construction cost.

On the French side, where the portal is on lower ground, it was found best to tunnel the descent under land, through the water bearing Middle and Upper Chalk, and to deal with the water by pumping and waterproofing.

The overall length between portals was established as 50 km, now reduced to 49.4 km.

Cross section. Special considerations applied to the selection of an appropriate cross section for the tunnel. It was very far from being comparable to an ordinary rail tunnel because it was intended also to function as a vehicle ferry, which would accommodate the largest of normal road transport vehicles–currently 40 tonne lorries with a maximum height of 4 m. This required the height within the tunnel above rail level to be 5.18 m after allowing for: wagon deck height 0.94, vehicle 4.00, clearance 0.10, wagon roof structure 0.14, to which must be added space for overhead line equipment. In comparison, the Berne loading gauge height of the Union Internationale des Chemins de Fer is 4.28 m with a proposed future increase to 4.65 m for new lines; the British gauge is smaller in height and width. Fig. 10.2(b).

For a circular tunnel this abnormal height of 5.18 m determined the diameter for a single track to be 7.1 m, with the use of special stock to carry heavy goods vehicles. It made possible the use of double decker wagons for cars, and the carriage of twice as many cars in a train of standard length.

In the current scheme the Berne UIC gauge has been adopted allowing all but abnormal loads to be carried within a tunnel diameter of 6.02 m, which effects a substantial saving in tunnelling volume and cost.

A pair of single-track running tunnels was preferred to a double-track tunnel as it minimises the risk of collision and dangers from derailments. There appeared to be no substantial difference in construction cost between twin tunnels and one double-track tunnel, in relation to which the comments in Chapter 4 on problems of scale may be referred to.

Studies have shown that a single one-track tunnel as the link between surface lines can be operated with a three-hour cycle, in which up to ten trains with a journey time of 35 minutes pass in one direction and clear the tunnel during a period of 1½ hours, and are succeeded by a similar flow in the reverse direction. Sixty trains a day in each direction can thus be handled, leaving a six-hour maintenance period.

In both schemes an additional service tunnel, of 4.5 m diameter, is to be driven. In operation, it gives access to the running tunnels for maintenance, and acts as a fresh air supply duct, and is also available for emergencies. During construction, it would be driven ahead of the main tunnel, functioning as a pilot, to prove the ground and make possible any necessary ground treatment. Fig. 10.2(c).

Loading loops. Another consequence of the abnormal loading gauge of the ferry trains was that they could not be used on track with normal structure gauge limits on either side of the Channel, and therefore that special terminal areas as close as possible to the portals became necessary for the loading and unloading of the ferry trains. Continuous loops of track were provided to handle the traffic without reversing the trains. Such elaborate provisions are unnecessary in the smaller scheme; some sidings would be required near the portals, and a simpler road/rail interchange wherever convenient to serve local traffic.

High speed rail links. Normal rail traffic–passenger, goods and container trains –could, of course, use the tunnel without difficulty. The connection to London

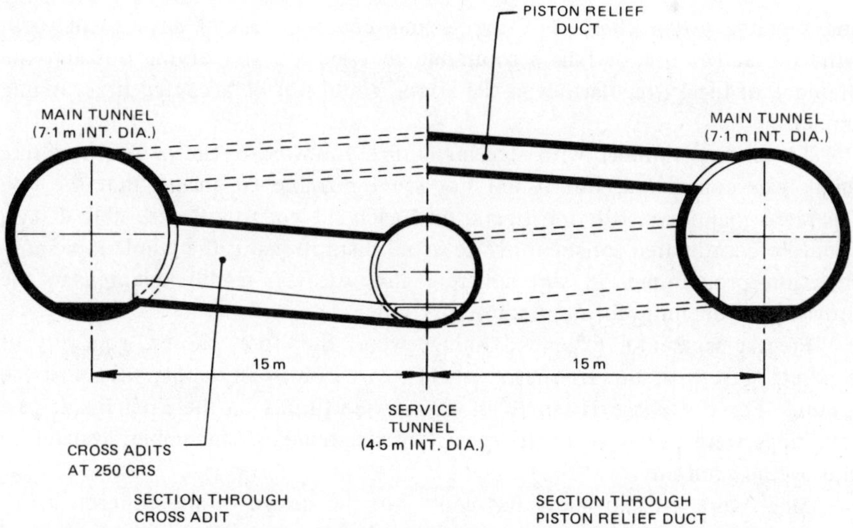

Fig. 10.2(c) – The figure shows the cross sectional arrangement of the running tunnels, service tunnel, cross adits and piston relief ducts adopted for the 1973 scheme. The 1980 proposal is for a single running tunnel of smaller diameter, but with a similar service tunnel and cross adits at 500 m centres.

is not built to the Berne gauge and could not handle the full volume of traffic anticipated, nor the proposed high speed intercity trains envisaged. Substantial new works were therefore contemplated, virtually duplicating much of the existing track and including new terminal facilities in London. It would also have involved further tunnelling, some orthodox hill tunnels and some to avoid urban congestion.

These extensive works have been abandoned in the current proposals, except for limited improvements of the existing track, and probably not giving the Berne gauge clearances.

Excavation. Machine tunnelling with a full face excavator was decided on as the best system for rapid and efficient excavation; the chalk could be relied on in general to stand unsupported for at least twenty-four hours, by which time the lining should be in place. Spoil disposal would probably be by rail, the surplus going to fill disused gravel pits; much material was required as fill in preparation of the terminal loading areas. Proposals for pumping out surplus chalk as a slurry into the sea bed received serious consideration and seemed likely to be advantageous. It would not be detrimental environmentally as the chalk would be rapidly and widely dispersed by swift and varying currents, and chalk in suspension is, in any case, a naturally occurring constituent in the straits.

Lining. A structural lining is essential to the safety in operation of such a tunnel. The headings of 1882 had stood up very well with minimal support and much exposed chalk, but the serious consequences of any fall of rock, with the added risk of falls attributable to vibration and drying out and the presence of local irregularities in the strata, could not be accepted in a running tunnel.

Temporary support with steel arch ribs, followed by an *in situ* concrete lining was considered, but it did not seem possible to ensure that the two separated major construction operations, each subject to unforeseeable delays, could be coordinated satisfactorily to avoid disruptions in the whole rapid construction programme. *In situ* concrete also was less readily adapted to the provision of openings for numerous cross adits.

Precast segmental expanded linings were, therefore, chosen, generally of concrete, but with the alternative of S.G. cast iron in any apparently unstable ground. For the Phase II length of the service tunnel on the French side cast iron rings were preferred in the special circumstances of timetable, availability and overall economy.

Much thought and study has gone into the design of the concrete rings, because even a small saving in manufacture, handling and erection is multiplied up to a very large economy in so long and uniform a tunnel. Finality had not been reached on the details of design in 1975, but new methods of calculation had been evolved, and tests have been made in the short lengths of service tunnel completed on the stress and deformations in ground and lining. These studies have allowed a substantial saving to be effected in the current scheme, whereby the thickness for the service tunnel segments has been reduced by about 0.1 m with corresponding savings in the running tunnel. Fig. 10.2(d).

Ventilation and aerodynamics. With electric traction the problem of fresh air supply to passengers does not usually call for any special provision, but in the Channel Tunnel the great length of tunnel without shafts and the high peak frequency of trains meant that many thousands of passengers might be in the tunnel and that the air at midway would not be readily replaced. The service tunnel was, therefore, utilised as a fresh air supply duct supplying the average requirement, while there were supplementary fans to supply peak and emergency requirements direct through the running tunnels. These fans might also be used in smoke control in case of fire.

The aerodynamic problems of rail tunnels are discussed in Volume 2. The particular problems of the Channel Tunnel were the long single tunnels, high speed working, and proportion of loading gauge to clear area. To provide relief from the pressure surges ahead of trains and to reduce aerodynamic drag, cross adits were to be constructed between the two running tunnels at 250 m spacing. They were separate from the cross passages connecting running tunnel and service tunnel because this gave a better environment for men and equipment in the service tunnel and left it available as the air supply duct.

With a lower operating speed in the current scheme, and with smaller

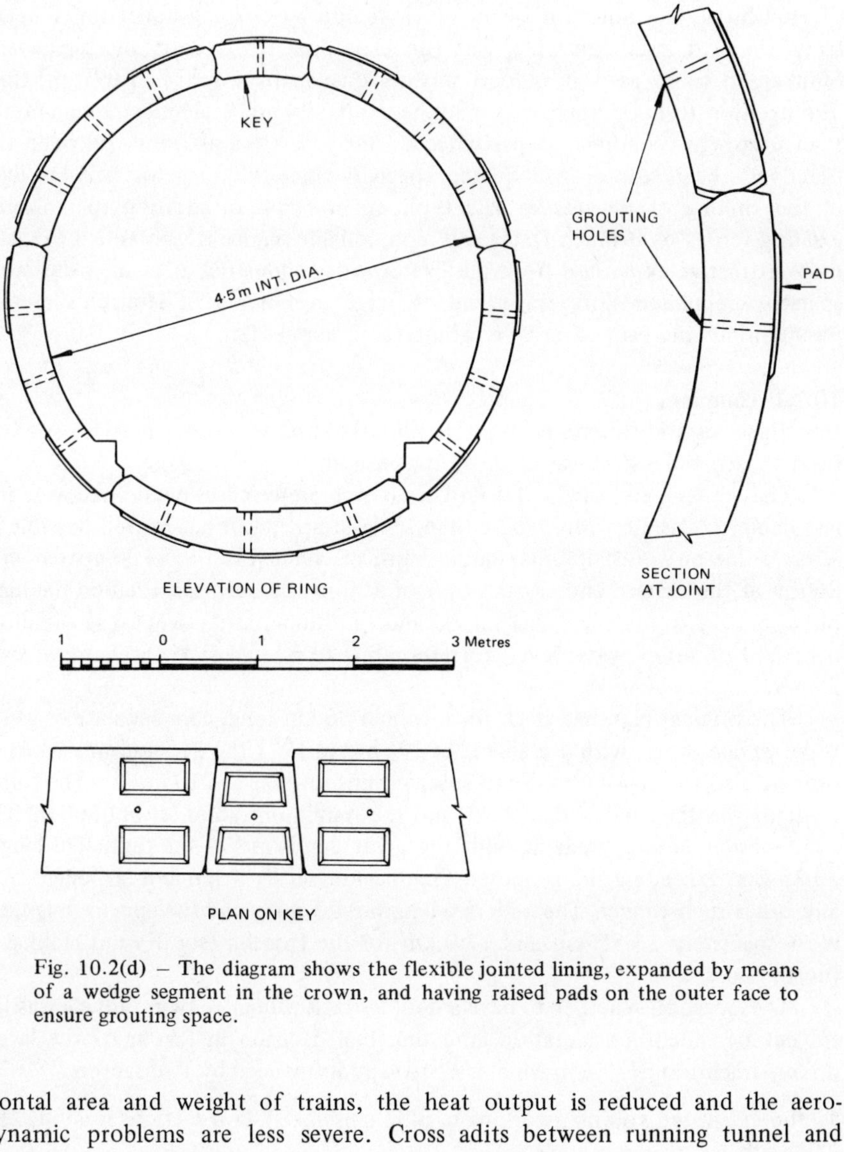

Fig. 10.2(d) – The diagram shows the flexible jointed lining, expanded by means of a wedge segment in the crown, and having raised pads on the outer face to ensure grouting space.

frontal area and weight of trains, the heat output is reduced and the aerodynamic problems are less severe. Cross adits between running tunnel and service tunnel are to be provided at a spacing of 500 m to serve for access and for ventilation and relief of aerodynamic surges.

10.2 SEIKAN TUNNEL

The Seikan tunnel in Japan provides an interesting comparison with the Channel tunnel. It also is an undersea railway tunnel, connecting the main Japanese island of Honshu to the northern island of Hokkaido by crossing under the Tsugaru strait where it is 23 km wide. The tunnel is designed to carry the high speed Shinkansen trains as well as local services on a narrow gauge (1.097 m) line. The decision to build appears to have been broadly political and strategic rather than narrowly economic. Dependence on ferry services as a link between the islands has been seen as inadequate, especially since the tragic loss of 1414 lives in the sinking of five ferries in a typhoon in 1954. In addition to replacing existing ferries by a much faster and more reliable service it is intended to make more attractive expansion from the overcrowded industrial areas into the more sparsely populated northern island. A large proportion of Honshu's area is mountainous and cannot provide for increased population.

10.2.1 Planning

Investigations and surveys were started in 1946 and in 1964 a possible site had been chosen and a start was made on the ground.

The western end of the Tsugaru strait was preferred as being shallower and less liable to seismic disturbances than the eastern end. It has proved possible to select a line on which the maximum depth of channel is 140 m compared with 260 m at the eastern end. Rock cover of 100 m has been specified on the basis of experience in undersea coal mines, and the tunnel at its deepest is therefore over 240 m below water level, corresponding to a hydrostatic pressure of over 24 bar.

The original plan had been for a tunnel 36 km long, carrying narrow gauge train services only, with a gradient of 2%, but in 1971 the decision was taken to provide also for high speed (Shinkansen) trains running at 210 km/hr. The ruling gradient was then reduced to 1.2% and the minimum radius set at 6500 m. The combination of this gradient with the great depth makes the tunnelled length 53.85 km, exceeding the proposed Channel tunnel by 4 km and far longer than any other such tunnel. The undersea length is 23.3 km and the approach lengths are respectively 13.55 km and 17.0 km for the Honshu (south) and Hokkaido (north) sides.

The question whether to have a double-track tunnel or twin tunnels was the subject of much consideration, and the final decision in favour of the larger double-track tunnel, 9.6 m wide, was strongly influenced by two factors:

1. the excavated volume, working to Shinkansen standard sections is about 25% less than for twin tunnels;
2. the undersea section requires extensive and elaborate grouting of fissured and faulted rock, and it was considered to be safer and more efficient to concentrate ground treatment on one tunnel.

Information is lacking on the operational ventilation and aerodynamic problems in the two-way tunnel, but it must be supposed that the experience gained in Shinkansen lines is an adequate guide. This applies also to the other problems to which so much study has been applied in the Channel tunnel design, such as evacuation of passengers from a breakdown and fire and emergency precautions generally.

In addition to the large running tunnel there are two subsidiary tunnels, namely:

1. a deep level drainage tunnel about 3.6 m diameter, initially driven as a pilot tunnel and serving also during construction for access, exploration, drainage and ventilation;
2. a service tunnel about 4 m diameter parallel to the running tunnel at a distance of 30 m, and driven in advance of it, also used initially for exploration, ground treatment, access, transportation and ventilation, and ultimately for services, maintenance, ventilation and emergencies.

The two subsidiary tunnels merge in the mid-tunnel section of 5 km. Fig. 10.3 shows plan and section of the crossing and a cross section of the three tunnels.

The geological structure is of Miocene strata, through which the tunnel passes, overlying at greater depth palaeozoic and granitic rocks. A wide variety of beds occur, many of volcanic origin. In the south half there is a predominance of andesites and basalts which are associated with much fissuring and therefore require extensive grouting to seal off inflow of water. In the north half there are volcanic tuffs, andesites, breccia, shale, siltstone, sandstone. Most of the rock is competent, but some of the mudstones swell following excavation. Across the line of the tunnel under the strait there are known to be nine principal faults with brecciated zones. In the middle third of the strait lie the most recent 'Kuromatsumi' strata of weak mudstones.

In designing the tunnel and its lining the principle adopted to deal with the very heavy water pressure is to grout as tightly as possible a cut-off ring of rock clear outside the tunnel and to drain off continuously from behind the tunnel lining any residual water seeping through. The loads resulting from water pressure and rock pressure are then carried by redistribution of stresses in the rock. The immediate support of the tunnel excavation is normally provided by arch ribs and sprayed concrete, followed up at the appropriate time by an *in situ* concrete ring up to 900 mm thick. In lengths where the ground is bad a circular section is adopted with 300 mm steel pipes reinforced and filled with mortar as ribs at 0.7 m centres.

The normal linings for the drainage pilot and the service tunnel also comprise steel ribs and sprayed concrete up to 200 mm thick. For the former no additional *in situ* concrete is thought necessary.

10.2.2 Surveys

Progressively more detailed topographical and geological surveys were made over the period 1946–1966. Coast lines were surveyed in 1946–1949, followed from

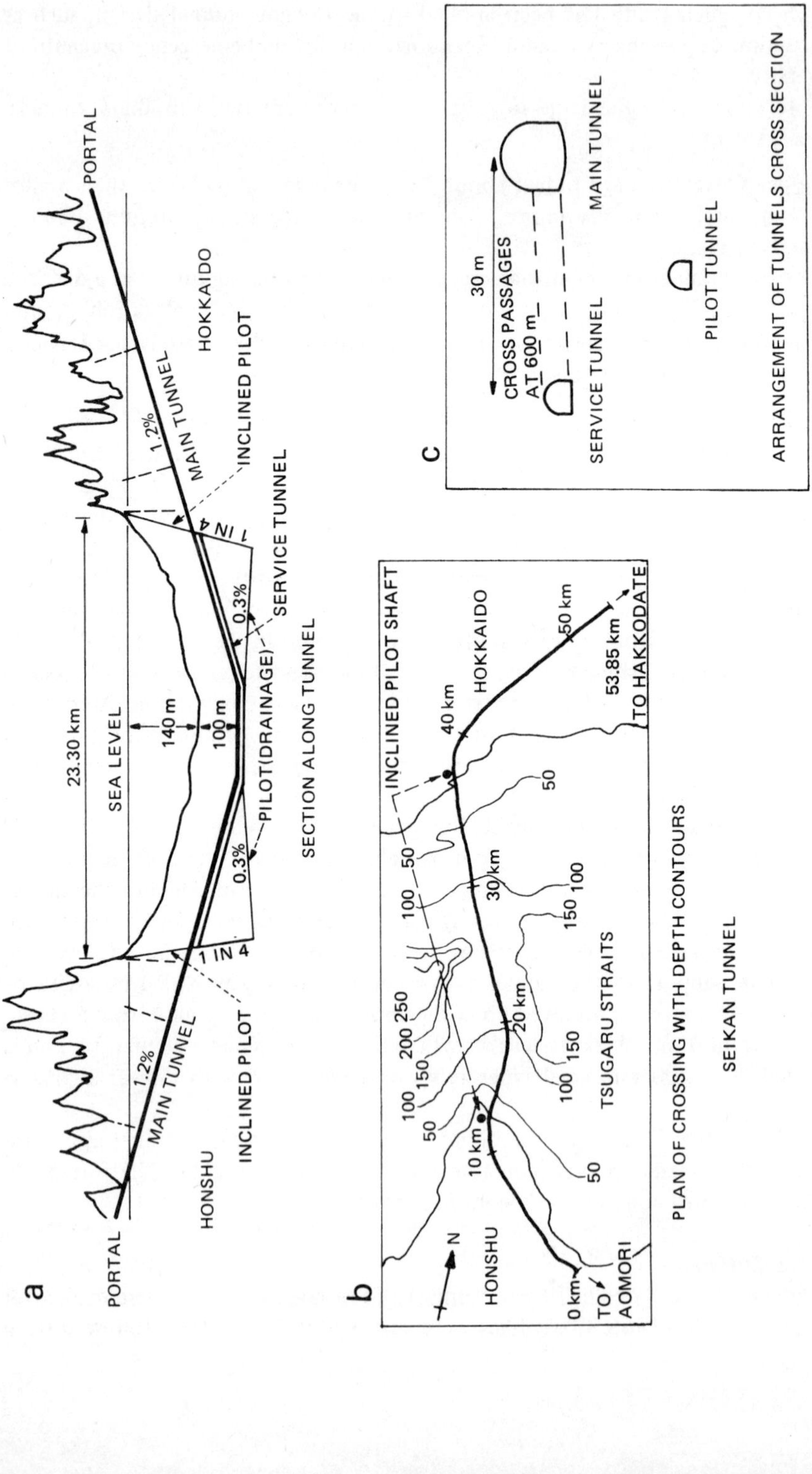

Fig. 10.3 – Seikan Tunnel. (a) section; (b) plan; (c) cross section.

1953–1961 by aerial surveys of the two peninsulas to be connected, the mapping in the last year being on scales of 1/2500 and 1/10 000, followed in 1963 by 1/500 detail surveys, and linking surveys across the straits in 1965, 1966, 1967.

Acoustic soundings to establish sea bed contours with a 2 m interval were carried out in 1954.

The geology was studied similarly over the same period with borings on land from 1946–1955 and 1961–1965 and on the sea bottom in 1956 and from 1960–1966. Seismic profiles were obtained about the same periods and sonic and magnetic traverses of the sea bottom were made in 1959, 1961 and 1962. Tidal observations were also recorded.

This preliminary work was carried out under the Japanese National Railways until the Japanese Railway Construction Public Corporation (JRCC) was established in 1964 to take responsibility for construction of this and other major railway projects.

10.2.3 Construction

Exploratory construction started in 1964 with the sinking of a 1210 m inclined shaft at a gradient of 1 in 4 on Hokkaido over a period of 3 years. After construction of various service chambers at the shaft bottom, driving of the drainage pilot tunnel seawards began on a rising gradient of 0.3%. A similar shaft was sunk on the Honshu side from 1966 to 1970 and the pilot also started. At the end of 1968 the Hokkaido service tunnel was started seawards from halfway down the shaft followed in early 1971 by the Honshu service tunnel. The work up to 1970 had been considered exploratory but the decision that the line should be constructed was then taken and the project was authorised, to Shinkansen standards, the following year. The work was divided into nine lengths: four under land in Honshu, two of 13.0 and 14.7 km under the strait, and three under land in Hokkaido, contracts being let for all except the drainage pilot, where work continues to be by the JRCC itself. The service tunnel construction was transferred from the JRCC to contractors in 1972. The land approaches are major tunnels in their own right, but cannot be further described here.

In driving the pilots and service tunnels, full face machines were tried with some success for substantial lengths, but frequent variations in the ground and difficulties of access to the face, together with problems of steering, particularly in soft ground, resulted in a decision to discontinue their use and adopt drill and blast methods throughout. The fear of undue disturbance of surrounding rock by blasting was somewhat relieved by comparison of elastic wave measurements showing the loosened rock when blasted to extend 1.2 m to 1.4 m compared with 0.6 m to 1.0 m when machine excavated.

The service tunnels have been used to open up and serve a number of faces in the main tunnels by driving adits at ½ to 1 km intervals.

The problems of excavation and construction are formidable enough because of the length and depth of the tunnel, but above all the major hazard is inundation from the sea through faults and fissures. The full hydrostatic water pressure has been experienced and, as is the characteristic of subaqueous tunnels in contrast to mountain tunnels, there is no hope of relieving it by drainage of the reservoir. The strategy of construction has been to explore the structure of the ground as thoroughly as possible by means of the drainage and service tunnels functioning as pilots and by exploratory drilling ahead of these tunnels. Drilled lengths of up to 800 m have been attained, and equipment with a claimed range of 2000 m has been acquired. As soon as fissures are located extensive grouting at pressures up to 80 kgf/cm^2 (three times the hydrostatic head) has been done, principally based on a specially developed combination of very finely ground blast furnace cement with waterglass, using the appropriate water-cement ratio to achieve the required setting time.

Despite these and other precautions the tunnels have been flooded on several occasions, the worst being in 1976 when the Hokkaido service tunnel struck a fault about 4.6 km out from the shore. It was not in fact one of the major faults. Inflow rose in a few hours to a rate of 70 m^3/min but stabilised at 30 m^3/min after 1½ days. For this time the inflow exceeded the pumping capacity but it was found possible to direct the water into the main tunnel, whose storage capacity proved adequate, and thus to avoid flooding the pumping plant in the low-level drainage tunnel. Additional pumps were installed, cut-off gates and bulkheads constructed and grouting carried out so that the service tunnel was recovered to within 70 m of the face. From here a by-pass was driven through the fault at right angles. These operations took eight months.

This technique of diverting the service tunnel so as to get a shorter and safer crossing of a fault is obviously not one that can be repeated for the main tunnel. Advance grouting of the whole zone will have to be intensive and every move at the main face will have to be most carefully monitored.

The pattern and cycle of excavation of the main tunnels has varied according to the ground conditions encountered. In the best ground the procedure followed has been advance of the arch, excavating and supporting the upper half, followed by excavation and lining below axis. Where this is not practicable a bottom pilot heading has first been excavated ahead, from which ground treatment can be effected as necessary, followed by excavation of the upper half and fixing of the arch ribs and then by invert excavation and construction. In weaker ground, bottom side headings have first been taken out to allow the concrete side walls to be built up to axis; excavation and fixing of arch ribs has followed and finally excavation of the dumpling and concreting of the invert.

In the worst ground a circular form has been adopted, side headings at axis being excavated and filled followed by the excavation and fixing of reinforced steel pipes as ribs for the upper arch, pipe struts at axis, and finally excavation of dumpling and circular invert, and fixing of ribs in the invert. Final permanent lining has been left for six months to allow rock stresses to develop, but heavy deflections have been experienced.

Removal of spoil has been by mine cars along the tunnels to hoppers and crushers and then by conveyor up the inclined shafts to surface tips. The handling capacity is hardly adequate for the number of faces being worked and the problem is bound to increase as the distances lengthen.

Pumping during construction is a major item. Capacity at each end has had to be increased to over 100 m^3/min after the flooding experienced. The average has been 26 m^3/min on the fissured Honshu side and 7 m^3/min on the Hokkaido side.

Ventilation has been important not only for fresh air supply but because of heat and humidity. Water has been encountered at temperatures of 20°–30°C and there is a substantial addition of heat from machine operation and the setting of cement. A circulation has been established utilising the whole cross section of the tunnels and connecting passages as follows: inclined shaft → pilot tunnel → connecting vertical shaft → service tunnel → main tunnel → vertical shaft. With this the temperature in the tunnels has been held to 25°C.

Early in 1980 the stage reached in construction on the main contracts, which total 27.70 km between shafts, was that the excavation of the main tunnel was nearly 80% of the distance across between shafts, leaving less than 6 km to be excavated. For the pilot tunnel the gap was about 3 km. The problems remaining are formidable but there is great determination to complete construction successfully. Probing ahead, extensive grouting and cautious advance through fault zones and other difficult ground are obviously the essentials in meeting the hazards of the geological disruptions and of unlimited water at high pressure.

10.3 COMPARISON

There are many points of interest in comparing the two major undersea tunnels. The Seikan tunnel is under construction and fast becoming a reality, while the Channel tunnel having been designed in considerable detail and brought to the point of actual construction in 1975 is still in abeyance. It may be revived but not necessarily to meet the same demands.

The Channel tunnel is the senior project having been the subject of serious proposals and progressively more detailed investigation and survey for well over one hundred years. The Seikan tunnel took some twenty years from the initiation of the concept in 1946 to start of work on the ground, and another five years to the authorisation of the full scheme and placing of construction contracts in 1971.

The political and strategic aspects have throughout been a major handicap to the engineering concepts in the Channel, because two independent nations have had to be in accord simultaneously on the scheme and its financing, and to agree on the division of responsibilities, with legal and administrative arrangements for construction and operation. There has been no clear national interest in either country overriding narrower sectional interests. The railway interest has been the most dominant in favour, but the shipping interests have naturally been less than enthusiastic, and development of air travel has drawn off an important element of passenger demand. Despite these and other diverse commercial aspects it is probable that the enterprise would have been launched about 1930 if military thinking had not continued its earlier opposition to the supposed hazard of invasion. In the 1973 project all these conflicting interests appeared to have been resolved, the special interest of road transport being the subject of elaborate provision, but the whole was found unacceptably costly in a time of economic difficulty. A wide range of schemes are again being actively discussed and evaluated, and again the initially favoured proposal is for a bored rail tunnel.

The contrast of the Seikan tunnel is startling–a single national government, deeply concerned to link the separate major islands of the kingdom was also concerned to eliminate the hazards and delays of shipping. The multiple competing routes between England and different continental ports have no parallel; there is one ferry route between Aomori and Hakkodate. The scheme formed part of the great national effort to provide rail links of a high standard throughout all parts of the mountainous islands so that when the tunnel project was finally authorised in 1970, it was upgraded and substantially enlarged to form part of the major system.

The Channel tunnel would link the United Kingdom, with a population of 55 million, to the European continent, whose EEC population exceeds 200 million, as compared with the Seikan link between the 70 million population of the main island of Honshu and the 5 million of Hokkaido, whose only very large city is Sapporo with a population of about 1½ million. To complete the connection to Tokyo, the Shinkansen line, at present under construction to Morioka 500 km north of Tokyo, must be extended by another 550 km, or a new line of greater length must be built up the west coast from Niigata. The comparable need for a new high speed line over 100 km long between London and Folkestone and improvements, or possibly a new line of 300 km, from Sangatte to Paris provides an interesting parallel to the Shinkansen proposal, emphasised by the fact that the existing Folkestone to London line cannot accept the full Berne loading gauge although the rail gauge is unchanged.

The forecast of passengers to be carried through the Channel tunnel in 1980 was 9 million by rail only, plus 6 million accompanied by about 2 million passenger cars; freight was estimated at 5 million tons, leaving 9 million tons carried by sea. In Japan, the current annual traffic across the Tsugaru strait by sea between Aomori and Hakkodate is 5 million passengers, mostly on railway ferries, and 15 million tons of goods, less than half on railway ferries; passenger cars are not separately described. There are also air services between the two islands. It is to be expected that much of the freight would continue to be carried by sea, but it seems unlikely that a frequent passenger ferry service would be maintained.

The topography of the two straits differs in that the Tsugaru sea bed reaches more than twice the depth of the Channel, while the 100 m chalk cliffs of Dover make a more abrupt change from land to sea than in Japan.

The geology shows marked and important differences influencing every aspect of the tunnel construction. In the Channel it has been found possible to locate with some precision a stratum of grey chalk and to align the tunnel to remain within it for all of its length except at the portals. It is fairly homogeneous, and believed to be free from significant faulting, so that excavation by a full face machine followed by erection of segmental lining could be specified. While no tunnel of such length can hope to avoid some unpredictable ground conditions, any anomalies should be over limited lengths and should be found in advance by probing ahead.

In the Seikan tunnel the line runs through many different and disturbed strata, intersected by at least nine significant faults under the sea. The ground ranges from soft and swelling siltstones and volcanic ash in various forms through sedimentary rocks to igneous intrusions and lavas of geologically recent date (Miocene). The faults are marked by shattered belts of rock allowing fissures to develop connecting with the sea bed and therefore threatening inundation. Machine boring techniques have been found insufficiently adaptable to the variations in ground, and the familiar and well tried techniques of drill and blast have therefore been adopted with sequences of heading excavation, enlargement and support, continuously adapted to the conditions disclosed by exploration ahead. The lining likewise is adapted to the ground.

On the railway engineering aspects the first comparisons are perhaps gradient and curves. There is a significant difference between the 1.03% or 1.11% of the Channel tunnel and the 1.2% of the Seikan tunnel; adoption of the former for a rise of 250 m would lengthen the Seikan approach by 3.4 km. Conversely, if the rise of 82 m near Dover were taken at 1.2% it would be reduced in length by 1.3 km. The Channel tunnel gradient was related to the operation of very long and heavy vehicle ferry trains rather than to through running of fast trains or to lighter trains of mixed freight.

The Channel curve limit of 4200 m related to a speed of 200 km/hr compared with the Seikan curve of 6500 m which has been adopted as the latest Shinkansen standard for 260 km/hr, the first, Tokaido, line being built to 2500 m and the next, Sanyo, line to 4000 m.

The choice of twin tunnels for the Channel was determined by operating advantages in respect of safety, ventilation and aerodynamics in which context the ferrying of very large road vehicles imposes special requirements; twin tunnels would also be preferred constructionally. For Seikan it is clear that the dominant consideration was the complexity of tunnelling in a difficult and dangerous ground. A wide zone outside the tunnel lining has to be grouted with meticulous care to ensure that the ground itself is substantially watertight and structurally sound. To repeat these operations for a duplicate tunnel a short distance away has obviously been found unattractive, especially when the excavation may disturb the equilibrium already attained round the first bore. Nevertheless, the decision to cut the larger hole in this difficult rock can only have been taken after much thought.

The drainage systems adopted also differ. Low level drainage tunnels leading down from mid-channel to pumping shafts near the coasts were considered for the Channel but rejected in favour of discharging seepage through a sequence of pumps and sumps, using the service tunnel. The tunnel profile under the Channel is not of V form but a flattened W. Although the tunnel will not be fully caulked, to avoid build-up of pressure, large volumes of water are not expected in the finished tunnel, nor during construction. At Seikan the situation is quite different. Even apart from inundations large volumes of water have to be drained and pumped out during construction and there is no question of an ultimately watertight tunnel, because pressures must not be allowed to build up. It has been essential to the whole construction strategy to construct the low-level drainage pilot tunnel ahead of the other work which can normally be drained by gravity to the bottoms of the inclined working shafts. The normal pumping requirement during construction has been about 26 m^3/min at the south and 7 m^3/min at the north shaft. The ultimate pumping requirement in operation may well be of this order.

Ventilation of the Channel tunnel has been referred to and is further discussed in Volume 2, but comparable information on the Seikan tunnel is not known. The moderately high temperatures experienced during construction suggest that some system of heat extraction, possibly by fresh air supply, will be essential to prevent a build-up of heat even if passengers are protected by air conditioning during normal operation. Problems of accident and fire and the evacuation of passengers have no doubt been fully considered.

BIBLIOGRAPHY

There is an extensive range of publications on the Channel Tunnel covering various aspects of its history. This selection is intended to be particularly relevent to current aspects. More comprehensive references are cited in the general bibliography in Volume 2.

Channel Tunnel

Economic Advisory Council, Channel Tunnel Committee, *Report, Cmnd. 3513,* HMSO, 1930.

Slater, H. and Barnett, C., *The Channel Tunnel*, Wingate, 1958.

Bruckshaw, J. M. and others, 'The work of the Channel Tunnel Study Group 1958–60', *Proc. Instn. Civ. Engrs.*, 1961, **18** (Feb.). *See also* discussion, 1962, **21** (Mar.).

Ministry of Transport, *Proposals for a fixed Channel Link, Cmnd. 2137*, HMSO, 1963.

Pequignot, C. A., *Chunnel*, CR Books, 1965.

Grange, A. and Muir Wood, A. M., 'The site investigations for a Channel Tunnel, 1964–65', *Proc. Instn. Civ. Engrs.*, 1970, **45** (Jan.). *See also* discussion 1970, **47** (Dec.).

Department of the Environment, *The Channel Tunnel Project, Cmnd. 5256*, HMSO, 1973.

Department of the Environment, *The Channel Tunnel, Cmnd. 5430*, HMSO, 1973.

Haining, P., *Eurotunnel, an illustrated history of the Channel Tunnel scheme*, New English Library, 1973.

Gould, H. B. and others, 'The design of the Channel Tunnel', *Struct. Engr.*, 1975, **53** (Feb.). *See also* discussion, 1975, **53** (Dec.).

Morgan, J. M. and others, *The tunnelling system for the British section of the Channel Tunnel phase II works, Laboratory Report 734*, Transport and Road Research Laboratory, Crowthorne, 1977.

House of Commons, Second Report from the Transport Committee, Session 1980–81, *The Channel Link, Report and Minutes, (HoC 155 – I, II & III)*, HMSO, 1981.

Seikan Tunnel

Sources of material are limited to papers from conferences and periodicals.

Tanaka, T., 'Seikan undersea tunnel', *Civil Engineering in Japan*, 1970.

Hama, K., 'The Seikan undersea tunnel', *Japanese Railway Engineering*, 1972 (Jan.).

Mochida, Y., 'Long horizontal boring', *Permanent Way*, 1973, **14** (Apr.).

Shinohara, H. and Akita, H., 'Grouting in pilot tunnel heading', *Permanent Way*, 1973, **14** (Apr.).

Hayward, D., 'Tale of a tunnel, on the survival story of sub-sea Seikan', *New Civil Engineer*, 1977 (19 May).

Matsuo S. and Endo, K., 'Seikan undersea tunnel', *Civil Engineering in Japan*, 1977.

Fujita, M., 'Seikan undersea tunnel', *International Tunnel Symposium*, Tokyo, 1978.

Mochida, Y., 'Excavation of expansive ground formation for the Seikan undersea tunnel project', *Tunnelling '79, 2nd International Symposium*, London, 1979.

Mochida, Y. and Kikuchi, T., 'Exploration undertaken as a safety measure in the construction of the Seikan tunnel, *The Safety of Underground Works, International Symposium, Brussels*, 1980.

'Seikan Undersea Tunnel', *The Present Status of Tunnelling Activities in Japan, Part 1,* Japan Tunnelling Association, Tokyo, 1980, 60–69.

Index

Table of Contents

Authors' Preface

The authors covered in Volume 1 the general field of tunnelling, more particularly bored tunnels, and they hope that is a worthwhile book in its own right. The Preface to Volume 1 outlined the general contents of Volume 2 as follows: tunnels in trench under land and water, shafts, ground treatment, geological and geotechnical aspects, and specialised applications for road and rail transport systems. Inevitably, it has been necessary to limit severely the discussion of every subject treated, but it is hoped that the chapter bibliographies will provide a useful lead deeper into the various subjects.

Again the emphasis has been on describing methods and principles and indicating choice to be made rather than on development and application of theoretical analysis in which engineering geology, soil mechanics and rock mechanics have so much to contribute. The aim is not to give detailed instruction on any subject, but to show what factors have to be considered in the perspective of the whole complex art.

An extensive bibliography of publications in English has been included, of wider scope than the chapter bibliographies, although incidentally including most of their contents to make it comprehensive. A vast amount of information is to be found in the publications of the Institution of Civil Engineers and of the American Society of Civil Engineers. These each have been made the subject of separate lists, in order of date of publication, and chronological order has similarly been followed in the general list of books and articles. Ever more numerous specialist conferences are occasions of presenting papers, some of major importance, some of local and temporary interest only. The conferences have been listed with an indication where possible of any particular theme relevant to this book. Individual papers have been included in the general bibliography if thought appropriate but many papers of great value are not separately identified.

New developments

New techniques and equipment are being introduced so rapidly that no attempt to describe and evaluate them has been possible. A very few may be mentioned here, of which detailed descriptions must be sought in the periodicals and in conference and other papers.

The recent successful use of an earth pressure balance tunnelling machine by a Japanese contractor in San Francisco is just one of many indications of the rapid changes in the tunnelling world. Two achievements of particular note on that sewer tunnel project were that the cutting tools of the TBM lasted throughout the drives without replacement, and secondly a number of timber piles were successfully 'chewed through' by the cutting head in the course of the tunnelling. That opens up new possibilities of shallow bored tunnelling in many other cities (in the Low Countries of Europe for instance) provided of course that the piles are no longer in important use as supports.

A recent promising innovation is the British 'Unitunnel' system. Major disadvantages of long distance pipe jacking are the high jacking loads and steering difficulties. The Unitunnel system overcomes both difficulties by jacking the pipe sections (or tunnel rings) forward one at a time. That is achieved by successively inflating and exhausting pneumatic hoses between each pipe (or ring) in such a way that one pipe at a time is being pushed forward off the two pipes behind it, and so on in turn until an additional pipe can be inserted in the working shaft or pit at the start of the drive: for most proposed applications, a simple annular cutting edge with a few steering jacks is all that will be required in the way of a shield, but excavation and face support can be organised in a variety of ways.

Meanwhile research continues on the use of novel forces such as laser beams and high pressure water jets for the excavation of rock, the latter having reached the development stages for softer rocks.

The most intractable tunnelling problem remains our inability to picture exactly the ground and water conditions which will be encountered during the drive. For some tunnels one of the most helpful developments is the new ability to navigate a probe for considerable distances ahead of the working face. But the expense is considerable and much uncertainty of ground and water conditions is likely to remain to challenge the tunneller.

Whilst tunnels will progressively be constructed more economically and safely than in the past, the authors see no danger of this branch of civil engineering losing its fascination for those who enjoy dealing occasionally with the unusual and the unexpected.

Acknowledgements

We reiterate our gratitude to those who have contributed to the text and the production of the book, whether or not named in the previous preface. Illustrations from some new sources appear in this volume: we express our thanks to all contributing.

1

Cut and Cover

1.1 CHARACTERISTICS

Construction of a tunnel by cut-and-cover offers an alternative to boring where a trench of the required depth and width can be excavated from the surface. In its simplest form, a trench is excavated, the tunnel structure is built, the trench is backfilled and the surface is restored, but the support of soft ground and the maintenance of existing surface and underground facilities and services make most projects much more complex.

For shallow tunnels the direct cost of cut-and-cover is likely to be much less than the cost of boring, but the incidental costs can change the balance completely. These include such costs, not always ascertainable or necessarily chargeable, as provision of alternative facilities for traffic using the surface, safeguards against subsidence, protection or diversion of services and drainage systems and social costs of disruption and loss of amenity.

With increasing depth, direct costs of trench excavation and support increase rapidly. In water-bearing ground the water must be managed by containment, pumping or groundwater lowering, while, in soft clays particularly, heave of the trench bottom may lead to serious loss of ground and subsidence.

For metro tunnels in particular there is unending debate on the choice between bored tunnels and cut-and-cover. Direct construction costs and operating economics obviously favour shallow cut-and-cover, but very shallow tunnels are rarely practicable and all the incidental costs and benefits must be taken into consideration. There is no universal answer: each city, and indeed each line in that city, has its own special problems.

The first London underground lines were built as cut-and-cover but objections to disruption, and incidental costs made it impossible to obtain the necessary powers to construct except by deeper bored tunnelling, adopted for all later lines in the central area. An interesting recent development has been the extension to Heathrow Airport, partly in cut-and-cover and partly, within the airport perimeter, in bored tunnel. Toronto presents another example of a city where the first line was built in cut-and-cover, but for subsequent work in the city extensive shield tunnelling, some of it with compressed air, has been adopted.

1.2 AREAS OF USE

Cut-and-cover tunnels are characteristically shallow tunnels in situations where open cuttings are not acceptable in the finished project. They are most commonly employed:

1. in subaqueous tunnels to form a transition between an open cut approach and the main tunnel;
2. at the portals of mountain tunnels;
3. in urban conditions where the route must be covered in and the surface restored; and
4. for reasons of amenity.

1.2.1 Subaqueous tunnels

In subaqueous tunnels a transition may connect an open approach ramp with a shield bored or submerged unit tunnel. At the outer end the preference for a covered tunnel over open cut may arise from:

(a) construction problems and costs for open cut;
(b) need to restore surface use over the tunnel; and
(c) danger of flooding.

(a) In the vicinity of a river the ground is usually alluvial such as soft silts and clays, water-bearing gravels with perhaps peaty layers. Any trench excavation will require substantial lateral support and management of water with precautions against bottom heave. While temporary support for a working trench may be feasible, permanent retaining walls and their foundations may become unduly heavy and costly and require such top strutting that a closed box structure is preferable. Upward water pressure acting on the base of an open cut may have to be counteracted by downward loading, more economically provided by back-fill over a tunnel roof.

(b) In open country a road or railway close to the river bank may pass across the tunnel approach at ground level or a flood bank may have to be restored. Drainage systems, whether in open channels or pipes, might be obstructed by a cutting. In urban conditions, the surface requirements are as more fully discussed below.

(c) Protection against flooding by abnormal tides may be more satisfactorily given by roofing over than by high flood banks or walls. Where construction is in a cofferdam extending into the waterway the need for a roofed structure is obvious.

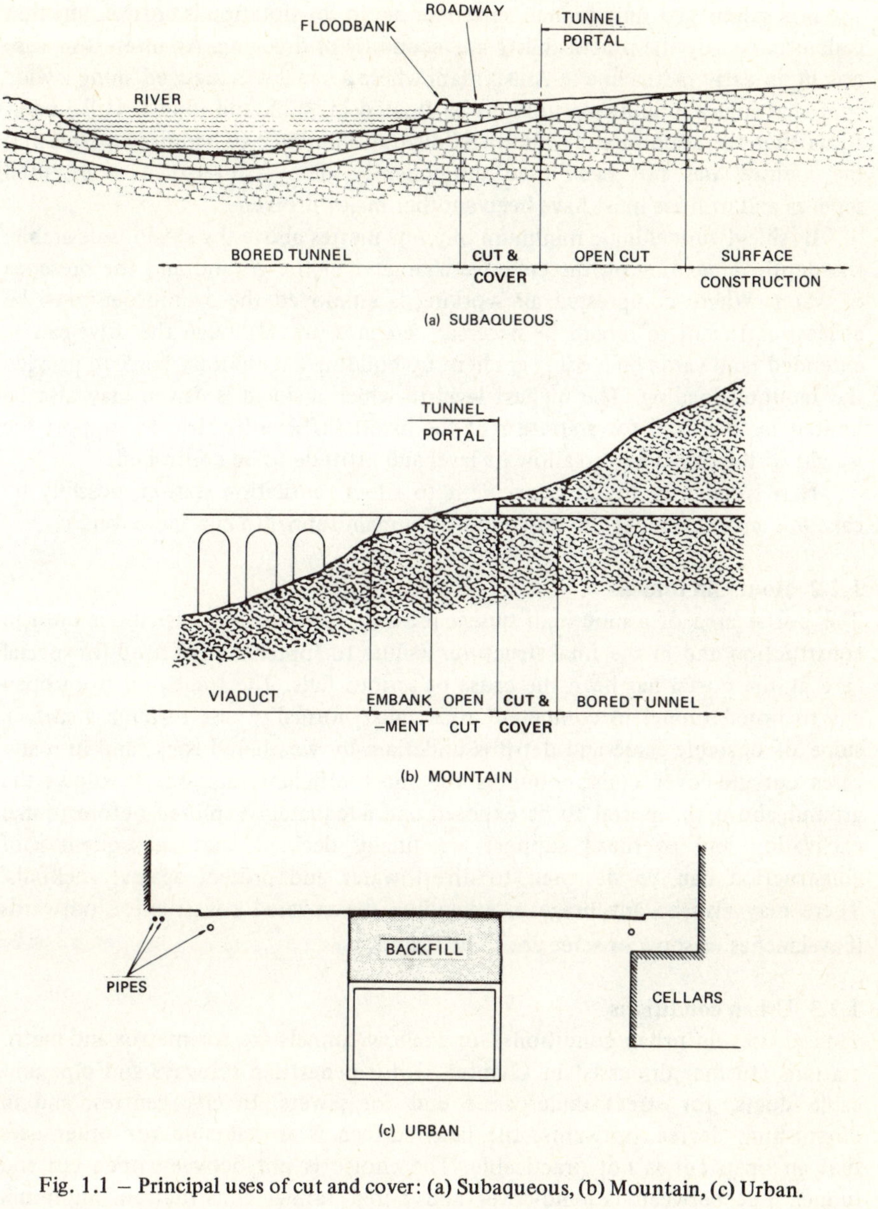

Fig. 1.1 – Principal uses of cut and cover: (a) Subaqueous, (b) Mountain, (c) Urban.

At the deep end of the transition the appropriate point of change may be determined by physical limitations defining a minimum depth for the main tunnel technique or a maximum for the cut-and-cover excavation. Comparative costs and access for construction may also be decisive.

For submerged unit tunnels sufficient depth for flotation is critical, but that is determined by the practicability and economy of dredging. An interesting case was in an early metro line in Amsterdam where a canal was dredged along a wide approach highway allowing units to be floated in, sunk and joined. It is understood that the scale of excavation and disruption proved to be unacceptable and the method has not been used subsequently. The maintenance of essential services and utilities must have been another major problem.

In shield tunnelling a minimum of a few metres above the shield is desirable, the depth depending on the cohesive character of the ground and the presence of water. Where compressed air working is employed the overburden must be at least sufficient to retain the necessary air pressure, although the drive can be extended landwards on a rising gradient by building a temporary bank to provide the required loading. The highest level to which a shield is driven may also be limited by the need for a stratum at the invert sufficiently firm to support the weight of the shield and to allow its level and attitude to be controlled.

It is frequently found convenient to site a ventilation station, possibly by caisson construction, at the transition from main tunnel to cut-and-cover.

1.2.2 Mountain tunnels

The portal area of a mountain tunnel presents considerable difficulties both in construction and in the final structure. Failure to appreciate the need for special care at the portal has been the cause of serious falls. The transition from open cut to bored tunnel in competent rock must normally pass through a surface slope of unstable scree and detritus underlain by weathered rock, and in many cases cut-and-cover construction is the most efficient method. It allows the ground above the portal to be exposed and adequately explored before tunnel excavation and overhead support are finally decided, and subsequent roof construction can be designed to divert water and protect against rockfalls. There may also be advantage in extending the covered construction outwards if avalanches of snow or scree are to be feared.

1.2.3 Urban conditions

Typical uses, in urban conditions, for shallow tunnels are for metros and metro stations (further discussed in Chapter 7), for pedestrian subways and pipe and cable ducts, for street underpasses and for sewers. In city centres, and in diminishing degree outwards, the land surface is so valuable for other uses that an open cut is not practicable. The choice is not between open cut and tunnel, but between cut-and-cover and bored tunnel construction. Minimum depth, and therefore normally cut-and-cover, is essential for pedestrian subways or street underpasses if they are to perform their functions. Throughout the history of metros the functional ideal has been to site them at the shallowest depth possible so that access between street and platform is quick and easy. The immediate obvious answer is cut-and-cover. In practice, bored tunnelling may be found necessary because street disruption, resulting from surface excavation, is unacceptable, or because existing underground obstacles, such as buried services, sewers or other tunnels crossing the line, force down the tunnel to depths where cut-and-cover ceases to be economic. As a city elaborates its underground network this factor of routes crossing one another becomes of increasing importance.

Fig. 1.2 – Metro construction by cut and cover in Rotterdam. Disruption to surface traffic is obvious.

Where the route is not so obstructed many ingenious methods have been adopted to minimise the time and extent of the disruption resulting from open excavation.

Even where metro running tunnels are at a depth where boring is necessary the complex station structures may, wholly or partly, be built in deep open pits. In some cases an 'umbrella' has first been built to carry the street traffic while excavation and construction proceed below.

Sewer systems range from small pipes laid in trench at the upper end of the system to progressively greater diameters and greater depths at which tunnelling may become necessary, either because of depth or because trench excavation occupying the surface is unacceptable in busy streets or under buildings. Even for small pipes excavation in heading may be necessary.

Fig. 1.3 – Metro 'umbrella'. (a) Trial assembly of material for use at crossroads above London Fleet Line station. (b) 'Umbrella' in place.

1.2.3.1 Street work

The essentials of cut-and-cover construction are: excavation of trench; building of floor, walls and roof; backfilling and restoration of surface. Particularly in urban conditions, complex procedures are usually found necessary. If the line of a street is being followed the principal disadvantages in construction special to this method, as compared to a deeper bored tunnel, are:

(a) Lengthy occupation of street with consequent noise and disturbance and disruption of traffic and access;
(b) restricted alignment;
(c) necessity of lateral support to minimise subsidence;
(d) interference with buried services–water, gas, sewers, etc.–which must be diverted or supported; and
(e) backfill material and restoration.

(a) The disturbance caused by opening up long lengths of a street can be greatly reduced, but never eliminated, by alternating excavation and construction for the separate elements of the tunnel. The side walls can be constructed first, occupying only a relatively narrow part of the street, the roof structure can then be built in short lengths which require only shallow excavation, allowing the street surface to be restored quickly, and providing strutting at the top of the walls. The excavation of the 'dumpling' can follow as in a bored tunnel without surface occupation or disturbance. The floor excavation may have to be done with particular care in narrow widths if there may be ground heave or if the side walls require to be strutted at the base.

Such a method does not require lengthy occupation of the whole street surface, but does still imply extensive obstruction of traffic as long lengths of wall construction obstruct lateral access and the whole width of the structure across the street must be occupied as each length of roof is constructed. In suitable conditions or for special lengths, it may be possible to start construction with the roof slab and to build the walls in tunnelled headings to underpin the edges of the roof slab.

(b) For metro and railway work in particular the need to follow an alignment within the limits of existing streets can be a serious disadvantage. Small radius curves may be unavoidable and the direct route between preferred sites for stations may not be possible. Basements and cellars extending beyond the building line impose further limitations.

(c) Lateral support. Open excavation with the sides at the natural slope of the ground is unlikely to be possible in street work, except in rock. Vertical sides reduce the width occupied to a minimum, but inevitably require lateral support, which may either be temporary or may embody the permanent wall structure.

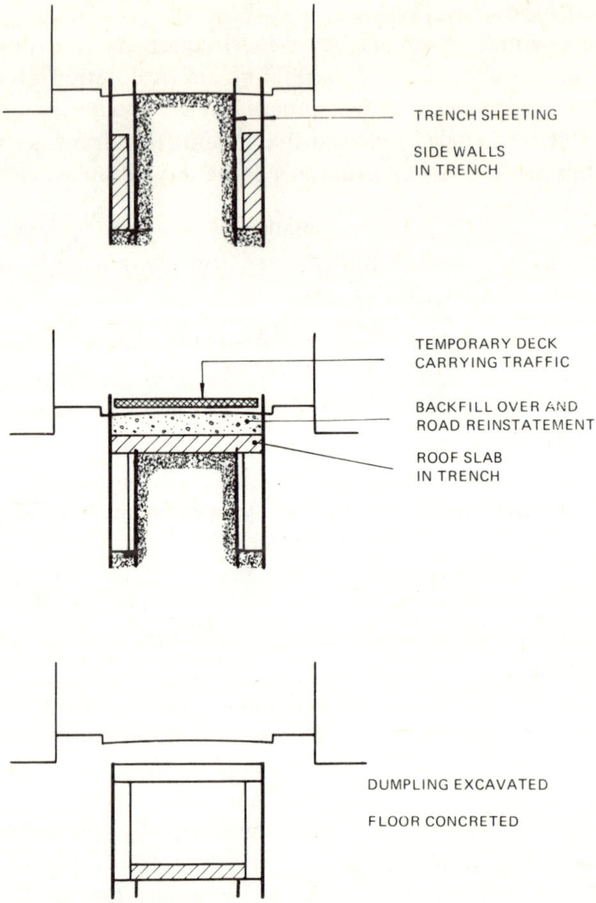

Fig. 1.4 – Cut and cover in street – normal sequence.

Although the tunnels under consideration are described as 'shallow', the trenches and pits required are deep by ordinary excavation standards and some discussion of settlement, and methods of keeping it to a minimum is required.

In any deep trench through cohesive soils the differential pressure on the side walls of the trench will result in strain displacement in the retained soil. There will inevitably be a tendency to plastic heave at the bottom of the trench and inward deflection of the retaining wall with resultant loss of ground and settlement at the surface. The magnitude of this depends on the nature of the soil, the depth of excavation, and the characteristics of the support system. Its importance will depend on the nearness and depth of foundations of adjacent structures and their sensitivity to movement. Strain movements can be minimised by strutting apart the sides at frequent vertical intervals as excavation proceeds. The prestressing of ground anchors and the incorporation of jacking devices in struts, and structurally stiff wall sections can probably reduce the deflections to acceptable proportions.

Ground water is of major importance. If it is retained by watertight piling, the water pressure must be carried by the lateral support system, and unless there is an effective cut off the full upward pressure will act on the bottom. If the water table is lowered by pumping, well pointing or otherwise there may be danger of consequent settlement in adjacent shallow foundations. Old buildings, supported on timber piles can be very sensitive to such changes in water level.

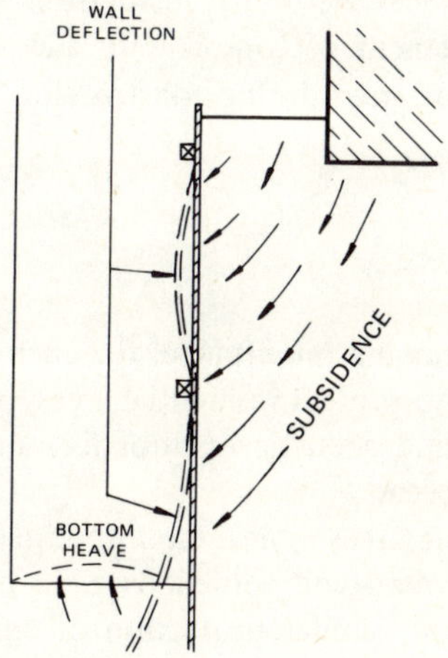

Fig. 1.5 – Pattern of subsidence adjacent to trench.

(d) Buried services are a major problem in cities where they have been laid —and not always accurately recorded—under the street surface at varying depths. Water and gas pipes up to 1½ m diameter are not uncommon; sewers may range from 100 mm house connections to trunk sewers of 2 m diameter upwards; high voltage cables, in ducts or armoured, and banks of telephone cables in ducts are likely to be encountered. Tramway conduits, hydraulic power pipes, and pneumatic despatch tubes are other hazards that may be present in some cities.

Where a cut-and-cover tunnel is to be built these services must be fully safeguarded during and after construction. Diversion beforehand out of the area of excavation is ideal for the tunnel work but is rarely possible for all of the services, although temporary diversion, or even cutting off, is sometimes practicable. Otherwise they have to be supported, protected and maintained across the open excavation and in due course must be very carefully reinstated and secured against subsidence. While so supported they constitute a continuing obstacle to excavation and to supports for any temporary decking. A fractured or leaky gas or water main can be a serious hazard in an enclosed excavation.

(e) Backfill. Restoration of the surface above the roof to its original use will require careful backfilling to avoid subsidence and to ensure that natural drainage is preserved.

In streets, as already noted, backfilling under and around buried services will require particular care. Reconstruction of the road slab and its surfacing and foundations will be necessary.

At greater depth it may be possible to utilise the excavated material as backfill, but unless it is suitable new material must be imported.

In parks and open spaces the proper restoration will demand careful attention to drainage and replacement of topsoil with adequate but not excessive compaction. Safeguarding of trees during construction and restoration may also be called for.

1.2.4 Amenity

A further reason, of growing importance, for enclosing traffic routes is that of local amenity in order to protect residential areas, parks, scenic areas and nature reserves against visual and acoustic disturbance, and interference with existing surface pathways and access.

Motorways in scenic areas or leading into cities across green belt land may be thus built below ground and roofed over. In by-pass construction skirting historic towns or villages, similar protection of amenity may be required. For such cases shallow cut-and-cover construction rather than boring is usually appropriate.

1.3 TECHNIQUES OF CONSTRUCTION

The key to the various methods lies largely in the support of the vertical sides. The principal techniques include:

1. Steel sheet piling with walings and struts or ground anchors.
2. Ground anchors.
3. King piles with struts or ground anchors and horizontal poling boards.
4. Slurry trench walls.
5. Large diameter bored piles contiguous or overlapping.
6. Concrete walls in heading.

The choice of method will be determined by the nature of the ground, the assessment of the difficulties listed above, available resources, cost of construction, land costs and social costs.

1.3.1 Steel sheet piling

This is the simplest engineering solution in soft ground provided it is free from obstructions natural or artificial, such as boulders or cellars and foundations or service pipes leading to and from buildings: the piling requires to be supported as excavation proceeds by walings and struts forming a framework wedged into place. Struts across the trench make access for excavation difficult: ground anchors are a possible alternative. The vertical intervals between frames are calculated initially to give acceptable stresses and deflections in the piles, but even when set and wedged tightly as early as possible some deformation of the ground will occur as noted above, because of ground stresses and plastic flow as the trench is deepened. Movement can be reduced by imposed loading in struts or prestressing in anchor cables, but the flexibility of the steel sheet piling makes unavoidable some loss of ground and resultant settlement, which may or may not be within acceptable limits at the distance from the piling of the nearest structure. Another possible source of settlement is the compaction of loose granular soil resulting from vibration.

Pile driving by hammer is a noisy operation but relatively quiet methods have been developed. Headroom for pitching a line of piles may not be available, particularly in city streets, where there are overhead wires, or where close to buildings.

Steel sheet piling provides an effective temporary barrier to water except where piles get distorted and out of clutch. In water-bearing ground this may necessitate remedial measures such as local ground injection treatment or, possibly, additional piles. Watertightness of the piling is rarely adequate for permanent tunnel requirements. The piling must frequently be left in place although the tops may be burned off at some stage of excavation and backfill.

1.3.2 Ground anchors

One great merit of ground anchors is that they avoid any obstruction of the trench by struts. Another is that they can be readily prestressed. They do, however, extend usually far behind the line of the wall, and very probably beyond the limits of acquisition of land. It will then be necessary first to verify that the anchorages are clear of basements or other structures, and then to obtain permission from the adjacent property owners. For tunnel construction the use of ground anchors will almost certainly be temporary, until the tunnel roof is in place and acts as a strut. If, for any reason, permanent anchorages are required, they would have to be fully protected, not only against corrosion, but against disturbance or damage caused by future site development.

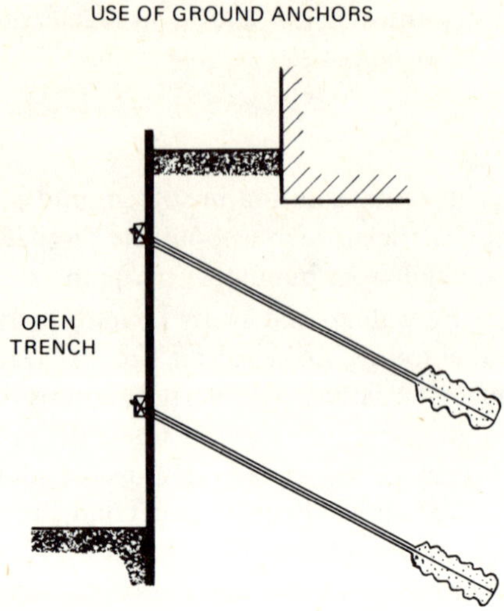

Fig. 1.6 – Section of sheet piling with walings and ground anchors.

1.3.3 King piles

An alternative is the use of bored king piles at regular spacing. They may be of heavy section steel lowered into a pre-bored hole. Where rock is present at suitable depth the toe can be grouted into the rock. The piles will require strutting across the trench as excavation proceeds, but ground anchors, pre-stressed if necessary, will minimise obstruction of the area of excavation. The

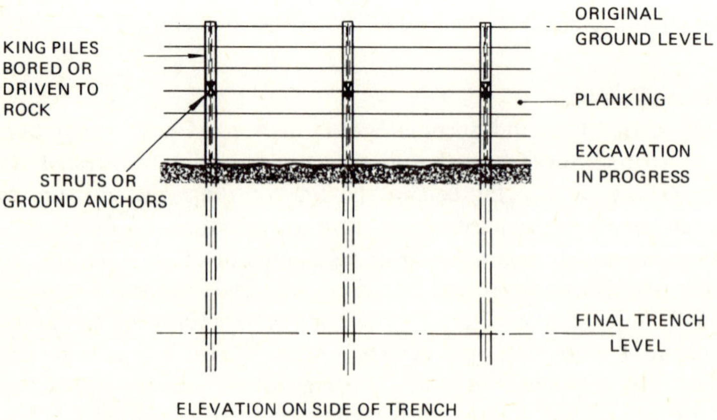

Fig. 1.7 – King piles.

Fig. 1.8 – Open cut approach to Mersey Kingsway Tunnel, supported by king piles and ground anchors.

ground between the piles is supported by horizontal poling boards set in place as excavation proceeds. A variant is where a length of wall is concreted in a slurry filled trench excavated between the king piles. The piles can be designed to carry a temporary road deck, beneath which excavation without further disturbance to traffic is possible.

As with steel sheet piling, possible settlement of the adjacent ground must be considered. Headroom, first for boring, and then for setting the piles, has to be ensured. Obstacles can be dealt with at the boring stage.

1.3.4 Slurry trench walls

Concrete walls cast in slurry filled trenches are a comparatively recent development. A trench is excavated without lateral support except that provided by bentonite or other dense clay slurry filling the trench throughout the process.

Fig. 1.9 – Cut and cover approach to second Dartford Tunnel: slurry trench walls.

Fig. 1.10 – Use of concrete bored piles in Liverpool approach to Mersey Kingsway Tunnel. (a) Contiguous pile walls completed and tops exposed. Arch shutter in place. Walls above with king piles and ground anchors. (b) Structure of cut and cover completed. *Note*: exposed shafts of bored piles (subsequently covered), *in situ* arch over, two main tunnel portals (11 m dia.) in headwall at rear.

Within this trench a concrete wall can be cast by tremie, with a reinforcement cage if required. The concrete wall so cast may either be temporary, performing the same function as sheet piling, or may form or be embodied in the permanent structure.

In cut-and-cover tunnelling this has the advantage of minimising surface disruption and obstruction, and of requiring less headroom than sheet piling, although some headgear is necessary. Also, the whole operation is quieter. The permanent wall can be thus constructed, but the finish, if cast against the ground, will be rough and will normally require to be faced up. This wall will be substantially more rigid than sheet piling, although struts or anchors will still be required during excavation of the main trench. Some surface settlement is inevitable but it can be reduced to a very small magnitude. A hazard with slurry trenching is that if unexpected voids are encountered the slurry may be lost and the trench sides may collapse. Possible voids in urban areas include unmapped basements, abandoned and fractured pipes and sewers, or old wells.

1.3.5 Concrete bored piles

Large diameter bored piles at close centres concreted *in situ* can be used to form a wall. They may be contiguous or may overlap ('secant' piling). The piles may be taken down to any required foundation level and can be reinforced. In the case of secant piling alternate piles, driven first, may be left unreinforced because they will be sliced down in boring the intermediate piles. With good workmanship substantial watertightness is possible, particularly so with secant piling. The concrete at the head of the piles can be finished to take the roof structure covering the tunnel.

In the Mersey Kingsway tunnel at the Liverpool approach the twin bored tunnels converge, where the depth to roadway is 18 m, to a single cut-and-cover tunnel 31.5 to 26 m wide at road level and about 244 m long. Contiguous bored piles, 2.4 m in diameter and keyed into rock, form the walls, and are strutted apart at the top by an arched roof with backfill over. In this instance the principal difficulty of open cut was that cantilevered retaining walls would have been excessively massive, and struts spanning across the top would have been long and heavy and exposed to temperature stresses. Apart from one street crossing the line of the tunnel, the land above was not built over.

1.3.6 Walls in heading

In special cases where access from the surface is particularly difficult, concrete side walls can be built in a heading, either so arranged that the roof slab will be seated on them subsequently, or to underpin a roof slab already in position.

BIBLIOGRAPHY

Cut and cover in general as a form of tunnel construction does not appear to be the subject of much separate writing. It is usually described and discussed as incidental to particular projects, especially metro systems and urban underpasses.

In Paris and Brussels a large number of street underpasses were built in recent years and are described in the P.I.A.R.C. documents noted below.

The following brief list includes general papers and descriptions of a few works; more appear in the general bibliography. Reference should also be made to Chapter 7 – Metro Tunnels.

Parsons, W. B., The New York rapid-transit subway, *Min. Proc. Instn Civ. Engrs,* 1908, **173**.

Humphreys, G. W., The London County Council Holborn-to-Strand improvement, and tramway-subway, *ibid*, 1910, **183**.

The main access tunnel at London Airport, *Concrete Constr. Engng*, 1952, **47** (Mar.).

Granter, E., Park Lane improvement scheme: design and construction, *Proc. Instn Civ. Engrs*, 1964, **29** (Oct.); *see also* discussion, 1966, **35** (Dec.).

Finn, E. V. *et al.*, Strand Underpass, *ibid*, 1966, **35** (Nov.); *see also* discussion, 1967, **37** (Aug.).

Road Tunnels Committee Secretariat:
Documentation and Studies, 13th World Congress, Tokyo, 1967
Documentation Digest, 14th World Congress, Prague, 1971
Documentation Digest Additive, 15th World Congress, Mexico, 1975
Permanent International Association of Road Congresses, Paris.

Peck, R. B., Deep excavation and tunneling in soft ground, *Proc. 7th Intnl. Conf. Soil Mech. Fndtn Engng,* Mexico, 1969, **State of the Art**.

Sverdrup & Parcel and Associates, *Cut-and-cover tunneling techniques,* **1**, *A Study of the State of the Art,* **2**, *Appendix*, U.S. Department of Transportation Reports FHWA-RD-73-40 and 41, (NTIS, PB 222997 and 8), 1973.

O'Reilly, M. P., Some examples of underground development in Europe, *Laboratory Report 592*, Transport and Road Research Laboratory, 1974.

Symposium, Cut and cover or tunnelling: the practical aspects, Concrete Society, May 1974.

Jobling, D. G. and Lyons, A. C., Extension of the Piccadilly Line from Hounslow West to Heathrow Central, *Proc. Instn Civ. Engrs*, 1976, **60** (May); *see also* discussion, 1976, **60** (Nov.).

Consulting Engineers Group, *Prefabricated structural members for cut-and-cover tunnels,* **1**, *Design concepts,* **2**, *Three Case Studies*, U.S. Department of Transportation Reports FHWA-RD-76-113 and 114, (NTIS, PB 273530 and 1), 1977.

Dasgupta, K. N. *et al.*, The Calcutta rapid transit system and the Park Street underground station, *Proc. Instn Civ. Engrs*, 1979, **66** (May); *see also* discussion, 1980, **68** (Feb.).

Coulson, C. R. and Stubbings, B. J., Alternative methods for cut and cover construction in congested areas, paper F2, *Mass Transportation in Asia Conference*, Hong Kong, May 1980.

McIntosh, D. F. *et al.*, Hong Kong mass transit railway modified initial system: design and construction of underground stations and cut-and-cover tunnels, *Proc. Instn Civ. Engrs*, 1980, **68** (Nov.).

2

Submerged Tunnels

2.1 DEFINITIONS AND APPLICATIONS

A submerged tunnel is essentially a subaqueous tunnel constructed by the prefabrication of units and their submergence and jointing in a prepared trench.

The descriptive term 'immersed tube' is also commonly used, but as 'immersion' has perhaps some suggestion of temporary submergence, and as a 'tube' is most typically of circular section, it seems preferable to adopt the wider terminology. The lengths in which it is prefabricated are here termed 'units', which are aggregated into the whole, rather than 'elements' which suggest diverse components of an integrated whole.

The construction techniques have very little in common with those of bored tunnelling, although the completed structure must meet the same fundamental requirements for traffic cross-section, gradients, alignment, ventilation, lighting and environmental amenity. The specialised skills employed are no longer those of tunnel mining–excavation and support–but of structural fabrication in steelwork or reinforced concrete, of handling floating units in a waterway with precise control of buoyancy and stress patterns, and of establishing uniformity of support and watertight joints.

Most submerged tunnels have been built for highways, some for railways and metros, and a few for sewers, water intakes and other uses. Many of the major tunnels are in estuarine waters under shipping channels and it is to such tunnels that the discussion principally refers, although the method is also widely used under rivers and canals and in harbours and offshore. Where the diameter is small, the techniques merge into those of pipe laying. It is normal for a submerged tunnel to be sunk below bed level, but it is possible for the structure to be partly or wholly above the bed, with a protective embankment covering it, provided it does not constitute a hazard to shipping and is not itself vulnerable.

2.1.1 Balance of advantage

Very sweeping claims are sometimes made for submerged tunnelling in comparison with bored tunnelling for subaqueous crossings, and there is no doubt that in many cases it is the right method. There are, nevertheless, important exceptions where submerged tunnels are impracticable and bored tunnels are appropriate. This subject is further discussed later in this chapter.

There are in particular two major aspects where this system will frequently offer substantial advantages, namely, in requiring shallower depth and in demand on less specialised resources.

2.2 TUNNEL DEPTH

Tunnel depth under a navigable waterway is determined by: (1) depth of channel and any intended dredging to greater depth; (2) depth of protective cover over the crown of the tunnel; (3) roof thickness; (4) headroom above the running surface or other design datum; and (5) construction depth below the running surface.

(1) The specified depth of channel for navigation is the same whether for bored or submerged tunnel. It is normally defined as the depth at low water and will apply over the whole channel width accessible to shipping.

(2) The submerged tube usually benefits in that the requirement for cover over the finished structure need not usually exceed about 1.5 m, which gives protection against ships' anchors. The nature of the existing bed is not very significant in this context, unless there is a particular risk of scour, because suitable material can be deposited to form the protective layer. In a bored tunnel, on the other hand, not only is it necessary to have at least similar protection against anchors and scour but there must also be a sufficient thickness of competent ground above excavation for safe working at the face. This depth is likely to be at least a few metres, and probably 10 m or more.

(3) The roof thickness of a circular tunnel is usually rather less than where a flat slab must span over two or three traffic lanes.

(4) For a bored subaqueous tunnel a circular section is almost invariable. A majority of American submerged tunnels are also circular tubes, but in most European practice a rectangular form has been adopted. In highway tunnels particularly, the circle implies a substantial extra height above the traffic clearance line, and even although the space can be profitably employed for ventilation, the total depth from crown down to roadway makes the road level that much lower.

(5) Similarly, the depth of invert below road level is greater in the circular form: but this does not affect road level.

These three factors – dredged depth, cover, and structure down to road level – therefore fix the level to which the tunnel must descend, and they normally show an advantage of several metres to the submerged tunnel. This in turn allows the approach gradients from the portals either to be made shorter with a reduction in the length of tunnel construction or to be made less steep with corresponding benefit to traffic using the tunnel. It may be remarked that for tunnelled metros there may be no such saving in tunnelled length and gradients.

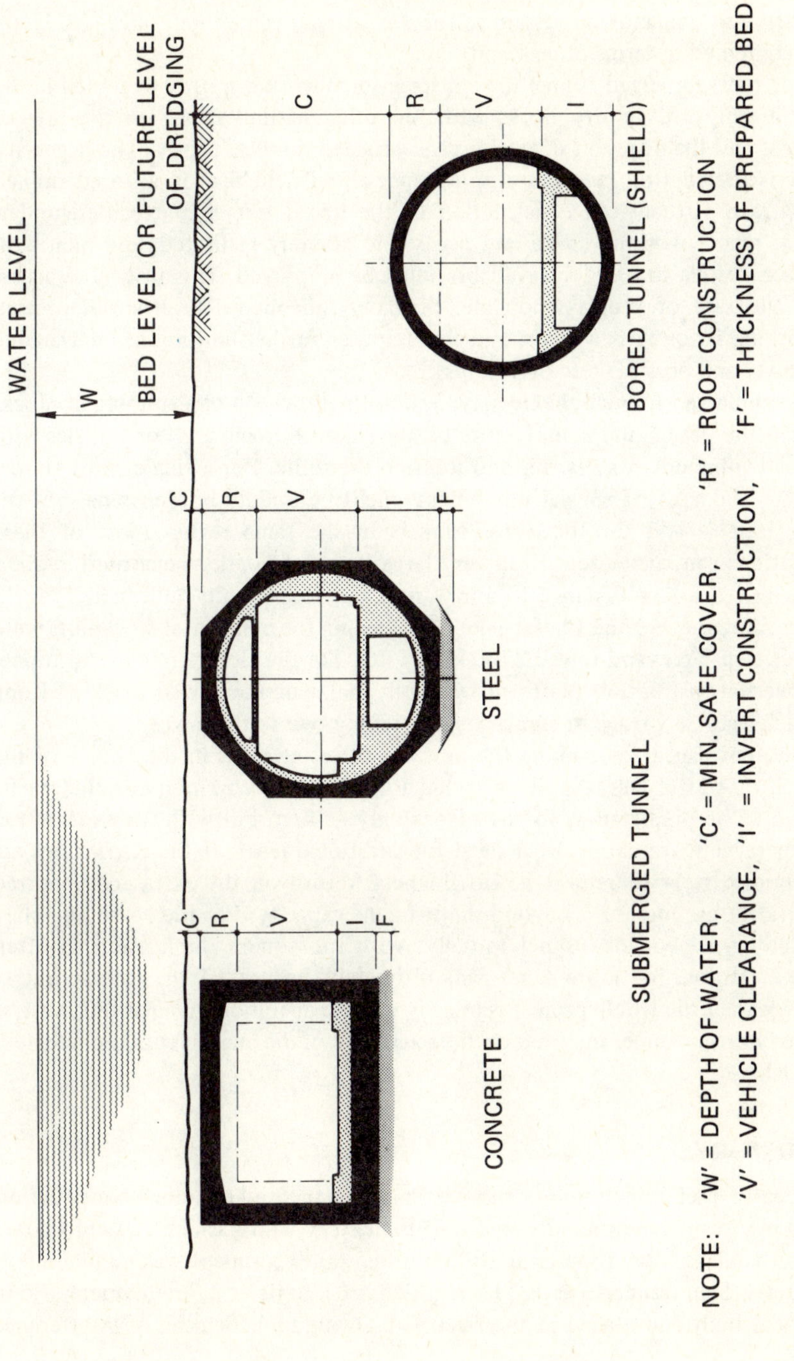

Fig. 2.1 – The figure shows the comparative depths required for tunnels of similar function but differing forms of construction.

2.3 RESOURCES

The different demands on resources for the alternative methods have ultimately to be compared in terms of cost and time.

For the submerged tunnel the constructional work is mostly executed in the open, in shipyard or dry dock, with few inherent limitations on the rate of progress, and the different operations can proceed simultaneously. The necessary labour is drawn from a wider and less specialised field than in a bored tunnel. In contrast, virtually all construction of the bored tunnel is executed underground, where working space and access are severely restricted, and men with specialised skills not widely available must be employed. This restricted access limits the rate of progress possible. In many subaqueous tunnels compressed air working is found necessary, and this imposes further limitations on available manpower and possible rate of progress.

A submerged tunnel scheme may require the provision of a suitable drydock, and also the design and manufacture of specialised floating craft or gantries with control equipment for placing and jointing the units. For a single short tunnel the time and cost of special equipment might be unduly burdensome, but for longer tunnels and for successive tunnels in the same region most of these installations can be re-used. Even very large items of floating construction plant can be towed a long distance to a new site, and so can units built in a drydock, or in a shipyard. For the Chesapeake Bay crossing the cylindrical steel units were fabricated in Texas and towed 2700 km to site. For the Hong Kong metro tunnel the special construction platform was built in Japan and towed to Hong Kong and back again, a voyage of similar length, taking two to three weeks.

Another aspect of making the best use of resources is in the choice of the number of traffic lanes to be provided initially. It is common for the traffic demand to be predicted as for two lanes only at first, but with the expectation that there will ultimately be a need for additional lanes. If the extra lanes are built immediately there will be no financial return on the extra cost incurred until traffic has increased beyond the two-lane capacity. For this reason a policy of building a two-lane tunnel initially, with subsequent duplication, is often adopted. If four or more lanes will ultimately be needed the capital outlay will be least if the whole project is built in a single operation. Against this must be set the interest charges incurred on the extra cost of the facilities not immediately essential.

2.4 HISTORY

The idea of prefabrication and submergence as a method of tunnel construction was discussed in London early in the 19th century, and a scheme was promoted for a crossing of the Thames at Rotherhithe. An experiment was conducted at Horseferry Stairs as early as 1811 when two brick units of 2.7 m diameter 7.6 m long were built and placed in the river bed. Owing to difficulties of interference with navigation in the limited period of slack water at high tide and problems of finance, the scheme proceeded no further, and Brunél's shield-driven Thames tunnel was eventually built, followed later by the development of the Greathead shield. The combination of shield and compressed air has been found effective for all subsequent tunnelling under the river Thames, but submerged tunnels are currently being considered.

Ideas for submerged tunnels were also put forward and discussed at various times during the 19th century in America as well as in England and France. The first substantial construction of this type appears to have been for a sewer in Boston Harbour. In 1893 sections of sewer, 2.7 m outside diameter and in lengths up to 20 m, were built on shore, using brick lining in a steel skin, and were floated out and sunk in a prepared trench, to form an inverted siphon across a channel, the units being connected with external bolted flanges. In the following year for an outfall the section was modified to comprise 200 mm brickwork, 150 mm concrete and a 100 mm timber skin.

In 1906–10 a tunnel for the Michigan Central Railroad was constructed to replace a ferry crossing of the Detroit river which connects Lakes St. Clair and Erie. The approaches were tunnelled with shield and compressed air, but for the river section of 812 m the softness of the clay and depth of 17 to 24 m made this method appear hazardous. A system of construction in trench was accordingly devised. Twin cylindrical steel shells, 7 m diameter x 80 m long coupled transversely with steel diaphragms and furnished, outside the proposed external wall concrete, with timber formwork, were prefabricated in a shipyard, closed at the ends with temporary bulkheads and floated to site. At site they were sunk, with the aid of pontoons and water ballast, onto preset foundation grillages in a prepared trench. End joints were made as steel lap joints carefully fitted at the shipyard and bolted up by divers. Concrete filling in the trench bottom and in the external walls and roof of the tunnel was deposited by tremie. Internal concrete lining was a subsequent operation.

It will be noted that the prefabrication was minimal, of the watertight membrane and external shuttering only, and that the concreting was all done *in situ*. It was perhaps more closely related to cut-and-cover, but under water, than to modern submerged tunnelling, but it did embody the germ of the method.

This set the pattern for cylindrical steel units and shipyard fabrication which was followed in the Detroit–Windsor tunnel in 1928–30 and subsequently at most other American east coast crossings. Circular tunnels were also adopted on the west coast, but as concrete structures built in drydock, for the Posey Tube (1925–28) and Webster Street Tube (1957–62) at San Francisco. The Bay Area Rapid Transit tunnelled crossing of the Bay was a concrete shell with two circular tubes and ancillary ducts within a steel hull. The typical American highway crossings are constructed with a single cylindrical tube of about 11 m diameter carrying two traffic lanes.

Rectangular submerged tunnel construction was initiated in Rotterdam for the Maas tunnel (1937–41). This was a highway tunnel incorporating dual two-lane carriageways and also cycle and pedestrian subways. The foundation problem of providing satisfactory bearing over a wide flat area was overcome by developing a system of sand injection beneath the units. The units, 61 m long, were partially built in drydock, floated to a jetty for completion, and sunk onto temporary supports. Steel plates were incorporated for waterproofing. Steel piling and concrete and compressed air working were used when closing the gaps. Jointing by the use of rubber gaskets between units was developed first for Deas Island tunnel, Vancouver (1959), and is now standard practice, largely dispensing with the need for compressed air or divers' work.

It is no accident that this particular form of tunnel has developed so vigorously in the Netherlands, stimulated by the geographical pecularities, although it was Danish engineers who initiated the special techniques. Three rivers, the Rhine, Maas and Scheldt, have created deltas through which they discharge to the North Sea by numerous channels, supplemented by canals, and with great cities and ports growing astride the rivers.

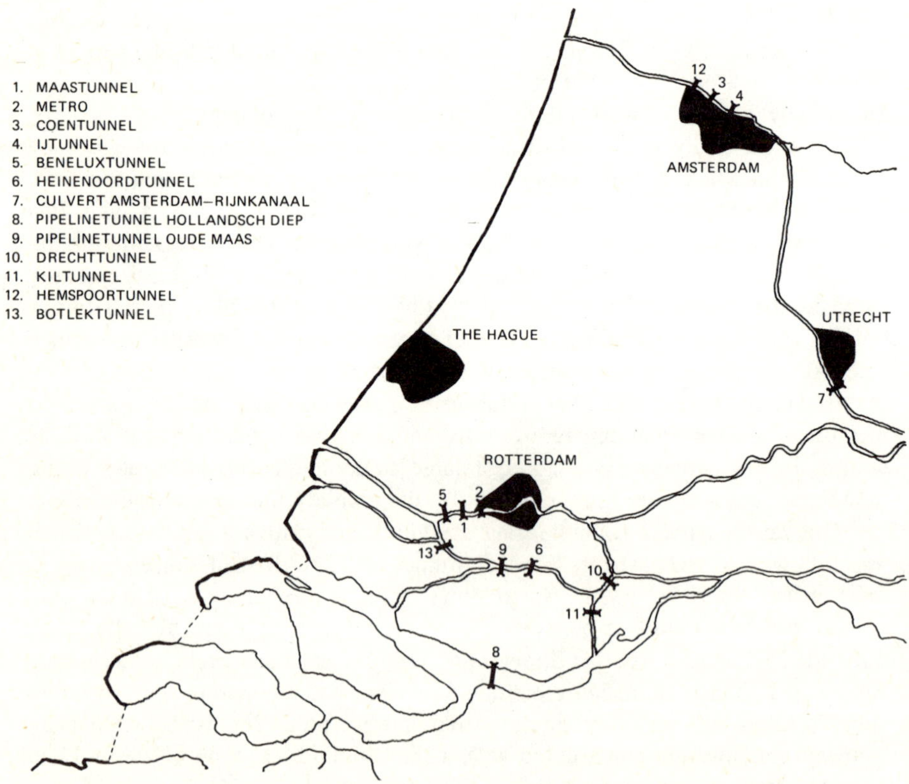

Fig. 2.2 – Principal submerged tunnels in the Netherlands.

Bridge construction has been handicapped by the need for wide and high spans for shipping and by the difficulties of foundations in soft alluvial ground. Deep bored tunnel crossings have likewise been considered unusually difficult because of the softness of the ground, but the relatively shallow submerged tunnel of rectangular form has shown great advantages, and the country's available resources in shipping and docks and navigational skills have been put to good use. The growth of these subaqueous tunnels in Holland may be compared with the sequence of Alpine tunnels in Switzerland nearly a century earlier.

The circular and rectangular sections have continued to be typical of American and European practice respectively, but each has benefited from devices adopted from the other, and the choice, in any particular case, is strongly influenced by available resources, and experience of the particular system.

2.5 SEQUENCE OF OPERATIONS

The principal operations in construction are:

1. Initial fabrication
2. Trench preparation
3. Fitting out, floating in and sinking
4. Jointing, bedding in and backfilling

The content and relative importance of the various operations differs according to whether the initial shell is of steel or reinforced concrete and whether it is of circular or rectangular form. There is no more a standard routine for this method of construction than for bored tunnels; each site presents its own problems and opportunities.

The differences in procedure between the use of steel cylinder and concrete rectangle are so marked that it is clearest to deal with the details of each method separately. Nevertheless, the general principles are clearly the same, and there is much common ground. The particular techniques of jointing and of bedding, although typical of one system or the other in practice, are not essentially so, and are now adopted as considered most suitable for the individual site. Steel cylinders and their traditional techniques are discussed first, followed by rectangular concrete tunnels.

2.6 STEEL CYLINDERS

2.6.1 Initial fabrication

The basis of design is shipyard fabrication of a cylindrical steel shell, but the finished tunnel usually comprises a reinforced concrete structural lining and a road deck, within a cylinder about 10 m in diameter of 8 mm thick steel, surrounded by concrete contained within a thin octagonal steel casing. The outer concrete, for which the steel octagon acts as formwork, provides external protection against damage and corrosion, and necessary ballast, contributing also to structural rigidity. (See Fig. 2.3.)

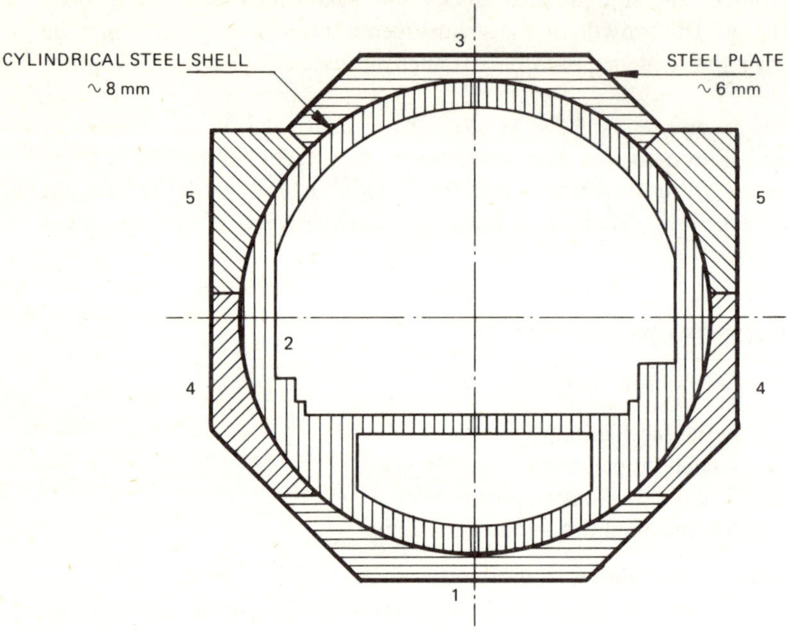

SEQUENCE OF CONCRETING

1. Structural concrete applied as a keel before launching
2. Structural reinforced concrete
3. Structural concrete added before sinking
4. Ballast concrete for sinking operation
5. Ballast concrete added by tremie after sinking

Fig. 2.3 – Typical cross section of American tunnel.

In the shipyard the steel cylinder is welded up and stiffened by transverse diaphragms connecting it with the external octagon. Watertightness of the welding is most important for the finished tunnel. A unit about 90 m in length is assembled and a concrete 'keel' is cast between the inner and outer steel casings. The open ends are closed off by temporary steel or timber bulkheads.

At this and all subsequent stages buoyancy of the units has to be carefully controlled. The weight of the steel cylinder, octagon and diaphragms may be of the order of 5 tonnes per metre and the concrete keel and bulkheads can bring this up to a launching weight of 15–20 t/m, but the water displacement weight of the cylinder is about 80 t/m. Initially, therefore, it will float one quarter submerged.

Before launching, reinforcing rods, for subsequent use, are placed within the cylinder and access hatches are provided. Sideways launching is preferred, but endways is practicable if stresses are properly allowed for.

After launching further external concrete is added including a cap to contribute to longitudinal stiffness, and concrete in the sides as necessary ballast for towing requirements.

Each end of the unit is furnished with devices for subsequent underwater connection between units as further described below.

Stresses have to be analysed for each step in construction. Longitudinal bending stresses are induced by uneven loading when floating. Initially, the end bulkheads must be considered as point loads on a beam, and subsequently the sequence of concrete placing must be planned to avoid excessive stresses. The thin steel cylinder must not be subjected to circumferential stresses which could cause buckling. Stresses must also be examined where loads may be imposed in towing and at points of support during sinking.

Fabrication of steel cylinders can be advantageous in effecting early completion of a project, because manufacture and delivery of units can be started quickly and arranged to comply with any reasonable sinking programme. If site work is delayed storage afloat of units is not likely to be difficult.

2.6.2 Trench preparation

The trench in the river bed must be excavated by dredging or by grabbing; if rock is encountered blasting may be required. Special dredging equipment is likely to be necessary not only because dredgers currently in use at a port normally operate only to the depth of the shipping channel, whereas an excess depth of 15 m or so will be required, but also because harbour dredgers are unlikely to be idle and available for any but brief periods.

It is usual to excavate the trench to about 0.6 m below the underside of the units to be sunk and with sides sloped back at a stable angle which may be about 1 : 3. The bottom has then to be filled up with gravel, screeded accurately to level. This screeding operation presents considerable difficulties at depth in a waterway. Heavy steel screeds must be controlled to follow the specified gradient and must be independent of tidal movements which affect all floating craft. Deposition of silt in the trench may make it necessary to clean out each length immediately before placing the gravel and again before landing the unit.

The cylindrical unit is relatively stiff longitudinally, and the narrow base of the external octagon can be bedded satisfactorily on the screeded surface without imposing unacceptable stresses on the structure.

Where ground conditions are variable and there is a risk of uneven settlement, piled foundations can be employed. In the Hampton Roads tunnel in Chesapeake Bay average settlement of 0.22 m has been measured but without unacceptable differential movement. It is attributed principally to rebound of soft silts, after trench excavation and before placing, followed by recompression.

2.6.3 Fitting out and sinking

The operation of fitting out will include completion of the inner concrete lining and road deck and of the external concrete, together with attachment of sinking equipment, such as masts, pontoons, winches and jacks. Some of this may be done at the shipyard site if it is reasonably near the construction site, but a special jetty may be necessary. The unit must continue afloat until required for sinking. The necessary balance of weight and buoyancy for the sinking stage may be maintained by the use of pontoons, the weight of the unit being slightly in excess of flotation, or the unit may remain self buoyant, requiring the addition of ballast, but finally it must be given a factor of safety against uplift.

At the fitting out site, the concreting of lining and roadway through hatchways will be completed, together with enough of the outer concrete shell to ensure minimum freeboard of 0.15 m or so. This brings the weight to about 99% of the displacement weight, which, for a 90 m unit, may be about 7000 tons with the final weight when in place exceeding this by about 600 tons. As much work as possible is done at the surface before sinking, and therefore pontoons or other means of temporary support are normally employed to keep the unit afloat while sinking weight is increased to about 300 tons net. Floating cranes, winches or block and tackle can be used for controlled lowering.

Position is carefully regulated by six anchor ropes, one fixed across the line from each corner and one each ahead and astern. Target masts are usually fitted on top of the unit so that level and position can be kept under continuous survey from fixed stations.

The net weight of the submerging unit is only about 300 tons and this load can be handled easily but the total weight is over 20 times as much and its inertial mass and that of the surrounding water may give rise to heavy dynamic loads in the supporting and anchoring ropes, particularly if any rapid movement develops and has to be checked. Very steady conditions are therefore essential during sinking.

Another factor which may be of great importance in tidal waters is the salinity of the water and any changes in it. Sea water has a specific gravity of about 1.025, which increases the buoyancy of a 7000 ton unit by 175 tons. Obviously the average salinity and any changes during handling and sinking of a unit must be studied.

2.6.4 Jointing and backfilling

The ends of each element are fitted with cylindrical steel sleeves projecting through square end plates and accurately made to match the adjacent units. Cover plates outside the sleeves project further, round the lower half of the circle at one end and the upper half at the other, so as to form a lapped joint when finally connected. Locating brackets project sideways at the ends of the horizontal diameter to be connected together with large diameter bolts. There may be further tensioning devices to pull the adjoining ends together (Fig. 2.4).

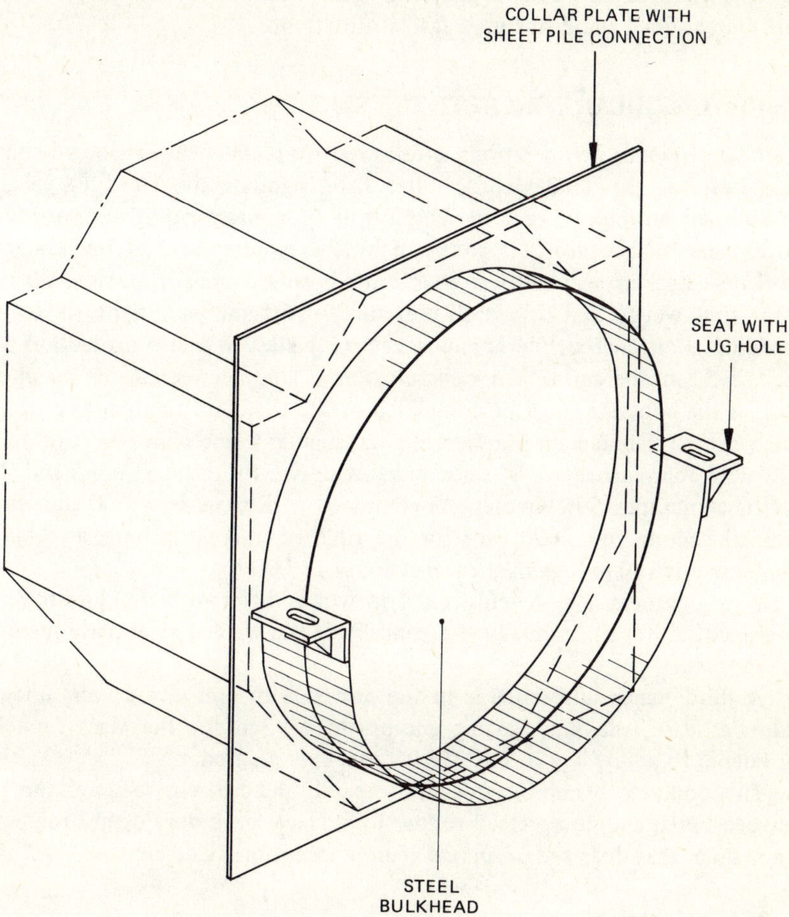

Fig. 2.4 – Perspective view of joint connection.

Divers are employed to make the connection, fit the bolts and caulk the gaps between cover plates and cylindrical sleeves. The unit is then further secured by careful placing of sand backfill under and against the sloping sides. The elaborate devices for the rectangular concrete units described below are not necessary. The vertical edges of the square end plates are fitted with sheet pile interlocks to take a vertical semi-cylindrical closer plate, within which concrete is placed by tremie to encase the whole joint. This completes the external sealing of the joint sufficiently for the bulkheads to be cut out at the appropriate time and for the steel tube to be finished off by watertight welding, followed by lining concrete.

The backfilling of the trench and protective cover over the top complete the operation. In addition to guarding against damage by shipping, the cover contributes added stabilising load against flotation.

2.7 RECTANGULAR CONCRETE TUNNELS

A similar sequence of operations is followed for rectangular reinforced concrete structures, but the detailed procedures differ significantly. The total weight of the finished structure per unit length will be approximately proportional to traffic capacity because it is governed by the requirements of buoyancy. The fabricated steel in a cylindrical steel unit, however, constitutes less than 10% of the final weight and is in itself structurally stiff and watertight, the concrete providing additional stiffness, and necessary ballast and also protection to the steel shell. In the reinforced concrete units, whether rectangular or circular, most of the concrete must be completed before the unit functions as a structure which can be handled. In practice, the rectangular tunnels have most often been built with four carriageways as compared with two for the steel tubes and also in lengths sometimes 50% greater. As compared with weights of 500 tons for the steel tube alone and 7000 tons for the finished unit 90 m long, a rectangular four-lane unit 140 m long might weigh about 28,000 tons.

Watertightness is a second aspect in which different techniques have been developed, although originally the concept of the welded steel plate membrane was adopted.

A third major difference is in the provision of uniform support under the wide flat base; the sand jetting equipment devised for the Maas tunnel was considered to be the key to this whole tunnelling method.

The concrete rectangle does not readily lend itself to steel lap joints between units, and compressed rubber joints have been developed progressively in a manner that does not ordinarily require attention by divers.

2.7.1 Initial fabrication

As already mentioned, a casting basin or dry dock is required for building of the units to the stage of floating out.

If an existing dry dock can be adapted the units can be constructed one or two at a time, but the operations of flooding and then pumping out again are time consuming and costly. It is more usual to excavate a special casting basin in which all of the units, or possibly half, can be built before any are floated. A very large flat casting area is necessary at a depth exceeding 8 m below high water. A sound foundation must be assured for each unit, preferably surfaced with gravel and possibly with other devices to ensure that water pressure has free access under the base at the floating out stage. Concreting plant has also to be installed on a very large scale.

In an existing dock, gates will be used for access to the main waterway, but if a basin is specially constructed for a single use, some more temporary device, possibly with sheet piling, will be required, and an access channel of adequate depth and width. Controlled flooding arrangements must be provided. In typical estuarine conditions ground water lowering throughout the construction period will probably be necessary to keep the dock dry and prevent heave of the bottom.

It is now usual to complete the construction of the unit except for some ballast concrete, in the dry dock. Buoyancy is of critical importance at all stages as has already been discussed for steel tubes. A weight of 99% of the total displacement weight with a freeboard of about 0.1 m is appropriate for the floating and fitting out stage, and finally after sinking an excess weight of up to 10% to give a factor of safety against uplift. These requirements determine closely the proportions of the tunnel cross section in terms of: (a) clear space; (b) wall, roof and floor thickness; and (c) ballast concrete. The specific weight of the finished concrete as actually constructed must be very accurately determined both for the high grade reinforced concrete in the structure, and the ballast concrete.

For two-lane and four-lane units at moderate depth these conditions can be met with a reinforced concrete cross section but, for a three-lane span and at depths exceeding perhaps 15 m or under high embankments, transverse prestressing may be necessary, because the structural depth in the unit cannot be increased without losing buoyancy. On the other hand, prestress in the unit before its working load is imposed may make necessary added reinforcement or other counterbalance.

Width of units is of course determined by traffic capacity required, and height primarily by traffic clearance plus concrete thickness. Additional space, if necessary, for ventilation or other purposes, may be provided laterally, but a small requirement for extra space to accommodate booster fans may sometimes be met by increased height. The length of units will be the maximum practicable for handling in the dock and waterway. In the original Maas tunnel the length was 61 m but in current practice 100 to 140 m is favoured and units up to 268 m have been used. Long units effect economy in jointing, but manoeuvrability in handling in the waterways must be maintained. Long units also present problems of stresses in handling and of shrinkage and cracking. They may lack flexibility in accommodating slight settlement on their foundations.

As already noted, overall waterproofing has developed from the original welded steel membrane. Designers have adopted combinations such as a steel plate for the base and a bituminous membrane in walls and roof, or wholly bituminous membranes. Systems have now been devised dispensing entirely with membranes and relying on the concrete alone, except that construction joints in the casting process are sealed with waterstops cast in.

In current practice the whole base slab is usually cast first in lengths with upstanding waterstops to seal the necessary construction joint between base and wall. The wall is cast in sections of 15 to 25 m to minimise shrinkage cracks, the construction joint again being crossed by a water stop. Where membranes are dispensed with, each concrete section is cooled artificially during the setting process of the concrete to extract the heat of hydration and minimise shrinkage cracking caused by the temperature differential between mature base and fresh wall concrete. Unless a very high standard of workmanship and control can be ensured, the bituminous or other membrane is still likely to be preferred.

The roof section may be cast at the same time as the walls, or may be a separate subsequent operation, the construction joints again being fitted with waterstops, for which special epoxy resin grouting techniques have been adopted.

The vertical construction joints between sections are designed, if necessary, to accommodate slight movement where settlement may be expected. Rigid units are necessary for handling in the towing and sinking operations and in the final adjustment of position. If a flexible design has been adopted it is made temporarily rigid until finally in place, either by longitudinal post tensioning cables or by reinforced concrete links across the construction joints in which the steel ties can subsequently be cut.

2.7.2 Trench preparation

There is no great difference from the procedures for steel cylinders previously discussed. The trench will normally be wider but a little shallower. An excavated depth of 0.6 m below the unit is considered appropriate. This space is not filled until after the unit is in place, when sand is jetted in. If the river carries much silt it may accumulate in the trench and will have to be cleaned out immediately before depositing the sand fill. The same jetting system may be adapted, first to blow out the silt and then to place the sand.

2.7.3 Fitting out and sinking

When the units, or batch of units, have been completed in the dry dock, they are fitted with end bulkheads, and, as appropriate, alignment towers and temporary access shafts, before filling of the dock. There are usually water ballast tanks provided in each unit to allow its buoyancy to be controlled by pumping. In many cases units are allowed to remain wholly or partly submerged in the building dock until required, only one at a time being floated and finally equipped for sinking. Pontoons are the most usual equipment for regulation of sinking, but floating cranes may be used or special jack-up platforms. Where the units are kept a little short of buoyant, pontoons may be floated into the dock over them, to pick them up with winches, the pontoons and units being transported as a single entity to the site of the trench, but this is dependent on ample depth of water in dock, exit channel and main waterway.

Sinking procedures are much the same as for steel cylinders; although the gross weight may be three or four times as much, the net weight is kept quite low, but dynamic and hydrodynamic loads must be very carefully monitored and kept under control. Motion of the unit cannot be changed rapidly.

In any but the simplest conditions of towing and placing, tests on models in a hydraulic laboratory are advisable to allow estimates of the necessary towing and anchoring power to be made and to predict the behaviour of the units in currents. Of particular importance are tidal currents possibly with varying salinity and therefore density. The final position of the unit will almost inevitably be across the flow of the current, and in a narrow channel, the unit will form a significant obstruction giving rise to substantial hydro-dynamic forces, and possibly causing scour in the channel.

At least six tugs are likely to be required to bring the unit into position for sinking. Lines from winches on the unit must then be fixed successively to the anchorages set in the river bed to allow the unit to be brought precisely into position above the trench. Lowering may be effected by first submerging the unit with the aid of water ballast, then lowering it from the pontoons very nearly to its final level and position, and after a final check landing it on its prepared points of support.

It is usual to land the unit on a preset beam or on foundation blocks at its outer end and preferably on a bracket on the previous unit at its inner end. Final adjustments both vertical and horizontal may be made by jacks working from the beam, but in recent practice the horizontal adjustment is often made by jacks acting longitudinally on the side walls at the joint under construction.

The net ground pressure from these tunnels is so low that a sand fill foundation is usually considered to give rise to little risk of significant settlement, although in soft silts there may be appreciable general subsidence. In a few exceptional cases piles have been adopted as the final foundation, and accurate level and bearing have been established by injection of grout between the heads of the piles and the bearing beam.

2.7.4 Jointing and backfilling

In connecting two units the initial water seal is established by use of a rubber joint gasket which allows the internal steel and concrete work at the joint to be completed. The rubber seal, having the so-called 'gina profile', is bolted to the face of the unit being placed and projects about 160 mm in all. The outer tip is in section a 40 mm triangle of soft rubber backed by the main body of the gasket about 160 x 120 mm of harder rubber. When the unit is positioned the nose of the gasket bears on a steel plate on the fixed unit and is pulled in

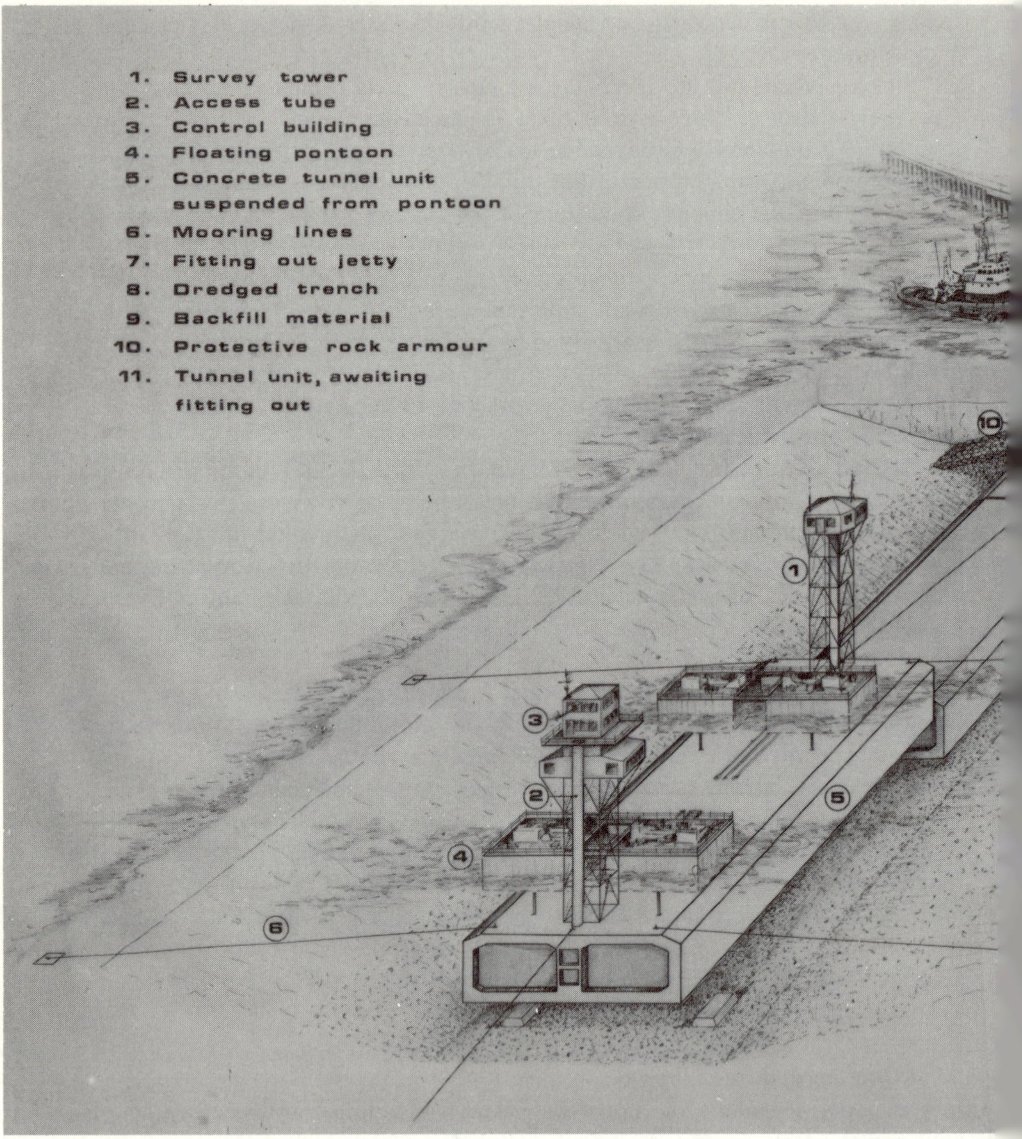
1. Survey tower
2. Access tube
3. Control building
4. Floating pontoon
5. Concrete tunnel unit suspended from pontoon
6. Mooring lines
7. Fitting out jetty
8. Dredged trench
9. Backfill material
10. Protective rock armour
11. Tunnel unit, awaiting fitting out
1
2
3
4
5
6
10

Fig. 2.5 – Sub

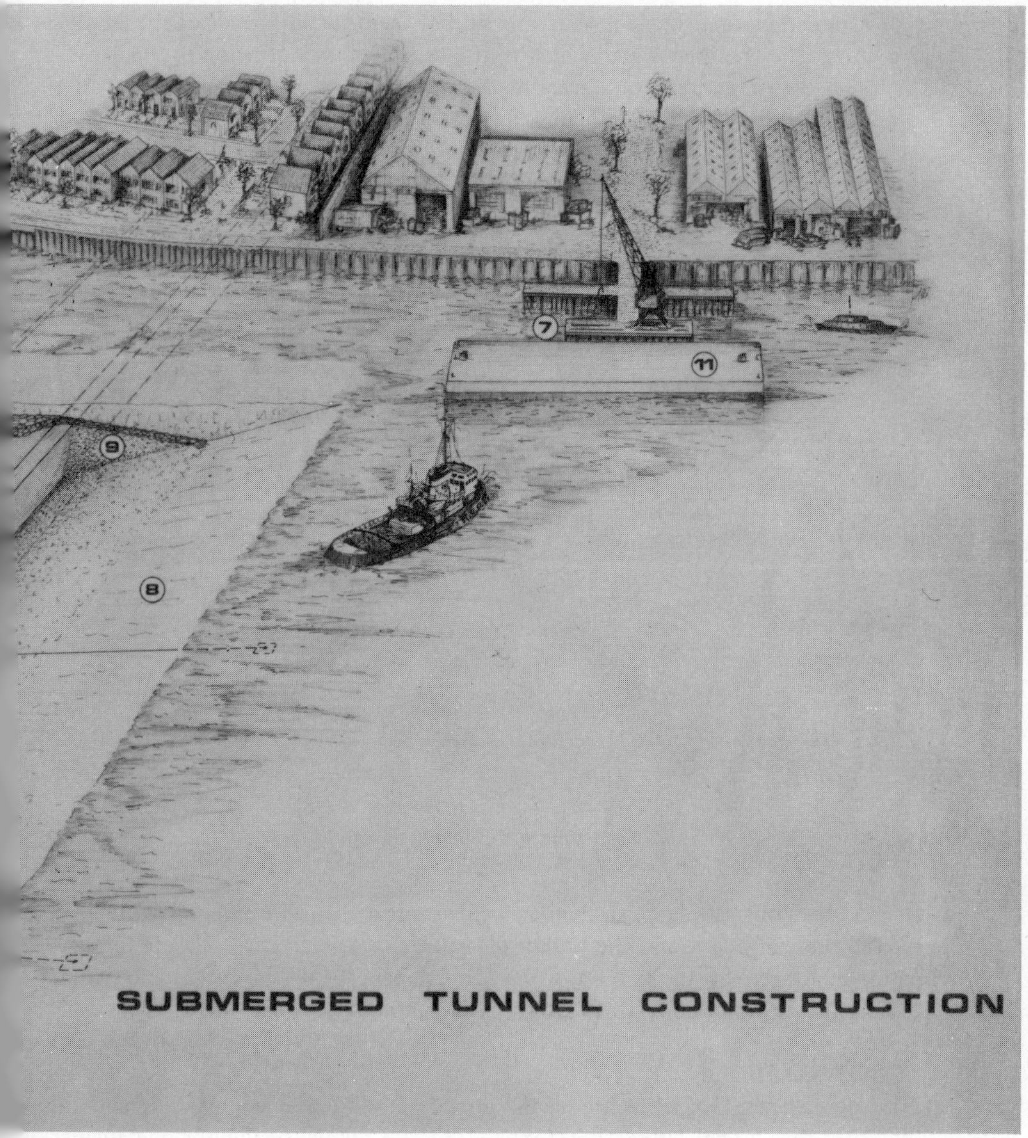

Construction.

towards it mechanically so as to compress the soft rubber tip and form a water seal.

A closed compartment is thus formed between the bulkheads of the adjoining units. The enclosed water which is at a pressure corresponding to the depth, is drawn off or pumped out, releasing the pressure and leaving unbalanced the pressure on the bulkhead at the far end. This forces the units into closer contact at the joint and compresses the main block of hard rubber. If necessary, to prevent excessive movement in compression, steel limit stops are fitted.

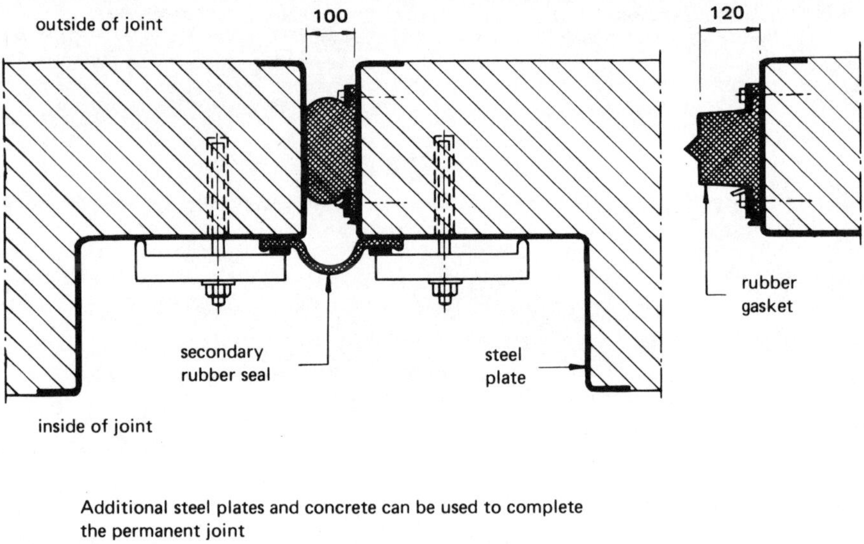

Fig. 2.6 – Joint seals for concrete units.

A steel 'omega' plate can then be welded across the gap inside the protection of the rubber gasket, and the bulkheads can be removed. The concrete lining is completed between the two units, the detail depending on the extent and nature of any future deflections which are to be accommodated.

2.7.4.1 Sand filling

The development of a technique to provide a solid well compacted sand fill beneath the tunnel units was the key to the whole system of rectangular construction. A turbulent mixture of sand and water is prepared on a barge, or elsewhere, and pumped down to nozzles beneath the unit. The jetting system can be adapted to remove accumulated silt before placing sand.

In its original form a complex special gantry was used to carry the jetting nozzles and to move and direct them so as to traverse systematically the whole area to be covered. At the same time as jetting sand in, excess water was sucked out through a suction nozzle, allowing the sand to settle as a densely compacted mass. One disadvantage of this system is that the gantry, located over each unit in turn, constitutes a continuing obstruction to shipping.

It has now been established that in favourable conditions fixed vertical nozzles, underneath the base of the unit and at a spacing of about 4 m, can inject sand satisfactorily. At the initial stage of injection the sand settles out as a conical deposit, spreading in all directions, until a central crater is formed where turbulence maintains the sand in suspension. A succession of narrow escape channels through the rim are eroded and, after contributing to the spread of the cone, become choked again. It is found in practice that virtually the whole base area can thus be given necessary support. The vertical jetting pipes may pass through the floor of the tunnel, being fed from within or, preferably, the supply pipe system may be carried round the side of the structure, being supplied from a mixing barge. It appears that the grading of the sand is not at all critical provided it is not so coarse grained and porous that water leaks away from the crater and turbulent flow is not maintained.

When support of the base is assured the sides of the trench are backfilled and the top of the tunnel is covered to the prescribed depth.

2.8 TOPOGRAPHY

The topography most favourable to the choice of a submerged tunnel is where the river crossing is wide and shallow with a sandy bottom, and with gently sloping banks and channel, and currents are not great. Depths have not hitherto exceeded about 30 m. The choice between bridge and tunnel for a new crossing has first to be the subject of careful study, which probably involves outline design, narrowed down progressively to details of type and technique. A precise balance can rarely be struck even in terms of cost: there are uncertainties as to actual circumstances encountered, and benefits and environmental value cannot be quantified with any precision.

Assuming that a tunnel is proposed, the relative merits of bored and submerged tunnels require to be assessed for the particular case. The general advantages of a submerged tunnel have already been outlined but some of the special difficulties need to be further examined. Each can usually be overcome, but if several factors are unfavourable the cost may be excessive.

2.8.1 Geology

The geology of the river bed is of primary importance. If it is of soft alluvial strata to a sufficient depth below the required clear channel level, the necessary trench can probably be excavated satisfactorily, but if a thin layer of silt overlies a rock bed the problem becomes more difficult. At the least, heavy dredging

Fig. 2.7 – Submerged tube in Australia. Single extremely long concrete unit 23 m × 5.5 m × 263 m carrying cooling water to Eraring power station under Dora creek. Unit ready to float out preparatory to sinking to form the crossing. Commission of New South Wales)

plant capable of working below normal depths will be employed, possibly with the aid of blasting. In such ground, a bored tunnel may not need so great a depth of protective cover as in soft ground, and free air or low pressure working may be feasible.

2.8.2 River profile

The transverse profile of the river bed in section along the line of the tunnel may be very significant. If the slope exceeds the acceptable tunnel gradient, then a trench which is of minimum depth at mid river, or at the edge of the shipping channel, must be made progressively deeper as the bed level rises more rapidly than the tunnel. It may be noted that if the tunnel crosses the slope obliquely the effect is to provide a gentler slope than at right angles.

Where the waterway is bounded by a quay wall the dredged depth may extend to the face of the wall, where a vertical profile is presented. Excavation of the trench up to the wall may imperil its stability and excavation behind the wall will necessarily be very deep.

A high flood protection bank will similarly add locally to the depth of trench and therefore to its width unless the sides are supported. At the site of the Dartford tunnel these conditions would have required a trench over 30 m deep through a flood bank recently raised, and having only a small margin of stability. This was one among several factors favouring a bored tunnel.

The Elbe highway tunnel at Hamburg (1971) presented a similar problem where the north bank rises steeply to a height of 35 m. The crossing is for six lanes of traffic. A submerged tunnel with reinforced concrete units was constructed under the waterway but boring was adopted for the long rising gradient under the north bank, the transition being effected at a ventilation station.

2.9 TIDAL WATERS

Tides in an estuary may make the handling and control of floating units very difficult and impose a strict timetable. Where the tidal range exceeds 5 m or so the shipping channels may not be usable at low water except for vessels of very shallow draught. Most shipping will therefore move up and down river near the time of high water which will also be the time for access through locks into and out of docks. This will also be the only period when adequate depth is available for towing tunnel units to site.

During sinking of each unit restrictions on shipping will be essential with complete closure of the channel for at least one tide and probably for 24 hours. In a tidal channel where there is a danger of a unit grounding while being towed to site, the operation should be effected on a rising tide, and to a strict and carefully devised timetable. Contingency planning is essential against the risk of grounding and thereby blocking the channel. Safe deepwater anchoring points may be identified, to be utilised in the event of unforeseen delays.

Once the unit is in place over the trench the operations of anchoring, sinking and final location may take about eight hours. Sand filling and subsequent removal of equipment may take several days, during which some restriction of shipping may continue to be necessary even if the channel is very wide.

Tidal currents in river estuaries may be very strong, and are likely to vary in direction and ultimately reverse as the tide ebbs and flows. Flood water in a river system may further complicate navigation. Powerful tugs adequate in capacity and numbers are essential for handling the units, which will be particularly difficult to control when turned across the channel ready for anchoring and sinking.

Salinity and therefore buoyancy may vary during the rise and fall of a tide, and it sometimes occurs that a front of advancing salt water pushes forward as a wedge beneath river water flowing downstream at the surface. A further phenomenon in tidal rivers may be the slow transport seawards of a relatively thick layer of silt at the bottom. This would be trapped by the tunnel trench and must be removed before backfilling under the tunnel.

In some estuaries the combination of large tidal range, convergent shore lines and a steeply rising bed, may result in the formation of a 'bore' like that of the river Severn in which the rising tide advances as a very steep wave front, which would obviously be a serious hazard in the handling of tunnel units.

2.10 ADVANTAGES

Provided that the difficulties can be overcome, the submerged tunnel of rectangular form gains added advantage when more than a two-lane tunnel is required. The cost of dry dock construction, if it is needed, is spread over the larger project and the trench is excavated with the same plant and less than double the volume of excavation. The longer the submerged tunnel, of course, the less significant is the heavy cost of a special casting dock and other special plant and equipment.

The advantage of shallow depth and short approaches up to ground level can be of overriding importance in city development where land is closely built up and in some cases space for access roads and ramps is more readily available in the immediate vicinity of the river.

Where the choice has been made in favour of steel cylinders rather than concrete it is obvious that a fundamental factor is the availability of low cost fabricated steel. The extra depth required may be quite acceptable, and speed of construction is likely to be much greater, partly because of the time required for preparing and equipping a suitable dock. Cylinders normally accommodate two traffic lanes only and therefore the rectangular form gains some advantage when four or more lanes are required as they can be embodied in a single unit with one joint, where the cylinders require two. Steel cylinders can more easily be made seaworthy for towing in exposed waters.

2.11 CHANNEL TUNNEL

The Channel Tunnel is in itself a subject of interest, but is of particular relevance in this chapter in exemplifying the limitations of existing methods and possibilities of development.

From early in the 19th century various forms of submerged structure have been suggested for a Channel tunnel, some laid on the sea bed and others in trench. It would certainly be essential that the structure should be in trench at a safe depth below possible damage by ships, and that any section above bed level should have very substantial protection. In the context of a rail tunnel the subject is discussed in Volume 1, but if a highway tunnel is to be contemplated, different considerations arise. Ventilation necessary for road traffic is not practicable for a full length tunnel without shafts, but much steeper gradients are acceptable for road than for rail, and bridge-tunnel schemes are suggested with shorter tunnelled lengths under the shipping lanes connecting with bridged lengths at artificial islands. In such a scheme the width required for multi-lane carriageways and ventilation ducts is a factor favourable to submerged unit construction. One scheme proposed embodies roadways on viaduct from each coast to offshore islands, with a central section in submerged tunnel ventilated by shafts up to artificial islands, all in combination with a through rail tunnel for the whole length.

The whole project, however, is on a scale so much greater than any existing submerged tunnel that every detail would have to be analysed critically. It would be very dangerous to accept as practicable any aspect because it is merely an increase of scale in a proved technique.

A factor favourable to the method is the great length of tunnel which offers the advantages of repetitive fabrication and handling, increasing experience and maximum use of equipment. In contrast, the great length to the working face of a bored tunnel is a severe logistical handicap. Access through the length of the submerged units would also be necessary for jointing and adjustment and the necessary ventilation might be difficult.

Among the principal adverse factors are the depth, the problem of gradients, the open sea exposure and the intensive international shipping.

Depth greatly exceeds that of any existing submerged tunnel for traffic, and the relationship between structural depth and maximum stress in slabs and the requirements of buoyancy might well preclude the use of a rectangular concrete structure, necessitating a circular shell, or arched form. On the other hand, temporary buoyancy might be provided by attached pontoons or special floating craft.

Depth also adds to the difficulties of making the structure watertight, particularly at the joints between units. Any access by divers for jointing or remedial work would be at pressures exceeding 5 atmospheres, requiring special modern diving techniques. Depth would also make more difficult precise control of placing by the long ropes needed for lowering and for anchors, so special plant would have to be devised and developed to function reliably.

For a railway tunnel, imposing strict limitations on gradients, the contours of the sea bed may present a problem. In normal surface construction in undulating country the required profile is based on a balance of cut-and-fill, working to the preferred gradients. For the submerged tunnel, embankments, except perhaps very locally and on a small scale, are undesirable as being difficult to build, vulnerable to shipping, and possibly subject to erosion and subsidence. Where the slope of any undulation exceeds the limiting gradient, increased depth of cutting becomes necessary with disproportionate increase of excavated volume. The surface technique of resorting to bored tunnelling where cutting depth becomes excessive is not available.

The disposal of material excavated from the trench and the procurement and placing of fill would present considerable problems of handling methods and cost.

The exposed conditions in the open sea are not altogether without precedent as in the Los Angeles sewer outfall and in offshore oil operations, but they are unprecedented in the handling and placing of units such as these. In the English Channel the tidal range is of the order of 5 m and tidal currents are strong and changeable. Stormy weather is frequent, with unpredictable wave action. The problems of precise location become formidable, even more so in fog. No doubt radar controls, laser beams and other devices could be adapted and developed, but at mid-channel, over 15 km from land, they would need to be very accurate and reliable, including projection downwards through 50 m of water.

Shipping in the Channel is international and includes almost the whole range of seagoing craft from Channel swimmers up to very large oil tankers. Regulation and control of such shipping in sufficient detail to safeguard anchored construction craft, in new positions as each unit was placed, would demand powerful and effective authority, not presently established. The hazards to shipping from any bridge-tunnel scheme would necessitate elaborate precautions, and also effective control.

It is not surprising that the estimates of cost and the uncertainties of feasibility have hitherto ruled out this technique for the Channel tunnel, but undersea oil exploration and exploitation have led the way in developing new methods and equipment, and it may well be that such a scheme will become practicable and competitive.

BIBLIOGRAPHY

For up-to-date details of submerged tunnel construction, and for historical records, it is necessary to refer to papers published in the proceedings of engineering institutions and of specialist conferences and also to articles in the technical press.

The earliest examples in the United States and Holland respectively are described in:

Kinnear, W. S., The Detroit River Tunnel, *Trans Amer. Soc. Civ. Engrs*, 1911, **74**.

Wilgus, W. J., The Detroit River Tunnel, *Min. Proc. Instn Civ. Engrs*, 1911, **185**.

Van Bruggen, J. P., The Road Tunnel under the River Maas at Rotterdam, *Engng*., 1940, **150** (Aug. 9 and 30, Sept. 27).

Important general papers on the subject, which include historical material and more extensive bibliographies are:

Brink, A., Recent developments in the design of submerged tunnels, *Struct. Engr*, 1966, **44** (Feb.); *see also* discussion, 1966, **44** (Aug.).

Bickel, J. O., Trench type subaqueous tunnels: design and construction, *Struct. Engr*, 1966, **44** (Oct.); *see also* discussion, 1967, **45** (Feb.).

Pequignot, C. A., Selective bibliography on immersed tubes, *Tunnels and Tunnelling*, 1969, **1** (July).

Brakel, J., Some considerations of submerged tunnelling, *Proc. Instn Civ. Engrs* 1971, **48** (Apr.); *see also* discussion, 1972, **51** (Jan.) (esp. Prangnell, K. J., pp. 142–144, on J. I. Hawkins' experimental tunnel under the River Thames at Rotherhithe in 1811.)

Williams, G. M. S. and Innes, K. W., Structural aspects of submerged tube tunnel construction, *Struct. Engr*, 1972, **50** (Feb.); *see also* discussion, 1972, **50** (Aug.).

Palmer, W. F. and Roberts, K. C., Developments in trench-type tunnel construction, *J. Constr. Divn., Proc. Amer. Soc. Civ. Engrs*, 1975, **101** (Mar.).

Glerum, A. *et al*., Motorway tunnels built by the immersed tube method, *Rijkswaterstaat Communications*, 1976 (25).

Brakel, J., *Submerged Tunnelling*, Technische Hogeschool, Delft, 1978.

Immersed Tunnels, *Delta Tunnelling Symposium*, Amsterdam, 1978.

Many of the tunnels have been described individually in some detail. References to them will be found in the above general papers and in the general bibliography in this volume.

3

Shafts and Caissons

3.1 WORKING SHAFTS

For many tunnels sinking of a shaft to provide working access is the first operation. Even where access for progressive excavation from a portal or from an adit is possible, shafts are usually advantageous in any but short tunnels in providing additional points of attack and more favourable working sites, as well as avoiding initially the complexities of portal construction.

A working shaft may be sunk as such and subsequently abandoned, but most often the shaft will have a permanent function and its siting and design will be such as to serve both purposes.

Working shafts are typically vertical but may be inclined, in which case special safety precautions against runaway vehicles are necessary. Neither in terms of construction nor of function is there a clear demarcation between an inclined shaft and an inclined tunnel. For escalators, at a slope of 30°, the term 'shaft' is widely used, perhaps because it replaces the old lift shaft.

The fundamentals of construction of shafts are very similar to those of horizontal tunnels, and may be compared briefly. Excavation is safer in that there is no overhead ground to be taken out, but all spoil has to be lifted and hoisted, without delay, to clear the working surface. Immediate support of the exposed ground generally presents less difficulty than in a tunnel, while for permanent support the same systems are available and linings are more easily placed. Management of water is of major importance: all water entering finds its way quickly to the shaft floor unless intercepted, and unexpected water presents an immediate threat of flooding. Another water hazard is heave of the bottom due to upward water pressure from below an impervious stratum. It may be noted that shafts have to be sunk through whatever ground happens to overlie the tunnel they serve; there is usually some choice of stratum for the tunnel, but not for the shaft.

A working shaft is ordinarily used for all construction access, but special shafts may sometimes be provided for such temporary functions as conveyor disposal of spoil, or for ventilation. Where compressed air working from the earliest stages is intended the shaft must accommodate an air lock. Shafts for tunnel construction do not often extend to the great depths of hundreds, or even thousands, of metres required for mining. The greatest depths are probably in hydroelectric work, including pumped storage schemes, to convey water at high pressure to the turbines. In long rail or road tunnels in mountain areas, intermediate shafts for ventilation may have to be deep but depths may be excessive, as in the Mont Blanc highway tunnel where ground cover exceeds 1000 m over 80% of its length of 11.6 km, and there are not intermediate shafts.

In many of the early English railway tunnels speed of construction was sought by sinking shafts at short intervals so that many tunnel faces could be worked simultaneously. From each shaft two faces could be worked, by miners and bricklayers alternately. In the very early Kilsby tunnel (1838) no fewer than 18 shafts were sunk to drive a tunnel 2.2 km long. Wider spacing, up to 1 km, was more usual subsequently. Of course, smoke clearance was another reason for provision of intermediate shafts.

For metro systems, which are constructed as near as practicable to the surface, shafts are correspondingly shallow. Permanent shafts are usually necessary in the vicinity of stations and can be sunk at an early stage with adaptations as appropriate for the tunnelling operations. Suitable sites for intermediate working shafts are often difficult to obtain in built-up areas.

Subaqueous tunnels are usually driven from a shaft or pit on each side of the waterway. Brunel, for his Thames tunnel, sank a single 15 m diameter brick shaft to a depth of about 17 m and launched the shield from the bottom. For the first tube of the Mersey Kingsway tunnel (1971) shafts were sunk on each side of the river at the sites of the ventilation stations and were used for four pilot tunnel drives, riverwards and landwards, but the main tunnel was driven from the south portal, cutting straight through to the site of the north ventilation station.

3.2 REQUIREMENTS

The principal requirements for an all-purpose working shaft are:

1. Availability from an early stage of construction until completion. The sinking of a shaft, from the bottom of which tunnelling can be started, is often the first site operation. If the shaft is ultimately intended for permanent use, its use for construction access may have to be given up at some stage as work proceeds, but adequate alternative access can then usually be secured.

2. The minimum acceptable area depends on the length and diameter of the tunnel to be driven but, except for a short length of small diameter tunnel, a shaft diameter of about 4 or 5 m is the least that will accommodate reasonably hoisting space for materials of construction and equipment, and a safe ladder bay for man access. A diameter of 6 m or more is much to be preferred,

especially where the depth and extent of the work calls for a separate man hoist. For a shield driven tunnel the shaft must provide space for lowering the shield components, and possibly for their assembly, although a separate shield chamber as an enlargement of the tunnel is often preferred, and is of course unavoidable where the shaft does not lie on the centre line of the tunnel.

The shaft must in any case be of sufficient diameter for the construction of a breakout at the bottom, whether for the actual tunnel or an access passage. Its diameter should preferably not be less than 1½ times the diameter of the breakout, because it is very difficult to support adequately the opening formed by the intersections of two nearly equal cylinders.

3. The working site must be reasonably level and have space for a crane, whether a fixed derrick or a travelling crane, and for storage of materials and of spoil. The materials to be stored include a stock of timber and steel for temporary support, tunnel segments or other permanent lining materials, cement and aggregate. Heavy ground loads, as from stacks of segments, should not be close to the perimeter of the shaft. For spoil disposal from a restricted working site a minimum stock pile is necessary to maintain the balance between output and disposal. An overhead hopper on a gantry is often used in city sites.

4. There must be good road or rail access to the shaft for delivery of materials and disposal of spoil.

5. The first working shaft has to be used to carry down the required tunnel alignment from the surface survey network.

In very many cases, and almost invariably in urban and subaqueous tunnelling, alluvial deposits, often water-bearing, overlie the chosen tunnelling stratum which may be clay or rock. The upper part, at least, of the shaft must therefore be excavated and supported through soft ground. In urban conditions it is likely to be of particular importance to minimise ground movements and strains which could result in settlement and cracking of adjacent buildings. The construction techniques must be decided with this in mind. In urban conditions, also, noise and other inconveniences during shaft sinking, tunnelling and in permanent use, must be kept within acceptable limits.

It should be appreciated that such relatively shallow shafts, required only during the construction period, are not comparable in magnitude with deep mining shafts, where heavy capital costs are necessarily incurred. Savings in time, permitting earlier mine development, and in winding costs, for large volumes of material over a long period of use, can justify substantial expenditure on special sinking techniques and equipment. The permanent headgear may be installed and utilised for the sinking operation. Diameters, commonly 6 m to 8 m, allow for ventilating air for the mining work as well as hoists for men and materials. The choice of methods and expedients for tunnel work must be adapted to dates of site access, availability of equipment and suitability for the ultimate permanent use of the shaft.

3.3 PERMANENT SHAFTS

In addition to provision of temporary construction access, shafts may be permanently required as elements in the tunnel system for many purposes, such as:

1. Lifts: (a) originally common for passenger access to metros, but now largely superseded by escalators; (b) access to underground power stations.
2. Escalators: inclined shafts, usually at 30°, for passenger access to metros and also to pedestrian and cycle subways.
3. Stairs and ladders: principally for emergency access and for maintenance.
4. Ventilation: (a) natural ventilation of main line and metro rail tunnels, and draught relief; (b) forced ventilation of highway tunnels, metros and rail tunnels.
5. Conveyance of liquid: (a) as part of the main flow in hydroelectric pressure tunnels; (b) collecting surface water for conveyance by tunnel; (c) drop shafts in sewers, where steep gradients are not acceptable; (d) undersea sewer outfalls; (e) undersea intake or outfall for cooling water for power stations.
6. Pipes and cables in river crossings.
7. Drainage and pumping, particularly from subaqueous tunnels.

Shafts are also constructed for purposes not directly ancillary to tunnelling:

8. Mining shafts (which have already been referred to).
9. Temporary storage and treatment of sewage.
10. Bridge and other deep foundations.
11. Hydraulic lift pits.
12. Wells.

3.4 WATER IN SHAFTS

Many shafts, perhaps a majority, are sunk through water-bearing ground and the management of this water is a most important factor in sinking technique. In urban tunnelling alluvial deposits must usually be penetrated to reach the chosen tunnelling stratum which may be clay or rock. In subaqueous tunnels soft water-bearing strata are likewise to be expected, and the shafts may be started through open water in a cofferdam or caisson. The principal methods of handling such water problems within the shaft are: (1) pumping, from a sump in the shaft kept ahead of the main excavation; (2) ground water lowering, either by well points or deep well pumping; (3) grouting of pores and fissures; (4) freezing; and (5) compressed air working.

The principles of these methods are discussed in Vol. 2, Chapter 5, Ground Treatment, and Vol. 1, Chapter 5, Compressed Air Working, but particular applications are further referred to below. High water pressures at depth impose a limitation on pumping, groundwater lowering and compressed air working.

3.5 SINKING TECHNIQUES

Techniques of sinking differ principally in the manner of providing support for the ground. More than one method may be utilised for a single shaft. In sound dry rock immediate ground support may be unnecessary, particularly if smooth walls can be ensured, as by reaming. The principal methods of shaft construction and lining include: (1) timbered pit; (2) sheet piling and secant piling; (3) concrete walling in slurry trenches; (4) precast segmental linings; (5) *in situ* concrete lining; (6) caisson sinking, in free air or compressed air; and (7) rotary drilling.

Whatever method be adopted for the deeper part of the shaft, the initial work, on land, is nearly always the levelling and preparation of the working site, the laying down of a kerb defining and protecting the perimeter of the shaft and the provision of a crane. Where the shaft has to be sunk through open water the first step will be construction of a staging, if practicable, and a sheet piled cofferdam or, alternatively, the use of a caisson built from a staging *in situ* or floated out.

3.5.1 Timbered pit

The principles of timbering in support of the ground in sinking a shaft vertically are very similar to the support in a horizontal heading. In soft ground excavation may be taken to considerable depths even when wet by the use of piling or forepoling. Poling boards are driven down ahead of excavation and secured by heavy walings as space is excavated. A pumping sump is maintained ahead of general excavation. Excavation thus proceeds in stages governed by the length of the poling boards in each setting.

Where the ground is suitable the sides of the initial excavation from the surface are often supported by timber runners or steel sheeting set vertically behind a timber or steel frame and driven down as excavation proceeds. This allows a depth of several metres to be taken out as the initial operation and simplifies the timbering compared with forepoling. Such a pit may serve as a starting point for a deeper permanent shaft of circular form with cast-iron or concrete lining. Apart from the advantages of an early start to work with readily available resources, complex permanent works may ultimately be required at the head of a shaft and the temporary timbered pit is readily adapted to requirements for later completion.

3.5.2 Sheet piling and secant piling

A shaft of limited depth may be wholly accommodated within a steel sheet piled

Fig. 3.1 – *In situ* concrete lining forming collar at top of shaft. (*E. Nuttall Ltd.*)

excavation whether on land or in water. The cofferdam can be made of any desired size and shape to accommodate headworks for the shaft. Depths exceeding 20 m or so make it progressively more difficult to ensure straight driving and secure interlocking of the piles. If the shaft has to descend into rock there is liable to be leakage of water at the soil/rock interface, and a sufficient margin of space between shaft and cofferdam must be left for appropriate water sealing devices in addition to allowing for possible deviation of the piles.

Fig. 3.2 – Sheet piled shaft with timber framing.

Steel sheet piling is obviously very useful where soft water-bearing ground of limited depth overlies a stratum into which the piles can be driven, for example in initiating a shaft to be driven through water-bearing gravel above a bed of clay in which the main tunnel is sited. It is also a most important method for intake and outfall shafts to be sunk under water.

Secant or contiguous bored concrete piles may serve the same function as steel sheet piles in preforming the shaft walls in soft ground.

In metro work the whole area of a station is sometimes excavated from the surface in a single pit, the running tunnels being then driven out from or into the bottom.

3.5.3 Slurry trench walls

This is an alternative to sheet piling or a timbered pit in relatively shallow shafts or shaft heads through soft ground, particularly granular soils. Narrow trenches are excavated, where shaft walls are required, by accurately controlled grabs, the sides being supported by the use of bentonite suspensions or other mud slurries. No timbering is required to retain the sides of the trench. Concrete to form the shaft wall is deposited by tremie pipe in the slurry-filled trench and displaces the slurry, which can be recovered for re-use. Shaft walls can thus be constructed, with reinforcing steel if necessary, before the main excavation is started, and the time taken and obstruction caused by timber strutting is avoided. Straight walls are usually implicit in the trenching techniques, and an octagon may be constructed to accommodate a finished circle.

The method is perhaps most useful in urban conditions. There is less danger of settlement in adjacent ground and structures and the noise and vibration of pile driving are avoided. Sound concrete requires careful supervision, and if a good surface finish is required a secondary lining will probably be necessary.

3.5.4 Precast segmental lining

Lining with circular cast-iron tubbing is one of the historic methods of shaft sinking and can be used to great depths, with increased thickness of lining where necessary for higher pressures. Precast concrete, pressed steel or welded steel segments are used similarly. The relative merits are discussed in Vol. 1, Chapter 8 on Permanent Linings.

A great advantage of segmental lining over *in situ* concrete is that in wet and difficult ground each segment can be set in place with a minimum of excavation and the whole ring will immediately develop its full strength in resisting compression as soon as all its segments are in place.

The description following is of a C.I. lined shaft in soft ground, sunk by underpinning from a prepared level which may be at the surface or at the bottom of a pit or cofferdam. It is assumed that excavation is by hand or by grab. Open excavation to a depth of a few rings is frequently possible with the bottom accurately levelled off and prepared for the building up of C.I. rings to above starting level. A supporting collar embodying if necessary a steel joist or other platform is constructed to secure the rings against any settlement as excavation progresses. The rings may be built up to above ground level to form a safety barrier. Shaft sinking proper then starts by the technique of underpinning. If the ground continues dry and firm, excavation for each ring may be taken out completely, possibly with a little timbering locally at the sides, followed by assembly and bolting up of a ring. It may be noted that, unlike tunnel rings, a special set of 'key and two top' segments is usually unnecessary. The shaft segments are identical and closure can be effected by taking out a little extra excavation behind the last pair of segments, which allows them to be fitted in from the outside. It is then important to seal the bottom of the

Fig. 3.3 – Ventilation shaft with concrete walls cast in slurry filled trenches.

Fig. 3.4 – 8 m shaft with concrete segmental lining being sunk through clay by underpinning.

ring and grout up the back of the ring, not only to seal against water but to fill any voids and prevent settlement of the ground and uneven pressures on the ring. It is essential at all times to ensure that the weight of the completed lining is adequately supported and cannot slide down.

Where the ground is loose or water-bearing, particularly in granular soils, it becomes undesirable to excavate the whole area of the shaft for a depth of one ring before erecting any lining. By local excavation of a small pit each segment successively can be set in place and bolted up, strutted back if necessary from the 'dumpling'–the ground left undisturbed in the centre of the shaft. A drainage sump can be kept ahead of the general excavation with the water level controlled by pumping or baling. Early and thorough grouting is even more important where water is present.

In a cast-iron or concrete lined shaft compressed air working can be adopted if necessary by provision of an air deck at the head of the shaft and an air lock for entry. With compressed air the water level can be pushed down to the bottom of the completed lining. It is of great importance in such a shaft to ensure that the air deck is designed to resist the maximum working pressure, and is sealed against leakage which may be through honeycombed concrete, or joints affected by shrinkage. In the shaft itself all joints will require caulking and bolts will require grummets.

Cast-iron is lacking in tensile strength and the shaft should not be subjected to tension from the uplift pressure of the compressed air. Kentledge should be added above the deck accordingly. Circumferential tension should also be minimised by grouting.

The general principles of compressed air working are described in Volume 1, Chapter 5.

In deeper shafts water pressures may be too high for such measures to be adequate. Pumping from external deep wells or from boreholes sunk through the floor of the shaft may effect adequate lowering of ground water levels. Advance grouting, or freezing, may be found necessary.

3.5.5 In situ concrete lining

In dry rock which is, at least temporarily, self-supporting, *in situ* concrete has substantial advantages in avoiding the problems of manufacture, delivery and storage of segments, and in being readily adaptable to varied conditions of overbreak. It has the disadvantages of not providing immediate strength, and of requiring formwork, which has to be erected, filled, left in place for the time required by the concrete and then stripped out and lowered for the succeeding lift. *In situ* concrete is probably the most widely employed lining for deep shafts. Excavation and lining have to be executed from a set of stagings, lowered as the work advances, equipped for drilling, for loading and hoisting and for shuttering and concreting. Excavation is commonly by drilling and blasting, the spoil being loaded into hoisting skips by a grab or mechanical

Fig. 3.5 – Deep rock shaft, prior to *in situ* lining. Working platform suspended. (*E. Nuttall Ltd.*)

shovel. The operation is cyclic and not continuous, and it is of great economic importance to ensure a properly balanced cycle of operations minimising loss of time. Where rock conditions permit, lifts of about 8 m for the lining are common, against which there may be four or more cycles of excavation. Ventilation adequate to clear the blasting fumes quickly is important.

Water poses the usual difficult problems and may be controlled by grouting or freezing. It is, of course, necessary to protect freshly placed concrete against flows of water, which might wash out the fines. Separation of lining from rock and provision of drainage channels is sometimes effected by insertion of corrugated sheeting against the rock. There is also the problem of thermal insulation where concrete is to be placed against frozen rock, both to avoid frost damage to the concrete while setting and to avoid thawing of the surrounding ground by the heat of the concrete.

3.5.6 Caissons

A caisson is essentially a watertight box, of which the base at least is prefabricated, set in position and sunk there by excavation from within. The walls and any required structure may be built up progressively as the caisson is sunk. The caisson may be open and be sunk by open grabbing, or it may be decked across at any desired level to provide a working chamber capable of retaining compressed air and equipped with an air lock. An open cylinder such as that used by Brunel to form the working shaft for the Thames tunnel is the same in principle. So is a 'monolith', the term being generally descriptive of a substantial concrete box, as used for quay walls, sunk into place by open grabbing.

A caisson may be used to form a shaft either on land or through water. It may also be constructed to embody a section of tunnel of special complexity, such as the ventilation supply and exhaust shafts and their connections into the tunnel, as in the Dartford and second Blackwall tunnels. In the Amsterdam metro a difficult length of tunnel in the city was constructed by sinking a series of caissons subsequently connected with the aid of nitrogen freezing; one reason for use of the method was to minimise settlement of the adjacent ground.

An open caisson is built up from a cutting edge set in position at the surface or in a pit. The walls are built up vertically together with any necessary cross walls. In dry ground men can excavate at the bottom out to the cutting edge, and the caisson sinks under its own weight and that of any added kentledge. Control of sinking may be aided by jacks or winches. In wet conditions excavation may be entirely by grab until a firm bottom is reached and penetrated by the cutting edge, which allows the caisson to be pumped dry. Sinking in these conditions is more difficult to control, and the caisson may finish out of position and out of level. Some allowance for this is necessary in the original design.

The use of compressed air can greatly assist precision and control and minimise loss of ground causing settlement in the vicinity.

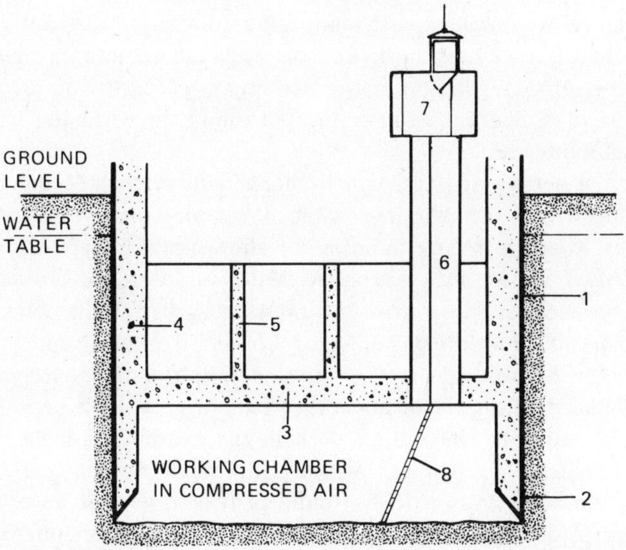

1. Steel shell
2. Steel cutting edge
3. Ait tight deck
4. Walls built progressively during sinking
5. Internal structure as required
6. Steel 'figure 8' access shaft with –
7. Ait locks for men and materials
8. Access ladder

Fig. 3.6 – Compressed air caisson.

The caisson must be provided with an airtight deck, usually 2.5 m or so above cutting edge level, through which passes a working shaft capped by an air lock. This forms a working chamber within which men can excavate, the spoil being hoisted in skips up the shaft and through a muck lock. Initially, excavation in free air may be practicable but, as the head of water externally increases, compressed air working becomes necessary. Water pressure can be balanced precisely at cutting edge level, unlike the conditions at the vertical face of a tunnel. The structure within the caisson will be built up as sinking proceeds, but careful regulation of weight is always necessary. There must be sufficient weight to overcome skin friction on the sides of the caisson, but there must be no risk of it descending out of control. In tidal waters the necessary balancing of water pressure by air means that at high water there is added uplift. Kentledge is often necessary at some stages of the work, or water ballast, which can be pumped in and out, may serve. Lowering of the air pressure, after a shift has completed excavation and come out, is known as 'blowing down' and may be a regular procedure. It can greatly assist sinking but needs to be done under control if the caisson is not to lurch suddenly and move out of position. For deep narrow shafts in particular, frequent careful surveying is needed to know what is the exact position and attitude of the cutting edge and to allow instructions to be given for excavation to be regulated accordingly.

Resistance to sinking, particularly skin friction, is not easily predictable, and may be unexpectedly high. Caisson walls are sometimes built to a slight inward taper, or stepped in above the cutting edge with the intention of reducing friction. Bentonite slurry injected round the perimeter has been found to be a useful lubricant.

The final setting to level can be done with reasonable precision by first constructing within the working chamber suitable concrete blocks finished to the proper level, with any appropriate allowance for bedding in. The final blowing down allows the caisson to settle on to those blocks, but it will probably be necessary to restore the pressure again for the concrete filling of the remainder of the working chamber, followed by grouting up.

It has been assumed in the above description that a separate working chamber beneath the level of any other structural work is provided, but it is, of course, possible to site the air deck in the caisson at a higher level, and to construct whatever is required within this much deeper working chamber. This may allow the caisson to be founded at a somewhat lesser depth but it makes control during sinking much more difficult with an unnecessarily large volume of air and a more top-heavy caisson.

Fig. 3.7(a) – Working chamber in caisson at Dartford Tunnel. Exploratory drilling in progress.

Fig. 3.7(b) – Caisson at Dartford Tunnel. Structure above working chamber forming shield chamber for reception of tunnel shield.

Fig. 3.7(c) – Excavation at cutting edge of caisson.

3.5.7 Drilling

Various drilling methods for shafts have been developed in recent years, including mud slurry methods and offshore techniques used in drilling for oil. Large diameter augers have been used in clay, at relatively shallow depths, but probably more for foundation work and for site investigation than for shaft construction. In mining work shaft raising in rock by boring and reaming machines has become established but diameters have usually been limited. Experience with TBMs can usefully be applied to such shaft work.

The provision, installation and removal of any such boring machines is a costly operation and unless the required shaft is a deep one or there are special circumstances, it is unlikely that the expenditure can be justified. Except for undersea outfalls or intakes, few tunnel access shafts have therefore utilised these methods. An exception is the St. Gotthard highway tunnel, where two of the four deep ventilation shafts were constructed by machine boring. The shafts in question were inclined at about 40° to the horizontal and were bored upwards at a diameter of 3 m and then enlarged downwards to the required 6.6 m, the lengths being 860 m and 500 m through paragneiss and granite.

In drilling shafts downwards, a major problem is the clearing of the broken rock from the face being drilled. Provided that satisfactory fragmentation of the rock can be obtained, circulating mud slurry can be used to remove the spoil. The mud slurry can also serve the purpose of supporting the sides of the excavation until excavation is complete, when a steel cylindrical lining can be lowered in and sealed and grouted into place.

The merit of raise boring is that the fragmented rock falls away from the face, but the method depends on having proper access at the base of the proposed shaft for drilling and removal of spoil. In raise boring it is common first to drill a central pilot hole and then to ream this out to the full required diameter. The pilot may be drilled either up or down and likewise the enlargement. If the enlargement is drilled downwards, spoil can be discharged through the pilot. It is obvious that if uncontrolled water, or unconsolidated strata are encountered in drilling up, difficulties may be very severe.

In shaft sinking to a depth of about 200 m, a method employed has been mud drilling of the required shaft followed by lowering with the aid of water ballast of a cylindrical steel liner sealed with a base plate. The cylinder was extended in 5 m lifts as it sank and was finally grouted in.

3.6 SHAFT BOTTOMS

There are various special considerations in constructing the bottom of a shaft, such as precautions against heave, sealing of the base, provision of a sump and pumps, and provision for breaking out for tunnel construction.

Even in a relatively shallow shaft water pressure at the bottom level has to be considered during the final operations and after completion. The base slab must be designed to resist uplift by its own weight and by anchorage into the lining of the shaft. Although the shaft may finish in a stratum which is impervious this may well be underlain by one which carries water under pressure, even a very thin seam of sand. If there is any reason to suspect this, a test boring through the bottom of the shaft is advisable. It may be feasible to relieve the pressure during construction of the base plug by allowing the water to rise through the test bore, with a gravel and sand filter if necessary, and to handle the water with the ordinary sinking pumps. A more elaborate deep well pumping procedure might be needed, or grout injection. Where uplift pressure is only marginally in excess of loading, the shaft bottom could be concreted in a series of trenches so that ground weight is not lost except over a narrow strip until replaced by concrete.

The base slab, already mentioned, has to be capable of resisting uplift and should normally be made watertight by grouting. It may also have to accommodate various items of construction plant and permanent equipment. In water-bearing ground provision of a sump and pumps is almost invariably essential both for the construction stage and in service.

The appropriate sump capacity may well be such that the whole shaft is sunk to below the level of the adjoining tunnel, and this will allow subdrains to be provided below the tunnel invert to assist in construction. In river crossings where the tunnel dips to its deepest level at mid river, drainage shafts are sometimes sunk to below this level and connected to the lowest point by special drainage headings rising gently from the shaft.

3.7 TUNNEL EYE

Another important feature of shaft construction is provision of a tunnel eye from which to break out. Where the shaft lining is segmental the cross joints in the rings at tunnel level should be arranged to suit the intended opening and minimise any cutting of segments or unnecessary exposure of ground. The operation of breaking out requires great care, especially because the excavation has to be done in ground already disturbed by the sinking of the shaft, and because water from higher levels may well follow down the outside of the shaft. A practice adopted in soft ground is first to make a small opening in the shaft and from this to drive a short timbered heading. At a distance of a few metres a 'break-up' for the full size tunnel is excavated. In this break-up chamber two or more rings are built and grouted and the tunnel is then built one ring at a time back to the shaft. Alternatively, two rings may be built within the shaft followed by advance one ring at a time through the shaft wall. The shaft rings when built carry the external ground pressure and in breaking them out the loads must be redistributed, initially to temporary steel or timber framing, and ultimately to the new tunnel lining.

Where a single segment is omitted, or removed, redistribution of load into adjacent rings may be expected, but as the opening is enlarged timber struts, wedged into each ring before removal of a segment, can take the load temporarily. The final junction between tunnel and shaft is likely to comprise steel joist framing embedded in concrete.

3.8 SUBAQUEOUS SHAFTS

There is an increasing demand for intake or discharge of water in rivers, lakes or the sea for power station cooling water, hydroelectric intakes, or sewer outfalls. In water of moderate depth the use of a cofferdam allows normal methods to be adopted, but special problems in constructing shafts and connecting them into tunnels arise in deeper water or in exposed situations or on a hard rocky bottom. A combination of expedients is likely to be necessary, dependent in some degree on whether or not the shaft is to be provided with headworks and to be capable of subsequent closure for maintenance.

For example, at the Wylfa Power Station cooling water intake the coast was exposed and the sea bed at −11 m was rocky with a tidal range of about 6 m. The

Fig. 3.8 – Tunnel eye. Framing at bottom of circular shaft to accommodate access tunnel.

shaft headworks were constructed from a jetty by sinking a pit, by blasting in the sea bed and concreting in a cylindrical shell and bulkhead. The lower part of the shaft was completed by raising a 1.2 m diameter shaft from the end of the tunnel and then enlarging to 4 m diameter downwards. In contrast, the Dublin outfall sewer at Howth Head, where strong currents were expected, was constructed almost entirely from the end of the tunnel below. In sound rock the shaft was raised from the end of the tunnel nearly to the sea bed and, after flooding the top of the shaft, was blasted with a carefully prepared pattern of charges, a sump having been prepared to absorb the debris. In fact, subsequent construction of headworks by divers proved necessary because of progressive silting up with sand. A similar method was employed for a cooling water outfall in the Bay of Fundy.

Other such shafts have been driven by employing offshore drilling rigs and boring down and capping off at or a little above bed level, then driving the tunnel into the shaft bottom and removing the cap.

BIBLIOGRAPHY

Much of the general material published relates to very deep shafts for mining development. The less deep shafts associated with tunnelling are mostly described as incidental to specific projects. They include shafts for tunnel construction, access, ventilation, hydroelectric power, marine outfall and intake. Caisson sinking is also used extensively for bridge and other deep foundations. A few relevant references are listed here, with further papers in the general bibliography.

Brown, E. O. F., *Vertical Shaft Sinking,* Benn, 1927.

Wilson, W. S. and Sully, F. W., Compressed-air caisson foundations, *Instn Civ. Engrs Works Construction Paper 13*, 1949.

Symposium on Shaft Sinking and Tunnelling, London, 1959.

Chapman, E. J. K. *et al.*, Cooling water intakes at Wylfa Power Station, *Proc. Instn Civ. Engrs*, 1969, **42** (Feb.); *see also* discussion, 1969, **44** (Nov.).

Haswell, C. K., Thames Cable Tunnel, *ibid*, 1969, **44** (Dec.); *see also* discussion, 1970, **47** (Oct.).

Collins, S. P. and Deacon, W. G., Shaft sinking by ground freezing: Ely Ouse – Essex scheme, *ibid*, 1972, **Suppl.** p. 129; *see also* discussion, p. 319.

Various authors, Shafts, *Proceedings Rapid Excavation and Tunneling Conferences*:
2nd, San Francisco, 1974, Session 16, Chapters 68–72;
3rd, Las Vegas, 1976, Session 3, Chapters 8–11;
4th, Atlanta, 1979, Sessions 14 and 15, Chapters 68–76.

Lander, J. H. *et al.*, Foyers pumped storage project: planning and design.

Land, D. D. and Hitchings, D. C., ditto: construction, *Proc. Instn Civ. Engrs*, 1978, **64** (Feb.); *see also* discussion, 1978, **64** (Nov.).

4

Geology

4.1 RELEVANCE

Of the natural sciences geology is that with which tunnelling is most intimately involved – from the inside of the rock mass. Tunnel engineering is concerned with those aspects of physical geology which are descriptive of the ground surrounding the tunnel, its response to stresses imposed and existing stress patterns. Geotechnology is a branch of civil engineering dealing particularly with this field, and relating the structural and mechanical behaviour of soils and rocks as affected by engineering construction to the geology in its widest sense. Its principal subdivisions are: soil mechanics, rock mechanics, hydrology and seismology.

There is inevitably an overlap between the subjects discussed in this chapter and elsewhere, more particularly in Chapters 3, 4, 5 and 6 of Volume 1.

It is mainly in geological terms that the ground can most accurately and concisely be described as an integrated and intelligible whole, both in surveys and reports, and in records of construction. Standard geological maps provide much valuable information, but inevitably need to be supplemented by site studies, to locate precisely and evaluate the significance of important features such as discontinuities, and by quantitative assessment of mechanical properties of the rock.

A brief discussion of the structures and vocabulary is therefore appropriate here, with a description also of the subject matter of soil and rock mechanics, not attempting to expound their principles.

4.2 SURVEY REQUIREMENTS

The tunneller wishes to know what strata he may expect to encounter, and how ground water is likely to manifest itself. He requires an assessment of the probable behaviour of the ground as and when it is excavated, including the effects of disturbance by the processes of excavation and drainage. In considering disturbance, the time factor is very important both in the initial response of the ground during construction and the subsequent relaxations and stress changes after the permanent lining has been installed. The requirement is to have defined in as much detail as practicable the nature of the strata and their geometry in all directions round the tunnel. The detail survey should cover a width at least several times the tunnel diameter and a depth from the surface to well below the lowest invert level foreseen. The description of the ground in general geological terms needs to be supplemented by analysis of its small scale structure in the terms of soil mechanics and rock mechanics, and by identification of the probable and possible locations of boundaries and discontinuities as precisely as may be.

The construction problems of excavation, support and water management, discussed in Vol. 1, Chapters 4 and 6, are all closely dependent on the composition and granular structure of the soils and rocks encountered and on the pattern of division by bedding planes, joints and faults. The broad division into soft ground and rock tunnelling is a useful one; at the extremes they are very different but an infinite variety of tunnel conditions and construction methods lie between. The geologist uses the term 'rock' for all natural deposits, although he does classify as 'unconsolidated sediments' much of the tunneller's 'soft ground'. In this chapter 'rock' is sometimes used in the comprehensive sense, but where discrimination is relevant 'rock' is applied to consolidated strata in contrast to 'soils' for unconsolidated sediments.

Much urban tunnelling is in soft ground, because cities are frequently situated on rivers with wide areas of alluvial deposits, and most tunnels, for urban transport and for sewers in particular, are constructed at as shallow a depth as practicable. In contrast, mountain tunnelling is likely to be through hard and varied rock which has been subjected to the stresses of mountain building earth movements. In geologically recent mountain systems, such as the Alps, and most other major mountain ranges, some strata may not be fully consolidated and may be under heavy continuing stress so that difficult and dangerous 'squeezing rock' is encountered.

There are two essential aspects of a geotechnical survey and report. These are:

1. Collection of all available information from site and other sources, and preparation of maps and sections to forecast what rocks and structures and ground water may be encountered.
2. Description and classification of the ground through which the tunnel will pass in terms of soil and rock mechanics so as to predict its behaviour in response to tunnelling operations.

It is of the greatest importance that the information should be presented in a form which can be utilised by the tunnel engineer in planning the project, if subsequent major changes required by unforeseen conditions are to be avoided. An adequate, clear and accurate statement of the available information and of

Position of Tunnel Face Set No. 9

1. Degree of weathuring in the Silurian

	MUDSTONE	
Complete	Zone 1	✓
High	Zone 2	✓
Moderate	Zone 3	✓
Slight	Zone 4	
Fresh	Zone 5	

SANDSTONE (Grade, Strength, Weathering)			
1)	Very weak	Complete	✓
2)	Weak	High	✓
3)	Mod. strong	Moderate	✓
4)	Strong	Slight	
5)	Very strong	Fresh	

2. Ratio of Sandstone to Mudstone 1:10
3. Thickness of Sandstone Beds 50–300 mm
4. Details of Discontinuities

(continues overleaf)

		Joints		Bedding Planes	
Spacing					
Very Close	$<$ 5 cm				
Close	5–30 cm	✓		✓	
Moderately Close	30–100 cm				
Wide	1–3 m				
Very Wide	$>$ 3 m				
Orientation		Strike	Dip	Strike	Dip
Tightness					
Tight		✓		✓	
Medium					
Open		✓		✓	
Irregularity					
a) Degree of Planeness					
– plane		✓		✓	
– curved				✓	
– irregular					
b) Degree of Smoothness					
– slickensided					
– smooth		✓		✓	
– rough		✓		✓	
Filling Material					
Type					
Thickness					
Degree of Cementing					
Not cemented					
Weakly cemented		✓		✓	
Strongly cemented		✓		✓	

Fig. 4.1(a) – Example of geological log for rock tunnel face.

(continued from previous page)

5. Groundwater
6. Description of Dykes, Faults, Folds, Synclines etc
   Description of Foliation
7. Comments

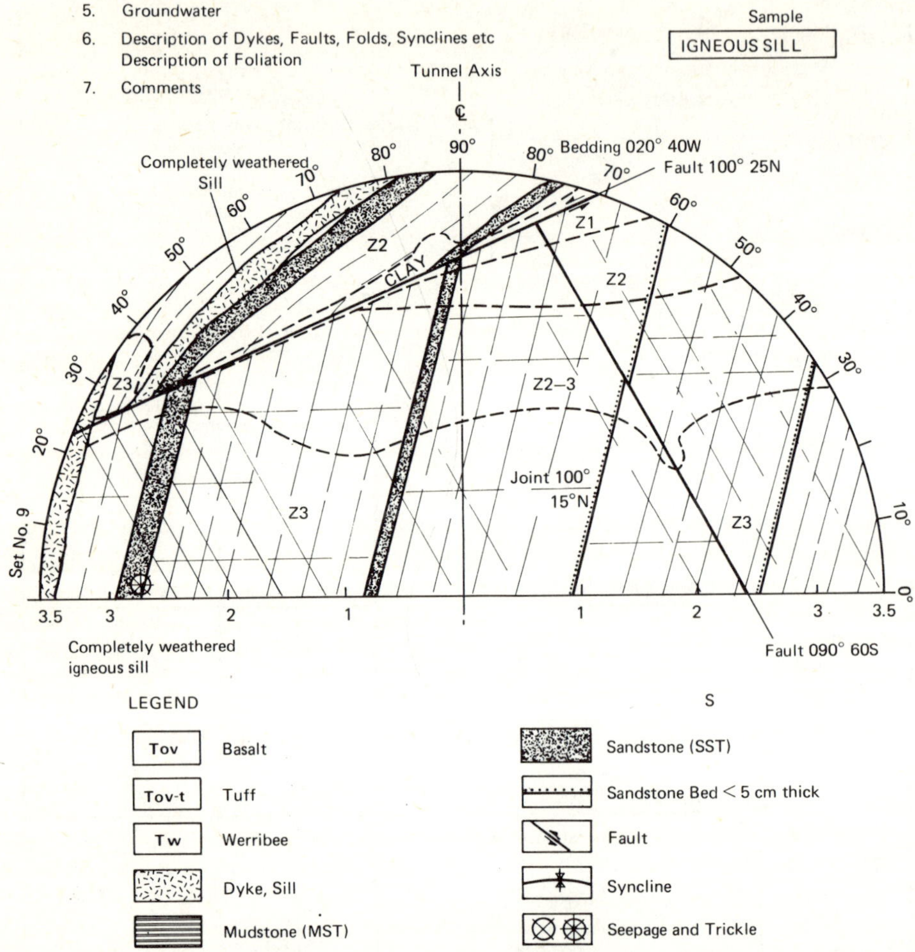

Fig. 4.1(b) – Example of geological log for rock tunnel face.

the uncertainties is essential to the production of an efficient scheme, including the proper choice of methods and plant to be used.

The geological information may be summarised as:

1. Lithology—type and characteristic of soil and rock.
2. Stratification and jointing.
3. Discontinuities—including faults and fissures.
4. Water.
5. Existing stresses.
6. Mechanical properties: (a) soil mechanics; (b) rock mechanics.
7. Seismic activity.

The manner in which some of these items appear in a tunnel face is exemplified by Fig. 4.1 which shows a tunnel log recorded for the Melbourne Loop. The form records lithology, weathering and discontinuities in sedimentary strata: steeply dipping weathered mudstone and sandstone. It exhibits also an intrusive sill and a normal fault. Frequent records of such detail at the face allow accurate plotting of the geological structure, which can be of the utmost value both currently and for future reference.

4.3 LITHOLOGY

Lithology describes the physical, chemical and textural characters of rocks and soils. These vary over so vast a range that classification and systematic nomenclature are difficult. Even the broad division into sedimentary, igneous and metaphoric is by no means clear cut, but is usually helpful, if occasionally confusing.

4.3.1 Sediments

Sedimentary rocks are mostly derived from the past erosion of other rocks and deposition of their fragments and particles, often on the sea bed, followed by consolidation.

Most sediments have been transported by water, but many surface deposits, particularly in the higher latitudes, have been carried by ice, and deposited unsorted as boulder clay or till, or as fluvio-glacial sands and gravels. The terms 'boulder clay' and 'till' for unsorted glacial deposits are largely interchangeable, but the latter is certainly preferable if boulders are absent. Sediments include also organic deposits of vegetable or animal origin, and salt beds resulting from evaporation of lakes and lagoons. When not fully consolidated and cemented they constitute the soils of the soft ground tunneller—gravels, sands, silts, clays—in which particle sizes and water content are of great significance. In consolidated form arenaceous rocks such as sandstone are constituted principally of quartz particles and are correspondingly hard, but may be weakly cemented. Argillaceous rocks are based on clay minerals and are fine grained and soft, forming mudstones and shales.

Limestones are largely of marine organic origin with a high proportion of calcium carbonate; they are typically fine grained and well cemented, but with a well developed joint structure. Their special characteristic of solubility in acidic water may result in enlargement of joints and creation of caverns traversed by underground streams and rivers.

Coal seams are organic strata, originating from swamp vegetation. A particular hazard from coal and other carbonaceous strata is the possible generation of methane gas. Another, man-made, hazard to tunnelling in coal fields is the presence of old worked out mines, possibly unrecorded.

Age of rocks is not of direct importance in tunnelling, but it does give some idea of their probable degree of consolidation and stability. Also, where strata of different periods are found the boundary between them constitutes an unconformity and is likely to mark an important change of conditions and possible plane of weakness. Fig. 4.2 shows some of these features in a section of Tyne Tunnel.

4.3.2 Igneous rocks

Igneous rocks, including volcanic and plutonic rocks, are probably the primary source of the earth's continental crust. Volcanic rocks include lava flows, which have been ejected in a molten state and spread over the existing surface, solidifying by rapid cooling and forming a fine grained crystalline rock such as basalt. They usually fit normally into the stratified pattern of sediments. Sills are intrusive lavas injected beneath the surface in more or less horizontal sheets between pre-existing sedimentary strata. Dykes are typically vertical sheets intruded into joints and fissures; they have sometimes formed feeders for lavas welling up to the surface. Sills and dykes are formed of relatively homogenous crystalline rocks of coarser texture than surface lavas because of a slower rate of cooling. In areas of volcanic activity ash beds, tuffs and ignimbrites may occur in considerable thickness and in various states of consolidation.

Plutonic rocks are formed in large masses at greater depth, not necessarily in a fully molten state, but they are of coarsely crystalline texture. Granite is the typical plutonic rock, but may also be the end product of metamorphic action on sediments.

Many of these igneous rocks are normally dense and impermeable but may be decomposed due to deep weathering at surfaces and at joints. They are further classified according to their chemical composition and crystalline structure. The content of silica, in the form of quartz, is used to divide them into 'acid' and 'basic', with further subdivision according to their feldspar content. The typical acid rocks are light in weight and in colour ranging from rhyolite lava to granite, while the basic rocks are dark and heavy like basalt, dolerite and gablro.

4.3.3 Metamorphic rocks

Metamorphic rocks are in origin sedimentary or igneous rocks which have been substantially altered by heat, pressure or chemical action. At one end of the range they grade into sedimentaries subjected to heavy stresses, and at the other into plutonic rocks. Shearing forces acting on rocks are likely to produce a foliated or laminated structure, as in gneiss and schist, or cleavage as in slate. Marble is a metamorphic rock generated from limestone by recrystallisation at high temperature. Schist derives from clays and shales, and quartzite from sandstones.

In general, metamorphic rocks are ancient and of massive structure, but with

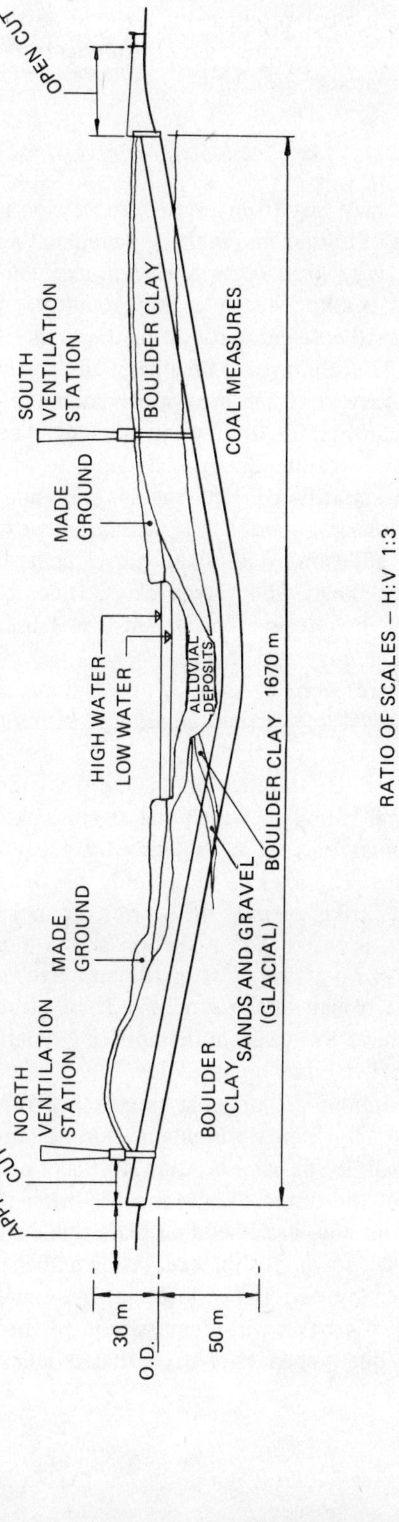

Fig. 4.2 – Tyne Tunnel – longitudinal section.

planes of weakness which may produce unexpected difficulties. In the less ancient mountain belts formations which are not fully consolidated may become unstable when disturbed by excavation.

4.4 STRATIFICATION

The arrangement of strata may vary from a simple succession of nearly horizontal beds to a complex pattern of folding and faulting inclined at any angle. The strata may include deposits of very local occurrence and irregular thickness. What is required from the survey is a knowledge of the geometry, in three dimensions, of each bed everywhere in the neighbourhood of the tunnel. The basic elements in the description, after the lithology, are thickness, dip, and strike.

The thickness may vary over the area surveyed, more particularly where recent unconsolidated deposits fill hollows in an older land surface, or have been subjected to surface erosion. Superficial deposits of glacial and fluvio-glacial origin may not be stratified in any regular way, and may have been so disturbed and eroded during a period of periglacial weathering as to defy prediction. Deposits, in addition to tills and gravel beds, include such forms as moraines, esker ridges, kames, and varved clays. Older beds may also have been thinned out locally by erosion or cut into by buried stream channels. Lateral variation in lithology may also occur.

Dip and strike define respectively the downward slope of the plane of a bed and the orientation of a level line at right angles to maximum slope and lying in the plane of the bed.

In undisturbed beds, which are uncommon, the dip will be zero. The angle of dip measures the local tilting to which the strata have been subjected by earth movement, usually as part of a pattern of folding into anticlines and synclines over a wide area. The axes of these folds may be horizontal, lying in the same direction as the strike, or may themselves dip more or less steeply and produce a more complex geometry. In a section across a simple fold the dip varies continuously from zero at the crest of the anticline through a maximum slope back to zero at the trough of the syncline. These structural features may be reversed topographically, the weak anticline being weathered into a valley and the strong syncline left standing up as a ridge.

The dip and its variations are of great importance in tunnelling through stratified deposits (Fig. 4.4). A tunnel running level in the direction of the strike may be expected to remain in the same stratum, if it is of adequate thickness. If the dip across the line of the tunnel (Fig. 4.4(a)) is other than small, the roof will tend to break away on any weak bedding planes, and there is likely to be a weak side where rock slides down into the excavation and a strong side opposite, but this may be modified by patterns of jointing. If a tunnel runs level in the direction of the dip (Fig. 4.4(b)) it will progressively cut through higher beds of the succession with consequent changes of strength and behaviour; any weakness

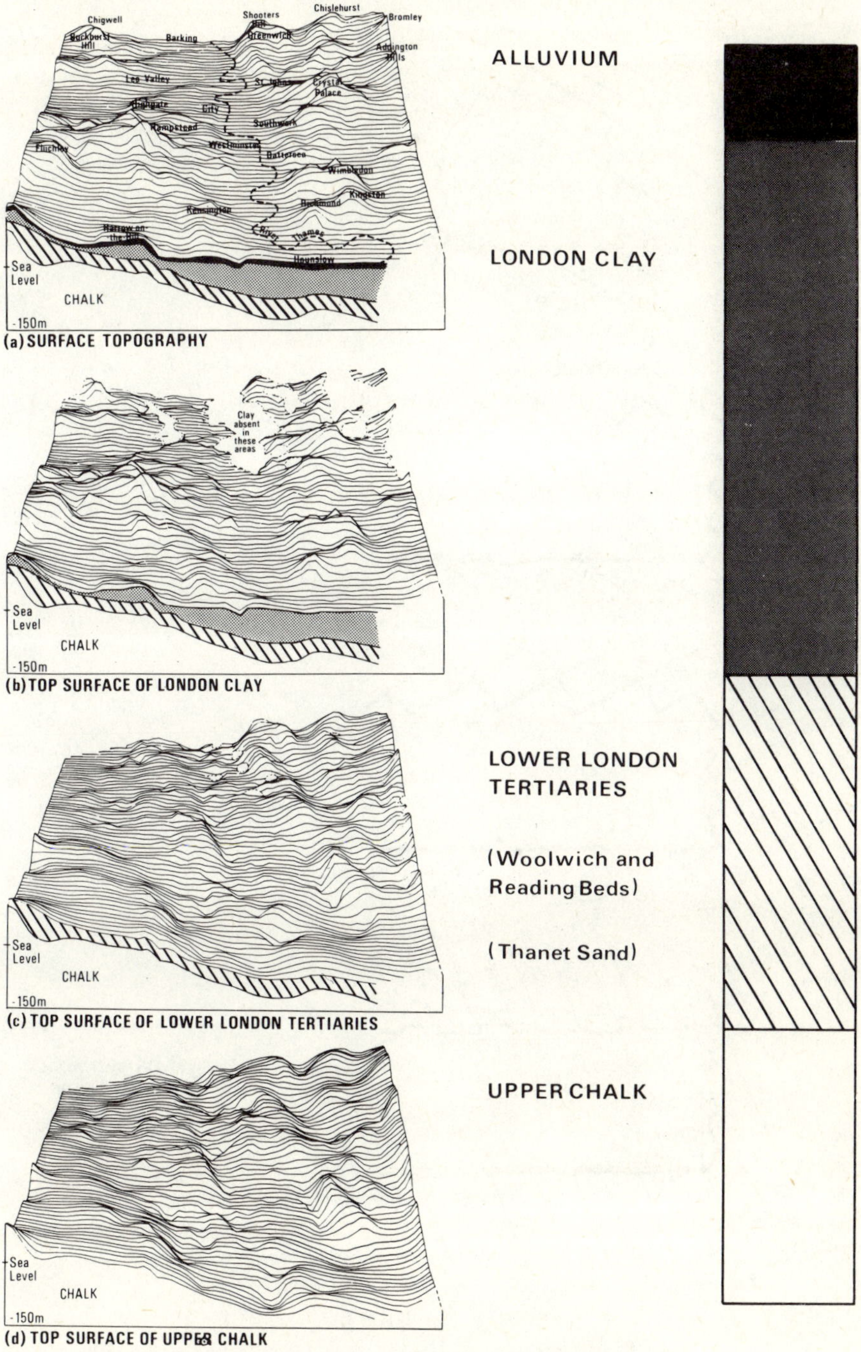

Fig. 4.3 – Geology of London Basin.

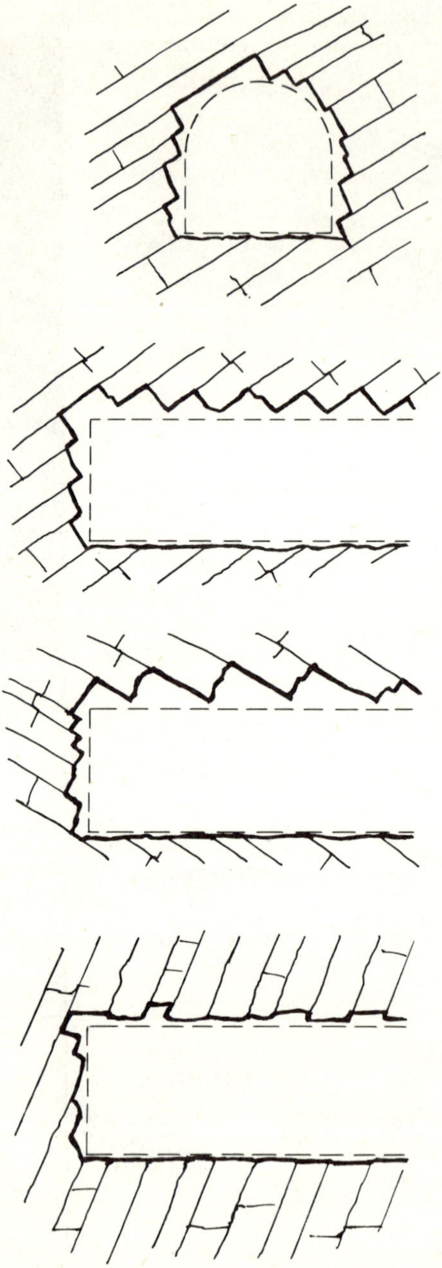

(a) DIP ACROSS TUNNEL

(b) TUNNEL FOLLOWING IN DIRECTION OF DIP

(c) TUNNEL DRIVEN AGAINST DIRECTION OF DIP

(d) TUNNEL DRIVEN WITH NEAR–VERTICAL DIP

Fig. 4.4 – Dip of strata in relation to tunnel excavation.

in shear on the bedding planes will result in weakness of the roof as support is cut away at the face. Conversely, a tunnel driven against the direction of dip (Fig. 4.4(c)) will progress into lower beds and is less likely to be weak near the face.

If the dip is nearly vertical the tunnelling conditions are very different according to whether the tunnel is driven along the line of the strike or at right angles to it. If along the strike, the same beds and bedding planes will persist, and any weakness in shear or any water-bearing stratum may make support very difficult. If the tunnel is along the line of the dip (Fig 4.4(d)) it will cut across the succession of beds at right angles and each bed which is itself competent can function as a transverse arch; a weak bed is only likely to require support while its thickness is penetrated.

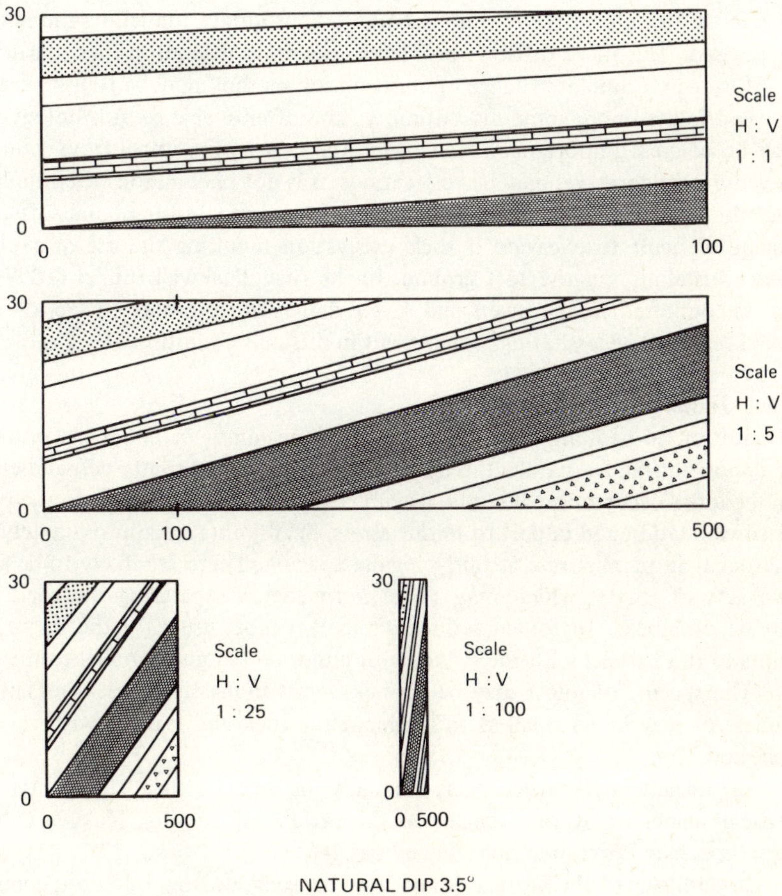

Fig. 4.5 – Scale distortion in sections.

In anticlinal structures the rock tends to be more lightly stressed as compared with compression in synclines. Water in fissures therefore may drain more freely from anticlines and accumulate in synclines.

The importance of dip and its variations will be apparent. It may be noted, however, that longitudinal sections of tunnels can be misleading where they are drawn to such different horizontal and vertical scales that they give the impression that all the strata are nearly vertical. For example, with scales in the ratio of 1 : 100, a dip of 5° is exaggerated to 83°. Such a section, therefore, does not give a clear picture of the actual transitions as encountered in driving the tunnel, although showing correctly the successive strata.

4.5 DISCONTINUITIES

In almost any uniform continuous ground, appropriate tunnelling methods can be devised. The more difficult problems arise from discontinuities, particularly if unforeseen. Complete change of construction method may be found necessary.

Bed joints imply some discontinuity, and often a change of lithology which may be of great importance. A change in the crown of a tunnel from sound clay to water-bearing gravel may be very serious; it is not uncommon when tunnelling near the upper surface of London Clay. A 'mixed face' of hard and soft ground can be difficult to excavate if rock excavation requiring the use of explosives causes instability in the soft ground. In the first Blackwall tunnel (1897) rock was encountered in the invert and severe damage to the bottom section of the shield resulted, necessitating replacement in difficult conditions.

4.5.1 Joints

In addition to bedding planes, which arise from minor or major discontinuities of deposition and separate the individual strata, sets of joints perpendicular to the bedding occur in most strata, usually attributable to shrinkage in the process of consolidation and in part to tensile stress. Such joints are approximately plane and occur in parallel sets at fairly regular spacing. There are likely to be at least two sets of joints, which may be at right angles, separating the rock into a mosaic of blocks. In folded sedimentaries it is most usual for the major set of joints to run parallel with the strike, with minor cross joints parallel to the dip.

The spacing of joints may be several metres in massive sandstones and limestones, or may be so small as to fragment the rock and leave it with very little cohesion.

In volcanic rocks there may similarly be sets of shrinkage joints in the vertical plane, or at right angles to the surface of cooling. Massive columnar structures may develop in dolerite and basalt.

In addition to the vertical joints flow layering, or flow foliation, commonly appears in volcanic rocks, and roughly horizontal sheeting joints develop in granites where vertical loading has been relieved by erosion.

Joints may be minimal cracks or relatively open. They may be filled with clay or with crystalline deposits. They provide a network of fissures through which water, or other fluid can travel, contributing to the weathering and erosion of the rock mass. The jointing is of major importance in assessment of the rock quality in relation to excavation and support.

4.5.2 Unconformities

An unconformity is a major structural feature where the sequence of strata is interrupted by an old surface of erosion, which cuts across tilted bedding planes and is buried by a new succession of conformable strata. Dip and strike are likely to change abruptly at this surface, with a complete change of lithology.

The unconformity may be marked by deep weathering of the underlying strata.

4.5.3 Faults

A fault is a fracture surface where relative shearing movement of the adjoining masses of rock has occurred. The extent of movement–the 'throw'–may be very large, or may be so small as to differ little from a joint. Most ordinary faults have ceased movement and are stable, but some major faults, particularly in earthquake zones, remain potentially active. The fault surface is ordinarily nearly plane; its orientation is defined by the dip (or by the complementary angle 'hade') and strike.

There are three typical patterns of fault movement: (1) A normal fault lies in a steeply inclined plane on one side of which the strata have dropped down relative to the identical strata on the upper side. (2) The converse is a thrust fault in which compressive forces have caused movement up the dip slope, which is usually less steep than in a normal fault. With a very low angle the fault becomes an overthrust. (3) The third type is the wrench fault in which the plane is near the vertical and the movement is mainly horizontal to left or right along the strike.

Overthrust faults are particularly associated with the process of mountain building and may mark sliding movements of many kilometres over wide fronts. They are likely to be associated with intense folding and subsidiary faulting and with metamorphism of the strata involved. The classic example is the Moine thrust in the north-west Highlands of Scotland. Such faults also occur on a spectacular scale in the Alps, Himalayas and other mountain ranges. Groups of smaller and steeper reverse faults are often associated with overthrusts.

Wrench faults also may be on a very large scale. One of the best known is the San Andreas fault in California which is over 1200 km in length. Movement totalling 600 km in the course of 140 million years has been identified, and currently the rate has accelerated to about 50 mm per year. A similar fault known as the 'Alpine Fault' in New Zealand extends from SW to NE over most of the length of both islands. Movement has been, and is, at similar rates. Movement at these faults is associated with earthquakes.

A minor fault may have sheared cleanly through the rock, producing polished faces with slickenside, but nearly all faults are accompanied by some degree of distortion and shattering of the adjoining rock in the shear zone, where broken rock may be reduced to rock flour or replaced by clay. The shattered rock is fault breccia and the fine and soft filling is known as 'gouge'. These shear zones provide difficult or hazardous tunnelling conditions. The fragmented rock is difficult to support, and unpredictable in its incidence; gouge may be washed out by water with loss of support to the adjacent blocks and danger of flooding, and clay may provide lubrication in the joints of the rock mass, and consequent instability.

The orientation of a fault and the width of its shear zone in relation to a tunnel being driven through it are obviously very important as are the magnitude and direction of residual stresses in the zone and the presence of water and the extent of its reservoir. A narrow vertical fault through which the tunnel crosses at right angles is likely to be much less difficult than one oblique to the tunnel or with a wide shatter zone. Undersea faults encountered by the Japanese Seikan tunnel have proved very difficult, as described in Volume 1.

4.5.4 Dykes and sills

Dykes and sills have already been referred to above. Dykes may be encountered as vertical walls of intruded igneous rock cutting across sedimentary beds, having followed and enlarged the course of a joint plane or a fault plane. Sills are horizontal intrusions which have followed a bedding plane, forcing apart the strata.

4.6 WATER

Site investigation of surface water and ground water has been discussed in Volume 1, where reference has been made to the danger of inundation. The hazards from internal flooding require some further consideration.

In subaqueous tunnels the major danger is almost always that of inundation by failure of the ground at or immediately behind the face before construction of the lining is complete. The source of water from sea or river, in the event of a breach at the face, is unlimited. Any flow of water, unless immediately stemmed, tends to erode a channel of increasing width and accordingly itself to increase rapidly, although it may eventually become choked with material carried in. Safety precautions and remedial action, apart from the use of a shield, close timbering and use of compressed air, may comprise reduction of permeability and strengthening of the ground by grout injection or freezing, or the use of a clay blanket in the river bed.

In mountain tunnels there are also special risks. Springs fed from fissures may discharge at high pressure and high temperature but the rate of flow is likely to diminish with time as the immediate reservoir becomes exhausted. Such inflows of water, however, may also carry with them solids, washing out the material filling joints, or may carry in sand from a weakly cemented stratum which is thereby disrupted.

In all tunnels the problem of an inflow of water has got to be considered in terms of the extent of the reservoir feeding it, and its erosion and transport of solids.

At very high pressures the stressing of the lining in the finished tunnel by the build-up of full hydrostatic pressure has to be examined. Where acceptable seepage with appropriate provision of drains and pumps allows external pressure to be limited, and in part controlled, this may be preferred to a fully watertight lining. In the case of hydroelectric power tunnels this is unlikely to be acceptable and the lining must be designed for both the alternative cases of high internal pressure in working conditions and high external pore pressure when the system is drained down.

Considerations of watertightness are also of major importance in sewers to ensure that the ground is not polluted by leakage outwards, and in water supply tunnels where contamination by inward leakage must be prevented.

4.6.1 Ground water

Ground water originates as rainfall or snow by soaking into the subsoil and bedrock where it becomes particularly relevant to tunnelling. Its absorption, movement, storage and discharge are regulated by the porosity and permeability of the soils and rocks, and by their geological structure and by the level of the water table. The water passes down through, or is retained in voids in the ground. These voids range from subcapillary dimensions in clay and fine grained sediments, through coarser sands, gravels and conglomerates and open joints and fissures to the extreme of vast caverns in limestone.

The water table is a critical level in the flow pattern. It is the buried surface below which the ground is saturated, and usually lies at some depth below the surface, varying from time to time with rainfall. In boreholes it is displayed as the standing water level, and near lakes and rivers it is nearly at the free water level, rising gently as distance from the open water increases. It marks the surface of a hidden reservoir of water, discharging through springs wherever the water table is above ground level, and sometimes through submerged springs in a river bed or even under the sea.

Above the water table soils and rocks are only partially saturated, and any free water moves downwards, at the rate that permeability allows, until it reaches the water table where it replenishes the reservoir and may raise its surface level. All pore water is not free, much being retained in fine grained material by capillary action. In addition there is adsorbed water bound to particles by molecular action. This adsorbed water is fundamental to the particulate structure and characteristics of a clay and provides shear strength in a way that capillary water does not; it is not readily displaced by consolidation or otherwise.

Below the water table movement through the saturated ground is necessarily much slower, being regulated by permeability and hydraulic gradients to the eventual points of discharge. As already noted, there may be more than one water table level because of the confining effect of impermeable strata.

Porosity measures the voids in rock or soil as a percentage of the total volume and is therefore substantially less than unity; it is not to be confused with 'voids ratio', which is the volume ratio voids : solids. Porosity is a characteristic of a laboratory specimen of rock and does not take account of the joints and fissures and other cavities; with the inclusion of these the mass porosity may be much higher.

Permeability measures the rate at which water can pass through the rock. In particular the coefficient of permeability, k, is, for present purposes, defined by the equation $q = ki$, where q is the rate of flow of water across a unit area and i is the hydraulic gradient in the direction of flow. As with the porosity the mass permeability utilising the joint structure may be much greater than that of an unfissured specimen. The mass permeability may also be strongly directional, following the dominant joint pattern.

Aquifer is the term used for a stratum of permeable rock which may hold water and through which water may pass.

4.6.2 Water in soft ground

The importance of pore water in soft ground lies in its action and effects on the soil. This indeed provides much of the subject matter of soil mechanics. In granular soils and silts the advantages of cohesion provided by dampness, and the disadvantages of either saturation or drying out which result in the loss of cohesion are further discussed below. The effect of any concentration of flow through such soils is to wash out the fines and set the loosened sand and gravel in motion.

In clays and in silts pore water content and pressure are determining factors in the shear strength and plasticity of the soil. Water content is not readily altered in clay, and residual pore water pressures may cause swelling of the clay, creating active pressures on support structures.

Impermeable layers have already been mentioned as giving rise to perched water tables. They may be of variable extent and thickness forming lenses which make further difficulties in the treatment of ground by grouting. In subaqueous soft ground tunnelling the face is extremely vulnerable to the action of water and the danger of inundation. It was to meet these conditions that the Greathead shield was devised, with supplementary equipment such as the hood and face rams. It is also for such ground that compressed air working is adopted, and ground treatment by grouting to reduce permeability and provide greater cohesive strength.

4.6.3 Chemical action

One aspect of ground water which may be of importance is chemical activity. Acidity in the ground water and sulphate content may attack cement used in grouting and pointing or concrete linings. The problem of attack on cement is principally one of flowing water which can carry away the reaction products and expose new surfaces to attack. Strong acid action has been observed in a cast iron lining in London Clay, probably attributable to oxidation of iron pyrites with resulting sulphate formation. Swelling ground attributable to chemical reaction is further referred to below.

4.6.4 Flotation

Another aspect of ground water in shallow tunnels is the possibility of flotation. An immersed tube tunnel, or one rising from a subaqueous crossing through soft unconsolidated ground, may be subject to nett uplift, requiring anchorage or ballasting against uplift.

4.7 ROCK STRESSES

In the normal configuration of stress in undisturbed rock the major principal stress is vertical and approximates to the direct pressure of the overburden. Accompanying horizontal stresses are about one quarter to one half as great. The significance of horizontal stress has been discussed in Volume 1. Abnormal stresses may be present in the rock or may be developed as a consequence of tunnelling operations. The principal phenomena to be considered are:

1. Stress exceeding rock strength.
2. High values of horizontal stress, or inclined stress, which have resulted from tectonic activity in the past during folding and faulting of the strata.
3. Inclined and irregular stresses of topographic origin, as in the steep side slopes of a valley.
4. Squeezing rock, where the disturbance resulting from excavation allows a stratum to yield in a plastic manner.
5. Swelling rock where physical or chemical action causes a stratum to swell and to impose heavy loading on tunnel support structures.

1. Direct stress of normal pattern may exceed locally or generally the unsupported strength of the rock during excavation. Rock bursts, popping rock or spalling may result. It is a familiar phenomenon below 600 m in deep gold mines where controlled de-stressing can alleviate it. It was noted in the chalk excavation preliminary to Channel tunnel construction in the walls of an arched heading with rib supports. It did not appear for some time after excavation and could be controlled by providing lateral support immediately after exposure. In a New York water tunnel under the East River at a depth of 130 m below ground popping rock was common in dolomite and metamorphic

rocks of Cambrian and pre-Cambrian age. Rock bolting was ineffective but steel ribs with timber lagging protected the work satisfactorily.

2. Abnormal horizontal stress may be difficult to ascertain from preliminary investigations, but the possibility of occurrence may be predictable if the geological history of the zone is known.

3. In hilly or mountainous country sloping ground is likely to have special characteristics as one principal stress may be inclined in the direction of the slope, but the pattern will depend on the relationship between the dip of the strata and the slope of the hillside. If both dip in the same direction, a tendency to slip is probable while if the dip is against the slope the ground will tend to be more stable.

The slope surface may be unstable, or of low stability against sliding if it consists of scree and hill wash above rock or if the rock is deeply weathered. These factors become of great importance in tunnelling either for construction of a portal entering the hill slope at right angles, or for a tunnel running nearly parallel to the slope.

4. Squeezing rock may develop in mountain areas where heavy stresses deform weak rock and cause it to yield and flow. When excavation exposes the squeezing rock it is extruded into the tunnel without visible fracture and without change in water content. The rock behaves like plastic clay, the rate of squeeze depending on the overstress, measured by the ratio of overburden pressure and shear strength.

The pressure on supports may reach very high values.

Squeezing rock can develop from metamorphism of shales or schists by the chemical action of deep weathering breaking down the original structure.

5. Swelling ground is attributable to change in water content causing expansion. One form is overconsolidated clay in which pore water content has been determined by former overburden pressures from strata long since removed by erosion. London Clay is a well known example. Excavation at a tunnel face releases pore water pressures and slow migration of water from the body of the clay towards the face causes swelling at the face.

Other types of swelling ground include beds of anhydrite ($CaSO_4$), in which access of water results in chemical change to gypsum with an increase of volume.

Magnetic iron pyrites, pyrhotite, decomposing to sulphuric acid and ferrous sulphate when exposed to oxygen and water, can result in swelling accompanied by corrosion problems.

4.8 GEOTECHNOLOGY

In order to make the best use of the information embodied in the geological survey and to assess the tunnelling conditions, it must be further analysed in terms of the mechanical behaviour of the ground during and after excavation.

As in all civil engineering, the old practical and empirical assessments of soils and rock have been largely replaced by the quantitative analytical methods, first of soil mechanics and then also of rock mechanics. In the application of these methods it remains essential to exercise at all stages engineering judgement based on experience. The siting of boreholes, the selection of samples for testing and the applicability of those test results all have to be considered and kept under review in the light of the actual ground encountered and its response to excavation.

It may be worth while to quote Terzaghi's words from his *Theoretical Soil Mechanics* (1943): 'The prospects of computing accurately the effect of a change in the conditions of loading or of drainage on the soil in advance of construction are usually slight–particularly (where) the action of water is involved. The action often depends on minor details of stratification which cannot be detected by test borings.'

4.8.1 Soil description

Soil mechanics is particularly related to the interaction of fine particles and water in the soils encountered in soft ground tunnelling, namely: gravel, sand, silt and clay, and their modification with changes of water content and under stress. In ground of this type jointing patterns are unlikely to be relevant,

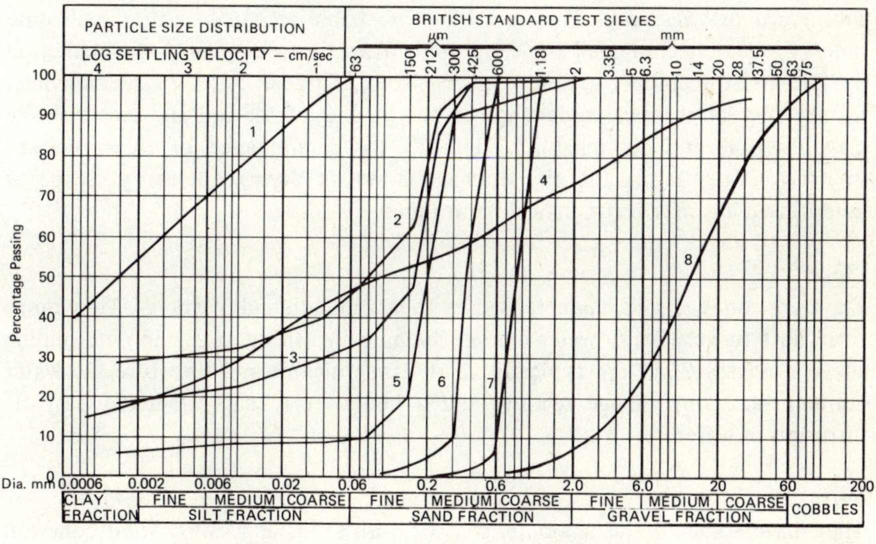

1. LONDON CLAY
2. FINE CLAYEY SAND
3. FINE SILTY SAND
4. BOULDER CLAY
5. FINE SAND
6. MEDIUM SAND
7. COARSE SAND
8. MEDIUM GRAVEL

PARTICLE SIZE ANALYSIS – TYPICAL GRADING CURVES

Fig. 4.6 – Soil grading curves.

except possibly in very stiff fissured clay. The presence and pressure of water, adsorbed, capillary, in pores and in fissures, can be of dominant importance as has been discussed above and in other chapters.

The primary mechanical classification of soils is by particle size distribution, followed by mineral composition and cohesive and plastic properties. The results of sieve analysis are best displayed by plotting a curve which shows the percentage of the sample passing each standard sieve against a logarithmic scale of particle diameter. Figure 4.6 shows some typical curves and shows also the size limits and the terminology. For the finer part of the range, below about 0.06 mm, sieves cease to be usable and sedimentation rates or other methods of estimating size are adopted. This is size grading only and it must be recognised that, for example, clay size particles do not necessarily have the characteristics of clay if their mineral constitution differs: finely ground quartz, as in rock flour, will behave quite differently from a true clay.

From sieve analysis a soil is classified descriptively by the name of its principal constituent, qualified by naming any other substantial fraction as, for example, clayey sand, silty clay. These soils fall into three main subdivisions: granular soils, silts, clays.

Granular soils

Such soils lack any appreciable fine fraction and have little or no cohesion when submerged or when dry. They are therefore liable to run in either such state unless adequately contained and supported. When damp they have some cohesion attributable to capillary action of the moisture. This can be of considerable help in tunnelling, but its value depends on the grading of the soil and presence of a clay fraction and its reliability depends on maintenance of the necessary moisture. Water levels and any drainage flows, or drying out as by escape of compressed air, may be of critical importance.

Clays

Clays are cohesive soils characterised by shear strength and plasticity. They range from soft to very stiff, water content being a significant factor in determining shear strength. Clays are typically nearly impermeable and therefore the water content does not change readily, and a bed of clay can effectively seal off intrusion of water.

Silts

They have some of the characteristics of both sands and clays. Their cohesion is dependent on water content remaining between narrow limits. Their permeability is such that they may become saturated, or dry out, fairly rapidly, losing all cohesion and running into an excavation unless retained by very close-fitting timbering. On the other hand, their permeability is too low to allow grout injection except with extremely penetrative fluids.

4.8.2 Soil properties

In terms of soil mechanics the principal properties of soils to be obtained from samples include:

Particle size grading, as already discussed
Mineral composition
Density
Porosity
Moisture content, from an undisturbed sample if possible
Shear strength, in clays
Angle of internal friction in granular soils
Plasticity–plastic limits and liquid limits in clays and silts

Other properties, important in most circumstances, must be determined by *in situ* measurements. These include:

Permeability
Pore water pressure
Flow of ground water
State of compaction
Fissuring

4.8.3 Quality of rock

For tunnelling in rock a similar list of properties can be given, again with the obvious proviso that not every sample needs to have all these tests made. What will be sufficient is very much a matter of judgement, influenced by time and cost. The information obtainable from core samples includes:

Lithology
Density
Porosity
Water content
Strength, including crushing and tensile strength and shear or any identifiable plane of weakness
Abrasiveness, in relation to cutting by drill bits or by machine cutters
Rock Quality Designation (RQD), being the percentage of core borings recovered in lengths not less than 100 mm

Site investigations are necessary to ascertain:

Permeability
Pore water pressure and flow of water
Joint and fissure pattern

Attempts have been made in recent years to establish quantitative assessment of the tunnelling qualities of the rock mass. One that is widely used is the 'rock mass quality index' of the Norwegian Geotechnical Institute, which provides an example of the complexities necessarily involved even although it still includes a large empirical element. Experience and sound judgement remain essential in evaluating the index and interpreting the result.

The index, Q, is compounded from numbers selected to represent RQD (as defined above), number of sets of joints, joint roughness, joint alteration, water pressure in joints, and a stress reduction factor allowing for such features as zones of weakness, high rock stress, or squeezing or swelling rock.

Q may lie in the range from .001 for the worst tunnelling conditions in squeezing rock up to 1000 for massive rock requiring no support. This range of values is not very convenient for description and ready appreciation, and might well be translated to a logarithmic scale, perhaps from zero to + 60.

It may be noted that, despite the complexities and the judgement required in estimating most of the components of the index, there remain other properties of the rock to be considered, including compressive strength, orientation of bedding planes and joints in relation to tunnel excavation, hardness, abrasiveness, and fragmentation in relation to the use of cutting tools, blasting methods and spoil handling.

4.9 EARTHQUAKES

Earthquakes are hazardous to all surface and underground structures. They arise typically from sudden release of stresses at depths ranging up to several hundred kilometres, often by movement in the plane of a fault. They manifest themselves, in diminishing degree as distance from the focus increases, by horizontal and vertical vibrations of the ground, by vertical or horizontal displacements across a fault plane, by opening of fissures and by consolidation and slipping of unstable ground.

Earthquakes may occur almost anywhere but the major zones of activity are the margins of the Pacific Ocean and a belt from the Mediterranean to the Himalayas and the East Indies.

Substantial surface structures are principally damaged by the horizontal component of acceleration in the vibrations at the surface, particularly in unconsolidated strata. No structure can, of course, resist a shearing movement in the plane of a fault if the structure is continuous across that plane. Tunnels in solid ground are unlikely to be severely damaged by vibrations unless the ground itself is sheared. There is, however, obvious danger of landslips at tunnel portals or in the sides of a mountain valley or gorge.

The B.A.R.T. tube 6 km long crossing San Francisco Bay was provided at both ends with joints designed to allow for earthquake movements. These joints allow for a longitudinal relative movement of 38 mm in either direction and a transverse movement of 100 mm in any direction.

BIBLIOGRAPHY

The subjects of geology, soil mechanics, rock mechanics, hydrology and other specialisations have vast bibliographies of their own ranging far beyond the aspects considered here. A brief list only, of items of particular relevance, is appended.

Terzaghi, K., *Theoretical Soil Mechanics*, J. Wiley, 1943.

Glossop, R. and Skempton, A. W., Particle size in silt and sand, *J. Instn. Civ. Engrs*, 1945, **25** (Dec.).

Terzaghi, K., Rock defects and loads on tunnel supports: Introduction to tunnel geology in Proctor, R. V. and White, T. L., *Rock tunneling with steel supports*, Commercial Shearing and Stamping Co., Ohio, 1946, 2nd edn. 1968.

Todd, D. K., *Groundwater hydrology*, J. Wiley, 1959; 2nd edn., 1980.

Chow, V. T. (Ed.), *Handbook of applied hydrology*, McGraw Hill, 1965.

Terzaghi, K. and Peck, R. B., *Soil Mechanics in engineering practice*, J. Wiley, 2nd edn., 1967.

Stagg, K. G. and Zienkiewicz, O. C., *Rock mechanics in engineering practice*, J. Wiley, 1968.

Wahlstrom, E. E., *Tunneling in rock*, Elsevier, 1973.

Barton, N. *et al.*, Engineering classification of rock masses for the design of tunnel support, *Rock Mechanics*, 1974, **6**, 189–236.

Blyth, F. G. H. and de Freitas, M. H., *A geology for engineers*, E. Arnold, 6th edn., 1974.

Blyth, F. G. H., *Geological maps and their interpretation*, E. Arnold, 2nd edn., 1976.

Manual of applied geology for engineers, Institution of Civil Engineers and H.M.S.O., 1976.

Jaeger, J. C. and Cook, N. G. W., *Fundamentals of rock mechanics*, Chapman and Hall, 2nd edn., 1976.

Cedergren, H. R., *Seepage, drainage, and flow nets*, J. Wiley, 2nd edn., 1977.

Holmes, A., *Principles of physical geology*, Nelson, 3rd edn., 1978.

Jaeger, C., *Rock mechanics and engineering*, Cambridge U.P., 2nd edn., 1979.

West, G. and Lake, L. M., *Site investigation for tunnels*, Int. J. Rock Mech. Min. Sci., 1981, **18**(5).

5

Ground Treatment

5.1 OBJECTIVES

Preparatory treatment of the ground to modify its behaviour during excavation for tunnelling can take various forms. The principal objectives are control of water and strengthening of cohesive forces, and the means include: (1) injection of grouts into pores and fissures; (2) freezing; and (3) lowering of ground water levels by drainage.

It is in tunnels and shafts below the water table, through granular soils which are permeable to water and lacking cohesive strength, that the usefulness of such methods is greatest, but they may also be employed to control water inflows and stabilise the ground in fissured and jointed rock or even clay. Compressed air working, described in Volume 1, is a very special technique for control or exclusion of water, and its application may be reinforced if necessary by most of the methods of ground treatment described herein.

Normally ground treatment methods are most effectively applied from the surface in advance of tunnel excavation, but they can also be applied from within a tunnel although such operations are almost inevitably in cramped conditions and greatly delay tunnelling progress, unless executed separately from a pilot or parallel tunnel.

When the Thames tunnel of the elder Brunel suffered a collapse, and he received innumerable suggestions from amateurs, he wrote: 'In every case they made the ground to suit their plan and not their plan to suit the ground'. In some cases, nowadays, ground treatment techniques can do just that; running sand can be cemented into a sandstone, wet silt can be frozen to a coherent solid.

5.2 GROUTING

The term 'grouting' describes the injection under pressure of a fluid grout, which may be a solution or a particle suspension, to occupy the pores, or the voids in the ground, or between a structure and the ground. Where the ground is water-bearing the grout is designed to expel and to replace the water, or, in the absence of water to fill the voids. When thus injected the grout reacts, physically or chemically, and may set to form a gel or a solid, reducing the porosity and usually increasing the cohesive strength of the ground, possibly to a very substantial degree.

An increasingly wide range of materials and of grouting techniques and equipment are available to suit particular requirements.

The two principal objectives are, as already stated:

1. to reduce the permeability of the ground, and
2. to strengthen and stabilise the ground.

In tunnel and shaft construction both are usually required. When working in compressed air the reduction in air loss may be just as important as the reduction in water inflow. Suitable filling material in the voids will contribute an increase in the shear and compressive strength of the soil at the same time as reducing permeability, but the appropriate materials and technique of application to suit the ground must be employed. If the volume to be treated contains a variety of soils, more than one type of grout may be required and even in a single stratum successive injections of finer grouts may be necessary.

5.2.1 Grout types

Grouts may be classified very broadly into:

1. Particle suspensions. Portland cement in water is the most familiar: this is only applicable where pore dimensions are sufficiently large to admit the cement particles. Bentonite or other clay suspensions are finer grained and therefore penetrate more freely.
2. Chemical grouts. These are aqueous solutions constituting pure liquids. Their penetration properties depend on their viscosity, which may increase with time after a physical or chemical action of setting to a gel is initiated. A 'one-shot' time-dependent fluid may be used, or a 'two-shot' process in which two different materials are injected in succession and react in the ground.
3. Others, such as bitumen emulsions, asphalt, etc., are not much used in tunnelling.

The choice and design of the most appropriate material and technique and the sequence and pattern of injection can only be made after a comprehensive site investigation with particular emphasis on permeability and sieve analysis. Specialists in grouting are likely to be required for all but the simplest cases, preferably with an ability to modify their techniques as experience is gained in the response of the particular ground.

It is seldom that grouting is completely effective, because grout may fail to penetrate equally all of the volume and some residual permeability is likely. A balance has to be struck between the cost in time and money of increasingly elaborate detailed treatment and acceptance of a sufficient improvement of ground stability and ground water control to make possible satisfactory permanent construction.

5.2.2 History of grouting

Ground improvement by grouting for civil engineering works was introduced early in the 19th century largely for marine and river works. Various materials were tried such as clays, pozzolana mortar, lime mortar or bitumen emulsions. In the last quarter of the century the use of Portland cement mortar became widespread. The use of grout by Greathead in shield tunnelling to fill voids between lining and excavated ground was an important advance, but is not directly relevant to this chapter, which is concerned with preparatory ground treatment.

In the middle of the 19th century grouting in the U.S.A. was developed in dam construction to strengthen foundations and form cut-off walls in pervious ground.

Chemically reactive grout comprising aluminium sulphate solution and water-glass which combine to form a soft silica gel was first used in 1911 at a colliery near Doncaster. In the well known and widely used Joosten process, patented in 1925, successive injections of sodium silicate and calcium chloride penetrate sands and gravels well and react to bind the sand very effectively.

In the last 30 years a wide range of different grouts have been introduced, and concurrently the necessary high pressure pumps and other injection equipment have been developed. Their applications to tunnelling have been extensive and they have become indispensable in difficult conditions.

The first Dartford tunnel (1963) marked the introduction of ground treatment to control losses of compressed air by injection of a range of grouts, including bentonite/cement, into a bed of open gravel through which the tunnel penetrated under the doubtfully stable river flood banks. The complex pattern of injections from an array of tubes at the surface, and from jetties in the river utilising the 'tube a manchette' (sleeved tube), was effective in limiting air losses.

5.2.3 Principles of grout injection

The principal factors governing the choice of grout mixtures and methods of application are:

1. Ground characteristics
   (a) Permeability of the ground
   (b) Flow of ground water
   (c) Ground water pressure
   (d) Depth and accessibility of zone to be treated

2. Grout properties
   (a) Size of particles in suspension
   (b) Viscosity
   (c) Setting time and rate of setting
   (d) Grouting pressure and injection rate
   (e) Chemical compatability

The overall permeability of the ground derives in part from the pore structure of the soil or rock and in part from the pattern of joints and fissures. What follows relates more particularly to unconsolidated granular soils, which are not characterised by joints or fissures.

Permeability, in this context, is defined by D'Arcy's Law relating to percolation of water through a uniform medium, such as sand: $v = ki$, in which v is the discharge velocity of water (cm/s), i is the hydraulic gradient (No), and k is the coefficient of permeability (cm/s).

While knowledge of k will give an immediate value for passage of water, the penetration of a grout will be restricted by its viscosity and by the action of solids held in suspension.

In soils in which k exceeds 10^{-1} cm/s slurry grouts of cement, or bentonite or other clay can be injected at low pressure. Down to $10^{-2.5}$ cm/s chemical grouts of relatively high viscosity, such as those used in the Joosten process are effective, but where permeability is lower resin grouts are likely to be necessary. The practical limit for these grouts is about 10^{-5} cm/s, which excludes fine silts.

Down to this level most soils can be stabilised and rendered impermeable, but the finer the pore structure the more difficult, slow and costly does effective grouting become. Where the strata are variable the grout injections will initially follow the path of least resistance, and lenses of fine silt or even fine sand may remain untreated and may cause trouble subsequently.

If ground water is flowing through the ground the viscosity and setting time of the grout must be such as to ensure its retention in the zone being treated. The pore pressure in the ground is obviously significant, as the injection pressure must be sufficiently in excess to expel the ground water throughout the desired radius of action.

The depth and accessibility of the treatment zone may range from a shallow bed easily reached from the surface to a site where a tunnel is to be driven below a river bed or under a built-up area, and where a pilot tunnel may be the only means of access. In shallow working vertical drillings can be made at close centres but wider spacings become necessary for both technical and economic reasons as depth increases, and the radius of effective penetration of the grout may become of critical importance. If grouting is to be done radially from within a pilot tunnel, as further discussed below, the problems of balancing spacing, penetration and setting time may be difficult.

Grout formulations are examined in more detail below. Where a particulate suspension is to be used the relationship between grain size of grout particles and pore dimensions in the ground is of critical importance. The thickness of the grout must be adequate to give a set to the appropriate strength in the time available, but not such as to demand excessive injection pressure. In liquid grouts viscosity determines the rate of penetration at a specified working pressure. The process of setting implies an increase of viscosity with time,

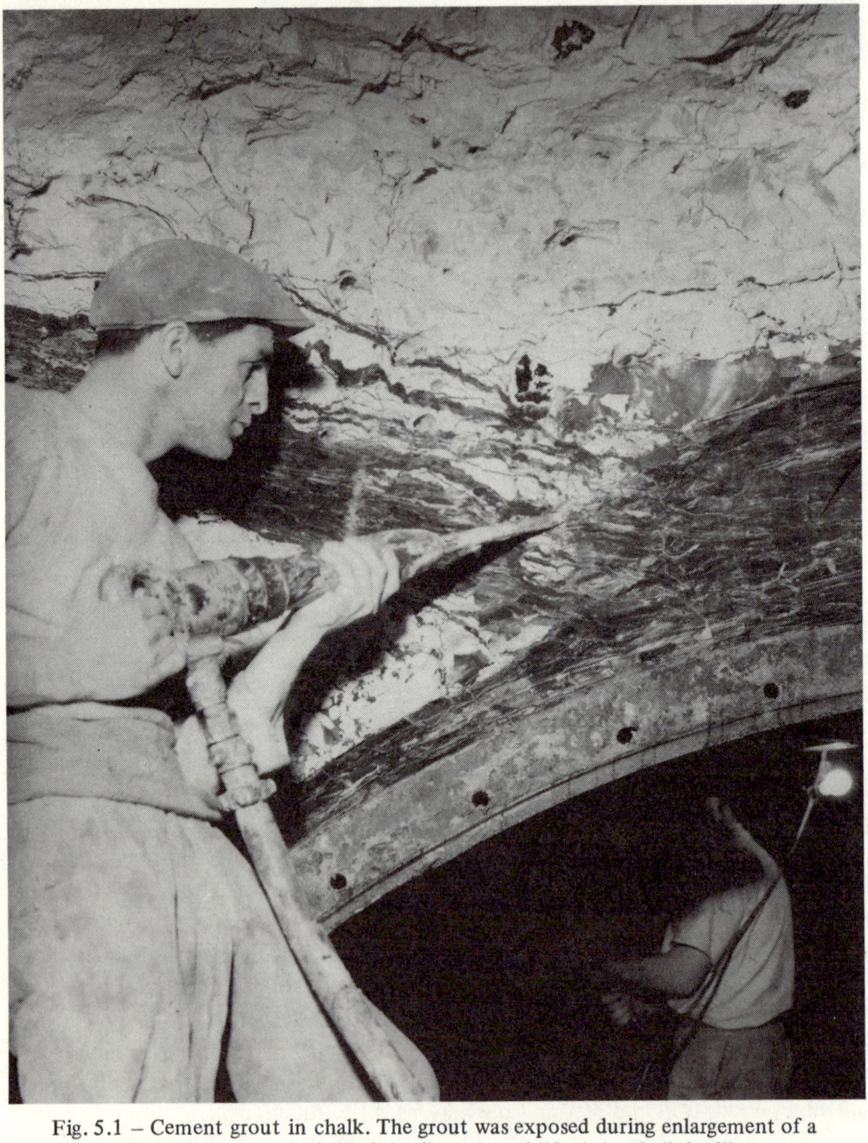

Fig. 5.1 – Cement grout in chalk. The grout was exposed during enlargement of a pilot tunnel. The cement had filled the fissures, and lifted the chalk bodily.

and this may be either slow and continuous or more abrupt. Time is therefore of great importance in determining the extent of penetration.

Grouting pressure has to be carefully calculated and controlled. An attempt to accelerate the injection process may simply result in penetration of grout locally along a plane of low resistance and its escape beyond the treatment zone, or the build up of a high hydraulic pressure, disrupting the ground and possibly causing heave at the surface without effectively treating the finer sands. This process can be deliberately employed to stabilise a fissured clay or other impermeable material by establishing a network of fissures filled with a strong grout.

The question of chemical compatability between grout and ground may have to be considered, as for example in ground of high acidity, or alkalinity, or in the presence of sulphates. Some grouts have toxic constituents, the handling and ultimate dispersal of which may have to be very carefully controlled. Any possible dangers of poisoning water in the ground also have to be considered.

Fissure grouting is aimed at the filling of fissures and joints in rock, and possibly in stiff clay, to reduce the intrusion of water and to minimise movement of blocks disturbed by tunnel excavation. The same principles apply but details of application and technique differ. Grouting of shattered fault zones ahead of excavation may be essential to safe construction. Pilot drilling ahead of the face as far as practicable is required to locate major water-bearing fissures and to allow them to be sealed by grouting. Whatever pattern of drilling is adopted for exploration and for treatment, it is impossible to ensure that every fissure and joint is located and treated; a boring may run nearly parallel to the plane of a major fissure and fail to intersect it. In some conditions fissures may be apparently blocked with soft 'gouge' which may be washed out by inflowing water when exposed. A useful treatment may be the deliberate washing out of the gouge by high pressure water, followed by refilling with grout.

5.2.4 Grouting materials

5.2.4.1 Particles in suspension

Cement Ordinary Portland cement is used in gravels and coarse sands, and also in fissure grouting. The average particle size of ordinary commercial Portland cement is 30 μm, but individual particles may be up to 100 μm. It is useful where passages for grout flow are of the order of 10 times the particle size. This corresponds to ground having a permeability greater than $10^{-0.5}$ cm/s.

To obtain maximum penetration by the cement a technique of progressive thickening for successive injections is used. An initial thin mix, such as 100 l water to 50 kg cement, allows the grout to reach and penetrate and seal the finer voids, excess water being expelled into the surrounding ground by maintained pressure. Successive stages with reducing water/cement ratio ensure also the filling of the coarser voids. The procedure ensures maximum strength and minimum permeability. Initial setting time is about one hour, but hardening is a slower process.

This progression with a single relatively cheap material contrasts with the technique of initially using cement grout and subsequently infilling the finer zones with expensive low viscosity grouts.

Rapid hardening cement is more finely ground and the smaller particle size increases penetration, with the possible added advantage of shorter setting time and high early strength.

High alumina cement gives early strength and resists sulphate and acid attack but cannot be used in alkaline conditions. When mixed with ordinary Portland cement there is a very rapid set, useful for plugging leaks or other emergency action.

Supersulphated cement is a finely ground mixture produced from granulated blast furnace slag and ordinary Portland cement with added calcium sulphate. It has good penetration and is highly resistant to sea water, acids and sulphates.

Other cement based mixtures may be adopted for special purposes. Addition of sand will provide a cheaper grout for filling large cavities.

Cement/Clay Mixtures of cement with clays can offer a cheaper filler grout than pure cement. The strength is lower but watertightness need not be impaired. Not all clays are suitable: Kaolinite and Illite are commonly used.

Bentonite is a very special clay, which has thixotropic properties when mixed with water, that is to say it behaves as a fluid while in motion or if agitated, but when stationary forms a gel, with properties of low permeability and some cohesive strength. It may be used alone to reduce permeability in gravels and coarse sands, if substantial strength is not required.

Cement/bentonite mixtures are used in fairly open soils and in fissure grouting. The bentonite improves the flow characteristics and does not significantly reduce strength if it is used sparingly.

Pozzolans Pozzolans are finely divided materials insoluble in water, but chemically reactive, named from a typical volcanic tuff occurring near Naples. They have themselves some cementing properties and can be used as additives to Portland cement. Pulverised fuel ash (PFA) is a waste product of coal-fired power stations, having similar properties, and is readily available. It has largely replaced natural pozzolans. Fly ash, also from power stations, has similar properties and uses.

PFA mixed with about 35% water forms a slurry, and further dilution up to 50% makes it flow freely, but the amount of water may be rather critical for particular pumping and injection requirements. It will react with lime and water to form a stable weak cement, but its particle size is in the silt range and its use in soils is restricted to gravels and coarse sands.

PFA and fly ash can provide economical fillings for large cavities.

PFA/cement in proportions up to 20 : 1 produce a cheap material with good strength characteristics, but large particle size limits its applications.

5.2.4.2 Chemical grouts

Those currently used may be grouped as silicates, lignins and resins.

Silicates The oldest and best known silicate injection technique is the Joosten two-shot process, in which sodium silicate is first injected followed by a solution of calcium chloride, which sets up an immediate chemical reaction resulting in a calcium silicate gel which binds sand into a weak sandstone. The consistency of the gel may not be uniform, being greatly influenced by variations in ground porosity. The process is well tried and is effective if expertly applied, but its use is restricted by the high viscosity of the silicate, because of which satisfactory penetration cannot be expected in gradings finer than medium sand.

The Guttman process is similar but the viscosity of the silicate is reduced by dilution with sodium carbonate, which improves penetration but gives lower gel strength.

One-shot processes have been developed by mixing the silicate prior to injection with appropriate reagents. The setting time for the reaction allows for the injected grout to penetrate sufficiently, but strength may be low and these are mainly used where reduction of permeability is the principal objective.

Examples of these additives are: weak gels–sodium bicarbonate, lime water, sodium aluminate; strong gels–ethyl acetate, amide, citric or oxalic acid plus a salt of aluminium or iron.

Lignins Low viscosity grouts with penetrating power adequate for fine sands can be developed from lignosulphite, a by-product of the wood pulp industry, in combination with bichromates. The mixing of these liquids results in formation of a firm gelatinous mass, with a setting time ranging from 10 mins to 10 hrs, dependent on the concentration of bichromate. The viscosity and final gel strength also vary. Some of the bichromate grouts may be toxic at certain stages and appropriate precautions in handling them are required.

Resins The grouts of greatest penetration are those produced by polymerisation of organic chemicals, including resins, which can be used at permeabilities as low as $k = 10^{-4}$ cm/s. They are mostly proprietary formulations, and are adapted to the particular conditions expected. They include:

AM-9 (American Cyanamid Co.) which is a mixture of acrylamide and a methyl derivative. Its setting time may be affected by the presence of salts in the groundwater and by its pH value.

Resorcinal with formaldehyde (Cementation Co.) which are injected together in aqueous solution, producing a polymerised resin with properties partly elastic and partly plastic.

Geoseal Resin Grouts (Borden Chemical Co.) are based on water-soluble resins in powder form, activated by caustic soda, with various modifications and additions to control viscosity and gel time. The grouts are described as MQ4, MQ5, etc. A two-shot grout is included in the range.

Other materials are sometimes used for special purposes. Bitumen emulsion can be used in medium sands to reduce permeability, but the cohesive strength is low. Hot bitumen can sometimes be injected to seal large cavities in the presence of flowing water.

Polyurethane is an oil based polymer with relative high viscosity.

Cyanloc 62 is a resin grout not water-soluble which forms a very stiff gel when mixed with sodium bisulphite. Being injected as a particle suspension the penetration is limited.

5.2.5 Equipment

The essential plant for ground treatment by grouting comprises:

1. Drilling gear
2. Batching and mixing equipment
3. Pumps
4. Grout injection tubes

The operations may range from relatively simple processes, such as cement or bentonite grouting from the surface. to treatment of varied and difficult ground with a succession of specialised grouts, and injections at high pressures from a confined space in a pilot tunnel. In all cases careful control and accurate records by skilled operators are required, and in the more complex applications a site laboratory and expert supervision by specialists are likely to be essential to a satisfactory outcome. Many of the chemical and resin grouts involve patented processes, and full details of formulation and use may not be published.

Percussion and rotating drills are both used. Casing will be provided or drilling mud used only if necessary to prevent collapse. There are specialised techniques which allow controlled change of direction during drilling.

Batching and mixing plant must be carefully planned and sited. It may be no more than a tank in which to mix cement and water, or a tunnel grout pan employing compressed air for injection, but, with the special chemical grouts, very precise proportioning of the ingredients and accurate timing of the mixing and injection processes may be necessary to ensure the correct viscosity and setting time; the plant may be very elaborate with extensive instrumentation. One important aspect is to ensure that the grout does not set prematurely and solidify in the pumping and injection system.

Pumps must allow control of quantities injected and pressures employed, and are therefore normally of the positive displacement type. With a two-shot system the use of separate pumps to avoid mutual contamination is obvious but also requirements for injection rates and pressures may be different. In some operations the larger voids are first filled with a relatively coarse and cheap material and followed up by infilling with a more penetrative chemical grout.

Grout tubes may be simple open-ended tubes, possibly fitted with an expendable tip to prevent blockage in driving, or perforated tubes which allow grout to be injected over a length may be used. The 'tube-a-manchette' makes possible successive injections of different grouts and choice of the level for injection. Perforations at appropriate intervals are closed by an external elastic sleeve which can be opened by the internal pressure of the grout; the level of injection at any time is regulated by pistons above and below the injection point (see Figure 5.2). In mixed ground initial injection of cement/bentonite in gravel and coarse sand can be followed by chemical and resin grouts in finer sands which the particulate grouts cannot penetrate, and injections can be repeated until the required reduction of permeability is effected.

Spacing of grout tubes normally is within the range 0.5 m to 4 m depending on the nature and extent of the required modification of the ground and the permeability and viscosity.

Where the surface is accessible modern drilling equipment will reach any depths likely to be required for tunnelling and associated shaft sinking.

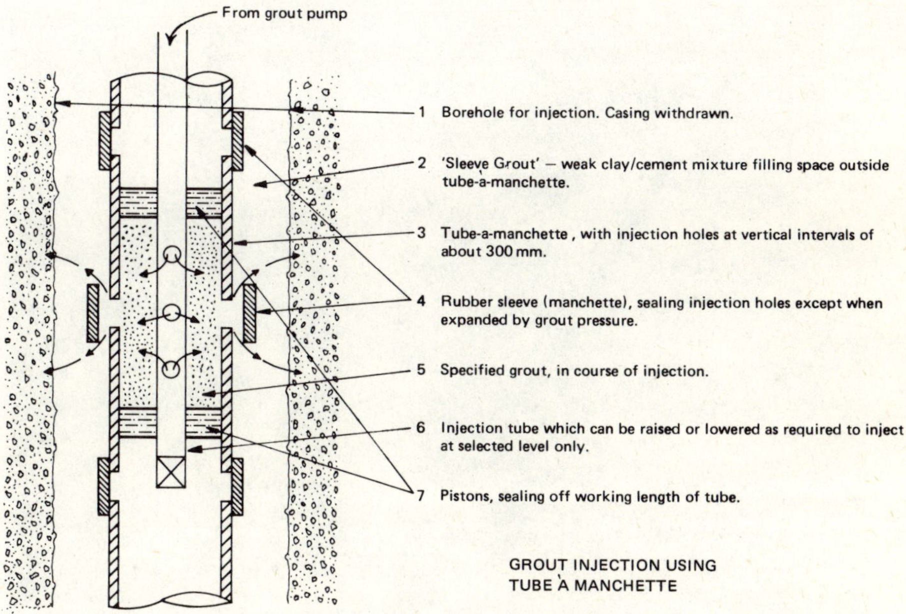

Fig. 5.2 – (a) 'Tube-a-manchette'.

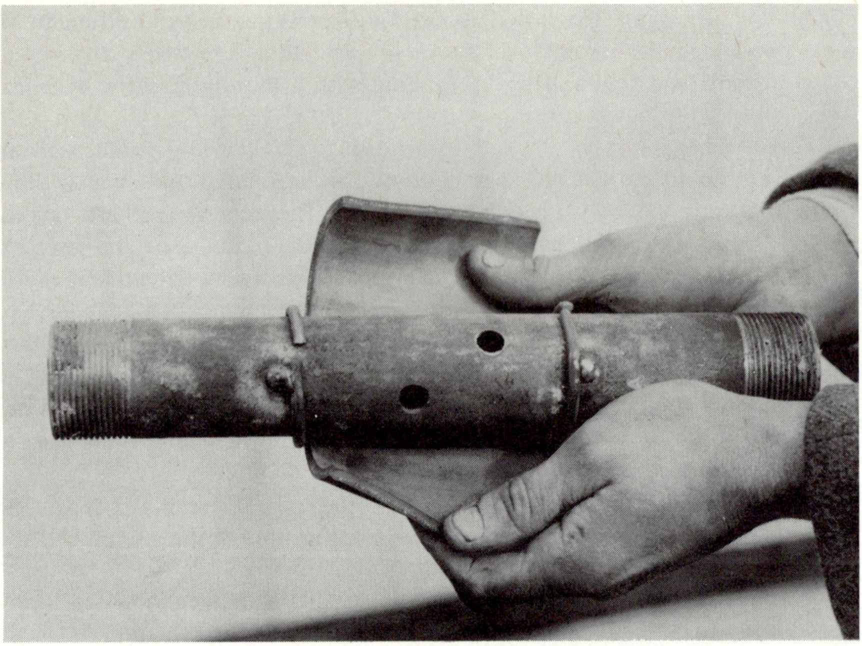

Fig. 5.2 – Tube-a-manchette (b) Perforated tube and rubber sleeve.

Fig. 5.2 – Tube-a-manchette (c) Internal piston.

5.2.6 Application

The treatment may be applied to the whole of a permeable or incoherent stratum affecting a tunnel, more particularly the area immediately outside the excavation. There may be an advantage in treating also the ground which is in fact to be excavated, as the added stability of the face will simplify face support, or may allow it to be dispensed with, and control of water will be improved. Selective treatment of critical zones can be economical in time and materials, but more precise control and supervision are necessary.

Access from the surface may be impracticable because of the presence of buildings or other obstacles, or in subaqueous tunnels. There are then the possibilities of treating the ground ahead of the face from the length of tunnel already excavated, or from a pilot tunnel, or a parallel tunnel already completed.

In working ahead through the tunnel face in normal conditions the treatment of ground outside the tunnel must be by a diverging cone of grout pipes. The length which can be treated is then limited largely by the accuracy with which pipes can be driven, and also by the angle of divergence of the pipes. In special cases a sufficient enlargement of the last length of tunnel will allow a ring of parallel pipes to be driven in a cylindrical pattern.

It may not be necessary to treat the whole circumference of the tunnel: a 'hood' over the crown of the excavation may be adequate, or other limited arcs where a thin layer of permeable ground intersects the excavation.

One major disadvantage of working from the tunnel itself is that ground treatment and tunnel excavation cannot proceed simultaneously, the appropriate plant and the specialised labour for the two operations alternating, perhaps on a two-week cycle. The tunnelling is then slow and costly.

Ground treatment in advance from a pilot tunnel has the advantage of continuity of operation and less disruption when the unexpected is encountered. Where a large tunnel is to be built a pilot may be driven inside its area and may be concentric or at high or low level. Grout pipes are driven out through prepared holes in the pilot tunnel lining and must be more or less radial. This imposes two limitations: the length of pipe which can be inserted without a joint, and working space generally, are limited by the diameter of the pilot tunnel, and the radial pattern of pipes means that spacing increases rapidly outwards.

The second Blackwall tunnel provided an interesting example of such ground treatment. The tunnel carries a highway under the tidal River Thames and is cast-iron lined with a diameter of 8.6 m. As much as possible of the tunnel lies within the London Clay, but there is a buried channel in mid-river, cut down in glacial times below the present bed level right through the stratum of clay and subsequently refilled, mainly with sands and gravels. The main tunnel, to be driven with the aid of compressed air and a hooded shield, had to pass through a full face of sand and gravel with a cover of about 6 m. The gravel generally had a permeability of about 10^{-1} cm/s and improvement to an average of 3 x 10^{-4} cm/s was sought. Two 2 m diameter pilot tunnels were

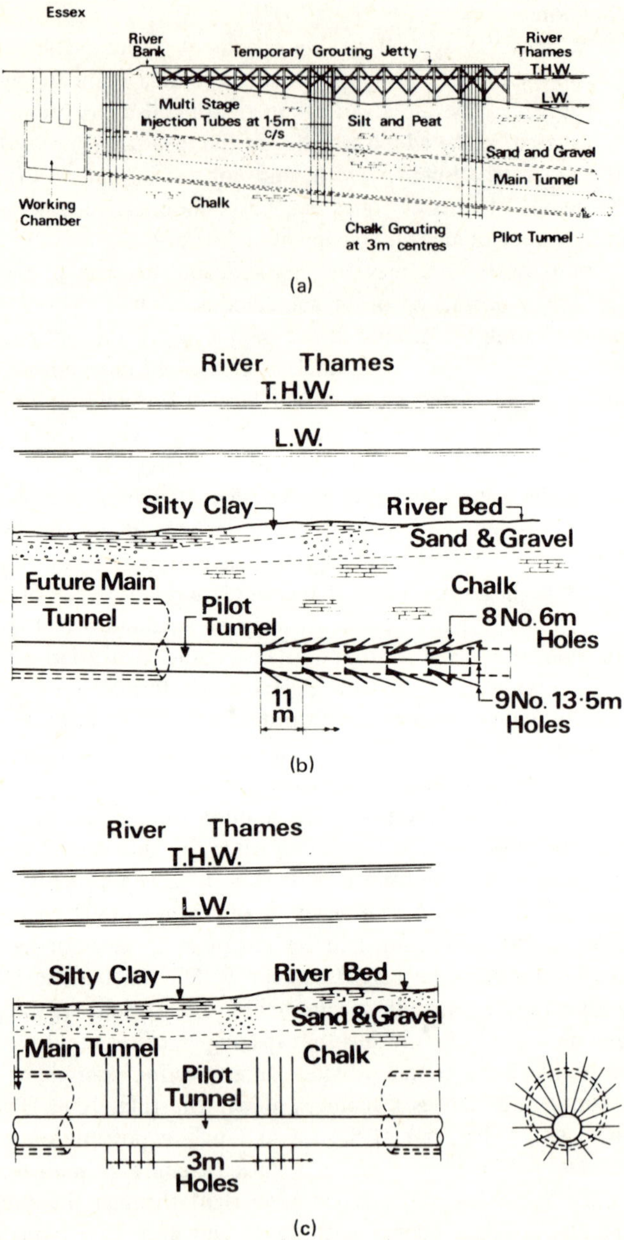

Fig. 5.3 – Three arrangements for ground treatment below river, for second Dartford Tunnel. (a) Under foreshore, from surface by use of jetty. (b) In a conical array, from face of pilot tunnel, which was constructed in 11 m lengths. (c) Radially, from pilot tunnel preparatory to excavation of main tunnel.

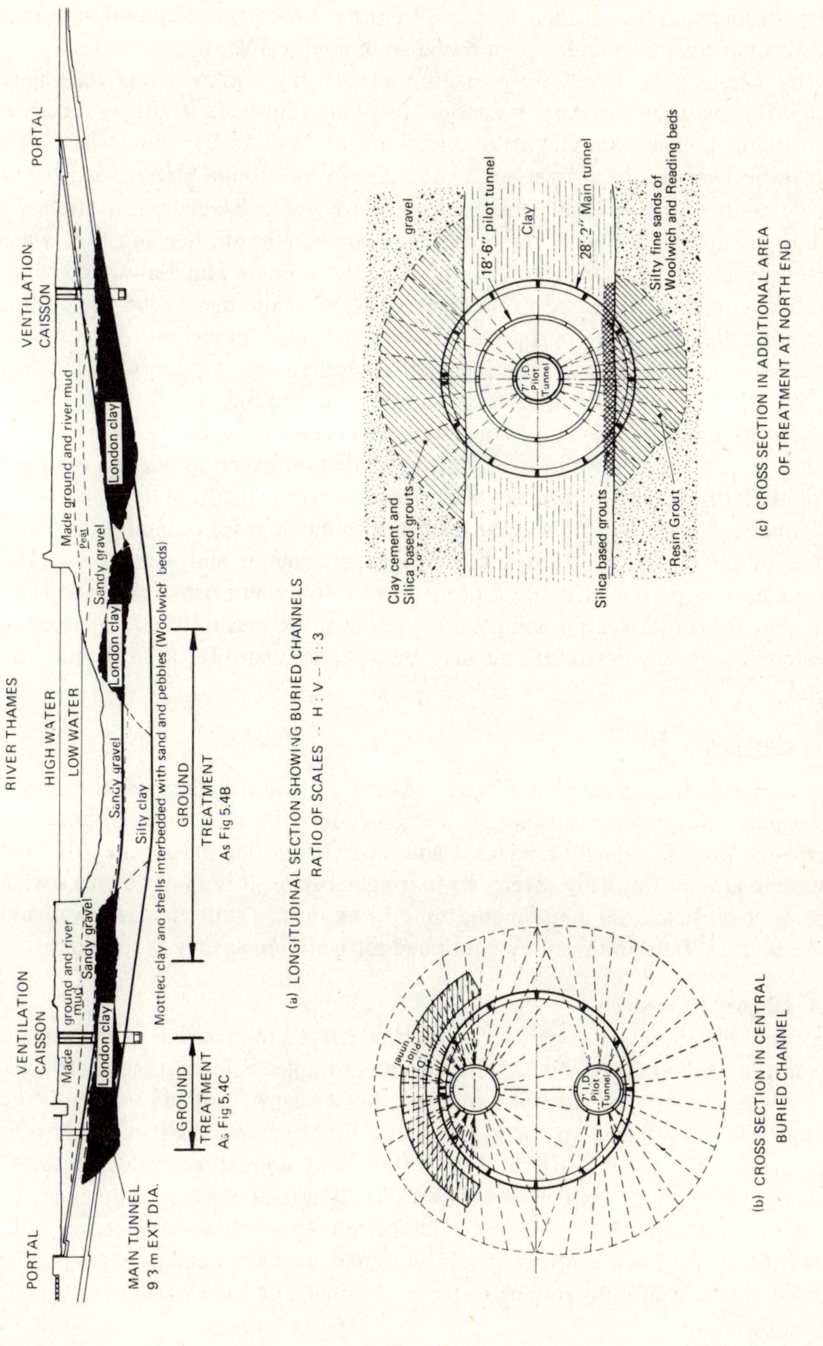

Fig. 5.4 – Blackwall Tunnel Duplication – ground treatment.

driven in advance as shown in Figure 5.4 to allow the grouting to be carried out. The small diameter was chosen to minimise risks of a blow or inundation during the driving of the pilots under compressed air and with shields.

The whole area surrounding the tunnel was injected with clay/chemical grout effecting the necessary reduction in permeability. In addition, because the working pressure in the main tunnel might have to be raised above the hydrostatic head at the crown, an arched canopy of ground above was further injected with a low viscosity chemical grout to give enhanced shear strength, resistant to uplift. Mixing of the grouts was done at the surface and they were injected by means of proportioning pumps in the tunnels which allowed gelling times and pressures to be controlled. In addition to the main zone there were other areas treated with appropriate grouts. The main tunnel was successfully driven; air pressures were a little higher than desired, but air consumption was satisfactorily low and there were no serious losses of ground.

Ground treatment by grout injection is necessarily a costly operation in which the cost per unit volume of treated ground will vary so widely, even for use of the same grout, as to make comparisons very difficult. Material costs for acrylamide and resin grouts may be 10 times as much as for cement grouts, and require more specialised plant and highly skilled labour and supervision. The case for using a particular method needs careful study, but use of an inadequate grout that fails to effect the requisite reduction in permeability and increase in cohesion is certainly wasteful and may be positively harmful in disrupting the ground.

5.3 FREEZING

Successful freezing of permeable water-bearing ground effects simultaneously a seal against water and substantial strengthening of incoherent ground. No extraneous materials need be injected and, apart from the contingency of frost heave, the ground normally reverts to its original state. It is applicable to a wide range of soils. It takes a considerable time to establish a substantial ice wall and the freeze must be maintained by continued refrigeration as long as required.

5.3.1 History of freezing

Uses for shaft sinking through water-bearing strata are recorded in the second half of the 19th century, one of the earliest examples being a mine shaft sunk using cooled brine in South Wales in 1862. In Germany from 1883 onwards the method was developed also for mine shafts. Freezing was used in repair of a shaft on the St. Omer canal in 1893. Tunnelling and other civil engineering applications were not extensive up to 1940. It was applied to tunnelling for the Paris metro in 1907; and over the period 1932–40 in connection with ventilation shafts for the Antwerp vehicle tunnel, escalators and passages for the Moscow metro, shafts for cooling water at Swansea, and for a water supply shaft in West Cheshire.

Uses in connection with tunnelling have since then increased, as refrigeration methods and plant have been improved, and freezing is widely used in many countries.

Freezing has been employed in the construction of difficult elements of metro systems in London, Leningrad, Tokyo and Helsinki, and for sewer tunnelling in New York City and for railway tunnels in Italy. The first application of liquid nitrogen as the freezing agent seems to have been for a sewer tunnel in Paris in 1961.

5.3.2 Principles of freezing

The effectiveness of freezing depends, of course, on the presence of water to create ice, cementing the particles and increasing the strength of the ground to the equivalent of soft or medium rock. If the ground is saturated or nearly so it will also be rendered impermeable. If the moisture does not fill the pores and an ice wall is nevertheless required, either because water is temporarily absent or because it is desired to cement with ice dry running sand locally, it may be necessary to add water. The strength achieved depends on freeze temperature, moisture content and the nature of the soil. Freezing can be particularly effective in stabilising silts which are too fine for injection of any ordinary grouts.

On freezing, water expands in volume by about 9%, which does not in itself impose any serious stresses and strains on the soil unless the water is confined within a restricted volume. With water content up to about 30% the direct expansion may be up to about 3%. Frost heave, which may occur in fine silts and clays, is a slightly different phenomenon described in more detail below. In rock or clay ice lenses may build up and enlarge fine fissures so causing increase in permeability after thaw.

If there is a flow of water through the ground to be frozen the freezing time required will be increased by reason of the continuing supply of heat energy and, if the flow is large and the water temperature high, freezing may be completely inhibited. Flowing water is also liable to result in an eccentric oval rather than a circular section round the freezing tube.

As in all ground treatment techniques, adequate site investigation is necessary to allow the best system to be chosen and to design the appropriate array of freezing tubes and select plant of adequate power.

Because freezing can be imposed uniformly on a wide range of soil types in a single operation, it may offer greater security in mixed ground than treatment by injection of various grouts.

The versatility of the principle in its application to tunnel construction problems is demonstrated by the variety of its applications, such as:

1. Shaft sinking through water-bearing strata.
2. Shaft construction totally within non-cohesive saturated ground.
3. Tunnelling through a full face of granular soil.

4. Tunnelling through mixed ground.
5. Emergency local soil stabilisation.
6. Temporary underpinning of adjacent structures and support during permanent underpinning.

The cost of ground stabilisation by freezing has been generally considered to be more than for other methods of ground treatment, but this is not necessarily so and comparative estimates are often worth while. Costs of chemical injection may be difficult to estimate with any precision, because the 'take' of the ground may be unpredictable and also because small pockets of soil may require special treatment. Freezing costs in terms of plant hire and time can be estimated with considerable reliability, provided that flowing water is not present and operations are not affected by unforeseen lenses of silt or pockets of gravel.

5.3.3 Processes

Freezing may be:

1. Indirect, by circulation of a secondary coolant through tubes driven into the ground.
2. Direct, by circulation of the primary refrigerant fluid through the ground tubes.
3. Even more direct by injection into the ground of a coolant, such as liquid nitrogen.

5.3.3.1 Indirect cooling

Primary refrigeration plant is used to abstract heat from a secondary coolant circulating through pipes driven into the ground being frozen. The primary refrigerant most commonly used is Freon ($CHClF_2$) which is non-toxic, non-flammable and non-corrosive. Its boiling point at N.T.P. is -40.8°C. Other primary refrigerants are ammonia, NH_4 (-33.3°C) and carbon dioxide, CO_2 (-78.5°C), not now commonly used. The secondary coolant, circulated through the network of tubes in the ground is most usually a solution of calcium chloride. With a concentration of 30% such a brine has a freezing point well below that of Freon.

The primary refrigeration process is basically the Carnot cycle of compression and expansion reversed. The heat developed by compression of the primary coolant must be extracted by air cooling or water circulation. When water cooling is practicable the plant is relatively quiet and can be run continuously in urban areas. Efficiency is necessarily governed by the ambient temperature.

The time required to freeze the ground will obviously depend on the capacity of the freezing plant in relation to the volume of ground to be frozen and also on the spacing and size of the freezing tubes and the water content of the ground.

With a tube spacing of 1 m, cylinders of 1 m diameter or more have to be frozen, for which a typical time is 4–8 weeks. Small zones, of less than 500 m^3, can be frozen more quickly, and large volumes, over 20,000 m^3 may take up to 3 months.

5.3.3.2 Direct cooling

In these systems the primary refrigerant is circulated through the system of tubes in the ground, extracting directly the latent heat of boiling, and therefore having a higher efficiency than the indirect process.

Freon R22 is most commonly used, and it has the incidental advantage that it is non-toxic and non-flammable. It is, however, asphyxiant if it displaces air. Leakage into the ground, which is readily detected by fluid indicators or gas detectors, is normally harmless, unlike leakage of brine which makes freezing more difficult by lowering the ground feeezing temperature. Direct freezing time is similar to that for the indirect process. The choice will depend on plant availability, estimates of cost and, perhaps, personal preference rather than on any differences in ground characteristics.

5.3.3.3 Liquid nitrogen (N_2)

Availability of liquefied nitrogen in bulk is dependent on the existence of a regular supply to heavy industry. The major differences in this ground freezing technique are that it is not dependent on site refrigerating plant and that the temperature is very much lower and therefore quicker in application. The nitrogen under moderate pressure is brought to site in insulated containers as a liquid, which boils at -196°C at normal pressure and thereby effects the required cooling. It can be stored on site.

There is a particular advantage for emergency use, namely, quick freezing without elaborate fixed plant and equipment, and this may be doubly advantageous on sites (accessible by road) remote from power supplies. In such conditions the nitrogen can be discharged directly through tubes driven into the ground being treated, and allowed to escape to atmosphere. Precautions for adequate ventilation are important.

Where there is time for preparation, an array of freezing tubes is installed for the nitrogen circulation, including return pipes exhausting to atmosphere. The speed of ground freezing is much quicker than with the other methods, the time required being days rather than weeks, but liquid nitrogen is costly. The method is particularly appropriate for a short period of freezing, up to about 3 weeks. It may be used in conjunction with the other processes with the same array of freeze tubes and network of insulated distribution pipes, in which liquid nitrogen is first used to establish the freeze quickly and is followed by ordinary refrigeration to maintain the condition while work is executed. This can be of particular help when a natural flow of groundwater makes initial freezing difficult.

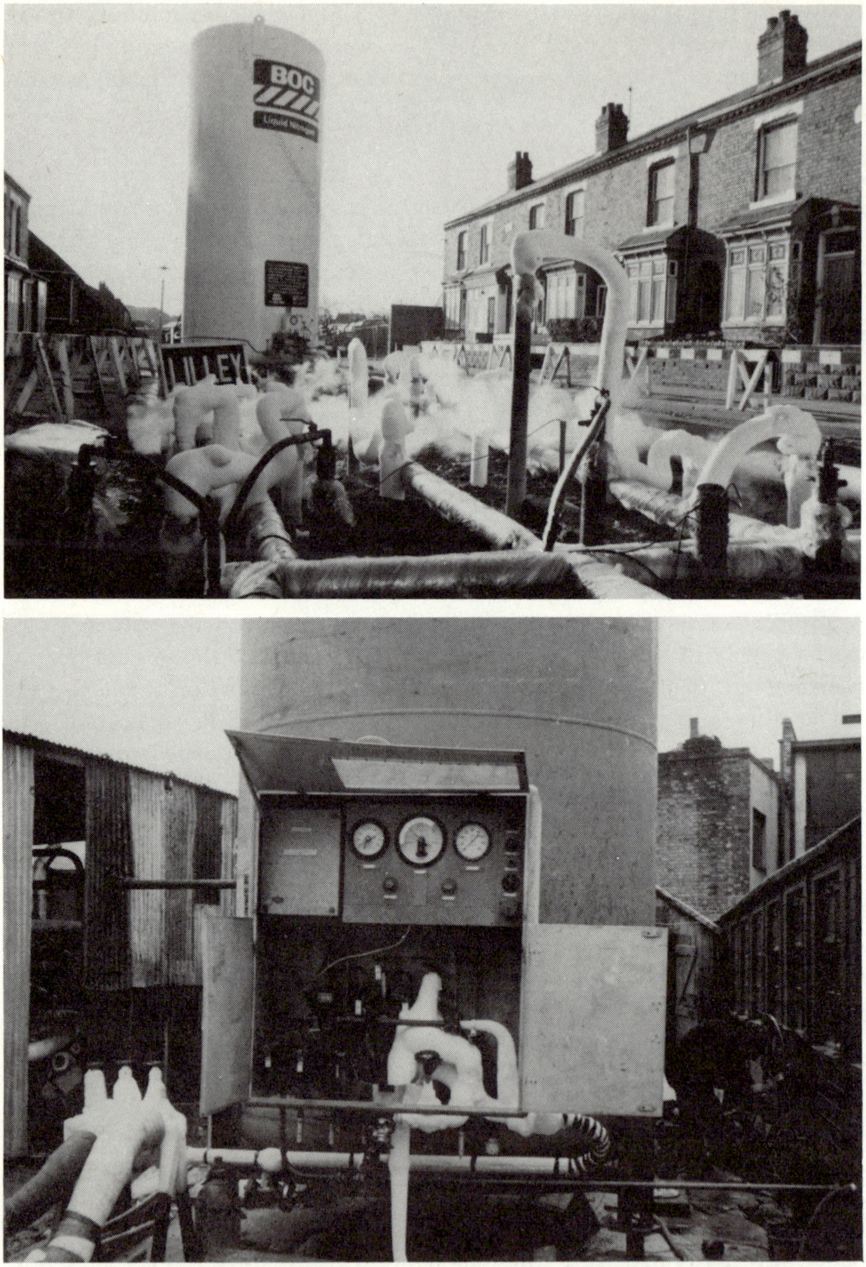

Fig. 5.5 – (a) Use of liquid nitrogen for ground freezing for a sewer tunnel. *Note*: Nitrogen tank feeding array of vertical tubes (*Lilley Construction Ltd.*). (b) Control panel regulating liquid nitrogen supply from tank behind (*A. Waddington & Sons Ltd.*).

5.3.4 Site investigation

A decision to use ground freezing methods may be taken as a result of the information obtained in the original investigation, but additional borings, samples and tests will probably be necessary to determine the best plant and arrangements. The aspects special to freezing include:

1. Extent of ground to be frozen.
2. Water content of the various materials, more particularly fine silts and clays liable to heave.
3. Salinity of groundwater as affecting freezing point.
4. Movements of groundwater.

Thermal capacity and conductivity of varying types of ground may be estimated from experience or established by laboratory tests, but the water content is the principal element, and minor variations have little effect on plant and freezing times. An ample margin of capacity is essential.

The site investigation should, of course, provide guidance in selection of the site for the plant and the distribution network, as well as the array of freezing pipes in the ground.

5.3.5 Shafts

Where a shaft is to be sunk through bad ground into firm impervious ground, an annulus only need be frozen by means of one to three rings of vertical freezing pipes installed by drilling from the surface. Excavation of the unfrozen core through the protective frozen annulus is followed quickly by building of the lining and any waterproofing, after which the ground may be allowed to thaw.

Where the shaft is entirely in water-bearing ground lacking cohesion, a plug at the bottom may need to be frozen, by means of further freezing pipes within the core. To avoid undesired and wasteful freezing of the core, these pipes can, with advantage, be insulated above the plug level. With mixed strata, pipes can be insulated at levels where freezing is not required.

Either precast or *in situ* linings can be used, and, if the freeze pattern is so designed and maintained, the whole volume can be excavated before building the shaft from the base upwards, but it is considered safer to work in units of 30 m depth or so.

Inclined shafts may be frozen either by use of inclined tubes parallel to the axis or by a carefully planned pattern of vertical tubes. Drilling of inclined holes requires specialised drilling equipment and skill to achieve the necessary accuracy.

In excavating a shaft the frozen annulus is relied on to carry the external ground and water pressures, and its thickness and temperature must be designed and ensured accordingly. The freeze must be maintained until the appropriate structural lining is completed and competent to carry the load. Any failure of the refrigeration plant allows a thaw to be initiated, with progressive danger of inundation and collapse. Where so planned and designed, however, an operation for which access is difficult in the presence of refrigeration plant but which requires only a limited time, up to perhaps 36 hours, may be executed after completion of the freeze, immediately following discontinuance of refrigeration.

5.3.6 Tunnels

The choice of drilling patterns is very similar to that for grouting. Vertical or inclined tubes so laid out as to provide a secure frozen envelope for the tunnel excavation may be driven down from the surface. Some tubes must pass through the tunnel excavation if the invert zone is to be frozen, and it is almost invariably

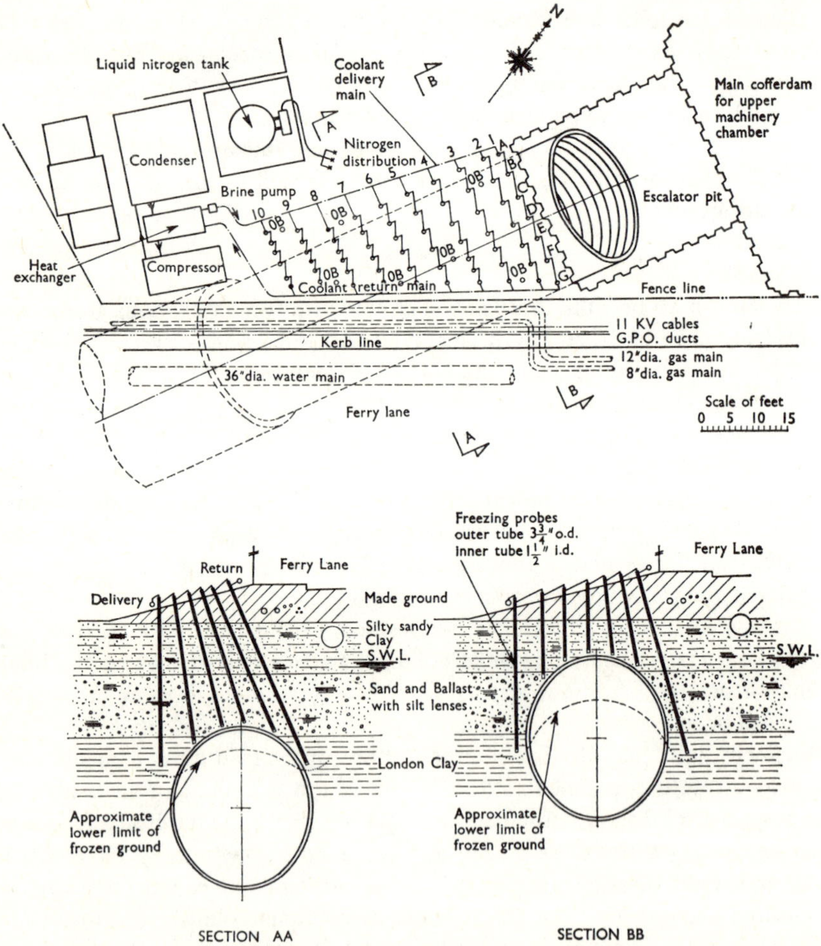

Fig. 5.6(a) – Freezing in 1966 for escalator tunnel on Victoria Line. Both liquid nitrogen, for rapid initial freezing, and ammonia-brine system were used. Site layout and sections. (*A. Waddington & Sons Ltd.*)

Fig. 5.6(b) – Freezing in 1966 for escalator tunnel on Victoria Line. (*A. Waddington & Sons Ltd.*)

necessary to withdraw or abandon these immediately before excavation, which has therefore to be completed and lined without delay. The face to be excavated will be frozen unless elaborate precautions are taken, but this is usually acceptable even though blasting may be necessary. Where treatment below the invert is unnecessary, work is much simplified. Difficulties of access to the surface area above the tunnel are likely to limit severely the application of the method in cities and under rivers.

Fig. 5.7(a) – Use of freezing for tunnelling through Kluuvi cleft in Helsinki, where a gorge in the rock is infilled with waterbearing glacial deposits. Rock tunnel enlarged to permit drilling of cylindrical array of freezing tubes, preparatory to tunnelling with segmental cast iron

Freezing from the tunnel ahead of the face is possible either by a conical array of pipes from the lined length of tunnel or by a cylindrical array from an enlargement or from a working pit or shaft, as for example a manhole shaft for a sewer. The former method can only be applied to a very limited length because of the spread of the pipes and the inaccuracy of horizontal drilling. it results in a very congested working area near the tunnel face, and the necessary freezing time makes the alternating progress of freezing and driving extremely slow.

The latter method of working with parallel tubes from a larger pit has substantial advantages because a longer length can be drilled and frozen and there is more working space, although this also is a slow method. Beyond about 30 m deviations of tubes are likely to increase rapidly, but it is now practicable to work up to 50 m, or in special circumstances 100 m.

Work from a tunnel face is therefore an expedient for special situations. As in shafts, the consequences of failing to maintain the freeze until the permanent lining is complete can be very serious.

Special difficulty may arise in freezing the crown of a tunnel in a river crossing where cover is shallow, because of the continuing supply of heat from the flowing water. A mat of freeze tubing installed on the river bed and covered with an insulating blanket has been used, the edges of the area being carefully sealed against hydraulic erosion.

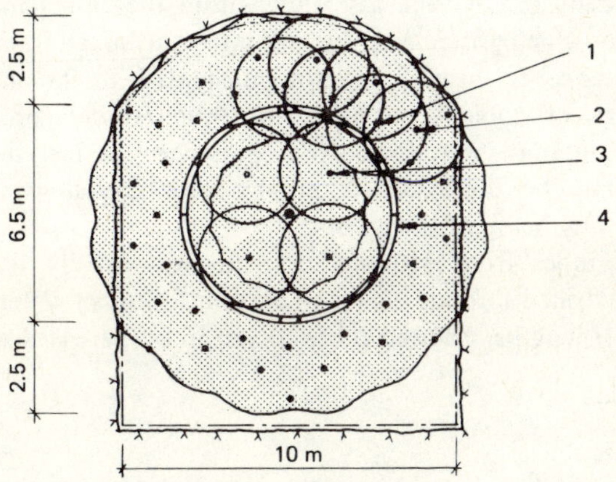

CROSS SECTION IN ENLARGEMENT
SHOWING PATTERN OF PIPES

1 INNER RING OF FREEZING TUBES
2 OUTER " " "
3 CORE " " "
4 CAST IRON TUNNEL LINING 6.1 m INT. DIA.

Fig. 5.7(b) – Kluuvi Cleft, Helsinki. Cross section in enlargement showing pattern of pipes.

The construction of tunnel junctions may be assisted by freezing the surrounding ground by tubes from the converging tunnels.

In an emergency when unexpected water is encountered locally, the merits of liquefied nitrogen freezing will be apparent.

5.3.7 Construction procedure

5.3.7.1 Tube installation

Freeze pipes consist of two concentric tubes, the outer being closed at the bottom. The coolant is circulated by pumping it down the inner tube to return through the surrounding annulus, where it extracts heat from the surrounding ground.

Spacing of about 1 m between tubes is usually appropriate, but may be made closer to provide a more uniform thickness of ice wall, or in critical areas.

Tube diameter will be chosen to suit the type of ground, length and thermal transfer requirement. Freezing will be quicker with a larger surface area for heat transfer and greater cross-section for flow of coolant. For the outer tube, the diameter is usually 50–100 mm.

Where practicable, uncased holes will be bored for subsequent insertion of the freeze tubes, but in unstable ground casings may be required, and these may be utilised as the outer coolant tube after the end has been sealed. The insertion of an independent freeze tube assembly within the casing may be preferred, especially if some lengths are to be insulated, although there will be some increased resistance to heat flow, not substantial if the gap is small. The possible leakage of coolant outwards into the ground, or of groundwater entering and contaminating the coolant must be considered. In deep shafts freeze tubes may be lubricated externally with special greases to minimise stresses imposed by heave of the ground.

Horizontal tubes from a tunnel or a pit will usually require casing and may have to be drilled through glands to prevent inflow of water. For emergency use or for short lengths, pointed tubes driven in by percussion may sometimes be practicable.

5.3.7.2 Accuracy

Spacing between tubes is determined by the need to close the ice wall as a continuous formation without 'windows'. A cylinder of frozen ground is built up round each tube as heat is extracted and a diameter of 1 m is a normal basis, which leaves some margin for deviations in the positions of adjacent tubes.

Deviation of long holes is difficult to control but it is important to ascertain the actual spacing at the ends of the borings and to ensure that tubes do not pass through the excavation zone unnecessarily.

It is therefore sound practice to survey completed holes accurately before refrigeration is started so that unsuitable holes may be abandoned, and any necessary supplementary holes may be drilled. An accuracy of about 1 in 100 is required, for which three types of instrument are used:

1. Pendulum device for vertical holes.
2. Mechanical deflection type.
3. Optical type, which employs a light beam or a laser.

These surveys are frequently carried out by specialists, using patented equipment.

5.3.7.3 Monitoring

Monitoring of the progress of freezing and its maintenance is effected by temperature readings at selected positions. The proposed outer boundary of the ice wall is likely to be chosen for points of measurement. Additional cased holes are drilled where required to accommodate thermocouples or thermistors, whereby temperature changes are measured and recorded as required. A typical provision might be five or more tubes round a shaft or tunnel, with thermistors at 3 m intervals down each tube. Observations show the progress of freezing, and circulation in particular tubes can be increased or reduced as found appropriate. Records should be maintained until the thaw is complete.

In an enclosed area, such as a shaft sealed at the bottom into clay, the water level and pressure will rise when the circumference is sealed, and this will indicate the effectiveness of the ice wall.

The refrigeration plant should be equipped with thermometers, pressure gauges and flow meters to evaluate heat extraction, and with integrating meters for power consumption to allow a check on overall performance.

It may be advisable to coordinate temperature records with levelling at the surface, to monitor any ground movements resulting.

5.3.7.4 Excavation

Soft ground frozen to a temperature of -10°C to -15°C has strength similar to weak concrete or soft rock, about 10–20 N/mm^2.

For hand excavation the heavier types of pneumatic tool are required, fitted with points. Soft ground tunnelling machines may be inadequate, and other mechanical excavators must be matched to requirements.

Excavation by drilling and blasting is often considered best, with charges carefully designed to avoid damage to tubes and minimise vibrations, with due regard to the limited zone of ice.

Excavation can be extended in advance of lining if a sufficiently thick zone is maintained fully frozen, but frozen ground can creep under load if support by lining is too long delayed.

5.3.8 Frost heave

Expansion and frost heave have already been mentioned. The immediate volumetric expansion of 9% which occurs as water solidifies is a limited action only relevant in an enclosed volume from which the liquid water cannot flow out freely. Frost heave proper is typically a surface phenomenon occurring naturally in cold climates where an actively freezing boundary in fine grained water-bearing soil can draw to itself a continuing supply of water, by capillary flow, whereby a progressively thickening layer of ice builds up by crystalline growth and forces up the overlying soil. In artificial freezing for shafts and tunnels it may likewise occur, but the supply of water need not be restricted to capillary flow. Enlargement of fissures may result, contributing to heave.

These phenomena are particularly related to silts whose analyses show a few per cent to have grain size less than .02 mm. From coarser soils water can be expelled more freely and finer clays become impermeable to the movement of moisture. Chalk can show heave, as can fissured clays.

Water pressure build-up may be observed if a zone being frozen is enclosed within an impervious envelope, which may be the ice envelope itself, or may be clay. Bursting action can be prevented by provision of pressure relief pipes, provided they are not allowed to freeze solid prematurely. The development of water pressure within the zone is a positive indication of the continuity of the ice envelope.

On termination of freezing the actions of expansion and heave are reversed as the ground thaws, which may occur naturally over a period of 6–20 weeks, depending on volume frozen, groundwater flow, ambient temperature. Thaw can be accelerated by circulation through the freeze tubes of steam, heated water, or brine at normal temperature. It has been noted in some cases where heave has been recorded that subsequent settlement has been greater by up to 20%.

Services such as water mains which might be damaged by freezing and heave can be protected by excavation to below their level.

5.4 DEWATERING

The management of water during excavation has already been discussed in Volume 1, but the problems arising at that stage can be minimised by preliminary dewatering of the ground, so as to improve the stability of the tunnel face and provide drier and safer working conditions.

In permeable strata where k exceeds about 10^{-3} cm/s, or where an aquifer can be dewatered below less permeable strata, the level of the water table over a wide area can be drawn down by pumping from boreholes and deep wells. These processes are widely used in open excavations and are suited also to cut-and-cover tunnels and shallow bored tunnels, but less so to deep bored tunnels.

The two principal methods are well points and deep filter wells. There are also tunnel drainage techniques of less general application such as pilot tunnels or headings excavated below and in advance of the main tunnel, and sub-drains and sumps maintained below the invert as the work advances.

In special circumstances in very fine grained soils vacuum consolidation and electro-osmosis might be considered.

5.4.1 Well points

Where an area of permeable ground is to be dewatered to a depth of less than about 6 m–limited by the effective vacuum lift of a pump–a ring of tubular well points of about 125 mm diameter can be sunk by wash boring round the perimeter of the area at intervals of a metre or more. The lowest part of each tube is perforated and fitted with a gauze strainer; at the surface the tubes are connected through a ring main to a suction pump. Some days of pumping may be necessary to draw down the water within and immediately outside the ring, followed by continued pumping to maintain the low level by intercepting inflowing water. Greater depths can be attained in conjunction with excavation in stages of about 5 m by installation of further inner rings of well points at the successive lower levels, while maintaining pumping in the original ring.

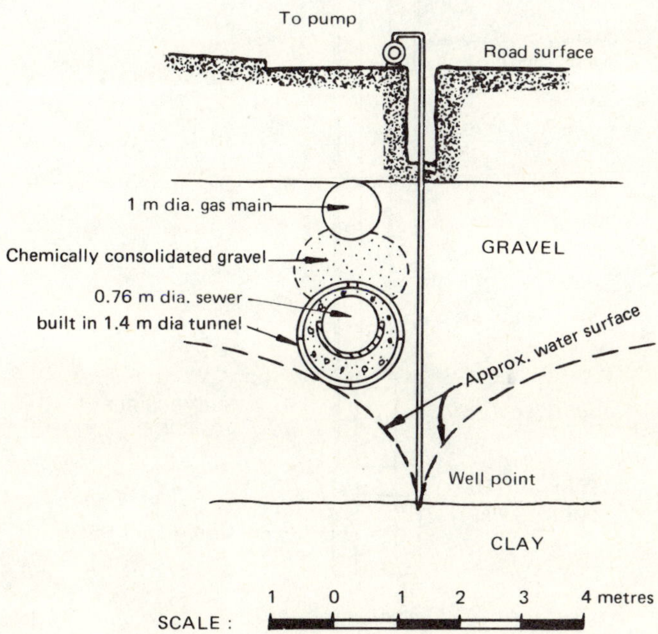

Fig. 5.8 – Example of well points used in combination with grout injection: Extension of London Central Line 1939. Diversion of 0.76 m dia. sewer was effected in a 1.4 m dia. C.I. lined tunnel, driven in free air through water bearing gravel, beneath an obliquely lying 1 m dia. gas main. Chemical grouting (to secure the ground and support the gas main) was injected progressively ahead of the tunnel working face.

5.4.2 Deep wells

As the name suggests, this method is for work at greater depths. A number of larger diameter wells, 300 mm or more, are sunk at spacings of 3 m or more in a ring to below the level required. Properly graded gravel filters are formed at the bottom of each bore around perforated suction pipes, and submersible pumps are installed at the appropriate level within the casings. In principle the operation is similar to that for well points, but the wells are larger, fewer and more widely spaced and are not limited in lift by considerations of vacuum suction. Because of the wider spacing the local draw down at the well must be further below the required working level.

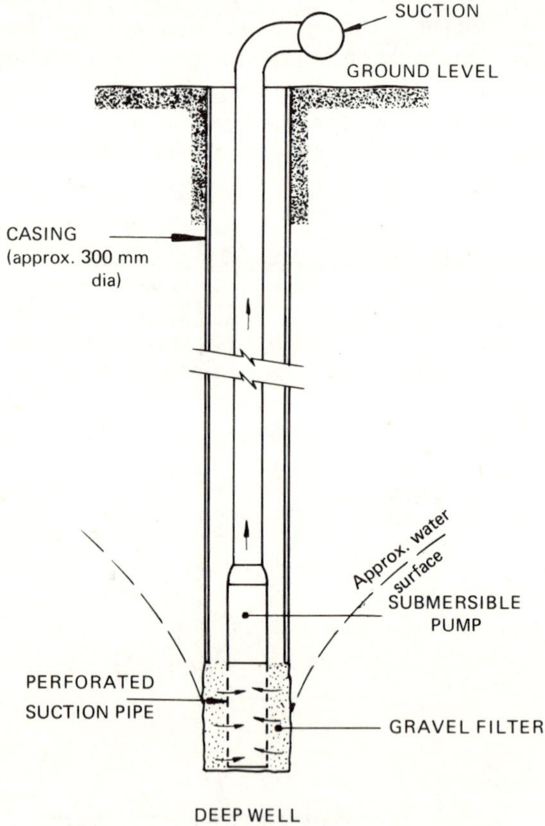

Fig. 5.9 – Deep well for ground water lowering below about 6 m.

5.4.3 Settlement

In both systems the filter arrangements are important to avoid settlement caused by drawing out fines from the soil. Even without any loss of fines there may be consolidation resulting in settlement of the ground affected. There is potential damage to foundations of buildings and structures over a wide area. Wells, springs and other sources of groundwater may be detrimentally affected, either temporarily or permanently. Near the sea, saline water may be drawn in. In such cases it may be possible to avoid the effects by recharging, through feed wells, important zones outside the area of excavation simultaneously with the operation of pumping down.

5.4.4 Pilot tunnels

Drainage pilot tunnels are an old and familiar expedient particularly relevant to subaqueous tunnelling both to improve conditions for excavation in the main tunnel and possibly to provide a permanent drainage system. The Seikan tunnel (Volume 1, Chapter 10) is an outstanding modern example of this method.

5.4.5 Vacuum consolidation

The method is applied in fine grained soils, such as silts, of low permeability. It is an extension of the well point system, in which a vacuum is created in the well points and the surrounding silt which is consolidated by the differential effect of atmospheric pressure at the ground surface.

5.4.6 Electro-osmosis

An electric current between electrodes driven into saturated fine grained soil will cause pore water to move from anode to cathode. If the cathode is constructed as a well point, seepage water can be pumped away progressively and consolidation of the soil will result.

BIBLIOGRAPHY

The subject is one continuously growing in its application to all forms of civil engineering construction. In soft ground applications it is almost a specialised branch of soil mechanics. The special aspects relevant to tunnelling are mostly discussed in papers descriptive of individual projects.

Mussche, H. E. and Waddington, J. C., Applications of the freezing process to civil engineering works, *Instn Civ. Engrs Wks Constr. Div*., 1946, (5).

Kell, J., Pre-treatment of gravel for compressed air tunnelling under the River Thames at Dartford, *Chart. Civ. Engr*., 1957 (Mar.).

Glossop, R., The invention and development of injection processes, *Geotechnique*, 1960, **10** (Sept.) and 1961, **11** (Dec.).

Ischy, E. and Glossop, R., An introduction to alluvial grouting, *Proc. Instn Civ. Engrs*, 1962, **21** (Mar.); *see also* discussion, 1962, **23** (Dec.).

Mansur, C. I. and Kaufman, R. I., Dewatering, Ch. 3 in Leonards, G. A. (Ed.), *Foundation Engineering*, McGraw-Hill, 1962.

Kell, J., The Dartford Tunnel, *Proc. Instn Civ. Engrs*, 1963, **24** (Mar.); *see also* discussion, 1963, **26** (Dec.).

Grouts and drilling muds in engineering practice, Butterworths, 1963.

Perrott, W. E., British practice for grouting granular soils, *Proc. Amer Soc. Civ. Engrs, J. Soil Mech Fndtn Div*., 1965, **91** (Nov.).

Kell, J. and Ridley, G., Blackwall tunnel duplication, *Proc. Instn Civ. Engrs*, 1966, **35** (Oct.); *see also* discussion, 1967, **37** (July).

Dempsey, J. A. and Moller, K., Grouting in ground engineering, *Proceedings Ground Engineering Conference*, London, 1970.

Cashman, P. M. and Haws, E. T., Control of groundwater by water lowering, *ibid.*

Sanger, F. J. and Golder, H. W., Ground freezing in construction, *Proc. Amer. Soc. Civ. Engrs, J. Soil Mech Fndtn Div*., 1968, **94** (Jan.).

Collins, S. P. and Deacon, W. G., Shaft sinking by ground freezing: Ely Ouse – Essex scheme, *Proc. Instn Civ. Engrs*, 1972, **Suppl**., p. 129; *see also* discussion, p. 319.

Powers, J. P., Groundwater control in tunnel construction, *Proc. 1st Rapid Excavation and Tunneling Conf*., Chicago, 1972, **1**, p. 331.

Shuster, J. A., Controlled freezing for temporary ground support, *ibid*, **2**, p. 863.

Bowen, R., *Grouting in engineering practice*, Applied Science, 1975. 2nd edn., 1981.

Bell, F. C. (Ed.), *Methods of treatment of unstable ground*, Butterworths, 1975.

Clough, G. W., *A Report on the practice of chemical stabilization around soft ground tunnels in England and Europe*, Dept. of Civ. Engng, Stanford University, 1975.

Herndon, J. and Lenahan, J., *Grouting in Soils*, 2 vols., U.S. Federal Highway Administration Reports FHWA-RD-76-26 and 27 (NTIS PB 259043 and 4), 1976.

Jones, J. S. and Brown, R. E., Temporary tunnel support by artificial ground freezing, *Proc. Amer. Soc. Civ. Engrs, J. Geotech. Engng Div*., 1978, **104** (Oct.).

International Symposium on Ground Freezing, Bochum, W. Germany, 1978.

Jones, J. S. and Brown, R. E., New advancements for ground freezing for tunnel construction, *Proc. 4th Rapid Excavation and Tunneling Conf*., Atlanta, 1979, **1** p. 722.

2nd International Symposium on Ground Freezing, Trondheim, Norway, 1980.

Powers, J. P., *Construction Dewatering, a guide to theory and practice*, J. Wiley, 1981.

6

Highway Tunnels

6.1 HIGHWAY REQUIREMENTS

The special construction features for highway tunnels include geometrical configuration, road construction, lighting, ventilation, traffic control, fire precautions, general facilities for cleaning and maintenance, and often provision for the pipe and cable services of public utilities. The requirements will vary according to the tunnel situation and character; whether urban, motorway, subaqueous or mountain, and whether long or short.

A highway tunnel is part only of a larger highway system. Its function in providing a crossing under an obstacle may be of major importance in linking up areas with poor communications, as under an estuary or through a mountain range. In some instances, tunnels may be relatively minor elements as in a road following a rocky gorge.

Road systems are designed on the basis of their traffic capacity, taking account of lane widths and clearances, gradients and traffic composition. The same formulae can be applied to tunnels, but it has been found that in tunnels where regular commuters constitute a substantial proportion of peak hour traffic the peak hourly flow may exceed the nominal capacity by 50% or more.

For a number of two-lane one-way tunnels actual flows of from 3000–4500 vehicles/hour have been observed as compared with nominal capacity from 2000–2700. In contrast, rural and mountain tunnels, in whose peak traffic the proportion of regular users is small, may have an actual capacity of 80–90% of the nominal capacity.

The principal factor determining the traffic capacity of a tunnel is the geometry of its alignment and cross section, but its lighting and ventilation may impose lower limits. Special factors influencing traffic flow include freedom from side turnings, absence of cyclists and pedestrians, and the regularity of use by a substantial proportion of drivers, or, alternatively, unfamiliarity. If capacity is found less than adequate the geometry cannot be readily altered, but deficient lighting can usually be improved to any necessary extent. Ventilation may be boosted, but major improvements are likely to be costly in construction and operation.

In all cases the geometry of the tunnel alignment is determined by the siting of the portals, and by the curves and gradients adopted to connect them. As part of a highway system the alignment must conform to the line of the highway as a whole, and must be compatible with the standards adopted. A tunnel, other than a very short one, is likely to be a substantial element in the functioning of the highway and in the construction cost. The highway standards must then be formulated in the context of the practical possibilities of tunnel construction. A short motorway tunnel through a ridge will, as nearly as practicable, be laid out to the ordinary motorway standards, but a subaqueous crossing under an estuary, or a long mountain tunnel will be sited primarily to suit tunnelling conditions and will largely determine the layout and standards of the approach roads connecting it to other parts of the highway system.

6.1.1 Urban underpasses

In urban conditions, for relatively short underpass tunnels, the points of connection into the road system are likely to be very narrowly limited, and further limitations on the tunnel line will be imposed by existing buildings and by buried services and structures. For underpasses following existing alignments, wide roadways are needed in order to provide space for separate descending approaches to the portals. The tunnels will usually be shallow and may often be constructed by cut-and-cover methods, but occasionally, as when at greater depth under a hill, by boring. The straightest practicable line will be adopted and gradients will be restricted if possible to less than 3%, because steeper gradients give rise to congestion when large heavily loaded vehicles are ascending.

The concept of a network of major traffic routes deep beneath a city has been advanced but presents exceptionally formidable problems of cost of construction, ventilation, intermediate access, fire and flood, and will not be further discussed here.

6.1.2 Motorways

Motorway tunnels are relatively new. In open country the obstacle requiring tunnelling may be a ridge, or hill spur, or occasionally the crossing of another road or rail route where viaduct construction is impracticable, or even the crossing of park land where the highway has to be sunk for reasons of amenity. In all these the tunnel line and gradients must conform with the standards specified for the motorway: sight lines appropriate to the design speed may require particular care, especially where vertical curves are necessary. Adequate lighting may be a particular difficulty. One possible relaxation for tunnel construction may be the omission of an emergency lane as provided on the 'hard shoulder', but in long tunnels lay-bys for use in emergencies might be a valuable contribution to safety and convenience. Limitations on speed may have to be considered.

Fig. 6.1 – Layby at mid tunnel. Mersey Kingsway tunnel was provided with an extra lane width over a length of 30 m, by construction of a special elliptical tunnel section.

6.1.3 Subaqueous tunnels

For subaqueous tunnels the line will usually be fixed nearly at right angles to the waterway to minimise the length of tunnel, unless the topography of the valley imposes another alignment. Roads tend to follow the line of the valley and right angled junctions are likely to be required between the existing roads and the tunnel approaches.

Steeply rising river banks where the climbing gradient out of the tunnel does not adequately exceed the topographic gradient make the approaches difficult. As an example, the Tyne tunnel (1967) on the north bank of the river gains the necessary height by making a turn of 200°, at the exceptionally small radius of 128 m, in rising to the portal at a gradient of 5%, and the emergent approach road completes the loop by crossing above the tunnel. As with urban tunnels any gradient exceeding 3–4% slows heavy traffic disproportionately. The Clyde tunnel, which is an urban subaqueous tunnel, has had to adopt gradients of 6% in order to connect into existing highways, close to and parallel to the river and therefore at right angles to the tunnel.

The profile of a subaqueous tunnel will usually comprise a descending gradient, a nearly level central length and a rising gradient. Vertical curves, required at the changes of gradient, should be as long as practicable, both to simplify construction if a shield is used and to avoid restricting unduly the sight line for vehicles in the tunnel approaching the change of gradient.

The depth of a subaqueous tunnel is determined by the bed level of the waterway and the geology of the river bed, together with the cross section of the tunnel. A circular cross section is usual for a bored tunnel, and this requires excavation at the crown substantially above the vehicle clearance level. In addition a safe thickness of undisturbed ground above is required during construction, depending on the character of the rock and soil and the construction methods adopted. In the case of navigable waterways, future dredging plans need to be considered. The first Blackwall tunnel (1897) was successfully driven with a specially designed shield and compressed air, with a mere 1.6 m of gravel and silt, but supplemented by 3 m of clay blanket, above the hood at mid-river (see Volume 1, p. 36). For the second Blackwall tunnel (1967) cover was about 6 m and the ground was extensively grouted. Again a shield and compressed air were used.

6.1.4 Mountain tunnels

Major mountain crossings also impose their own rules. A pass, or col, approached by a steep and narrow valley can only be crossed by a surface roadway by use of very steep gradients, sharp bends, and zig-zags. Such passes are also liable to be closed or made dangerous by snow and ice and possibly by rock falls and avalanches. The shortest practicable tunnel under the col may ameliorate the worst conditions, but a longer tunnel at a lower level makes possible a much better highway.

The cost of any such tunnel is so high that a substantial traffic expectation is needed to justify the expenditure, but if sufficient demand exists it is likely to be stimulated by improved facilities, so that a major project including improved approach roads and a long tunnel at a lower level may well be found advantageous. The whole layout of the highway system for such a crossing involves much study of topography and geology, and balanced judgement in the choice of alignment and design of viaducts and minor tunnels in the approach roads as well as in the major tunnel.

Although there is more freedom of alignment for a road than for a railway, because steeper gradients and sharper curves are acceptable, the problems remain difficult and complex. To the limitations of longitudinal gradient and curvature must be added in greater degree the transverse slope of the ground, which is of increased importance where wide carriageways are required. A narrow road can be accommodated on sidelong ground with a limited need for cutting and embankment, whereas a major highway with dual two- or three-lane carriageways cannot. With any specified gradient it is geometrically possible even on the steepest slopes to set out a centre line ascending by sharp zig-zags, but width of construction and minimum radius for curves make such routes quite impracticable if the slope is excessive. Tunnels through projecting spurs and under the summit col become necessary.

The comprehensive design of highway crossings for mountain routes is too large a subject to be dealt with in adequate detail here. Their geometry must be related to the topography and the geology in order to design, and ensure the stability of cuttings, embankments, viaducts and portals leading to tunnels.

Fig. 6.2 – Canopied entrance to Mont Blanc Tunnel – West portal, France.

6.2 GEOMETRY OF CENTRE LINE

The geometry of any tunnel has two aspects, that of the centre line and that of the cross section. They are not entirely independent but can best be considered separately.

The principal factors determining the centre line include:

1. The relative positions of the portals and directions of approach
2. Geology
3. Clearances from external obstacles
4. Gradients
5. Vertical curves
6. Horizontal curves

6.2.1 Approaches

The obvious initial approximation of aligning the tunnel in a straight line joining the portals will almost inevitably be modified in any but a very short and simple tunnel by introduction of curves to suit the approaches, and varying gradients to carry it under and around obstacles.

6.2.2 Geology

The choice of the most suitable strata for tunnelling may influence the alignment and level greatly, as may the avoidance of water-bearing ground or unstable rock. The subject is dealt with in Chapter 4.

6.2.3 Clearance from external obstacles

Reference has already been made to the clearance required below the river bed in a subaqueous tunnel. Adequate clearances from existing structures are also necessary in urban conditions. As a broad generalisation it is usually satisfactory if uniform undisturbed ground outside the tunnel extends for one tunnel diameter; if discontinuities or obstructions occur within this zone, more careful analysis and special precautions are likely to be required.

6.2.4 Gradients

Although road vehicles are normally capable, even when fully loaded, of climbing steep gradients, this is not an acceptable basis of design for highway tunnels. Heavy vehicles on steep gradients are reduced to the use of their lowest gears, and can only proceed at a road speed of perhaps 10 km/hr, which is extremely restrictive of other traffic and greatly reduces the traffic capacity of the tunnel and its efficient use, at the same time making abnormal demands on the ventilation system. While limitation of gradients to 3% is desirable, up to 5% of even 6% is acceptable, particularly if two traffic lanes in the direction of climb are available. In short urban underpasses steep gradients may be necessary where space is very restricted. If their function is relief of rush hour congestion and their traffic is mainly light vehicles, little delay may in fact be experienced. In long mountain tunnels it will usually be possible so to select portal positions that the gradients are within the desired limits. On the other hand, in subaqueous tunnels the length and cost of the scheme will be minimised if the maximum practicable gradient is accepted. One expedient adopted on the Mersey Kingsway and Dartford II tunnels has been the combination of a two-lane tunnel with a three-lane open approach on the ascent, to provide a 'crawler' lane for slow and heavy vehicles, the gradients being 4% both at Mersey and at Dartford.

A minimum gradient should also be specified, to ensure longitudinal drainage of the roadway. For most conditions 0.25% is probably acceptable.

6.2.5 Vertical curves

Changes of gradient are normally small in motorway and mountain tunnels and connecting curves are correspondingly short and can follow the ordinary highway specification.

In subaqueous tunnels and in urban underpasses, however, gradients of 5% or thereabouts are normal and the junction between this slope and the central level length should be eased by a curve, whose radius will be determined partly by the length available and partly by the method of construction. In a shield driven tunnel with precast segmental lining the steering of the shield must be gradual, and the segmental rings must either be designed as tapered rings to suit a specified radius or adapted to the curvature by insertion of tapered packings in the joints. In a tunnel of 9 m diameter using 1 m rings, a radius of 1000 m requires a taper of 9 mm per ring. If specially cast taper rings are employed they will probably have a taper of about 50 mm and therefore only one special ring in five or six need be used.

Such a curve is much sharper than is usual in highway practice, and would not be acceptable in a motorway tunnel, but may be appropriate in a subaqueous tunnel in which speed is limited, and may be essential in an underpass where length is very limited. It may be noted that, on the open highway, minimum radii for convex vertical curves are determined by sight lines appropriate to design speed, and for concave curves by vertical acceleration. In a tunnel the concave curve will also restrict the sight line because a low ceiling line will limit visibility ahead.

6.2.6 Horizontal curves

In plan, curves may be necessary to align the tunnel with its approach roads and to avoid obstacles in the ground. The same considerations apply in determining the radius as in surface roads: design speed, centrifugal force, superelevation, line of sight.

On very sharp curves, some extra lane width for long vehicles is desirable, but may be impracticable or prohibitively expensive. A special consideration, particularly where the traffic emerges on a rising gradient–as is normally the case in a subaqueous tunnel–may be the avoidance of an alignment directly towards the sun when it is low in the sky. A curve in the open cut approach, or a screen wall, may be found helpful in minimising dazzle.

6.3 CROSS SECTION

The cross section is determined by the space required for traffic, and other facilities, and by construction methods.

6.3.1 Traffic space

The traffic space required is defined by the width of lane and maximum load height of vehicle. The minimum normal tunnel will accommodate two lanes of traffic. Three lane tunnels are not uncommon where a rectangular section is used, in cut-and-cover construction or in immersed tubes. The Mersey Queensway tunnel (1934) was unusual in providing for four lanes of traffic in a circular bore, but the lane width is only 2.74 m and the side lanes have a height limitation at the kerb of 4.42 m. The more common pattern for bored tunnels is to construct them in two-lane widths, as in the second Mersey Kingsway tunnel with two tubes, or the Elbtunnel with three tubes.

For vehicular traffic the equivalent of the 'structure gauge' on the railways is a rectangle, defined by lane widths and height, plus such clearances as are appropriate for road vehicles, which are not under the same precise control as rail vehicles, in construction, loading, steering and suspension.

There has been a progressive increase in traffic lane widths and heights, and consequently in tunnel diameters since the end of last century which can be seen in the following table of dimensions of British subaqueous tunnels of circular section. In such tunnels the dimensions are more closely restricted to the minimum acceptable than in mountain tunnels or cut-and-cover construction because any increase adds disproportionately to cost and difficulty. In some cases, roadway dimensions have been determined by commitments entered into many years before actual construction. Thus the Dartford tunnel was planned and shields were built and installed in 1938 to the standards then current, but work on the main tunnel did not proceed until 1957 and 6.40 m was the widest roadway that could be fitted into the 8.59 m circle, even with slightly substandard headroom of 4.88 m. The same shields and cast-iron lining were then used at the Blackwall II tunnel, the width being reduced to 6.10 m, but with full headroom of 5.03 m.

BRITISH SUBAQUEOUS HIGHWAY TUNNELS

Tunnel	Construction	Internal diameter	Length between portals	Lane widths	Height at Kerb	Height at Max.	Max. gradient %	
Blackwall I	1892–97	7.52	1361	2x2.44			2.9	
	1967–68			2x2.97	3.23 4.35	4.41		Improvement
Rotherhithe	1904–08	8.43	1480	2x2.44	4.88	5.64	2.7	
Mersey Queensway	1925–34	13.41	3237	4x2.74	4.42	7.16	5	
Dartford I	1937–38							Pilot only
	1957–63	8.59	1429	2x3.20	4.88	5.33	3.6	
Clyde	1961–64	8.99	747	2x2x3.35	4.57	5.03	6.3	Twin tunnels
Tyne	1961–67	9.53	1684	2x3.65	4.88	5.26	5	
Blackwall II	1960–67	8.59	1174	2x3.05	5.03	-	4	Duplication Blackwall I
Mersey Kingsway	1966–71	9.63	2256	2x2x3.65	5.03	5.49	4	Twin tunnels
Dartford II	1973–80	9.54	1435	2x3.65	5.03	-	4	Duplication Dartford I

The current British standard clearance for height is 5.03 m, which is greater than the ordinary minimum of 4.5 m on the European continent, or 4.3 m in North America.

With the circular form the rectangular traffic area occupies no more than about 60% of the area of circle for a two-lane roadway, but footways for inspection and maintenance are usually accommodated at the sides, while the space beneath the roadway is very valuable as a ventilation duct and for various services, and space above any ceiling can also be used for ventilation and cables. Although surprisingly economical for two lanes, the circular form does not lend itself so well to three or more lanes, the traffic area falling below 50% of the circular space.

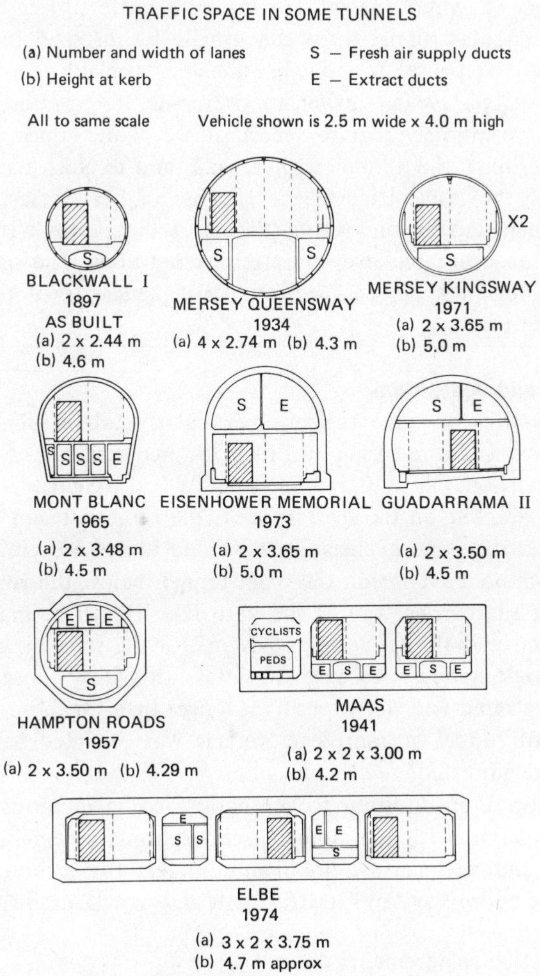

Fig. 6.3 – Traffic space in some tunnels.

6.3.2 Other space

Walkways are often considered essential for inspection and maintenance, and for emergency use both for access to the site of an accident and for escape. They have to be protected from impact by vehicles, and can with advantage be elevated above road level and also protected by a kerb and a safety margin. Additional space may also be necessary for ventilation ducts. In a circular tunnel the spaces beneath the roadway and above the clearance line are available without extra excavation, and in a horseshoe tunnel there is normally a substantial area in the crown, but in a rectangular structure extra width or depth must be excavated. This subject is more fully discussed in Chapter 9.

Other provisions often required, particularly in river crossings, are the accommodation of water mains, gas mains, electric power cables, or other services. In a circular tunnel these can usually be installed in the under-road space but at the expense of reducing the area available for ventilation and increasing the necessary fan power to overcome the friction and turbulence generated. Gas mains may require special study of the contingency of leakage or of a burst filling the tunnel with a toxic and explosive mixture. Electric power can only be transmitted at the expense of heat losses, and these may have a significant effect on raising the air temperature within the tunnel, possibly to an unacceptable level in extreme conditions. The risks of accidental damage to any of these services and the consequences to tunnel users need to be assessed carefully.

6.3.3 Cyclists and pedestrians

In the construction of some tunnels, particularly urban subaqueous tunnels, there is a demand for crossing facilities for pedestrians and cyclists. These can be disproportionately costly if incorporated in a vehicular traffic system. One solution, adopted on the river Tyne, is the provision of a separate smaller tunnel for pedestrians and cyclists. In the Clyde tunnel a cycle track and walkway were accommodated within the main tunnel, below the road deck, but the headroom over the carriageway at the kerb line had to be made substandard, namely, 4.57 m, and the fresh air provisions for the subsidiary passage had to be to a very high standard because the intake of CO by those taking physical exercise is more rapid and the exposure is longer than for those driving through. In the Dartford tunnel a transporter vehicle was provided for cycles and no pedestrians were admitted.

Other facilities, in addition to ventilation, to be incorporated within the tunnel are the services for the tunnel itself: lighting, emergency services such as telephones and fire alarms, fire mains, meters for carbon monoxide and visibility, public address systems, traffic lights and signals, drainage and pumping.

6.3.4 Construction requirements

The shape of a highway tunnel, whether rectangular, horseshoe or circular in form, is largely dictated by the method of construction adopted to suit the ground conditions. For cut-and-cover the rectangular shape is usual, for rock to be excavated by blasting, the horseshoe or other arched form, for excavation by full face machine the circle, and likewise for most soft ground subaqueous tunnels (other than submerged tunnels). In long mountain tunnels a rising gradient is preferred to simplify drainage during construction; in shield driven tunnels sharp curves, horizontal or vertical, present considerable difficulties in steering the shield and building the lining.

6.4 ROAD CONSTRUCTION

In a tunnel through sound rock, road formation is unlikely to present any unusual problems. Drainage of ground water seepage is, of course, important and the gradients and cross falls of the road surface must be accurately followed. Specified headroom may require particular care at the kerb where the tunnel roof arches down.

In a circular tunnel the roadway is necessarily suspended above the invert and the structural slab may be of reinforced concrete, precast or *in situ*, or of composite prestressed concrete beam and *in situ* slab construction. It must be very accurately positioned in relation to the tunnel lining.

The programming of road construction in a tunnel is not easy, unless it can wait for completion of the tunnel excavation and lining, because the road works obstruct access for any other supplies to and from the working face.

Temperature movement is not normally a serious problem in a long tunnel, because fluctuations are smoothed out, even round the year, in that the tunnel lining and surrounding ground act as a heat reservoir.

Because of the very limited clearances, usually minimal above the kerb lines, the kerb should be sufficiently high to ensure that it cannot easily be mounted by the wheel of a vehicle, and should also be robust. In setting out the kerb line it may be necessary to prepare and follow a 'wriggle diagram', as described in Volume 1, to accommodate unavoidable deviations in the tunnel alignment.

For the surfacing of the roadway, accurately laid concrete in the road slab without bituminous topping may be specified, but it must be remembered that ordinary highway paving trains cannot usually be operated within the confines of a tunnel. An asphaltic or bituminous wearing coat is most common in practice. It should be to a high standard. It may be noted that it will not be exposed to extreme temperatures and to sunlight. As in the case of concrete work ordinary paving equipment cannot be used, and also ventilation during laying must be ensured. Eventual resurfacing needs consideration; the old wearing coat will almost certainly have to be stripped off because headroom must be maintained and a shut down of the tunnel is likely to become necessary.

Fig. 6.4(a) & (b) – Road construction in first Dartford Tunnel. Precast units assembled in tunnel invert.

6.5 SECONDARY LININGS

It is usual to provide a secondary lining to the side walls of a highway tunnel to contribute to visibility and for aesthetic reasons, and sometimes to assist in waterproofing. The roof may also be similarly treated, but in many cases a false ceiling is constructed to form ventilation ducts and cable runs above, or to shed water or for acoustic absorption. In all but very short tunnels, artificial lighting is considered necessary and its efficiency is greatly enhanced by suitably light coloured reflective surfaces. In most tunnels some seepage of water through the primary structural lining is inevitable and even very slow penetration of groundwater can discolour large areas, both by direct deposit of dissolved salts and by assisting the adhesion of particles of soot and oil from vehicle exhausts. An air gap, or a waterproof membrane between primary and secondary linings can be helpful in preventing this seepage staining.

The principal criteria in the selection and design of such linings for tunnel walls are:

1. Costs of installation, maintenance and replacement.
2. High light reflection, but diffuse not specular.
3. Non-absorbent surface resistant to water, oil and dirt; washable with detergents and not readily scratched.
4. Resistance to aging and corrosion in tunnel atmosphere.
5. Rigid dimensionally and free from vibration.
6. Fire resistant and not contributory of noxious fumes in fires.
7. Replaceable when damaged and removable where required for access to services behind.
8. Minimum thickness, shaped to profile.

Many different forms of internal finish have been used, although in some tunnels bare rock, or unpainted or painted concrete or cement rendering are found adequate. Finishes include white glazed tiles, coloured tiles, vitreous mosaic tiling, vitreous enamelled steel sheeting, plastic coated aluminium sheeting, PVC sheeting. Higher aesthetic standards are to be expected in a major city tunnel than in an industrial or dock environment or in a mountain area.

The merits of ceramic or vitreous tiles are apparent, but they present problems of adhesion and cracking. They are often fixed direct to a primary concrete lining without an air gap, which makes it difficult to control leakage and staining at joints. Repair of damage should not be unduly difficult.

Coated steel sheeting can be prefabricated in large panels with some form of framing to provide rigidity and to make fixing and replacement easy. Access to cables and other services can readily be given. Vitreous enamelling provides a hard washable non-absorbent surface but it can be damaged by impact. Plastic coating is of course less brittle but more easily scratched and possibly more absorbent.

Fig. 6.5 – Examples of secondary linings.

Fig. 6.5(b) – Vitreous enamelled steel sheeting to walls in Tyne Tunnel.

Fig. 6.5(c) – Vitreous enamelled steel sheeting forming cable ducts in walls of Mersey Kingsway Tunnel. No other secondary lining is employed but epoxy resin coating and paint are applied directly to inner steel facing of composite steel and concrete segments.

Plastic panels are lighter, usually cheaper and more easily adapted to fit where there are irregularities. They are also more readily scratched, possibly more absorbent, and are sometimes rejected as an unacceptable fire hazard, particularly in respect of fumes generated.

Roof or ceiling linings present similar problems. Reflection of light is not considered to be of the same importance as for walls. Indeed, black ceilings have been installed, one argument being that there is created for the driver a greater impression of overhead space. Sound reverberation in a tunnel is sometimes considered to be so objectionable as to justify the installation of acoustically absorbent linings. A false ceiling can be used for this purpose, but absorbent materials, such as slag wool, are essentially open textured and not washable. Therefore such a ceiling will inevitably become darkened by smoke which will be least objectionable on a dark surface. Much of the sound reflected directly from the walls is not attenuated by acoustic treatment of the ceiling.

In the absence of a false ceiling the crown of the tunnel may be treated in the same way as the walls, but a less costly finish is likely to be adopted, possibly cement rendering and painting of the primary lining surface. The problem of cleaning has to be considered.

It is usually necessary also to accommodate in the ceiling or at its junction with the walls, the lighting units and their feed cables, and in some cases jet booster fans are installed above the vehicle clearance level.

6.6 LIGHTING

The inadequacy of natural lighting for any but a very short tunnel is obvious, and the design of lighting on a safe and economic basis is complex. It must be suited to the performance of the human eye, subjected to varying conditions of light, and must ensure that the roadway with guide lines and any obstacles on it can be seen and appreciated at a sufficient distance ahead, appropriate to the vehicle speed and braking distance.

The subject of highway tunnel lighting is dealt with in some detail in the periodic publications of the Permanent International Association of Road Congresses (P.I.A.R.C.), the latest of which relate to the XVI Congress held in Vienna in 1979.

The units employed below in describing lighting are for the *luminance* or brightness of any illuminated surface and *illuminance* or flux of light directed onto the surface.

For the purpose of tunnel lighting they are connected by the simplified formula

$$L = \frac{E \times P}{\pi}$$

where: L is the luminance measured in candelas/m^2
E is the illuminance measured in lux
P is the coefficient of diffuse reflection of the surface, which lies in the range 0.1 to 0.3 for various carriageway surfaces.

6.6.1 Optical adaptation

Levels of lighting, adequate at night on the open road, whether from overhead lighting or from vehicle headlamps, are totally inadequate for a driver entering a tunnel from daylight. The comparative luminances (measured in candelas per square metre) range from 8000 cd/m^2 or more in an open sunlit approach to about 2 cd/m^2 or less for good highway lighting at night. The inadequacy arises from the fact that the eye at any time is temporarily adapted to a certain average brightness and can discriminate only over a limited range above and below this average. It suffers from glare from much brighter light and cannot discern any detail below about 5% of its adaptation level. The eye does, however, change its adaptation level fairly rapidly in response to a change in lighting. At moderate levels of brightness it will accommodate to a decrease in the luminance of the field of vision by about half in every 3 seconds. In terms of vehicles travelling at a speed of 72 km/hr this corresponds to halving the luminance every 60 m. This progressive reduction cannot be continued indefinitely even in a long tunnel, the rate of adaptation at low levels being much slower. At the exit from a tunnel adaptation to brighter light is much more rapid.

At any point in a vehicle's entry to a tunnel two levels of lighting are significant: first, the general luminance of the driver's field of vision to which his eye has adapted, and second, the luminance of the roadway at a safe distance ahead so that the vehicle may be stopped if an obstacle is seen. The discussion that follows relates primarily to long tunnels, but the same principles apply to short tunnels, except those so short (less than about 40 m) that the exit is clearly visible before entry.

6.6.2 Entrances

The origin for calculation of lighting levels at entry to a tunnel is a point in the approach road at the safe stopping distance from the entrance. This may be of the order of 100 m or more depending on speeds, gradients and road surface. From this point the eye should be able to see any obstacle within the tunnel entrance, which should not simply appear to be a black hole. At this origin maximum luminance in sunlight may be between 2000 cd/m^2 where there is a hilly wooded background and 8000 cd/m^2 in very flat and open country. In selecting the appropriate figure for lighting design, the finished approach structure must be envisaged. The luminance may be kept low by such expedients as a dark roadway with dark retaining walls or other structures, or by screening with trees. From the value of the luminance at this point the necessary luminance

Fig. 6.6 – (a) Portal of first Dartford Tunnel, showing lighting transition effected by canopy over to reduce intensity of sunlight and intensified artificial lighting within the portal. (b) Canopy at Liverpool portal of Mersey Kingsway Tunnel, spanning five traffic lanes.

in the tunnel entrance is usually taken to be between 1/15 and 1/30 but sometimes as high as 1/10. A very high level of artificial lighting of 1500 to 2000 lux is required to provide luminance of perhaps 100-200 cd/m^2, even with the assistance of a light coloured roadway and walls. In some tunnels canopies have been constructed extending from the portal and admitting reduced daylight to the roadway while excluding direct sunlight. This certainly has merits in economising on cost of lighting, but at a substantial construction cost. It is found difficult to keep the screens clean and efficient, and in some climates snow and formation of ice create hazards.

6.6.3 Tunnel interior

Within the tunnel, which for this purpose includes any canopy or similar extension, four lighting zones are recognised, namely:

1. Threshold
2. Transition
3. Interior
4. Exit

Within the *threshold zone*, the luminance may be kept constant as calculated for the entrance or may decrease slightly in parallel with the natural decrease in the external approach zone, maintaining a constant ratio, such as 1/15 at the chosen safe distance ahead. The threshold zone need not extend the whole safe distance from the actual entrance, but from a point a short way outside at which the entrance and its level of luminance dominates the field of view.

The *transition zone* immediately succeeds the threshold zone and within this the luminance is progressively reduced to the value adopted for the interior. The rate of reduction is precisely defined by a curve of adaptation, which may very approximately be represented by halving the luminance every 3 seconds.

For the *interior zone* the recommended minimum luminances range from 10 cd/m^2 in urban tunnels carrying heavy traffic to 3 cd/m^2 in very long tunnels in open country or where speed is restricted or traffic is low. The corresponding illuminance ranges from 175 to 50 lux for average reflectance of the road surface.

The *exit zone* in a one-way tunnel does not usually require extra lighting. In special cases, such as very long tunnels, or emergence to the very high luminance of snow or sea, a short transition zone of increasing luminance may be helpful.

It will be appreciated that the lighting in threshold and transition zones has to be varied throughout the day in proportion to the natural luminance at the origin, more particularly at twilight, but also where daylight fluctuates greatly between sun and storm. The control may be manual, or by time switching, but regulation by photometer sited outside the tunnel near its entrance is likely to be most satisfactory for major tunnels.

At night the tunnel lighting should be closely comparable with that in the approaches. If too bright, the exit becomes a black hole. If roads are generally unlit the luminance within the tunnel should be between 1 and 2 cd/m^2.

Short tunnels. The threshold and transition zones designed as described will occupy a length of 300 m or more from the entrance so that in a shorter one-way tunnel the exit portal will be reached before lighting levels have been reduced to the recommended minimum interior levels, and in two-way tunnels the critical length is doubled. It will be appreciated that the cost of such lighting is proportionately higher than for long tunnels.

It is difficult to generalise about the appropriate installations for such short tunnels down to the length where there is through visibility and in which artificial lighting, except at night, can be dispensed with. Each tunnel must be considered separately in relation to approaches, length, curves, vehicle speed, traffic density. Illuminated kerb lane markers may be helpful, or reflectors where use of headlights can be ensured.

6.6.4 Lighting installation

The most suitable systems are usually found to be fluorescent tubes, low pressure sodium or high pressure sodium. Uniformity of lighting, concentrated principally on road and walls, is desirable, and avoidance of flicker which can result from spacing out of lamps, both from the direct effect of the lamps themselves and their reflection in other vehicles. Uninterrupted longitudinal rows are usually the most suitable for heavy urban traffic with multiple rows where enhanced lighting is required. When lower intensities are appropriate alternate tubes may be switched off. Spacing exceeding 12 m is conducive to flicker.

The lighting circuits have to be designed to allow lighting to be adapted to varied daylight levels. It is also important that lighting should be so planned that the tunnel is not plunged into darkness if one circuit fails, and also with minimum emergency lighting against the contingency of major supply failure.

6.7 TRAFFIC MANAGEMENT

Management and control of traffic is found necessary in many tunnels if they are to be used to the best advantage with safety. In the absence of any control the tunnelled highway should conform fully to the dimensions and standards, including clearances, of the open highway and should impose no separate restrictions. Even at best, there are differences of environment likely to cause many drivers to slow down as they approach, unless they are very familiar with the route. Although clearances laterally and overhead may be the same as under bridges, the convergence in perspective makes them appear more restrictive. Lighting must be of a very high standard if it is not to cause hesitation by some drivers, and uncertainty for all.

Such ideal standards are rarely practicable in any but very short tunnels, whether from difficulties of construction or excessive cost. These physical limitations do in themselves exercise some restraint on traffic, although, as noted earlier, nominal road capacity may sometimes be substantially exceeded.

Various forms of restriction and control are applied in most major tunnels. Apart from imposition of tolls or stopping of vehicles at a national frontier, the primary reason for exercising control is safety. Other reasons are to ensure maximum flow, and to integrate the tunnel traffic flow with city traffic.

The ordinary safety regulations applied elsewhere are adopted in tunnels as found appropriate. Limits on height and width and maximum speed limits, which may be special to the tunnel, or general to the highway, are usual. Lane changing is frequently prohibited, often with restriction of heavy vehicles to a single lane. In two-way tunnels overtaking is usually prohibited. These are all controls of common application on highways and may not necessitate separate enforcement, but the consequences of any accident or breakdown is potentially so dangerous, and likely to be so disruptive to traffic, that special patrols are often employed, and also TV coverage to identify location and cause of congestion as early as possible.

Important loads exceeding regulation width or height are sometimes routed through tunnels when alternative routes are not practicable, but the closing of part or all of the tunnel to other traffic during the passage of the load is necessary. Extra height may be available in the middle of the roadway.

The safety limitations on traffic special to tunnels are in respect of ventilation and carriage of dangerous goods. Ventilation, discussed in detail in Chapter 9, is usually adequate, even in congested traffic, in busy tunnels, but there are important exceptions, where the traffic capacity of the tunnel is limited by the ventilating air available. Such is the case in the Mont Blanc tunnel, which, even after an increase, only provides for traffic at a rate of 800 veh/hr at a spacing of 100 m, although the road capacity is much greater. Other major Alpine tunnels are similarly limited.

Supervision of traffic and ventilation must be adequate to ensure safety if congestion does develop. Automatic control of fan output in relation to measurement of pollution levels is usually too slow.

6.7.1 Dangerous goods

Materials inherently dangerous by reason of their flammable, explosive or toxic nature are widely transported by road. Although accepted as a necessary hazard on the open road, they may be quite unacceptable in a tunnel or admitted only with special precautions. Their passage on the highway is not in itself hazardous, but the contents of the load may be released by collision damage or overturning, or by a fire in the vehicle. The risk that a collision or other accident will occur in a tunnel is no greater than outside, but the potential consequences in the confined space of a tunnel are in many cases quite disproportionate. They will vary, depending on the characteristics of the tunnel and the traffic present in it. Regulations to be made and control to be applied should therefore be carefully considered to meet the particular circumstances and needs of each highway tunnel, and to contribute to its primary purpose of carrying highway traffic.

One of the important classes of dangerous goods is liquid fuel, whose transport in road tankers constitutes a substantial proportion of heavy vehicle traffic. The major danger from this and other flammable liquids is that of fire from spillage following collision. An added hazard, more particularly in subaqueous tunnels with their concave profile, is of the liquid flowing into the drainage system and sump, and possibly evaporating to form an explosive mixture. Such tankers are usually in charge of careful and skillful drivers, and in some tunnels are allowed through only at offpeak hours and when escorted by tunnel staff in a patrol vehicle.

The danger from explosives is obvious, but could be catastrophic in a subaqueous tunnel, or in one where structural damage might be followed by collapse. The use of alternative routes is to be preferred, but escorted passage through the tunnel with temporary exclusion of other traffic is possible. Toxic materials, more particularly in gaseous or liquid form, and materials becoming toxic in contact with water, also may be excluded, or may have to be passed under control.

In any such control system one of the greatest difficulties is to identify and intercept vehicles with dangerous loads. Where there is a toll barrier all vehicles can be inspected, but in many cases toll collection is concentrated at one end of the tunnel, half the vehicles having already passed through.

6.7.2 Means of traffic control

In a controlled tunnel the organisation will include any or all of the following:

1. Toll barriers
2. Traffic lights and illuminated signs
3. Staff patrol cars and breakdown vehicles
4. Closed circuit TV
5. Public address system
6. Instruments for monitoring carbon monoxide, visibility, numbers of vehicles, speed

6.7.2.1 Toll barriers

The collection of tolls in itself exercises a discipline on tunnel users but is something of an obstruction to free flow of traffic. When at full capacity, each traffic lane needs between two and three toll booths to ensure its full utilisation. A simple toll structure requiring payment with a single coin is ideal, and can operate with very little delay where toll collectors are equipped with suitable recording instruments and where drivers are familiar with the procedure. Tokens

Fig. 6.7 – (a) Wallasey approach to Mersey Kingsway Tunnel showing: (1) Access junctions. (2) Toll plaza with space for 12 toll booths. (3) Open cut approach to portals. (4) Ventilation towers on both sides of river.

Fig. 6.7(b) – Control room above toll booths.

or tickets can be helpful, but season tickets can be troublesome as they may be misused by drivers, or exploited by toll collectors, because they break the direct check between vehicles passed and tolls collected.

The passing of the toll booth is usually regulated by a traffic light which is operated by the toll collector's action in recording the receipt of the toll and the classification of the vehicle. It is usual for these records to be registered simultaneously at a central office. Automated toll collection has been tried, but has not yet been found entirely satisfactory. As compared with efficient manual collection from a regular user, no time is saved, and the driver requiring change must in any case use a manned entrance. A driver unfamiliar with the tunnel is likely to be delayed by the need to read the instructions about the use of the instrument and to find the correct amount of cash.

The space for 2½ lanes per tunnel plus the width of a toll booth plus safety clearances makes necessary a very wide area for the toll plaza. It may also be necessary to leave space for emergency access by fire engines or breakdown vehicles at a time when all traffic is halted and all toll lanes are congested. If all tolls are collected at one end of the tunnel space may be saved by making a number of the lanes and toll booths reversible, unless the peak flows occur in both directions simultaneously, which is most unlikely.

6.7.2.2 Traffic lights and illuminated signs

Within most tunnels traffic lights are undesirable. Their presence causes uncertainty to drivers, and they are unlikely to be of much use in emergency. The place for stop signals is in advance of the portal to exclude traffic from the tunnel when necessary.

Normal road signs cannot usually be accommodated within a tunnel and any instructions to drivers can best be given by special illuminated signs, operated when required. One which may be considered necessary for conditions of acute congestion, to minimise generation of fumes, is 'Stop. Switch off engine'.

Signs indicating fire points and emergency telephones may also be installed.

6.7.2.3 Patrols and breakdown vehicles

Car patrols are almost essential to supervise traffic and deal with emergencies. The duties may be carried out by traffic police but in many cases special tunnel staff are employed with police status and with a fleet of vehicles belonging to the tunnel. One of the duties of such patrols may be to escort dangerous loads through the tunnel.

Breakdown vehicles must be readily available, capable of clearing obstructions quickly. It is advisable to have them provided at both ends of the tunnel, as the scene of an accident may not be accessible from one end because the roadway is blocked with stopped vehicles.

It is necessary to ensure that the recovery vehicles can operate properly within the confines of the tunnel. For speed and efficiency it is advantageous to have heavy recovery vehicles, light recovery vehicles, and for the patrol cars to have some capacity for towing out broken down cars.

Fig. 6.7(c) – Heavy breakdown vehicles for use in tunnel. (*Mann Egerton & Co. Ltd.*)

6.7.2.4 Closed circuit TV

TV cameras in the tunnel displaying as required in the control room can be of great help in identifying and locating any incipient congestion or other trouble. They should have the capability of showing any part of the tunnel and also the traffic approaching the portals. A number of screens are likely to be kept on permanent display with the capacity to switch on and direct any one of the cameras. If there is a radio link with patrol cars–for which special devices are necessary within the tunnel–fire fighting or other emergencies can be directed from the control room.

6.7.2.5 A public address system

Such a system is only for use in emergencies. It could not normally be heard by drivers in closed cars, and against traffic noise, but could be used to give instructions and prevent panic in a fire.

6.7.2.6 Instrumentation

Levels of carbon monoxide must be monitored continuously to give guidance on ventilation. Instruments should be installed at strategic points in the tunnel air flow. Visibility, likewise, should be measured at a number of points. For both of these continuous recorders in the control room are required and some form of alarm to indicate critical levels.

The importance of knowing traffic flow, spacing and speed has been discussed. The use of a number of electromagnetic inductive loops under the carriageway coupled to a central computer allows these factors to be assessed and used. Older methods of counting by interception of a beam of light or by contact pads have not proved very reliable, if the aggregate difference between vehicles in and vehicles out is to be used to count the number of vehicles in the tunnel at any moment. Confusion arises from multiple axle vehicles and trailers and from vehicles not centrally in the lane.

6.8 FIRE PRECAUTIONS

Although serious fires in road tunnels are of rare occurrence the potential consequences are usually thought to make appropriate special precautions, more particularly in long tunnels carrying a heavy volume of traffic. Some of these requirements have already been noted above as an aspect of traffic management and the regulation of ventilating air is referred to in Chapter 9.

Two major examples of fires may be adduced:

The Holland tunnel in New York suffered from a major vehicle fire in 1949. A load of carbon disulphide in drums, which had entered contrary to regulations, caught fire and burned fiercely with emission of flame and thick smoke, which made close access impossible. No lives were lost but 66 were injured and 180 m of the lining was destroyed. The tunnel was closed for 2½ days.

In the Nihonzaka tunnel (opened in 1969) in Japan on the Tokyo–Nagoya highway there was a very serious fire in July 1979 with the loss of 6 lives. There are twin two-lane tunnels about 2000 m long. A heavy flow of traffic entered the tunnel when released after a long hold-up due to earlier collision. About 400 m into the tunnel there was a collision between two cars and four lorries, one of which was carrying liquid ether, which ignited. A sprinkler system operated but water supplies proved inadequate and the fire raged for 2 days, causing spalling of concrete from the crown with exposure of reinforcement, and burning out 160 cars. Regulations on transit of dangerous goods were not applicable to this tunnel.

In major tunnels it is usual to provide fire points at intervals of 50 or 100 m along the tunnel, at which are installed

1. press-button fire alarms and emergency telephones connected to the control centre, and if possible to the fire brigade;
2. hose reels carrying 50 m or so of 37 mm hose;
3. portable chemical extinguishers, and

4. outlets complying with fire brigade standards connected to a high pressure fire main of about 100 mm diameter or more.
   These fire points are usually on the walkway and should be readily accessible to any driver in trouble. They should be clearly marked with illuminated signs.

The dangers to life consequent on a fire are caused by heat, smoke, oxygen deficiency, carbon monoxide. The importance of a fresh air supply and an extraction system in combating these is obvious, except that continued combustion depends on air supply, which should therefore be carefully regulated if practicable.

Heat is a direct danger to life including the radiant heat from burning gases flowing under the ceiling. It also attacks and impairs the strength of structural materials which are not themselves combustible. In a major fire reinforced concrete may suffer spalling of concrete followed by loss of strength in the reinforcing steel. Smoke destroys visibility and may also contain toxic constituents; its extraction at roof level assists escape of persons at low level. Oxygen deficiency and carbon monoxide may develop in a major fire but more particularly in the absence of any fresh air supply.

Aspects of fire risks to be considered in design include, in addition to ventilation:

1. escape routes, possibly by cross passages to an adjacent tunnel, of particular importance in very long tunnels;
2. provision of emergency lighting adequately protected against fire;
3. choice of lining materials, particularly in ceilings and ceiling ducts, which will neither contribute to combustion, nor give off toxic fumes, nor collapse;
4. appropriate arrangements to drain and trap burning fuel spilling on the roadway;
5. protection of cables and wiring wherever potentially vulnerable.

6.9 PIPES AND CABLES

A tunnel itself requires a range of services supplied by pipes and cables. In addition, as already mentioned, public authorities may make arrangements for their mains to be routed through the tunnel, more especially in a subaqueous tunnel.

6.9.1 Water

A fire main in a tunnel is a normal provision. Water may also be required for washing down and these services may be separate, or combined. Trunk mains of larger diameter may also be installed to link the public systems at the two ends. Those water pipes can usually be accommodated without difficulty or danger, if safeguards are provided against the risk of flooding from a burst.

Frost protection may be necessary if the water pipe is exposed to fresh air being drawn in for ventilation.

6.9.2 Gas

There is no normal requirement for a gas supply to the tunnel. High pressure trunk mains are not usually acceptable, in particular because of the risk of explosion following a leak or burst, and the difficulty of providing adequate safeguards.

6.9.3 Power cables

Tunnel supplies for lighting must be carried through the tunnel; those for ventilation and pumping may also be so routed together with linking power cables to allow standby supplies to be available from either end. These will usually be in ducts giving some protection against mechanical damage and against fire. Spare ducts may be provided for subsequent developments or for replacements.

Public supply cables forming part of the grid may also be sited in the tunnel. If carrying heavy currents they may require to be cooled either by

Fig. 6.8 – Tunnel service cables behind secondary lining in second Mersey tunnel. (*B.I.C.C.*)

use of the ventilation air supply or by water cooling or other means. Automatic protection can be given against explosive cable rupture. Mechanical protection against damage by collision may be ensured by appropriate siting.

6.9.4 Communication cables

These are likely to be required for transmission of instrument readings and control relays and for telephones and alarms, also for any TV and public address systems. They will usually be carried in ducts or conduits, with accessible junction and distribution boxes at intervals. Watertightness is important and may be difficult to achieve.

BIBLIOGRAPHY

There is some difficulty in presenting a bibliography for this chapter in that most construction aspects of highway tunnels are not peculiar to their function, and have thus been dealt with and included in other chapters, including in particular Chapter 1 – Cut-and-cover, and Chapter 2 – Submerged Tunnels, to which reference should be made. The requirement for ventilation is so complex and specialised as to be treated separately in Chapter 9.

It has, however, been thought of interest to list here the whole sequence of U.K. subaqueous highway tunnels, and a few others, from 1897 to the present day, as they have been recorded in useful detail. For worldwide practice generally, the P.I.A.R.C. reports apply a necessary corrective to this insular perspective. References to tunnel lighting are also given.

Hay, D. and Fitzmaurice, M., The Blackwall Tunnel, *Min. Proc. Instn Civ. Engrs*, 1897, **130**.

Tabor, E. H., The Rotherhithe tunnel, *ibid*, 1908, **175**.

Anderson, D., The construction of the Mersey Tunnel, *J. Instn Civ. Engrs*, 1936, **2** (Apr.).

Van Bruggen, J. P., The road tunnel under the River Maas at Rotterdam, *Engng*, 1940, **150** (Aug. 9 and 30, Sept. 27).

Kell, J., The Dartford tunnel, *Proc. Instn Civ. Engrs*, 1963, **24** (Mar.); *see also* discussion, 1963, **26** (Dec.).

Schreuder, D. A., *The lighting of vehicular tunnels*, Philips, 1964.

Granter, E., Park Lane improvement scheme: design and construction.

Rayfield, F. A. and Clayton, A. J. H., ditto: planning and traffic engineering, *Proc. Instn Civ. Engrs*, 1964, **29** (Oct.); *see also* discussion, 1966, **33** (Mar.) and **35** (Dec.).

Morgan, H. D. *et al.*, Clyde tunnel: design, construction and tunnel services.

Haxton, A. F. and Whyte, H. E., ditto: constructional problems, *ibid*, 1965, **30** (Feb.); *see also* discussion, 1967, **37** (July).

Kell. J. and Ridley, G., Blackwall tunnel duplication, *ibid*, 1966, **35** (Oct.); *see also* discussion, 1967, **37** (July).

Road Tunnels Committee:
Report and *Documentation and Studies*, 13th World Congress, Tokyo, 1967;
Report and *Documentation Digest*, 14th World Congress, Prague, 1971;
Report and *Documentation Digest*, 15th World Congress, Mexico, 1975;
Report, 16th World Congress, Vienna, 1979;
Permanent International Association of Road Congresses, Paris.

Prosser, J. R. and Grant, P. A. St. C., The Tyne tunnel: the planning of the scheme.

Falkiner, R. H. and Tough, S. G., ditto: the construction of the main tunnel, *Proc. Instn Civ. Engrs*, 1968, **39** (Feb.); *see also* discussion, 1968, **41** (Nov.).

Lyons, A. C. and Schofield, J., The Great Charles Street tunnel, *Tunnels and Tunnelling*, 1969, **1** (May).

Margason, G. and Pocock, R. G., *A preliminary study of the cost of tunnel construction, Laboratory Report 326*, Transport and Road Research Laboratory, Crowthorne, 1970.

Parry, R. R. and Thornton, G. D., Construction of the Monmouth tunnels in soft rock, *Proc. Instn Civ. Engrs*, 1970, **47** (Sept.); *see also* discussion, 1971, **49** (July).

Lyons, D. J., Trends in research on motorway design and use in *Motorways in Britain today and tomorrow*, London, 1971.

Brown, C. D., Homes and highways, *Consult. Engr*, 1971, **35** (Nov.).

Muir Wood, A. M., Tunnels for roads and motorways, *Q.J. Engng Geol.*, 1972, **5** (1 and 2).

Megaw, T. M. *et al.*, The Mersey Kingsway tunnel (3 papers), *Proc. Instn Civ. Engrs*, 1972, **51** (Mar.); *see also* discussion, 1973, **54** (May).

O'Reilly, M. P. and Munton, A. P., Prospects of urban highways in tunnel, in *Transportation Engineering 1972*, London, 1973.

Sir William Halcrow & Partners and Mott, Hay & Anderson, *Report on Roads in Tunnel for the Greater London Council*, G.L.C., 1973.

International Recommendations for Tunnel Lighting, Pub. no. 26, Commission Internationale de L'Eclairage, Paris, 1973.

Muir Wood, A. M. and Sharman, F. A., Tunnelling as a social benefit, in *Transportation and Environment: policies, plans and practice*, Southampton, 1973.

Symposium on Tunnel Lighting, London, 1974, *Lighting Research and Technology*, 1975, **7** (2).

Shutter, G. B. and Bell, G. A., Design and construction of the second Dartford tunnel, in *Tunnelling '79*, London, 1979.

7

Metro Tunnels

7.1 CHARACTERISTICS

The term 'metro' is being increasingly used to distinguish railway systems specially adapted to cities and their immediate environs. The terms underground, U-bahn, subway, rapid transit, S-bahn, used in many cities are less distinctive. It is of interest that the earliest such system was London's Metropolitan Railway, followed by the Paris Metro.

The most important characteristic of a metro is the ability to transport rapidly large numbers of people on an exclusive right of way, free from any interruption by other types of traffic. Frequent, predictable and reliable services are of great value to the passengers, of whom a majority travel regularly between home and work, with many also travelling in and out of the centre for education, shopping, entertainment, and also within the inner city area.

Metros are to be distinguished functionally from main line and suburban railways by their use for shorter and more frequent journeys although no hard and fast line of demarcation is possible. A possible definition of a 'metro' is: an urban passenger railway with stations at frequent intervals to provide convenient and rapid transport over relatively short distances within a city and its environs.

Metro systems are almost invariably required to meet the needs of existing cities, whose centres are so closely built up already that surface railways or elevated railways are quite impracticable and tunnelling becomes essential. Tunnelling, either by cut-and-cover under existing streets or by boring under streets and buildings, minimises interference with existing traffic capacity, demolition of buildings and visual intrusion.

Tunnelling costs are necessarily high and therefore it is usual to tunnel only in congested areas and under high ground, and to run on the surface or as an elevated railway wherever adequate space is more readily available, possibly over or alongside an existing highway.

Correct siting of metro stations at close spacing in the inner city is an obvious necessity if maximum traffic is to be attracted, although in suburban areas stations must be more widely spaced to avoid the delays and costs inherent in frequent stops. A balance has to be established between these delays and costs and the extra traffic attracted. In central tunnelled sections the device of constructing double-ended stations can increase substantially the catchment area of a single station. Entrances, booking offices, and access stairs are provided at both ends of the station platforms and may therefore be more than 200 m apart which is often a significant proportion of the distance between stations.

Before planning a route in any detail it is essential to have an assessment of the number of passengers to be expected and their origins and destinations, and more particularly the number during a 15-minute peak. From this information alternative proposals for possible routes can be prepared and a single route selected for detailed study of its traffic potential and engineering aspects.

The enclosed nature of metro systems with their own right of way makes possible automation of many aspects, thereby reducing the labour intensive character of public transport generally. Train operation, signalling and ticket issue and control can all be regulated automatically and largely centralised, but the fact that the system handles human beings limits the extent to which machinery can function satisfactorily in relation to passenger convenience and safety and in meeting emergencies.

7.2 COMPULSORY POWERS

For the construction of a metro system in a city compulsory powers are usually necessary to acquire land and rights of way, and for operation and maintenance. Provision must also be made for appropriate compensation, the assessment of which should not be allowed to delay planning and construction.

In the United Kingdom a Private Act of Parliament is normally obtained for the purpose by the promoters, describing the project in sufficient, but not excessive, detail and prescribing the centre line within defined limits of deviation. Other forms of authorisation will be more appropriate elsewhere but the nature and limitations of the powers to be exercised by the promoters must be essentially similar.

The right to compensation of property owners adversely affected by construction or operation requires to be defined while the promoters must be given rights to enter property for surveys, for trial borings and soil investigation, and also for temporary and permanent underpinning and strengthening. A time limit is usually defined within which the rights must be exercised, but with possibility of extension. Claims for injurious affection by construction or operation are also normally limited in time.

On the other side of the balance it is usual that the provision of good rapid transit enhances property values where it serves a residential area, and also improves business and shopping areas. There may also be substantially reduced congestion in streets. Such benefits are dispersed throughout the community and are not received by the promoters to offset against the necessary items of compensation; it is usual and proper therefore that the central authority should pay a substantial part of the costs incurred. Greatly enhanced land values followed the construction of the Toronto system, but similar effects were not experienced in the early years of the BART system in California.

The negotiations with those affected by a new line are likely to be far more difficult for cut-and-cover construction than for a deep tunnel. Objections sometimes are given disproportionate publicity; the promoters have to justify their proposals in much greater detail.

The special problems to be examined include:

1. the establishment of a legal right of way;
2. siting of stations to suit prospective passengers;
3. layout of running tunnels linking stations within acceptable limits of gradient and curvature;
4. avoidance of underground obstructions such as deep foundations and existing tunnels and shafts;
5. adaptation to geological conditions;
6. maintaining appropriate distance from establishments particularly sensitive to vibrations such as laboratories, large hospitals or concert halls.

Each of these aspects influences the others and the structure finally adopted must be a compromise on the basis of estimates of feasibility and cost of possible alternatives.

Access necessary for construction must be determined as a compromise between conflicting ideals. Working sites are essential but are hard to find in densely built-up areas, and are often subject to restrictions in respect of noise, loss of amenity, access by heavy vehicles, or prospective building construction. Permanent acquisition of sites for stations, or for ventilation or electrical supply may provide many of the necessary points of access. For cut-and-cover construction the obstacles are much more severe: streets must be occupied for a long period, excavated and ultimately restored or, if streets are not followed, buildings must be cut through or demolished with little respect for property boundaries.

When the layout of the line and its stations has been settled, ventilation and associated problems of draught relief need to be considered. The extent and complexity of the system necessary depend on the climate of the city and the standards adopted, but siting of shafts and provision for fans should be an integral part of metro design and necessary land should be acquired before construction of the line begins.

Survey to a high standard of accuracy on and under the surface and including location of buried services, is an essential preliminary.

Planning must not overlook requirements for inspection and maintenance when the system becomes operational. The limited clearances in tunnels and the short intervals between trains, restrict these operations to those hours when train services can be suspended and current cut off from a live rail or other conductor. Except where a twenty-four hour service is maintained, a few hours are usually available each night, but such operations are necessarily very costly and should be kept to a minimum by careful design and choice of the most suitable systems. It is even more difficult and costly if all-night services are run.

7.3 GENERAL PATTERN OF PLANNING

The planning of a metro system is extremely complex and involves many different and conflicting interests. It is one element only of a city's long term planning but one which will affect the future growth and living pattern of that city. Such broad plans have necessarily and properly political aspects, but the specific selection of a route and its detailed design is properly a matter of engineering judgement. The intention here is to outline such aspects as are directly relevant to construction and use of tunnels.

Decisions in principle should be embodied as early as practicable into a full set of design criteria, progressively developed into precise specifications governing such matters as methods of construction, types of rolling stock and other plant and equipment, operating speeds, track geometry, tolerances, air temperatures, rates of air flow and very many others. One very broad decision to be taken early is the level of mechanical complexity and automation to be adopted for operation. This will be influenced by the ultimate availability, after training, of suitable operating and maintenance staff.

Capital cost is inevitably high, more especially in congested areas, and expenditure on construction must be justified by providing a transport system serving the needs and wants of a large number of passengers. In modern cities very many people have their homes as far as practicable from the centre, and travel to and from the centre for business and pleasure. This establishes a demand for transport on radial routes and success in meeting this demand conveniently and economically encourages more people to adopt this pattern of living and travel. Therefore both an analysis of existing wants and predictions of growth require careful study in planning a route. The principal routes are often apparent along corridors determined by the topography. Sometimes a line may deliberately be so laid out as to encourage development of a particular area, while catering also for existing demand from other areas.

A single radial route is not satisfactory as it involves reversing trains at a central terminal but a diametrical route crossing the centre, or a combination of two radial routes with through running is very much better. The two radial lines may even be nearly parallel but connected by a U-loop. The first lines of the Toronto Subway embodied this pattern; no lines can run out to the south because of Toronto's situation on the shore line of Lake Ontario.

A system comprising separate and relatively short lines has the advantage that any delays are confined to one line and thus affect a minimum number of passengers.

Radial routes are thus the normal pattern for city transport, but circular routes have also been adopted. The earliest metros in London were linked to form the 'Inner Circle' in 1884; its special function was to connect the main line termini which, for reasons of congestion, were widely separated and were sited round the perimeter of the city's core. Glasgow also in 1896 adopted a circular route, which has not been added to subsequently. Another example of a circle route is in Moscow where it is in addition to the numerous radial lines.

Circular systems are far less common than radial lines and are only justified in special circumstances. Where both coexist convenient interchange for passengers is essential but interworking is usually found to be operationally undesirable, because delays on one radial line can disrupt the circle service and in turn all the other radial lines. In addition, with interworking the train capacity of the circle line limits the use of the radial lines unless each has its own separate track round the circle.

An outer circle line is rare and is more difficult to justify, particularly so if tunnelling has to be considered. It would serve as a bypass to the city, or for communication between various suburbs or satellite towns but it is most unlikely to attract traffic of sufficient intensity to justify the capital and operating costs.

In addition to the running route a rolling stock depot is required for stabling of trains and day-to-day maintenance. Provision must also be made for major overhauls, whether at the same site or elsewhere. A large area of land is necessary which will obviously be most economically acquired as a surface site remote from the city centre, towards the outer end of the line.

The systems considered above are self-contained units, but there may also be direct links through a city centre between existing main or suburban lines having many of the characteristics of metros. The new Melbourne Underground Loop is such a system, comprising an inner circle carrying services from suburban lines and providing the advantages of frequent stations in the centre serving all radial lines, and the operating benefits of through running of trains without the need for reversals and stabling. The disadvantages noted above of interference between radial and circle services are avoided by the provision of four loop tracks each in a separate tunnel large enough to carry standard rolling stock, and also possible double-decker suburban stock.

On a smaller scale the new Liverpool 'Loop' line extends the cross-Mersey line as a one-way loop providing passenger interchange at the main line stations and eliminating reversal of trains. It is complemented by the 'Link' line connecting two local lines across the city and providing interchange with the 'Loop'.

The choices between cut-and-cover and deep bored tunnels are discussed in more detail below, as is the question of size of rolling stock to be accommodated.

Fig. 7.1 – King's Cross Underground Station in 1868.

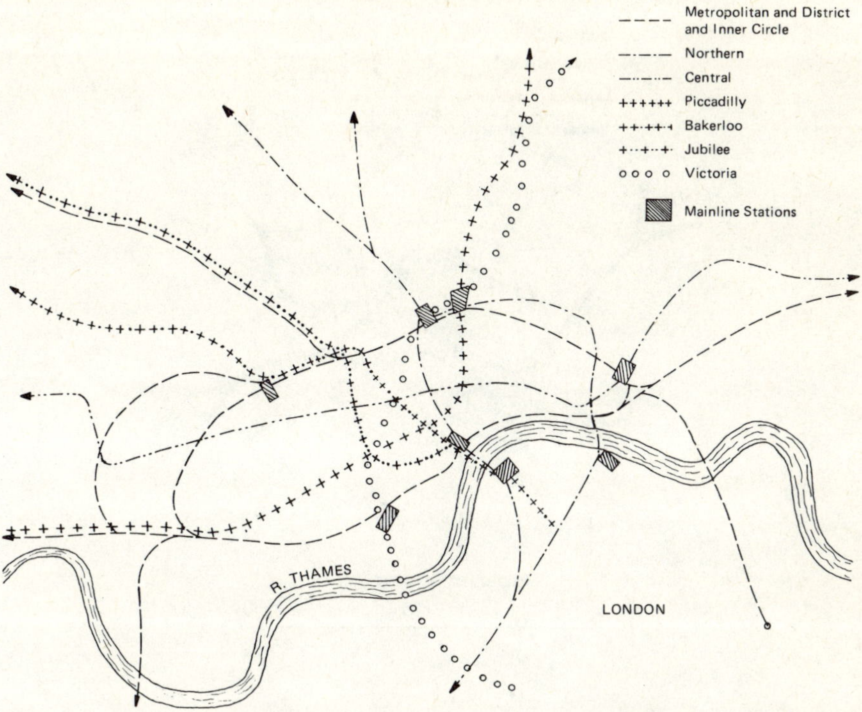

Fig. 7.2 – Line diagrams of metros: (a) London,

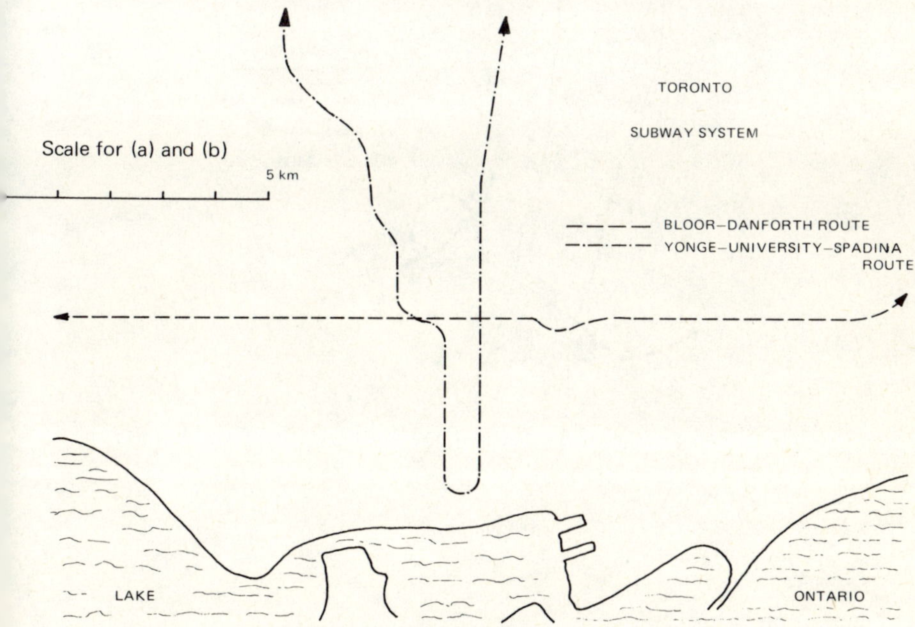

(b) Toronto,

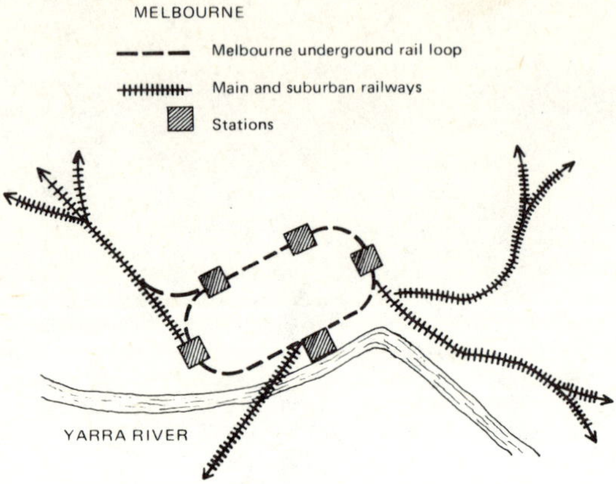

(c) Melbourne,

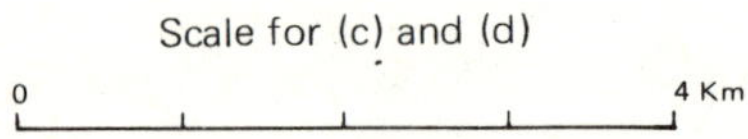

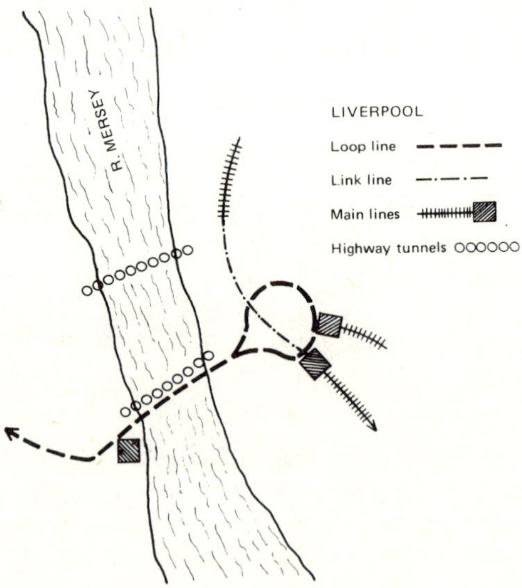

(d) Liverpool.

Fig. 7.3 – Opening of London Central Line, July 1900. (*London Transport*)

7.3.1 Control centre

A central control establishment is now becoming of major importance for the efficient operation of a modern metro. The primary function is the supervision and regulation of train services, for which purpose full information must be continuously furnished and displayed.

Other mechanical functions which can usually be centralised with advantage are control of the various electric current supplies, and the remote operation of plant and equipment such as escalators, control gates, ventilation, pumping, emergency flood gates, etc.

Closed circuit TV and other links can provide essential information from stations, both by viewing passenger movements and by transmitting particulars relevant to operation of equipment. Such a control centre is also the proper focus for relevant information from outside sources–weather forecasts, flood warnings, crowd movements related to sporting or other events, road accidents and diversions.

7.3.2 Establishment of routes

Having decided on the areas to be served by a new line its route must be laid out in detail with consideration of the necessary rights of way and land acquisition, as well as the engineering possibilities.

The balancing of costs against benefits is of major importance in justifying the expenditure to be incurred although neither costs nor benefits can be quantified with any precision. Costs of land acquisition and engineering construction, equipment and maintenance can be estimated reasonably well in financial terms, as can passenger use and income from fares, but other large factors such as environmental intrusion, disruption during construction, relief of road congestion, enhancement of land values, etc., cannot be evaluated with any precision, and are not likely to be of any direct benefit to the immediate promoter unless he is also a property owner, or a transport operator, public or private. In choosing the exact alignment, both in plan and in level, more detailed examination of the comparative costs and benefits of alternatives is necessary.

The two main elements of direct cost are land acquisition and construction. Acquisition of a right of way comes first. There may exist a surface right of way which can be used, following a major road or within the boundaries of a railway. Even in the absence of such a right of way over part or the whole of the line, it may be cheaper to acquire the necessary new land for a surface line than to tunnel. This is unlikely to apply in the central area of a city where tunnelling on a substantial scale is to be expected. Here too a right of way under the ground is required, and a line beneath existing streets has many advantages, provided sharp curves can be avoided. In many cities such street routes are natural traffic arteries and thus meet readily the needs of passengers. In the past, however, streets have sometimes been too closely followed and the railway has been constructed with undesirably severe curves, and even reverse curves. Acquisition of necessary land or rights to pass under private property at such points is likely to be fully justified. There are also routes within a city where the best line departs completely from the surface street pattern.

Siting of stations is a most important aspect of detail planning. Experience in London and elsewhere has shown that a station sited where inconvenient to the public will significantly fail to attract passengers to the metro. A station should be sited as close as possible to a major traffic focus, but even if the ideal site cannot be achieved, it may be worth while to provide pedestrian subway access to the station, incorporating escalators or horizontal conveyors where they can appreciably improve the time and effort of getting to the trains. It is very important that entrances to stations should be sited in the best places, at street level, and attractively designed and clearly signed; back street entrances of dull appearance and not readily identifiable will do much to discourage passengers.

The integration of metro stations in a city as part of a larger complex of public buildings, shops and surface transport is a wider subject discussed in more detail later in this chapter.

Stations providing interchange with other metro lines or with main line railways require particular consideration in planning. Changes in the course of a journey are a deterrent to use of the system, which must be minimised. The time and convenience of passengers needs to be carefully studied, keeping distances and differences in level as small as possible. To be able to change trains by simply crossing the width of a platform is obviously excellent, and considerable ingenuity has sometimes been exercised in arranging tracks to provide this layout. Where lines cross at right angles a substantial difference of level is inevitable and escalators may be provided. It is worth noting that where both lines operate on a through ticket the interchange passages should be entirely within the 'paid' area to avoid unnecessary checking.

At stations on the outer sections interchange with buses and private cars may be an important problem involving car parking and waiting facilities, but these are unlikely to involve tunnelling, except for pedestrian access subways.

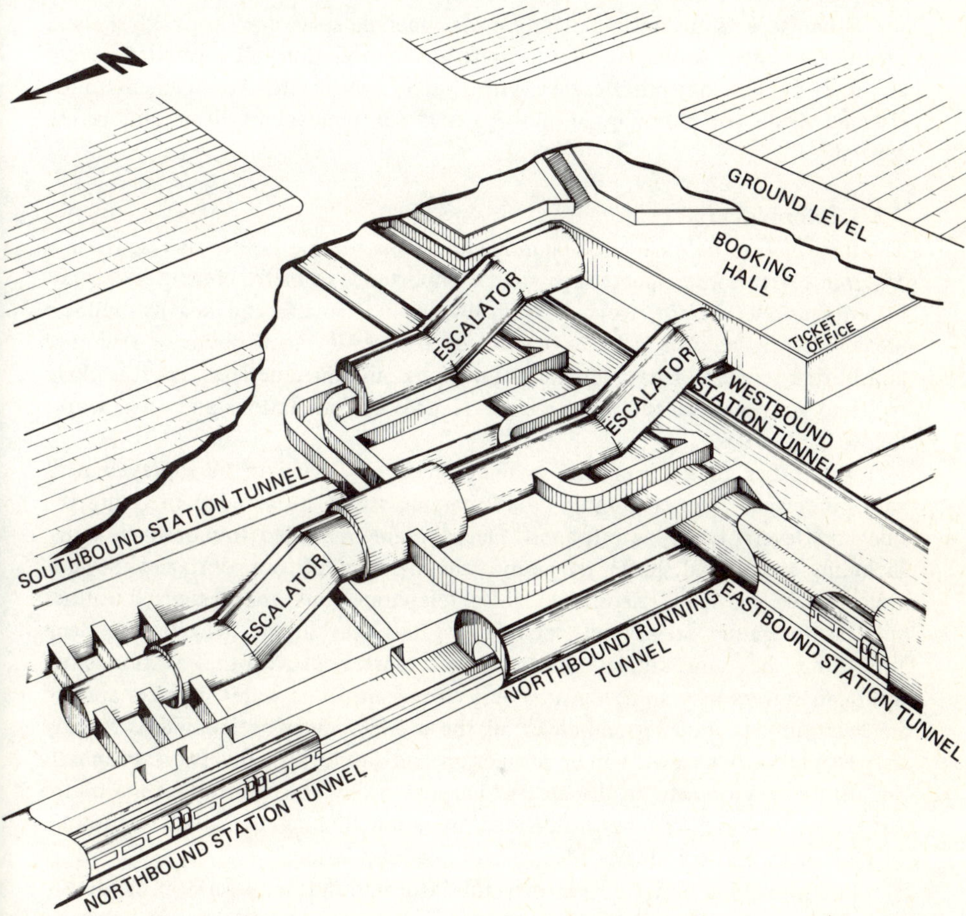

Fig. 7.4 – Station structures where two lines cross. (*After London Transport.*)

The design of a station will be governed by the forecasts of peak passenger flows upon which will be decided the area of ticket hall, circulating areas, numbers of escalators or lifts, widths of stairs and passages and the access provided to platforms. Booking hall size and shape will be adapted to the ticket vending and checking system proposed. Platform lengths must always accommodate the longest train required in peak periods, and this length must be provided even at the less-used stations and near the outer limits of a line.

Train movements in the tunnels cause air flows in a station which may be found objectionably high, particularly in passages, escalators or ticket halls, unless the station is designed to accommodate maximum flows by appropriate layout of all passages and adequate cross sections, assisted where appropriate by draught relief shafts.

Other aspects of station planning may include space for shops, kiosks and facilities for the public, as well as staff accommodation, all located to avoid as far as possible any interference with flow of passengers. Arrangements must also be made for handling of moneys and for removal of litter and refuse generally.

7.3.3 Information

The information necessary for planning and constructing tunnels is discussed in Volume 1, but some aspects of particular importance to metro construction may be emphasised. As the work is typically in city centres and heavily built-up areas, existing records may be expected to provide much detail of structure and buried services. The older the city and its buildings the less easy it is likely to be to get access to full and accurate records of foundations and other works below the surface.

Any other tunnels, in use or abandoned, may need to be surveyed with some precision if unnecessarily wide margins are to be avoided in fixing the line and level of the new tunnels. Deep basements, piled foundations, wells, including abandoned shafts, rivers and streams, may all form obstacles. Sewers, water mains, gas mains, electricity cables, telephones and other less usual utilities such as hydraulic mains and pneumatic tubes may all be present. For deep tunnelling they are unlikely to be serious obstacles except at station sites, although sewers may sometimes be very deep, and some water supplies are run in deep tunnels. For cut-and-cover all the services must be located as exactly as possible with a view to their protection and support, or diversion. Proposals for future development in the area which may affect the stability of the metro tunnel may sometimes be anticipated by strengthening the tunnel structure locally, or by constructing in advance suitable foundations.

For tunnelling under a river very full information is necessary on the depth, as existing and as it may be affected by any dredging, or by scour in floods or attributable to river training works. Available records of tidal levels and flood levels should be studied and continuing records preserved unless adequately maintained elsewhere. Piled structures for wharves and jetties and for bridge piers need to be investigated and, of course, any other bridge foundations and river walls.

7.3.3.1 Geotechnical investigations

General requirements for tunnelling are described in Volume 1, comprising initially study of existing information in maps, borehole records and records of previous construction, followed by preparation of a scheme for investigation of possible routes by boreholes, trial pits and shafts, and by geotechnical methods.

While existing records are likely to be extensive in a heavily built-up area, much of the information may be irrelevant to tunnelling, because it usually is related principally to the upper layers of the soil above tunnelling depth, and tunnel excavation problems are not considered.

The problem of obtaining sufficient suitable sites for boreholes is often difficult. Selection of positions may unfortunately be governed more by availability than by tunnelling requirements. In this respect tunnelling under streets has advantages, more especially cut-and-cover construction, because extensive surface occupation is implicit. Opportunities of boring on temporarily vacant sites sometimes occur briefly, and it is helpful if those planning the metro are able to take advantage of them quickly even if the full scheme has not been developed.

Parks and open spaces may offer favourable sites for borings, but severe restrictions on such use are often imposed.

The apparent merits of geotechnical investigations–electrical, magnetic, sonic and other techniques–are at a special disadvantage in urban conditions because of the presence of pipes, cables and other buried artefacts.

7.3.4 Type of construction and programme

While the selection of a route is primarily dependent on traffic needs, to meet which appropriate engineering techniques must be employed, the precise siting and detail layout of stations will differ according to whether bored tunnels, cut-and-cover, or surface or elevated tracks are adopted. The latter are rarely suitable near city centres and are not further considered in any detail.

The first possibility likely to be examined is that of cut-and-cover construction. A major operating advantage of relatively shallow stations lies in easier access from the surface, which is of great importance in attracting passengers to a system providing rapid transit, especially over short distances. The time taken in descending from the street to the station platform, and the waiting time determined by frequency of service may be critical in deciding whether the metro or other surface transport is used for a particular journey. Likewise, on journeys involving interchange, the time and effort necessary to change trains should be kept to a minimum.

There are two substantial disadvantages in shallow construction by cut-and-cover. First is the disruption inevitable during construction. Streets must be opened up for long periods with extensive disturbance of street traffic. Vehicle access to adjoining properties can only be maintained with difficulty and they are subject to inevitable noise and dust and other loss of amenity, often resulting in loss of business. Where streets are not followed extensive demolition of property is necessary or very costly underpinning.

Second is the necessity to pass under buried services—pipes, cables, sewers—and possibly under rivers and existing tunnels, which may force down the railway to a depth at which trench excavation is more expensive than bored tunnelling. This last factor is of increasing importance as a city elaborates its metro system with a number of crossing lines.

Although bored tunnelling appears initially the more costly method, the indirect savings in reduced disruption and loss of amenity and in reduced costs of property acquisition will frequently make it the preferred alternative. There is, of course, the handicap of deeper stations requiring longer escalators and therefore increasing the time for access by passengers.

In considering bored tunnelling the choice of the most suitable stratum will be important in deciding on depth. Information is particularly necessary on water table levels, and the possibility of tunnelling in impermeable strata. Depending on distance between stations and acceptable gradients, stations may be built at relatively shallower depths with the running tunnels descending and rising again between stations.

Another important consequence of the choice between cut-and-cover and bored tunnel is that of size and cross-section. Generally, a trench lends itself to a rectangular cross-section and can be made reasonably spacious without disproportionate increase in cost, whereas in a bored tunnel of circular or horseshoe section, extra space is disproportionately expensive.

7.3.4.1 Construction sites

At an early stage of planning it becomes important to find possible construction sites from which work can be executed economically. In cut-and-cover the whole line constitutes a working site progressively, but larger areas are required beyond the mere width of the tunnel for assembly, operation and maintenance of plant and storage and handling of materials. For bored tunnels similar areas are likewise essential with additional requirements of access to the tunnelling by shafts, pits or adits.

Although station sites are useful as working sites for tunnelling when available, the construction of the stations often cannot be subordinated to the tunnelling operations and other areas must be found for tunnelling operations.

Ventilation shafts, cable shafts and emergency exits may provide satisfactory temporary construction sites, and therefore should be determined as early as possible with this use in mind and with the intention of utilising working shafts for permanent requirements.

The spacing of such working shafts will depend on, and in turn influence, the depth and type of tunnel to be built, the use of a tunnelling shield or machine and the lengths of the drives; all being related also to the nature of the ground. The maximum practicable working area should be provided immediately adjacent to each shaft, although some of the requirement can be met by provision of a second area in the vicinity.

It is generally found that the construction and equipment of stations takes longer to complete than does the connecting tunnelling and it may therefore be appropriate to begin the building of stations first. Nevertheless, the installation of specialised railway equipment in the running tunnels can be a lengthy process and the time required should not be underestimated. (See Figs. 7.5a and b).

7.3.4.2 Staged construction

Where a long line of new railway is to be built it is advantageous to divide it into a number of sections which can be opened to traffic successively at intervals of a year or less. There are benefits both in construction and in operation. Contracts can be let progressively, avoiding peak loading of the construction industry and providing continuity of work over a longer period for those with special skills, while applying the experience gained on one section to the building of the next. Operationally, the staff responsible for bringing each section into use can be concentrated more efficiently, and likewise make use of experience gained. Electrical trackside and other equipment, liable to deteriorate in unavoidably humid conditions, can be operational more quickly.

If such staged opening is planned the early acquisition of a site for the rolling stock depot will be most valuable so that it may be used as the main depot for installation of track and other equipment in the tunnels. Where the depot is remote from the city centre this implies the opening first of an outer section of the line which again has the advantage of providing a training ground where the traffic is less intense than in the centre. A requirement of each stage is the provision of temporary facilities for reversing trains at the end of each section, and the stages may well be fixed by positions where an emergency crossover is a feature of the permanent design and can be utilised.

The staged opening affects all sections and aspects of the project: survey, acquisition of land, preparation and letting of contracts, supervision of the works, manufacture and supply of rolling stock and other equipment, and training of railway staff in operation and maintenance.

7.3.5 Track geometry

The basic geometry of a metro railway in terms of curves and gradients determines very closely the alignment of tunnels and of stations. Adequate margins of safety for stability of trains must be maintained, with curves designed to provide a comfortable ride for passengers.

Fig. 7.5(a) – Construction site for Victoria Line: – During construction.

Fig. 7.5(b) – Construction site for Victoria Line: – As subsequently restored.

The horizontal geometry is initially considered as comprised of straights and circular curves, which should have the maximum practicable radius. The curves are superelevated to match a specified design speed, usually somewhat less than the maximum. Transition curves are then inserted, modifying a length of curve and straight, so as to build up the curvature and superelevation at a comfortable rate. Unless sufficient length is available, adequate superelevation is not attained and speed restriction becomes necessary, although some deficiency from the calculated balance at maximum speed is acceptable.

Curves of small radius must be avoided, not only because of the long transitions required but also because of the excessive cant needed if normal maximum speed is to be maintained and because of the overhang of coaches on such curves, encroaching on tunnel clearances or making necessary a larger diameter of tunnel.

In some existing systems radii below 150 m have been employed but 400 m is a desirable standard as a minimum for new construction, although radii down to 200 m have been adopted for important systems in recent years.

Vertical alignment is regulated by the ruling gradient, minimum drainage gradient, station and sidings gradients and vertical curves. The maximum acceptable gradient for a long slope is limited by the motive power of the trains. Very steep gradients may be necessary in hilly cities, as for example in Montreal, but in less severe conditions 3% or 4% may be adequate as a maximum.

In order to ensure effective drainage, it is advisable that no length of tunnel between stations should be quite level. It is sometimes necessary to provide a low spot for a drainage sump and pumps. At stations it is desirable for safety that the train should not roll under gravity even with the brakes off. With modern stock this requires a level track, although previously a small gradient has been considered acceptable. Similar considerations apply to sidings, and certainly in no case should a siding fall towards a main line.

Vertical curves of large radius are required at changes of gradient, which should not be sited in the vicinity of crossovers because it is not possible to maintain satisfactorily switch blades in vertical curves.

Because stations are relatively closely spaced it is important that trains have braking and acceleration rates as high as is consistent with passenger comfort. The trains can be assisted by provision of a rising gradient approaching a station and by a falling gradient leaving the station; this constitutes a so-called hump profile. Some economy of current consumption and brake wear is also achieved.

7.3.6 Power supplies

Electrical power is required in a tunnelled metro system for traction, lighting, ventilation, pumping, escalators or lifts, and for the signalling and control systems as well as for various services in booking offices, staff accommodation and elsewhere.

Traction power is most usually DC, which has been found suitable in relation to the limited route mileage, generally less than 200 route miles. The widely used conductor rail system is generally DC within the range 600 V to 1000 V, while for an overhead DC system about 1500 V is common, and for an overhead AC system 6 kV to 25 kV. The driver of any train should be able in emergency to cut off the traction power supply in the section.

Station and tunnel lighting is usually fluorescent and may be supplied at 240 V AC through a local transformer substation. Emergency lighting from a separate energy source is essential. Ventilation and pumping, probably with remote or automatic control, may require a 240 V or 415 V supply between stations. Escalators and lifts demand substantial and reliable power. If duplication of their supply requires the running of power cables through the tunnels a high voltage AC connection is much to be preferred because of the fire risk from a cable carrying a heavy current at lower voltage.

Supply for the signal system must be available along the whole length of the line. It is preferable that this be a separate system so connected into substations that supply is maintained when any one substation is shut down. Voltage of about 600 is suitable and frequency of 50 Hz is acceptable although there is an advantage in having an independent frequency whereby interference is limited.

7.4 DESIGN

There is not a well defined boundary between planning and design; each reacts on the other, but broadly it may be suggested that engineering planning is done at scales smaller than 1 in 1000, while engineering design takes over at scales of 1 in 250, not ignoring the smaller scale layout plans.

The functional design of the various elements–structures, machines, equipment–is very closely restricted by the limitations of space in all three dimensions and precise geometrical calculation pervades the whole subject.

If existing rolling stock has to be accommodated, that must be the starting point for consideration of tunnel size, gradients, curvature and provision of appropriate track and trackside equipment. It is obvious that specially designed stock adapted to the proposed tunnel can provide great economies and advantages, provided that there is no requirement to accept other rolling stock from connecting lines. It is in bored tunnels of circular section that the disadvantages of standard railway stock are most apparent; carriages are usually rectangular in form and, in the circular tunnel which provides adequate clearance at the corners of a rectangle, space overhead and at the sides is likely to be larger than necessary. London's 'Tubes', including recent and current projects, exemplify great economy in the proportion between passenger carriage area and excavated cross section. Early standardisation of tunnel diameter as 12′6″ (3.81 m) external and 11′8¼″ (3.56 m) internal, increased progressively to 3.81 m internal, has governed design and development of rolling stock. Most new systems in other cities are of larger dimensions, as for example Toronto, Washington, Hong Kong and Brussels, either because the rolling stock was initially developed to suit cut-and-cover sections or because greater importance was placed on the spaciousness inside the cars, especially in hot climates. Some authorities insist on the ability to have staff in the tunnels during service hours and this naturally makes large clearances necessary.

In the tunnel design the aim must be to provide a structure requiring no major maintenance. Its alignment, horizontal and vertical, must be the best possible, because subsequent alterations are necessarily difficult and time-consuming and probably involve operational shutdown.

The lining's primary function is to support the ground and structures above, but it must also accommodate railway equipment such as cables and signals. Electrical equipment is sensitive to dampness, but moisture is almost unavoidable and the lining and fixings must be designed to minimise the effects. Even when the utmost precautions are taken to ensure watertightness in construction, slight ground movements may occur, resulting in seepage of water into the tunnel. An efficient drainage system should be provided, but there may remain the problem of atmospheric humidity and condensation of water on cold tunnel surfaces, more particularly in the early stages of commissioning before air movements and heating from traffic are effective. A fully operational ventilation system may therefore be an early requirement.

7.4.1 Tolerances

The size of the tunnel will usually be the smallest area within which the rolling stock and trackside or overhead equipment can be accommodated with appropriate clearances—further discussed later. In addition, tolerances for various errors of construction must be provided.

1. The first source of error may be in the setting out from the survey. No great errors should arise, but the transfer of lines and levels to the working face or to the shield is done in difficult conditions under pressure of time, and involves measurement of line, level, and shield attitude (see Volume 1).

2. A tunnel in soft ground, whether machine excavated or shield driven or without a shield, cannot be excavated and built to a perfect alignment and profile without intolerable delays. The more rapidly the excavation is advanced the more difficult it is to avoid departures from the specified line and level. If a shield or machine is found to be out of line it cannot be immediately adjusted to its correct position; it must be deflected so as to aim at coming on to line again a short distance ahead yet without overshooting into a reverse error.

Fig. 7.6 – Victoria Line. Brixton Station.

3. The lining introduces a third source of error, although with some opportunity for correction of minor errors of excavation particularly if relatively thick *in situ* concrete lining is used. Where precast segmental lining is to be used, errors may arise in the segments themselves from casting tolerances and also in erection to a true circle and in packing of joints. Imperfections in one ring add to the difficulties of erecting the succeeding ring correctly.
4. Following the lining of the tunnel, stresses in the surrounding ground readjust themselves and loading of the lining normally causes slight movement; the tunnel usually 'squats' by a small amount. Further movement is to be expected if another tunnel is built in the vicinity.

All these causes may accumulate to an error of several centimetres. The more experienced the tunnel miners and the surveyors and engineers, the more accurate the work is likely to be, but the more quickly the tunnel is advanced for economic reasons, sometimes for safety reasons, the more difficult it is to keep the deviations small.

In specifying a contractually acceptable tolerance all these factors must be taken into account, but even where the tolerance is exceeded it may be wise to adopt expedients which avoid rebuilding of the lining.

On completion of the lining a detailed survey is made from which a 'wriggle diagram' is produced (see Volume 1). A new alignment of the track centre line can then be calculated to utilise to the best advantage the clearances actually available, and to reduce or eliminate any necessity for rebuilding. The importance of experienced staff and their value to all parties will be apparent. Long experience on the part of all concerned allows the specified tolerance in London to be as little as 25 mm for hand excavation or 38 mm for mechanical excavation but with relaxations as suggested above.

The problem of construction tolerances has been described in the context of a bored tunnel in soft ground, but the same principles apply, although in lesser degree, for rock tunnels and to cut-and-cover; namely, allowances for setting out, excavation, structural lining, and ground movement. Operating tolerances are discussed below in terms of structure gauge and loading gauge.

7.4.2 Trackside equipment

Equipment within running tunnels should be kept to a minimum, because of the difficulties of access for maintenance and in the event of failure or accidents. As much as possible is therefore housed at stations. The necessary minimum in the tunnel, in addition to cables, is likely to include:

1. switchgear for isolation of traction feeds;
2. signal heads;
3. trainstops or similar devices for emergency braking of trains, and
4. disconnector boxes for track circuit and equipment connections.

The equipment must be designed to fit within the space available in the tunnel, leaving specified clearances which have to be accurately checked in relation to actual rail installation, allowing also for superelevation and curves horizontal and vertical. Design of covers for boxes housing equipment is of importance, to seal equipment against dust which is likely to have a high metallic content and to ensure that the covers cannot be left in a position encroaching on clearances.

7.4.2.1 Current collection systems

Metro systems, having relatively low route mileages, show a general preference for conductor rail systems. Overhead H.T. systems impose extra tunnelling costs because of necessary clearances.

If the conductor rail system is adopted there is a choice between 3-rail and 4-rail systems, the former being naturally the preference of engineers responsible for installation and maintenance costs of permanent way. For signal engineers there are advantages in the fourth rail system as this permits separation of the signalling currents in the running rail from the traction currents in the conductor rail; also, impedance bonds required in the third rail system can be eliminated. For close headway systems it may become more economical to provide a fourth rail.

The 4-rail system has a further benefit in permitting continuous monitoring of both poles of the traction system with earlier detection of faults and possible rectification before the service is affected. Recent development of audio frequency track circuits affect the balance of advantage.

7.4.2.2 Cable runs

Cable supporting systems usually consist of brackets, ducts or troughs. It is good practice to segregate low power signal cables and equipment from higher powered cables of the traction system by siting them on opposite sides of the tunnel. Fire precautions are important, particularly against the accumulation of dust round cables and in the adoption of cable insulation which minimises the risks of fumes, smoke and flame, while possessing electrical and mechanical strength. Thermal expansion and contraction is not a major problem in the relatively constant temperature of a tunnel.

Transition points between running tunnels and station tunnels require careful design of cable routes to preserve accessibility and avoid sharp bends. Segregation of signal cables from power cables should be maintained at these points, and also under platforms and in cable shafts.

The provision of suitable fixings for cable brackets or other supports must be considered in the design of the lining. Figure 7.2 shows the arrangement of brackets provided in the Victoria Line running tunnels. The frequent bolt holes of bolted cast iron linings offer a ready answer but even then it is necessary to provide adjustments which take account of the 'roll' of the iron and such local effects as superelevation which may cause the cables and equipment to encroach on clearances. Bolted concrete segments usually are less convenient in this respect, and with other types of lining, may make necessary the provision of special fixings for brackets, with scope for adjustment.

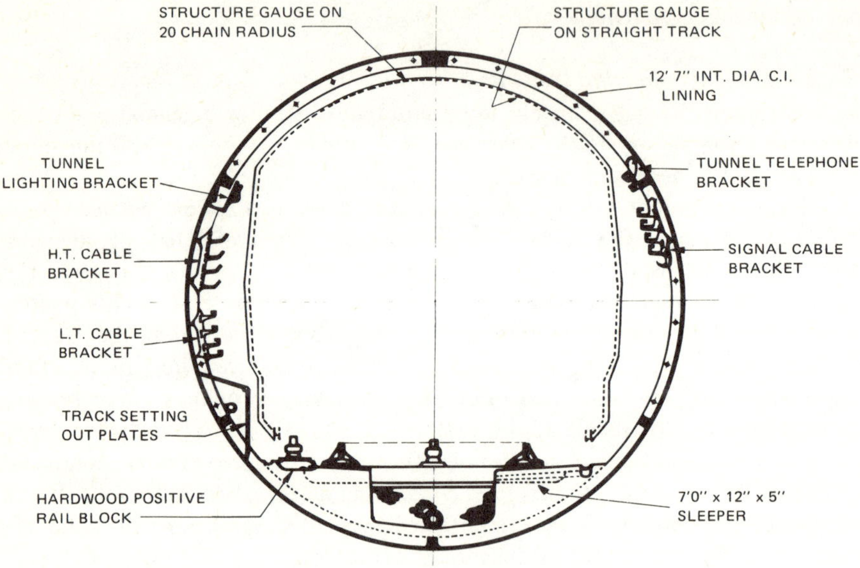

Fig. 7.7 – Victoria Line: cable brackets in tunnel.

7.4.2.3 Walkways

Emergency arrangements for safe discharge of passengers from a train immobilised in a tunnel between stations must be carefully planned. The emergency may be mechanical or electrical breakdown, or an obstruction in the tunnel, or fire. With decreased numbers of operating staff consequent on automatic operation in modern metros, the provision of easy means of escape becomes of increasing importance.

The ideal is an accessible platform on one side, at or near carriage floor height, but this is very expensive to provide both in tunnelling space and in its construction. While some metro systems are obliged by law to provide such walkways, in many there is no such platform and no space; the coaches are connected by a through corridor and passengers are expected to walk to the end of the train and there descend and walk along the track to the nearest station. For such circumstances the absence of a fourth rail is an advantage.

Where walkways are provided their design must provide for the accommodation of cables and pipes for the various services.

7.4.2.4 Lighting

It is not normally necessary to operation for running tunnels to be lighted, but reliable standby lighting for use in emergencies and also for use in night maintenance is required. The standard and quality of station lighting is of major importance in the contribution it makes to an attractive environment, over and above its narrowly utilitarian function.

7.4.2.5 Communications

An emergency telephone connection accessible between stations to drivers and guards of trains should be installed. It may be by wires carried on brackets fixed to the lining, or possibly by a suitably adapted radio system.

7.4.3 Noise and vibration

Reduction of noise experienced by passengers in tunnelled metros has been a problem actively investigated for many years. Noise originates in motors and in axles, wheels and interaction with the track. The hard surfaces of tunnel lining and track reflect sounds readily, unless treated with sound absorbing material (which should be non-toxic), and in the confined space of a tunnel the noise transmitted into the passenger coaches may be excessive.

Vehicle design can effect some improvement as can the use of rubber in the wheel mountings. Sound absorbent panels, acting also as baffles, may be fixed at wheel height as close to structure gauge as possible and the tunnel surfaces may be lined or sprayed with suitable material. Reduction of exposed area of concrete surfaces at track level and increase in ballast is also helpful. Careful sealing of openings in the body of the car together with careful design, construction and maintenance of track, wheels and suspension are even more important to reduce the noise experienced by passengers.

Large rolling stock allows greater opportunity of insulating the coaches from noise generated between wheels and rails. In the smaller London tube stock the tops of the bogies are below the passenger seats and little space is available for efficient sound insulation. Where stock is larger it is possible to design coaches without opening windows and to install air conditioning.

An underground railway may produce low frequency vibrations, or rumbles, which are occasionally heard or felt, particularly at night in buildings over the track. In Great Britain it has usually been found that this new traffic noise, initially considered obtrusive, is soon accepted as background noise which is rarely noticed.

The vibrations, transmitted through the ground, are not harmful to sound buildings. Nevertheless, there is a problem for which solutions are being sought. Rubber mountings for the tracks, plastic sleepers, heavy double sleepers and other devices to isolate the track from the ground are among the methods being tried. The extra expense is substantial and benefits are uncertain, more especially as it is only in special cases that objections are serious. Promoters of new metros are sometimes obliged to adopt such measures, but many cities maintain that the expense is not justified.

7.4.4 Drainage

Drainage is provided along the invert of the tunnel to dispose of all water from seepage or spillage. The gradient of tunnel, track and pipes should be at least 0.25%, leading to a sump and pump, which may be midway between stations. An important function of good drainage is to ensure reasonably dry conditions for track circuits, used to detect and register the presence of a train. Rail fixings to sleepers have to be designed with this in mind, particularly in damp areas. The vertical profile of a new line often makes it necessary to provide a drainage sump between stations. Such sumps should be pumped out automatically, but with appropriate warnings of pump failure. They will be discharged either to the nearest station sump, or direct to the surface, utilising a cable or vent shaft.

7.4.5 Track support

Conventional railway track support comprises cross sleepers of timber or concrete laid on stone ballast. This distributes loads and allows for efficient drainage. In the open it is easily installed and adjusted to the required line and level; regular maintenance is necessary and checking and readjustment. In tunnels these operations become much more difficult because working space and times of access are restricted and tolerances are small and therefore require more frequent checking and reinstatement; concrete support is therefore preferred.

The requirements to be satisfied as far as practicable by the track system adopted include:

1. Minimum tunnel size, and therefore minimum practicable clearances and tolerances.
2. Low cost of track installation.
3. Accurate installation requiring accurate and secure setting of rails before concreting in place.
4. Effective drainage, with particular reference to electrical controls.
5. Minimum operating maintenance with provision for vertical adjustment of at least 10 mm to allow new and worn rails to be matched up. Spring clip fastenings may be preferred to screwed fastenings in that the latter must be periodically tested for tightness.
6. Long term reliability.
7. Low noise level in coaches.
8. Minimal transmission of vibrations to ground.
9. Smooth track bed to prevent accumulation of dust which can ignite and smoulder.

Various designs of fastenings and concrete work are available, but the most suitable for any particular metro will depend on rolling stock, tunnel structure, skills and resources available, climate and other special factors. Whatever system is adopted involves some compromise between the differing requirements, and will be capable of progressive development in actual use.

Noise is greater with concreted than with ballasted track and increases significantly with speed. Below about 50 km/hr noise from a concreted track is not likely to be serious, but where speeds are higher, further measures become necessary as discussed above.

7.4.6 Structure gauge

The loading gauge is the profile beyond which no part of a train may project.

The structure gauge is the profile within which no part of a structure or fixed equipment may encroach.

It is the structure gauge that directly concerns the tunneller, but its profile is derived from that of the loading gauge by addition of safety clearances.

The loading gauge is set out in relation to the running rails as a base and is developed from the maximum car profile (including if necessary any projection caused by swinging car doors) by addition of allowances in width and height, for:

1. Track movement, tolerances and rail wear.
2. Train displacements in motion due to spring deflections, tolerances, defects and wheel wear.
3. On curves, the overhangs arising from the geometry of the coach in relation to the curved track.

Track movement to be allowed will depend on the form of track support adopted. Obviously, conventional ballasted track with cross sleepers is liable to a degree of movement much greater than when sleepers or other supports are encased in concrete and demands correspondingly greater tolerances.

Car displacement in motion may be both horizontal and vertical and has multiple causes: in the springing and suspension of the vehicle and between bogies and frame; in wearing down of original new surfaces of rail and tyre; and in variations of track gauge. These effects can only be estimated on the basis of experience and will be affected by the standards of maintenance imposed and the degree of wear allowed before renewal.

As a conventional rectangular coach travels round a curved track the bogies follow the curve, but the middle of the coach 'cuts the corner' and overhangs on the inside of the curve, while the ends of the coach overhang on the outside. Once the design of the rolling stock is settled, these overhangs are calculable. Their order of magnitude may be indicated by the example of an ordinary rectangular coach running on a 400 m radius curve for which the overhangs are about 50 mm.

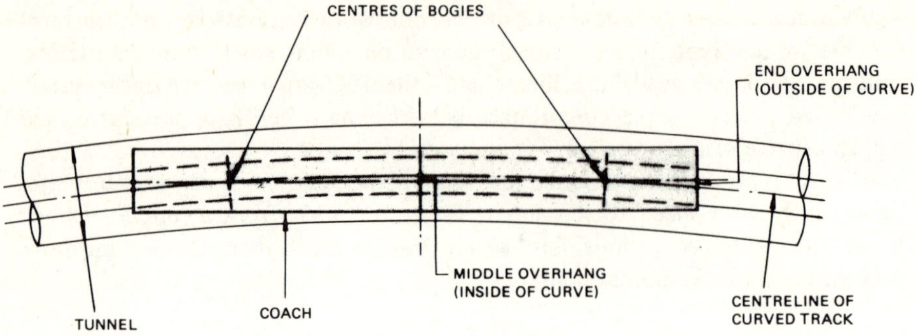

Fig. 7.8 – Coach displacement on curved track.

Minimum tunnel diameter, accommodating overhang, makes for economy of construction and can be approached by such measures in coach design as location of the bogies to equalise centre and end overhangs and rounding the ends and tops of coaches.

The structure gauge lies everywhere outside the loading gauge at minimum distances prescribed. Loading gauge and structure gauge for the Victoria Line are illustrated in Figure 7.9.

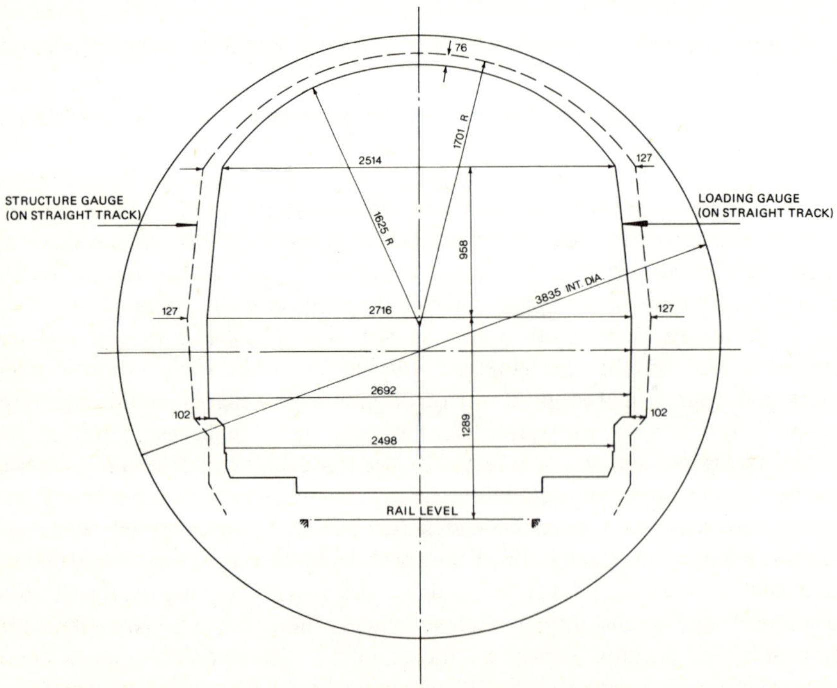

Fig. 7.9 – Structure gauge for Victoria Line. All dimensions are in millimetres.

Where the track is curved the allowances for overhang must be made, but in addition the structure gauge must be set out in relation to the rails which are normally superelevated or canted on a curve. The centre line of the tunnel structure and all critical points of its lining must then be offset by appropriate horizontal and vertical adjustments. The cant is built up gradually on the rails as a curve is entered and the tunnel centre line must correspondingly be progressively modified.

If the tunnel includes an emergency walkway it must, of course, be wholly outside the structure gauge but unless, exceptionally, it is to be used by staff when trains are running, it is unnecessary to provide the further clearances and safeguards which would be essential to give protection against passing trains.

7.4.7 Passenger flows and carrying capacity

A significant part of the cost of metros results from their having to carry peak passenger flows, which in most cities are limited to a few hours morning and evening five days a week. If they are to serve their purpose of reducing road traffic congestion all elements of the system must be designed accordingly: type of train, number of coaches, coach layout, length, frequency of service, together with station detail, length and width of platforms, widths of stairs and passages, numbers of escalators and lifts, ticket halls and turnstiles and ticket issue and collection. Some aspects of design have to be based on peak flows of as little as 15 minutes.

The limiting factor in train service is the duration of the station stop, which must be as short as possible if a close interval is to be maintained without the progressive delays resulting from trains held between stations. Numerous wide doors and clear space for free movement with cars are necessary for the quickest loading and unloading, but this means that seating capacity is limited and that coach designs are less suitable for suburban and longer journeys. Thus the highly specialised metro cannot satisfactorily be extended to distant outer areas. The practical carrying capacity of modern metro systems is between 30,000 and 50,000 persons per hour, or possibly more, in each direction, but light rail systems with a capacity of 15,000–20,000 per hour are now favoured in many cities.

Station design may be considered to start with the ticket hall, divided into the 'free area' outside the ticket inspection barriers, and the 'paid area'. Ticket control in and out may be either manual or automatic. Visual inspection may in practice be twice as quick as automatic gate control but is then likely to be less effective. A rate of about 25 passengers per automatic gate per minute is a fair average but may well be exceeded where most of the passengers are thoroughly familiar with the system. An escalator fully used will require about 5 gates handling 7,500 persons per hour. Between the top of the escalator and the gates at least 10 m should be provided for circulation to avoid dangerous congestion. Booking hall area and layout must also be designed to suit the ticket issuing and checking methods with careful study of pedestrian flows.

In passages a crowd density of 1.4 persons per sq m allows a free flow at about 3.8 km per hour which is 88 persons per metre width per minute. For fixed stairs the corresponding rates are 62 ascending and 69 descending, and escalators running at 44 m/min (2.5 km/hr) have a capacity of about 133 per min per escalator. At increased passenger densities the rate of flow in passages slows down and at a density of 4.3 person/sq m is reduced to a shuffle. The necessity is apparent for proper proportioning of passage and stair widths and circulating areas and avoidance of sharp bends and constrictions if a station is to handle peak flows satisfactorily.

7.4.8 Construction resources

The design of structures and methods of construction for a metro will be influenced by the availability of resources, both materials and human skills. The principal materials employed in any tunnelling are steel, cast iron, concrete and timber. Steel can provide temporary supports for trenches and tunnels, and permanent linings either as structural sections or as reinforcement. Cast-iron is the material which, in combination with the Greathead shield, made possible the development of soft ground tunnelling both in London and New York, and continues to provide a material of great merit for segmental linings, particularly in water-bearing ground. Spheroidal graphited iron is a form of cast iron having special structural advantages. Concrete has almost completely replaced masonry and brickwork for *in situ* linings in rock and in cut-and-cover tunnels, and is also preferred to cast-iron for segmental precast linings in many metros. Timber is not a material of permanent construction but is still most important for temporary support during excavation.

The principal choice to be made at the design stage is probably that of the material to be used for the permanent lining. The merits are more fully discussed in Volume 1 and briefly below. Availability of precast segments, iron or concrete, in the right quantities and sizes at the right times has to be ensured at the design stage. The initial use of concrete segments in place of cast iron in London in 1935 resulted from difficulties in securing adequate supplies of cast iron for the extension of the Central Line. The materials for concrete segments are usually available locally and a special casting yard can be set up, but a very high standard of workmanship is necessary if satisfactory segments are to be produced in quantity. Concrete segments are likely to cost substantially less than cast-iron, but for difficult tunnelling in water-bearing ground cast-iron may well have the advantage.

The choice between timber and steel for temporary support may be influenced by availability, but steel may be definitely specified at the design stage for some purposes, particularly where temporary supports are to be built in, and settlement may result if timber decays.

Tunnelling methods to be adopted must also be considered in relation to the availability of skilled men, familiar with particular techniques or capable of being trained. The degree of mechanisation to be adopted in construction will depend on many factors, but among the most important is the ability to assemble, operate and maintain the machines. Experts and skilled men will be necessary and it is virtually certain that some of them must be brought in from outside. It may be that the heavy capital outlay for mechanised work cannot be justified and that the better course will be to employ much larger numbers of less skilled men using simpler methods, which in any case are more adaptable to unforeseen ground conditions.

7.4.9 Tunnel lining

The lining for a tunnel must be selected at an early stage as a part of the whole metro system. Problems of ground support, methods of construction, performance in use and problems of maintenance must all be taken fully into account, in addition to cost of supply and installation.

Ground conditions may vary very greatly within short distances and although there might appear to be an economy in utilising cheaper linings where the ground is favourable, using more expensive linings, such as cast-iron, and S.G. iron only where difficult ground is encountered, there are very real disadvantages, even if the point of change is ascertainable in advance. A change involving alteration of the excavated area is not usually acceptable where a shield or excavating machine is in use, and even if the same excavated area is maintained the disruption of the system of progress can be very detrimental. With precast linings segments are manufactured well in advance and stocks are built up so that lining material is available in the tunnel at all times without delay. A decision to change cannot be taken quickly unless duplicate stocks of the alternative lining have already been built up. It is therefore generally advisable to use throughout a lining adequate for the worst conditions. Exceptionally, where a short length of particularly bad, or heavily loaded ground is foreseen, it may be of advantage to construct this length with special linings, but the converse modification of using lighter linings for a short length of favourable ground is most unlikely to effect any economy.

It is therefore best to design the lining for the worst sustained conditions in the length of tunnel under construction. This does not preclude the use of different linings on different sections of a line, as for example the use of cast-iron in an inner city area under buildings and of concrete further out where building loads are lighter or where a more favourable choice of ground is possible.

Watertightness is another factor to be considered in choice of lining. Cast-iron can more readily and effectively be made virtually watertight than can concrete, where some porosity is almost unavoidable and where cracks are likewise liable to occur and are very difficult to seal. Even in apparently impervious clay and with cast-iron lining some seepage of water is almost inevitable. Drainage must be provided to prevent accumulation of moisture; good ventilation is also helpful.

The requirement that there shall be an accurate relationship between track and trackside equipment must also be met, so that shape of lining and positioning of fixing holes must be precisely specified.

In a tunnel with *in situ* concrete lining all these considerations must be taken into account in the design of the shuttering. With prefabricated linings the choice between cast-iron, S.G. iron, concrete and steel depends on these and other factors, more fully discussed in Volume 1.

7.4.10 Ventilation and draught relief

There are two aspects of ventilation which have to be considered in the design of stations and running tunnels. The major air requirement is for cooling of the tunnels to keep within acceptable limits the build-up of heat arising from the motors and from braking. The ground provides a substantial heat sink, but in the long term the temperature inevitably rises until the ventilating air supply becomes warm enough to carry away surplus heat and reach a heat balance. The second aspect, and in immediate effects the more important, is passenger comfort. This requires an adequate supply of fresh air at an acceptable temperature and a limitation on air velocity in stations where the passage of trains can cause strong and unwelcome air movements.

The necessary air supply is that which, at the maximum acceptable rise of temperature, will remove the energy generated by the trains, averaged over a long period. The problems will be different in temperate or in cold climates where a substantial temperature rise is acceptable, and in hot climates where any increase may be unacceptable in passenger areas. This will influence the choice of points of supply and extraction of air. In the Victoria Line the air is normally drawn in as fresh air through the stations and extracted through shafts midway between stations, but the fans are reversible and can be used to draw in fresh air between stations. The piston effect of trains in tunnels can cause surges of air at stations. Reduction can be effected in part by draught relief shafts in the tunnels close to the stations. There may also be passages and entrances particularly subject to such draughts; much can be done in the design and if necessary special additional air passages can be provided. Seven metres per second air speed is considered to be an acceptable maximum for most areas, but a limit of 4.5 m/s is to be preferred if practicable, particularly for a flow of cold air.

Refrigeration of metros is a problem to be considered in hot climates. The fundamental difficulty is that waste heat is generated in the operation, and therefore plant requires further cooling from outside the system. If refrigeration is effected in the passenger coaches, there is added heat to be removed by the tunnel ventilation, while much of the effect may be lost when doors are opened at stations.

7.4.11 Environmental standards

The functional and aesthetic aspects of metro architecture cannot be separated. Apart from the trains themselves, most of the visual impact on users is in stations.

Construction problems and costs, particularly in tunnel, impose severe limitations on space both in plan and in height. Lighting, ventilation and noise have all to be considered.

In any but very shallow stations there will usually be two levels, with a booking hall at or immediately below ground level and deeper station tunnels, with escalators connecting them.

Ideally, the layout of a station should be such that passengers entering from the street progress directly past the ticket office to and down stairs or escalator and on to the platform, with a minimum of turns. The flow capacity of the access route should be almost uniform, except that the entrance gates at the top should have slightly less capacity than the escalators, and any exit gates or barriers should have a slightly greater capacity, so that the escalators themselves do not suffer dangerous congestion.

Platform length will be fixed by the length of the longest train plus the stopping margin; width should be sufficient to accommodate easily the maximum number of passengers entering between trains and also those detraining. Passengers entering and leaving must have room to move and sort themselves out. Emergency margins of space are also necessary for such circumstances as the emptying of a crowded train which has to be withdrawn from service, while incoming passengers continue to arrive on the platform. A reservoir area at the foot of escalators may be provided but closure of entrance barriers in the ticket hall should also be possible. Distinctive features and colours for different stations may be considered aesthetically pleasing, but have also operating advantages in enabling regular passengers to identify a station readily from a crowded train and thereby to alight without delay. This is of course additional to the provision of clear direction signs and station names which should be bold and simple and should be easily visible to passengers whether sitting or standing in the trains.

Another aspect of station design is the elimination of odd corners and recesses which can readily accumulate rubbish and cause a fire hazard. The same consideration applies also to the running tunnels between stations.

In deciding on the appropriate finishes for walls, floors and ceilings throughout a station, materials must be chosen which are available in suitable colours and are hard wearing and easily cleaned. Fire resistance and behaviour in a fire are of great importance. Where the station structure is below ground, and more especially below the water table, perfect waterproofing is almost impossible because of the inevitability of minor movements of ground and the ultimate porosity of even the best concrete. The very smallest penetration of ground water will, over a period, stain a decorative surface and deposit solids as it evaporates; it is therefore often advisable to make allowance for this by applying the decorative finish to a false lining with a small air gap behind it.

Lighting within a station should be of a high standard, but should not be higher than is adopted for the passenger coaches. Emergency lighting for stations to an adequate level should be automatically available in the event of failure of the main supply. Running tunnels need not be lighted ordinarily, but there should be emergency lighting provided for the evacuation of passengers from a stationary train.

Ventilation, to provide cooling, or heating, and draught relief have been discussed above.

Fig. 7.10 – Tyne & Wear Metro – Station Tunnel:
Secondary linings: Wall: Vitreous enamelled steel fixed to brickwork.
Platform: tiles on concrete beams.
Track bed: concrete.
Far wall: curved concrete slabs, advertising panels, corrugated 'Cellactite' to shed water.

7.4.12 Integrated stations

Apart from the detailed consideration of the best layout and environment within stations, the greatest shortcoming of many city stations is their separation from the activities which, in fact, they serve. In most cases they simply spew out and suck in passengers to and from busy and often narrow and crowded streets. The reasons are clear: minimal property acquisition and disturbance are required, and it is in any case likely that the constructing authority has no adequate powers to buy and develop, or that complex negotiations with adjoining owners will prove abortive. It is usually the cheapest and easiest solution simply to provide steps leading down to a ticket hall and attracting the passengers directly off the street. This may well be the best method in old cities close to historic buildings where disturbance should be minimal, but where an area is being redeveloped or in a new city or underdeveloped suburb, much better arrangements should be possible.

An urban station above an underground railway might well be served by escalators connecting to a covered shopping arcade at ground level and also serving a covered bus stop; other escalators might lead to an upper floor level of the shops and connect to pedestrian ways above traffic level with bridges across the streets. Other such connections should serve important buildings and offices and centres of entertainment. The benefits to the community should be substantial if costs could be properly apportioned.

Various metros have been able to incorporate exits through big department stores to the benefit of all. Montreal (perhaps encouraged by the climate) went further during their construction in the 1960's by incorporating some of its down-town stations in integrated developments covering several hectares; at the same time the different stations were made more attractive by varied architectural treatments separately commissioned. Some other new authorities in recent years have moved in the same direction–Melbourne, Helsinki, Hong Kong–but more success has been achieved in providing interchanges with other forms of transport than in the integration of station design and construction with other city development.

Fig. 7.11 – Montreal Metro – Station Tunnel.

It is probably generally true that metro promoting authorities lack the wide powers which would be necessary to use the opportunities occurring when a metro station is to be built in a city area ripe for reconstruction. It would certainly add to the complexities and delays of planning, but the problems are rather organisational and legal than engineering and architectural.

7.4.13 Passenger safety

One of the most important aspects of safety on underground railways is the avoidance of overcrowding, more particularly on platforms when there are service delays. When the designed capacity is likely to be exceeded, close supervision and early restriction on the entry of further passengers becomes essential, and it is important that the designed layout includes properly sited gates and barriers which can be shut as appropriate to the particular emergency.

In some stations a portion of the ticket hall below street level may also function as a public pedestrian subway; it should be possible to shut off access to the station proper while the subways still function normally. This consideration may also apply during non-traffic hours.

A vital element of station crowd control is obviously continuous monitoring and good communication between members of staff concerned. It was formerly necessary to have a large staff using an elaborate telephone system, but it is now possible to operate efficiently with a smaller staff using a closed circuit TV system combined with a public address system, and possibly remote control of escalators and barriers. This is an aspect of the central control system described earlier in this chapter.

7.4.14 Fire precautions

Any major fire on an underground railway is a serious matter, but they are very rare provided suitable materials and maintenance techniques are used. Those fires which do occur are mostly caused by accumulation of dust or inflammable rubbish, which is very vulnerable to ignition by cigarette ends blown about by draughts from moving trains. Such fires are not usually serious and can be extinguished by use of water or sand.

Very serious fires can result from overheating of equipment or cables due to an electrical fault, a major hazard particularly where plastic coverings can give rise to toxic fumes, although much work has been done to reduce this danger. Water cannot be used to extinguish fires in electrical equipment, while sand or other methods may be difficult to apply in some locations. Use of fans to remove smoke may be possible if the draught created does not intensify the fire.

With modern rolling stock fires are very rare and can usually be dealt with by use of portable extinguishers, or in electrical fires by cutting off the traction current.

It is a simple truth, not always appreciated, that the best precaution against fires is the avoidance of any materials which can burn, in the finishes and equipment of the metro and in the trains themselves.

Passenger emergency escape routes need to be fully considered in the design of the system, including contingency plans for exit to an adjacent station if an escalator link or other essential way out is blocked.

7.4.15 Flood protection

Underground railways, and particularly deep tunnels, are potentially vulnerable to flooding. In a city, long lengths are likely to be below the level of rivers, canals and sewers. The primary precaution must be to study carefully all records and to ensure that entrance levels at stations, and any openings for ventilation, cables or other purposes, are set above known and expected flood levels and are provided with barriers or other protection, to be used to raise the level if abnormal floods are experienced. Advance warning systems should be organised whereby information from all sources is directed without delay to a control centre from which instructions may be issued to stations for local action, or barriers may be closed by remote control. Drainage from the tunnel must be so designed that a hydraulic surcharge in the sewer to which it flows cannot cause a back flow. Flap valves by themselves do not give sufficiently reliable protection against major risks. In wartime, tunnels passing under waterways may be particularly at risk if the tunnel structure is breached. Floodgates may have to be installed at each end of vulnerable lengths to seal off the remainder of the system. Obviously electrical operation of gates would be provided but hand operation should also be possible to meet the contingency of power failure. At less vulnerable sites simpler forms of barrier might suffice.

7.5 CONSTRUCTION

Many of the problems of metro construction have already been discussed in considering planning and design. Tunnel construction methods have been described in earlier chapters, but some aspects special to metro tunnels may be examined in more detail. What follows is more particularly related to bored tunnels in soft ground.

7.5.1 Running tunnels

These comprise lengths of the order of 1 or 2 km between stations, usually for single tracks and therefore of minimum diameter, built to very small tolerances and requiring great accuracy of construction. Tunnelling may start from a shaft either in the vicinity of a station, or often midway between stations, in which case it may be advantageous to drive the tunnel right through the station site with subsequent enlargement. Shields or excavating machines have to be provided and assembled below ground in a shield chamber which may be formed as an enlargement of the future running tunnel. The special problems are those of fitting the work conveniently into an overall programme, working very accurately in line and level, and handling excavated spoil and lining material on a restricted working site, possibly subject to strict regulation to minimise inconvenience to street traffic and to neighbours. Full mechanisation of tunnelling is only found advantageous for the longer drives and when ground conditions are reasonably uniform, because of the time consumed in setting up machines, their capital cost, their lack of flexibility in varying ground conditions, difficulties of accurate control of the geometry and the elaborate back-up facilities needed to handle peak output. A Greathead shield ('hand' shield) drive with precast segmental lining is likely to progress more slowly, but can be started more quickly and will utilise disposal facilities more uniformly over a longer working life. It is also easier to introduce the modifications necessary if ground conditions change unexpectedly.

7.5.1.1 Crossover enlargements

For operational reasons track crossovers are required at places where it may be desired to reverse a train, either in normal operation or in emergency. Their relevance to the opening of a line in stages has already been referred to. With single track running tunnels the enlargement necessary to include both tracks and the intervening space changes the scale of the tunnelling and presents special problems of excavation and support during excavation. It is likely to be best to drive and line the running tunnels straight through in the first instance and to use them as pilot tunnels to be enlarged to the full crossover size. For London standards the minimum diameter of the crossover tunnel is 9.6 m.

7.5.2 Stations

The term 'station tunnel' is here used to refer to a tunnel encircling rail track and platform and not to other tunnels forming part of the station complex, for passages, escalators, circulating areas, etc.

It is usually advantageous so to design station tunnels that the running tunnel shield or other equipment can be driven through the station length either before or after the station tunnel itself has been excavated.

Special problems arise in having to drive two or more large tunnels side by side and in providing cross connections between, and access to escalators. A common pattern is that illustrated in Fig. 7.12 where the station platforms are connected to an intermediate circulating area in tunnel which forms the lower end of an inclined escalator tunnel. Numerous access openings have to be constructed at the back of the platform through the tunnel wall.

The sequence of operations has to be planned with great care to fit together the timing of running tunnels and station tunnels with the excavation for escalators and circulating chambers, allowing proper time for the installation of escalators and other equipment in relation to the opening date for the line.

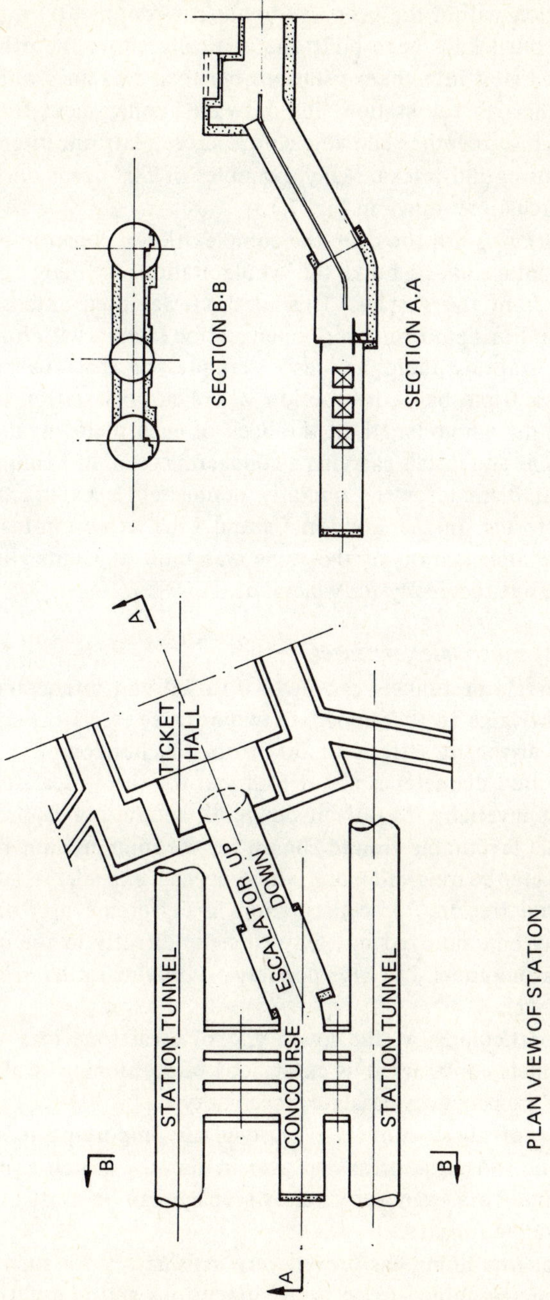

Fig. 7.12 – Typical arrangement of tunnelled station with escalators.

It is not, of course, essential that the station tunnels for the two tracks should be side by side on the same level. There are instances where, in order to keep construction within the property boundaries defined by a street at the surface, station tunnels have been built one vertically above the other. Another important variation is at interchange stations between two lines which converge and are parallel through the station: the outward bound tracks from each line may be brought close together allowing simple cross platform interchange, and likewise the inward bound tracks. Many examples of this occur on the London Victoria Line interchanges shown in Fig. 7.13.

Where the station is not too deep the complexities of construction in tunnel may make it advantageous to build the whole station, or a large part of it, in open excavation from the surface. This method was used extensively in the Montreal metro, and made possible very open station layout, with lofty ceilings.

In tunnelled stations there are also examples of spacious construction. Moscow adopted a triple bay construction where normal station tunnels were opened out along the whole length at the back of each platform and supported by a line of columns and lintels carrying a connecting arch. In Leningrad running tunnels of normal diameter were similarly connected by a much larger arch spanning the platforms. In the London Central Line extension in 1935–40 an elaborate multiple arch station of this type was built at Gants Hill, providing a spacious station, but too costly for general use.

7.5.2.1 Escalators, concourses, passages

Construction of escalator tunnels can involve varied and interesting problems. The tunnel must be large to accommodate two or three escalators and operating machinery and is driven at a slope of 30° from the horizontal. For a pair of escalators the required diameter is about 6 m and for three escalators 7 m. The tunnel will almost invariably be driven downhill, and, although the lower part may be in the most favourable ground chosen for the running tunnels, the upper part may be in water-bearing alluvium where special expedients such as grout injection, or ground freezing, or construction in cofferdam may be necessary. A pilot tunnel may be employed not only to assist directly in the construction of the main escalator tunnel, but also possibly to provide earlier access to other work at the deep level.

Concourses, particularly at the lower end of escalators, may also require large diameter tunnels constructed as extensions of the sloping escalator tunnels and used to house the necessary escalator machinery.

Watertightness of these tunnels is particularly important to protect the installed equipment and structures, and also in the upper arch exposed to the public, where stains from even very minor seepage can be most unsightly and damaging to decorative finishes.

Cast-iron segmental lining has proved very satisfactory for such tunnels, all the joint faces being machined to accurate dimensions and in addition caulked.

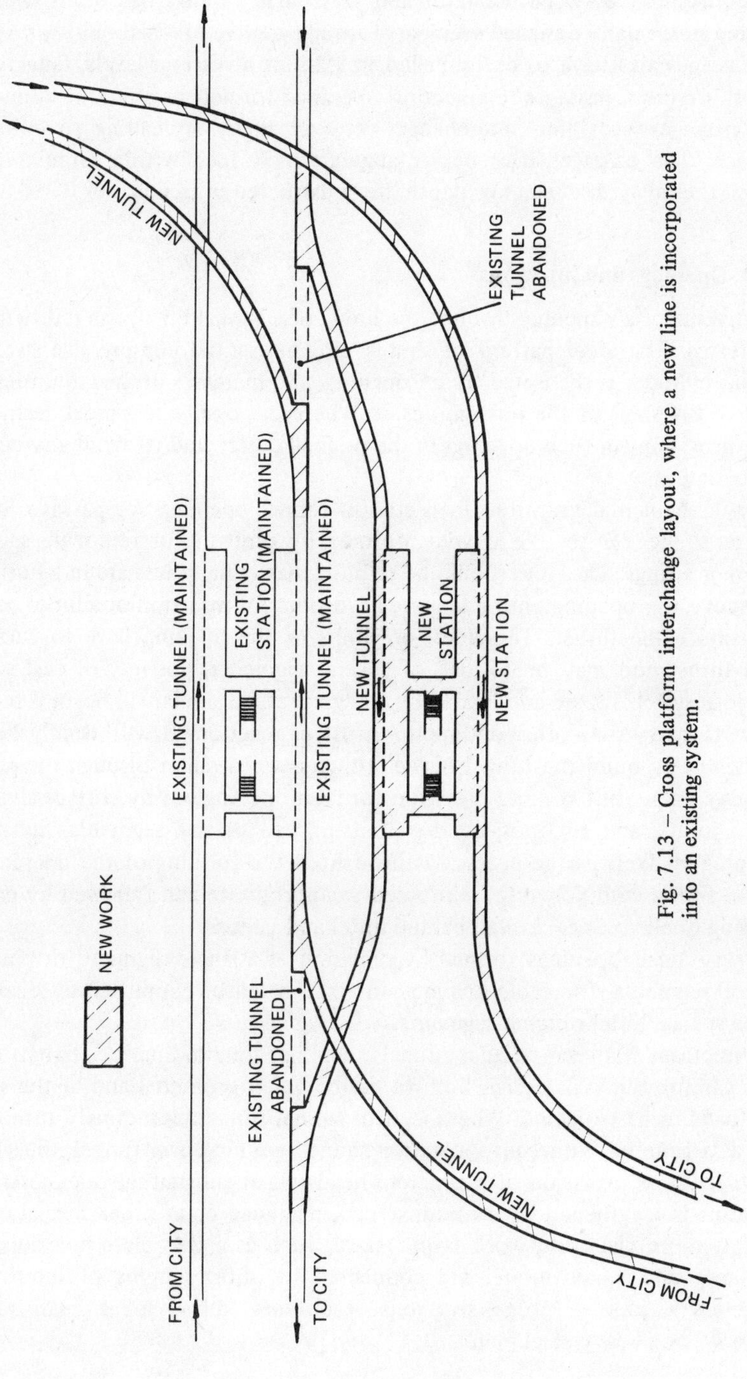

Fig. 7.13 – Cross platform interchange layout, where a new line is incorporated into an existing system.

The connection between escalator and concourse tunnels has often been constructed in specially designed segments forming a curve of small radius.

Passages also have to be tunnelled in stations at various levels, ranging from very short cross passages, connecting openings formed in adjacent tunnels, to long passages providing interchange between lines, or leading to exits at a distance. The passages may be rectangular or formed within circular tunnel linings; they may, if at shallow depth, be constructed in trench.

7.5.3 Openings and junctions

The making of an opening through the lining of a tunnel involves a redistribution of stresses. The ideal pattern of simple circumferential compressive stress in a circular cylinder is disrupted by an opening, and increases in the circumferential stress at the sides of the opening result. Wherever possible it is most desirable to make provision for such openings in the original design and to avoid any need for improvisation.

In a segmental cast-iron lined station tunnel openings for passages usually require to be 2½ to 3½ m wide, formed by omitting or removing segments from 4 or 5 rings. The rings should be so built that radial joints are in a horizontal line above the opening with adequate headroom for insertion of a lintel beam to redistribute the thrust. The sides, or jambs of the opening have to carry this extra thrust and may be formed of special segments, possibly of cast steel. A steel joist lintel is fixed and loaded by wedging, and a similar sill beam is required below to carry the corresponding up-thrust or reaction. It will usually be most practicable to build the tunnel in the ordinary way straight through the position of the opening, but making provision for later opening out by correct alignment of the joints, and also possibly by omitting one or two segments. Substantial propping is likely to be necessary throughout the forming of the opening, and careful filling with concrete, reinforced as appropriate and followed by pressure grouting of all voids excavated behind lintel and jambs.

Very small openings formed by omission of a single segment, or perhaps a pair of segments, for cable or pipe runs may often be simplified by dispensing with a special lintel or jamb segments.

Junctions between circular tunnels, whether longitudinal or at right angles, need careful study in design both in the complex geometry and in the stresses and loads to be expected. Where a small tunnel runs longitudinally into a large one, as a running tunnel into a station tunnel or a crossover tunnel, the simplest construction is provision of a concrete headwall of substantial thickness. Where a station is lengthened or a running track realigned or in other special circumstances, more elaborate work is necessary, such as a step plate junction where the large and small tunnel are connected by short lengths of intermediate diameters varying in progressive steps, each step embodying a small concrete headwall, or a heavy steel plate.

In some cases longitudinal junctions will be formed by use of taper rings of conical form either concentric or offset. These will be used where continuity of structure is particularly important due to heavy loadings and poor ground, and where waterproofing is very important as in escalator/concourse junctions.

For tunnels of different diameter joining at right angles an 'eye' may be provided in the larger tunnel within which the smaller tunnel is built. For this purpose the ratio of diameters should not be greater than 2 : 3 because the formation of such a junction between tunnels of the same size makes it necessary to open out a disproportionate area of ground, and the geometry of the connection has to be very accurate.

7.5.4 Construction programme

As with any other complex engineering project, if a metro is to be built efficiently and opened as soon as possible, a carefully prepared programme is necessary to ensure that each operation is started at the appropriate date and is so phased as to be completed in advance of subsequent dependent operations. This generalisation covers a very elaborate network of activities and interests. Legal powers must be obtained as soon as the line is decided, finance must be ensured to meet the anticipated expenditure, sites must be acquired, necessary materials and labour must be available and the work must be designed in detail so that contracts can be let and equipment ordered.

Key events and dependencies have to be identified, and times required for various activities estimated as closely as possible, so that a network incorporating those activities and the links between them may be established for use in forecasting and control. The general sequence of requirements for any section of running tunnel will include:

–acquisition of working sites
–detail design
–advance ordering of tunnel linings
–advance ordering of equipment
–letting of tunnel construction contracts
–completion of tunnel construction contracts
–installation of track
–installation of trackside equipment

Dates for these may be governed by priorities in relation to other units of work, in particular to station construction and also to access from adjacent lengths of running tunnel, and may also, where interchanges are involved, depend on modifications and diversions of existing tracks and tunnels.

Similar considerations are involved for station construction and depots, and for various aspects of equipment and also the supply of rolling stock. A very extensive and intricate network is built up such that computer control is likely to be necessary if it is to be fully utilised to monitor and regulate construction to the best advantage. Accurate reporting of departures from programme allows the network to be brought up to date; from this the ultimate consequences of delays can be seen and appropriate action can be taken by readjustment of the various activities.

Even without computerization flexible control of construction programmes is most valuable.

7.6 MAJOR SYSTEMS

The circumstances of every city vary so greatly that each produces a different metro system adapted to its own needs. When a system is initiated it is likely to be greatly influenced by existing public transport facilities; the first line may be an interconnection of main line or suburban subway termini, it may be the adaptation of tramways to a subsurface route, or a completely new system.

The factor precipitating a decision to construct a metro with underground lines is usually the congestion that develops in and about the centre of a city having a population of the order of one million people or more. It is, of course, difficult to make any precise comparison of city populations because traffic generating zones are likely to differ greatly from administrative areas, and in any case any metro line serves a narrow corridor rather than a city as a whole.

7.6.1 Route lengths

The major metro systems each having tunnelled routes in excess of 100 km are in cities with populations exceeding 6 million as follows: (The figures are from U.I.T.P. Handbook of 1975.)

City	No. of lines	Passengers per year x 10^6	Route lengths (km) Total	In tunnel	
New York	36	1266	407	232	Incl. P.A.T.H.
Paris	19	1225	248	178	Incl. Express
London	8	655	393	171	
Moscow	8	1628	150	132	
Tokyo	7	1498	150	125	Incl. TBTMG

It is further of interest to compare dimensions adopted for coaches in various systems and the tunnel sizes. The Glasgow system is included as showing the minimal size adopted in 1896. The limited height of the London tube stock, 2.88 m, is also of interest, in comparison with heights of 3.49 to 3.70 m adopted by many other cities and the intermediate height of 3.20 m in the new San Francisco (BART) system.

7.6.2 Dimensions of typical* coaches and tunnels

City	Length	Width	Height	Passenger Capacity Total	Passenger Capacity Seats	Tunnel
Glasgow	12.80	2.34	2.69	84	42	3.35 m dia.
Leningrad	19.70	2.70	3.70	170	40	5.10 m dia.
London (tube)	15.94	2.59	2.88	162	40	3.71 m dia.
Moscow	19.10	2.70	3.70	170	40	(7.6 m x 3.9 m (5.46 m dia.
New York	18.30	3.05	3.66	200	54	7.96 m x 3.89 m
Paris	15.15	2.40	3.49	157	24	7.10 m x 5.20 m
Tokyo	18.0	2.79	3.65	124	52	7.8 m x 3.9 m
Toronto	16.95	3.15	3.64	208	62	4.88 m dia.
Washington	22.86	3.05	3.30	175	81	

*Note that several of these cities have more than one type of coach in operation, and tunnels of various diameters.

BIBLIOGRAPHY

This chapter bibliography provides more especially references to British practice, for which very valuable and detailed descriptions of planning, design and construction problems are readily available. The meetings and publications of the International Union of Public Transport (U.I.T.P.) provide a world-wide forum, although concerned rather with operation than with construction. Further reference to the list of conferences in the main bibliography may be found helpful.

Baker, B., The Metropolitan and Metropolitan District railways.

Wolfe Barry, J., The City lines and extensions (Inner Circle completion) of the Metropolitan and District railways, *Min, Proc. Instn Civ. Engrs,* 1985, **81**.

Greathead, J. H., The City and South London railway; with some remarks upon subaqueous tunnelling by shield and compressed air, *ibid*, 1895, **123**.

Dalrymple-Hay, H. H., The Waterloo and City railway, *ibid*, 1899, **139**.

Biette, L., Underground railways: the metropolitan system of Paris;

Mott, B. and Hay, D., Underground railways in Great Britain.

Parsons, W. B., Underground railways in the United States, *Trans. Amer. Soc. Civ. Engrs,* 1905, **54**(F).

Parsons, W. B., The New York rapid transit subway, *Min. Proc. Instn Civ. Engrs*, 1908, **173**.

Gilbert, G. H. *et al.*, *The Subways and tunnels of New York*, J. Wiley, 1912.

Jones, I. J. and Curry, G., Enlargement of the City and South London railway tunnels, *Min. Proc. Instn Civ. Engrs*, 1927, **224**.

The Post Office tube railway, London, *Engng*, 1928, **125** (Jan. 27, Feb. 10 and 24, Mar. 2 and 16).

Hall, H., The new Piccadilly Circus station, *Min. Proc. Instn Civ. Engrs*, 1929, **228**.

Groves, G. L., The Ilford Tube, *J. Instn Civ. Engrs*, 1946, **26** (Mar.).

The Stockholm Underground Railway: a technical description, 1958, A technical description of the Stockholm Underground Railway, 1964, Public Works Dept. and Passenger Transport Co., Stockholm.

Turner, F. S. P., Preliminary planning for a new tube railway across London, *Proc. Instn Civ. Engrs*, 1959, **12** (Jan); *see also* discussion, **13** (Aug.).

Jackson, A. A. and Croome, D. F., *Rails through the clay: a history of London's tube railways*, Allen & Unwin, 1962.

Howson, F. H., *World's underground railways*, Ian Allan, 1964.

Dunton, C. E. *et al.*, Victoria Line: experimentation, design, programming and early progress, *Proc. Instn Civ. Engrs*, 1965, **31** (May); *see also* discussion, 1966, **34** (July).

Havers, H. C. P., *Underground railways of the world*, Temple Press, 1966.

Follenfant, H. G., *Underground railway construction*, London Transport, 1968.

Follenfant, H. G. *et al.*, The Victoria Line, *Proc. Instn Civ. Engrs*, 1969, **Suppl.**; *see also* discussion, 1970, **Suppl.**

Howson, F. H., *The rapid transit railways of the world*, Allen & Unwin, 1971.

Nock, O. S., *Underground railways of the world*, A. & C. Black, 1973.

Follenfant, H. G., *Reconstructing London's underground*, London Transport, 2nd edn., 1975.

Associated Engineers, *Subway Environmental Design Handbook*, U.S. Department of Transportation, 1975.

Edwards, J. T., Planning and design of the Hong Kong Mass Transit Railway, *Proc. Instn Civ. Engrs*, 1976, **60** (Feb.); *see also* discussion (Aug.).

Jobling, D. G. and Lyons, A. C., Extension of the Piccadilly line from Hounslow West to Heathrow, *ibid*, 1976, **60** (May); *see also* discussion (Nov.).

Railway and tracked transit system noise, Proceedings of workshop, Derby, 1976, *J. Sound Vibr.*, 1977, **51** (3), Proceedings of 2nd workshop, Lyon, 1978, *ibid*, 1979, **66** (3).

Bayliss, D. A., *The Post Office railway London*, Turntable Pubns., 1978.

Rogers, L. H. (Ed.), *International statistical handbook of urban public transport*, International Union of Public Transport (UITP), Brussels, 1979.

Dasgupta, K. N. *et al.*, The Calcutta rapid transit system and the Park Street underground station, *Proc. Instn Civ. Engrs*, 1979, **66** (May); *see also* discussion, 1980, **68** (Feb.).

Cuthbert, E. W., The Jubilee Line – 1, the project.

Lyons, A. C., ditto – 2, Construction from Baker Street to Bond Street and from Admiralty Arch to Aldwych.

Bubbers, B. L., ditto – 3, Construction from Bond Street station to Admiralty Arch, *ibid*, 1979, **66** (Aug.); *see also* discussion, 1980, **68** (Feb.).

Edwards, J. T. *et al.*, Hong Kong mass transit railway modified initial system: system planning and multi-contract procedures.

McIntosh, D. F. *et al.*, ditto: design and construction of underground stations and cut-and-cover tunnels.

Haswell, C. K. *et al.*, ditto: design and construction of the driven tunnels and the immersed tube.

Taylor, R. L. *et al.*, ditto: design and construction of above-ground works and trackwork, *ibid*, 1980, **68** (Nov.).

Smyth, W. J. R. *et al.*, Tyne and Wear metro, Byker contract: planning, tunnels, stations and trackwork, *ibid*, 1980, **68** (Nov.); *see also* discussion, 1981, **70** (Aug.).

Howard, D. F. and Layfield, P., Tyne and Wear metro: concept.

Bartlett, J. V. *et al.*, ditto: management of the project.

Tough, S. G. *et al.*, ditto: design and construction of tunnelling works and underground stations.

Nisbet, R. F., ditto: interchanges and surface stations, *ibid*, 1981, **70** (Nov.).

Roscoe, R. B. *et al.*, Melbourne Underground Rail Loop, (11 papers), *Fourth Australian Tunneling Conference,* 1981.

8

Railway Tunnels

8.1 GENERAL CONSIDERATIONS

It is probably the railway tunnel through high ground that is thought of as typical when a tunnel is spoken of without other qualification. The Shorter Oxford dictionary in defining 'tunnel' uses the phrase 'now most commonly on a railway'. This obviously relates to public experience of tunnels in the 19th and early part of the 20th centuries, and to the days before road traffic developed to its present importance and resulted in the proliferation of highway tunnels. The associations of the word 'tunnel' may well become different in future.

From the first railway onwards tunnels have been an essential feature, primarily to provide a track with a limited gradient. In Volume 1 the early history of railway tunnelling has been described and some general characteristics of railway tunnels have been outlined. The fundamental reasons for tunnelling are topographical, with two distinct types

(1) to pass through rather than over a natural obstacle such as a hill, ridge, escarpment or mountain range, and

(2) to pass under a river, estuary or sea channel.

(1) Hill or mountain tunnels are the most frequent and characteristic type. Their function is to minimise the heights to be climbed and to keep gradients within acceptable limits, without involving lengthy deviations.

(2) Subaqueous tunnels, on the other hand, impose descents and unwanted gradients on the railway. On main lines (excluding metro systems) they are comparatively few in number, but are usually long, difficult and costly. They are therefore only constructed where the rail link is of major importance and alternatives are impracticable, or excessively long.

At the extreme for mountain tunnelling are the crossings of high mountain ranges as in the Alps, and elsewhere, and for subaqueous tunnelling sea crossings such as the Channel Tunnel and Seikan Tunnel described in Volume 1.

8.1.1 Urban tunnels

A third category which might be adopted is that of tunnels under urban areas. These are usually dictated by topography and as such need not be considered separately. Examples include the very earliest main line tunnels at the Liverpool terminal of the Liverpool-Manchester railway (1830), and the Hudson river tunnel in New York. In so far as other urban tunnels are not topographically determined, they resemble closely Metro tunnels in concept and construction. The principal differences may include:

–larger cross section
–terminal stations
–no intermediate tunnelled stations
–less frequent services
–longer and heavier trains
–non-electric traction
–use by goods trains.

New York City provides an interesting example of the dominance of topography. The main line from west to east has to pass in tunnel under the Hudson River to reach Manhattan Island and, after crossing the narrow island, to pass under the East River, again in tunnel. The Pennsylvania Station, near the heart of

Fig. 8.1(a) – The first railway tunnels, at Edge Hill, Liverpool (1830): Terminal tunnels, operated by rope haulage, comprising a passenger tunnel up to Cross Street and a goods tunnel descending to Wapping dock (see Vol. 1, Ch. 1).

Fig. 8.1(b) – Brunel's Thames Tunnel adopted for railway operation in 1869 (not the same Wapping as in Liverpool!).

the city, is therefore at a low level, 12 to 20 m below the street and is a through station between tunnel approaches rather than a terminus.

Melbourne's new Underground Rail Loop provides within the city centre better suburban services by eliminating the delays and congestion of reversing trains at termini. It also provides new stations giving better and more frequent access. Except in respect of the large tunnel cross section which is about 6 m diameter, the new Loop has all the characteristics of a Metro: primary use by passengers, frequent stations under city streets with easy access, electric traction.

The Paris express Metro also has the large cross section and wider spacing of stations more characteristic of main lines.

The generalisation about a smaller cross section in metros is only relevant where they are self contained systems whose use is restricted to their own special rolling stock and loading gauges. Main line tunnels must normally be capable of accepting all stock in use and must often allow clearance for hinged carriage doors swinging open and overhead clearance for electrical equipment or for the ventilation of diesel fumes or smoke. This larger and deeper cross section increases the tunnelling difficulties and costs, not only in respect of the excavation and lining, but by making more difficult the safeguarding of surface structures and by the necessity to go deeper below obstacles.

Fig. 8.2 – Potters Bar tunnels through London Clay. Old brickwork tunnel and modern duplicate employing expanding segmental concrete lining. (*British Rail*)

Important terminal stations require so much space for numerous tracks and platforms with facilities for reversing and various shunting operations that tunnel construction is not practicable, but the lines radiating out from a city terminal are likely to be forced into tunnel if they run across the topographical grain of the country. New York City has already been mentioned. London also exemplifies this; the Great Western railway following the Thames valley could be constructed without any tunnelling near the city, but the North-South lines were immediately involved in tunnels. Stephenson's London–Birmingham railway out of Euston Station started northwards in deep cuttings and then had to be tunnelled under Primrose Hill and again further out at Watford. To the north east there had to be a sequence of three tunnels through London Clay at Potters Bar which were built as double track tunnels in 1849–50 and duplicated in 1959. The various lines southeastwards to the coast all had to pass through ridges, which required construction, among others, of the Penge tunnel through London Clay under Sydenham Hill in 1863, the Merstham tunnel through Chalk in 1841, the Bletchingly tunnel through Weald Clay in 1842. These are all within the general classification of hill tunnels and are not typically urban, because they do not involve the special problems of working in closely built-up areas and under streets and buildings.

In urban conditions construction of main line tunnels as compared with Metros is simplified by the absence of frequent intermediate stations having problems of access and restrictions on gradients. The lower frequency of service and greater clearances make minor maintenance during traffic hours a possibility. On the other hand ventilation problems may arise from use of diesel power, or more rarely steam.

8.1.2 Hill tunnels

There is a wide spectrum from the relatively short and shallow tunnel which cuts through a hill spur or ridge in order to minimise gradients and shorten the route and avoid long or sharp curves, to the crossing of a mountain range with no practicable way over or round. Such tunnels will first be discussed in general terms, followed by a description of the special railway aspects determining design. Subaqueous tunnels will be described later.

Just as it was the necessity for level lengths that compelled the canal builders to tunnel, particularly for summit levels, so the limitations of workable gradients forced the railways to do likewise. Where a ridge or escarpment is to be crossed the shortest tunnel is achieved by aligning it as squarely as possible at the narrowest width. There is some scope for keeping the tunnel short by climbing to the highest practicable level, but there must be a balance struck, taking into account both overall capital cost and operating cost and revenue earned. The shortest possible tunnel is not the best if it is attained only by unreasonable expenditure on long and deep cuttings, long and high embankments, steep gradients, and a climb to an excessive height.

Having fixed on the maximum acceptable gradient, the surface section of the line will be laid out to provide a close balance between the volume of material excavated from cuttings and tunnels and the volume required to build embankments, having regard to the factors for bulking and the suitability of the materials. Stability of embankments built with clay can be critical, and without the benefits of the science of soil mechanics, the early pioneers had many instances of continuing trouble in building high embankments, and also in stabilising cuttings.

In establishing such a line through hilly country the point of change from deep cutting to tunnel has to be decided with comparative cost as an important factor. In the early days, before development of major excavating machines, it was considered that the change should be made when cutting depth exceeded about 20–25 m, provided that the cutting could be relied on to stand at a slope of about 1½:1. Obviously, in unstable ground where a flatter slope becomes necessary, the volume of cutting increases considerably and a shallower depth justifies tunnelling. In rock which will stand up more nearly vertical, the excavated volume is correspondingly less at any depth, but tunnelling problems are different and a new balance of advantage must be sought.

The economics of excavation have changed somewhat with modern machine methods, but even where open cutting may appear relatively cheaper than tunnelling, the problems of drainage of deep cuttings, the stability of high side slopes, and the taking of a wide area of land are likely to be more decisive in the choice than the narrower considerations of construction cost.

8.1.3 Mountain tunnels

This balance between cutting and tunnelling costs is only one aspect of the much larger and more complex problem of the best route across a mountain range where there is no way round and a surface route would involve impossibly severe gradients and curves in addition to excessive exposure to snow and rock falls. A major problem is, again, at what level to drive the summit tunnel. A long deep tunnel offers the counterbalancing advantages of less height to climb at easier gradients and less severe exposure to the weather. The ultimate argument lies in the assessment of what capital cost can be financed by the expected traffic. The problem of connecting two established railway networks, as in the Alps, was very different from that of driving a new line such as the Canadian Pacific Railway through undeveloped country. The original Alpine tunnels were deep and long because that offered immediate operational savings, whereas the CPR across the Rocky Mountains used very steep gradients and sharp curves because traffic initially would not be of sufficient volume to effect large total savings. At Kicking Horse Pass, approaching the summit, the original line was at a gradient as steep as 1 in 25; this was later reduced to about 1 in 45 by looping the line in two spiral tunnels driven into the mountainside when traffic became adequate to justify the cost, even though these loops added 4 miles to the length of line. Similarly, on the original transcontinental lines of the USA, zig-zag layouts involving reversals were adopted; they were costly to operate and slow but saved heavy capital expenditure.

Fig. 8.3 – Swiss railway tunnel in approach to Lötschberg and Simplon Tunnels. Originally constructed for single track, but with provision as shown for widening to double track. (*Berne-Lötschberg – Simplon Railway.*)

As the systems developed and traffic grew, a number of long summit tunnels were built to improve operation. In the USA these include the Moffat tunnel (1928) 10 km long near Denver in Colorado which replaced a line at 3500 m over Rollins Pass, and the Cascade tunnel (1929) 12½ km long near Seattle in Washington replacing a shorter tunnel and reducing the summit level by 150 m.

The same pattern of long railway tunnels being constructed, as later improvements in mountain areas to reduce steep gradients and improve operationally expensive routes, has been followed also in New Zealand and in South Africa.

The original Trans-Iranian Railway from the Persian Gulf in the south to the Caspian Sea in the north via Tehran was largely built in the years prior to 1938 and required much tunnelling, particularly in the Bakhtiari mountains in the south and the Elburz mountains in the north including a number of figures-of-eight with spiral tunnels and with gradients up to 2.8%.

The line northwestwards from Tehran to Tabriz, a distance of about 700 km, was built as a single track with curves as sharp as 250 m radius.

It was to be largely rebuilt to modern standards for high speed traffic with double track minimum curves of 5000 m to 1350 m and with gradient limited to 1.0%, except in the mountainous area approaching Tabriz where gradients of 1.4% are allowed. Inside the mountain tunnel portals a gradient of at least 0.8% for the first 250 m is sought to ensure rapid flow of drainage water and minimise ice formation. The distance would be reduced by about 100 km by a new route through the mountains from Poledokhtar to Tabriz.

In Japan major railway developments have been in progress since 1960. A policy of linking all parts of the four main islands by modern high speed trains has been pursued very actively. Because the country has many mountainous areas bordered by narrow and densely populated coastal strips, tunnelling on a large scale has been necessary to ensure both the easy gradients and large radius curves demanded for high operating speeds. Initially, for the Tokaido line from Tokyo to Osaka the design speed was 210 km/hr and the ruling gradient 2% and curve 2500 m. On its extension first to Okayama and then through the Shin Kanmon tunnel to Hakata on the island of Kyushu, design speed has been increased to 260 km/hr and construction limits have been progressively modified to 1.5% gradient and 4000 m radius (but with an exceptional 1.8% and 3500 m in the Kanmon tunnel). Another line runs NW from Tokyo to Niigata. The proportion of tunnel to open line is extremely high in these lines and the tunnels are among the world's longest.

8.1.3.1 *Portal Cuttings*

In contrast to the lowland railway requiring a tunnel through a hill or escarpment which tends to approach the slope squarely, the high mountain railway, which makes use of a valley is to a great extent laid out on sidelong slopes. Deep cuttings in such ground are not usually practicable so that the problem of siting the tunnel portal is different. The ruling gradient determines the possible alignment up the side of a valley, cutting obliquely across the contour lines until they become too steep or too sharply curved. Sometimes a viaduct will be required, as across a side valley but eventually tunnelling becomes essential, initially perhaps only for a short length through a spur but eventually into the main mass.

These simple geometrical requirements of the topography are only the first step in laying out the centre line because the width of the formation is very significant on steep sidelong slopes, and is likely to require the building up of a bank on the low side while a notch is cut into the slope on the other side. The

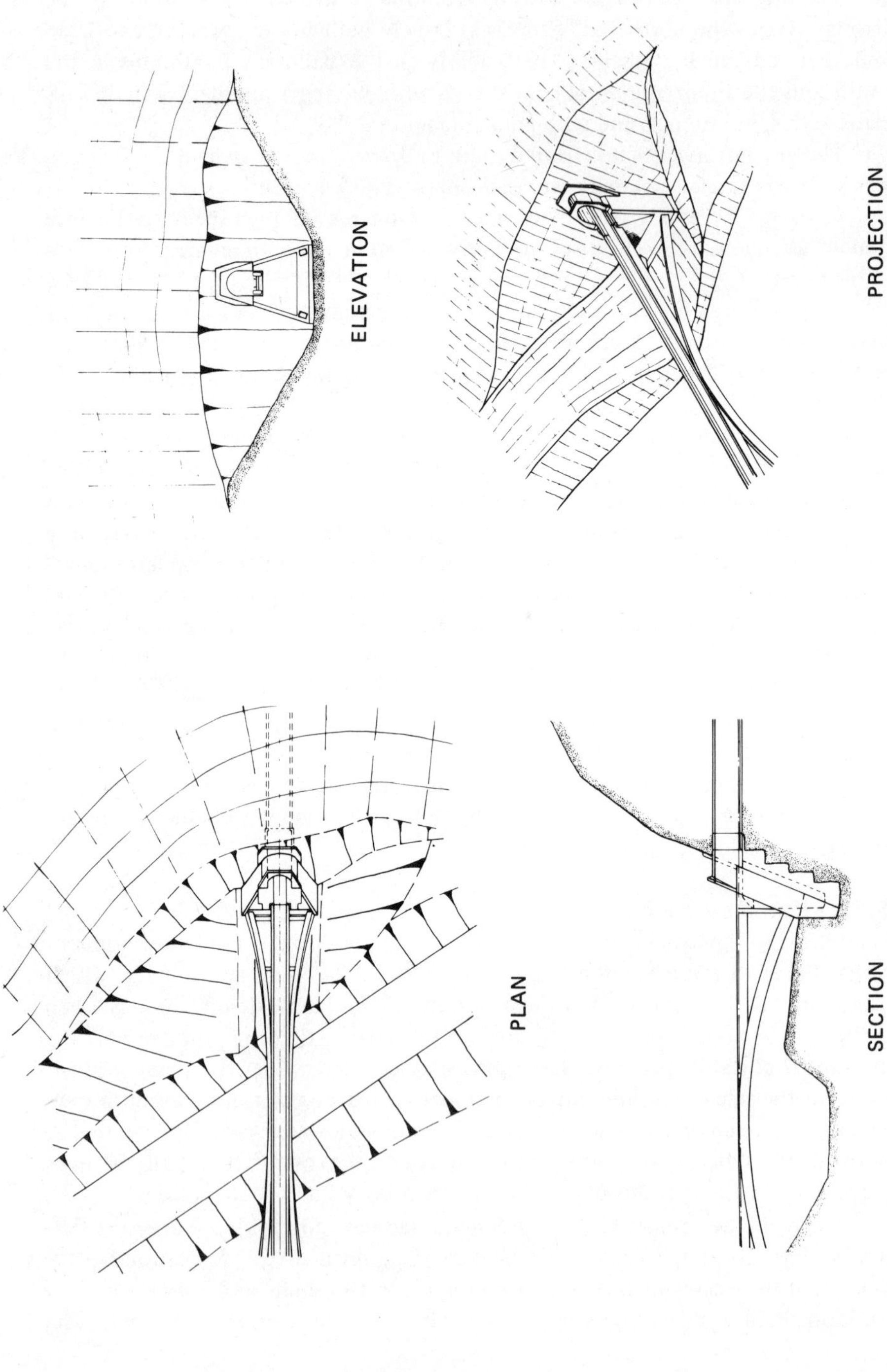

Fig. 8.4 – Portal and tunnel approached by bridge, proposed for Iran railways. Note transition problems at portal.

Fig. 8.5 – Avalanche protection on Austrian railway. (*Austrian Federal Railways*)

characteristics of the rock and the whole geology of the area together with its exposure to rock slides, avalanches, snow and other conditions are of vital importance. Another consideration in seismic areas is that high viaducts become extremely costly to safeguard, turning the economic balance in favour of the tunnel.

Siting and construction of tunnel portals also presents difficult problems in the angle of approach and the penetration of loose and weathered rock before reaching competent rock. There may be a need to protect the approach to the portal against rock falls or snow by an extension of the tunnel. The operating length of the tunnel will then exceed the bored length.

Such problems of siting are not peculiar to railway tunnels but apply also to highways, which are, however, much more flexible in that steeper gradients and sharper curves are acceptable.

8.2 ROUTE SELECTION

The promotion of a railway link through country where tunnelling may be required will first be determined in relation to its potential usefulness commercially and socially, and also sometimes in the context of strategy or even in terms of national prestige. The comparative merits of railway and highway must be weighed and the whole project studied in terms of its capital cost, operating cost, and benefits whether measured in terms of revenue or otherwise assessed.

Although the proposed railway may be seen primarily as a transport link between its termini, which may be junctions with other railways, it will almost certainly be required to serve also intermediate places. These requirements are constraints on the choice of route, as are the various topographic features, but they also provide opportunities for attracting profitable traffic and may attract sufficient additional traffic to justify expenditure on higher standards of construction and equipment.

Of first importance in selecting a route between points to be served is the adoption of the most direct alignment with minimum summit height, gradient not exceeding a selected maximum, and curves within a prescribed radius. Tunnels and viaducts may serve all these purposes by their independence of the surface topography. Most lines require both, but the proportions as part of the whole line vary greatly according to the ruggedness of the country, and the extent to which the line accommodates itself to the grain of the country or cuts across it. The whole economy of a line in construction and operation is dependent on making the best use of the ground, including the adoption of tunnels.

Where a tunnel is adopted its siting and alignment to the best advantage of the line as a whole may be a very complex problem. Apart from the choice between a shorter tunnel at a higher level and a longer tunnel lower down on a simple basis of construction costs compared with operational savings, the geology of the area may influence very strongly the preferred tunnel line and level. There may also be questions of intermediate access for construction or for ventilation or other operational reasons.

8.2.1 Topographical survey

The first requirement for planning a railway route and its tunnels is a topographical survey of as wide an area as possible. Adequate maps for initial needs may exist but progressively more detailed maps are required as planning and design proceed and the alignment is more precisely defined. The subject has been dealt with in general terms in Volume 1 but there are special aspects relevant to railway tunnels. At no stage can the scope of the survey be limited to the tunnelled length, but must be adequate to include long open approaches and potential areas of portal construction. In rugged mountainous country every detail of the alignment has to be studied as part of the whole. Viaduct, sidelong cutting, portal and tunnel interact and cannot be seen in isolation. Possible sites for vent shafts or access shafts may be important, also any watercourses or springs or standing water or drainage features. As for all tunnels, any buildings in the vicinity need to be located exactly.

For the preliminary stage of establishing technical feasibility and identifying possible alternative railway corridors, maps to a scale of 1:50,000 are suitable. Areas requiring more detailed mapping can be identified, and limitations imposed by gradients and curves can be examined.

For the next stage maps of the selected corridors at a scale of 1:10,000 allow feasible alignments to be confirmed, and unsuitable corridors to be eliminated, gradients and curves again being checked. Areas for more detailed mapping can again be determined, and the location and scope of major structures required can be identified.

Maps to a scale of 1:2000 allow the best alignment to be finally fixed and its geometry to be specified. The major structures can then be defined and quantities prepared for structures and earthworks.

For subaqueous tunnels similar procedures are necessary but there are additional aspects of prime importance which include

–accurate survey of bed levels
–determination of nature and stability of bed
–possibilities of deepening by scour or by dredging
–flood levels past, present and prospective
–tide and surge levels, normal and abnormal
–presence of waterbearing fissures and fault zones

8.2.2 Geological survey

The geological mapping of the route is just as important as the topography, in some cases more important. The same process of progressively more detailed study of narrower areas as for the topographic survey is desirable, but may not be practicable.

A general study of available information covering the possible routes, supplemented by a visit to the area, will help in establishing the preferred corridors but much will depend on the detail and accuracy with which general mapping has been done, and also on the rock exposures available for inspection or the extent to which they are masked by superficial deposits.

At the 1:10000 mapping stage, field reconnaissance of the alternative corridors is essential to assess the character and structure of the rocks and to prepare proposals for a detailed exploration of the preferred route with the aid of boreholes and geophysical survey.

As the 1:2000 mapping develops the need for further boreholes and tests, particularly at the sites of major structures will become apparent and should be organised, preferably as an extension of the main borehole programme.

Apart from the aspects of rock nature and structure essential to all tunnelling, as dealt with in Chapter 4, special railway problems arise, for example, at the tunnel invert, where squeezing clay may be a long term hazard unless an invert arch is provided, and where good drainage can be vital to a sound road bed.

Ground water can be detrimental to the electric transmission system unless proper precautions are taken, and also can reduce rail friction seriously, especially in conjunction with clay or if frozen, and thus reduce the haulage capacity of locomotives particularly on a gradient. In some climatic conditions dripping water can freeze into dangerous icicles and can coat the rails with ice.

The geological conditions at and near the portals are likely to require special study where there is danger of rock slides and of the presence of badly weathered rock.

Subaqueous tunnels demand even more detailed investigation of the possible effects of ground water both in the construction phase and in subsequent operation. Porosity of the strata and the presence of fissures can be critical to the feasibility of tunnelling methods and may influence the choice of lining to minimise seepage and consequent pumping installations.

8.3 GEOMETRY OF ALIGNMENT

The topography and geology of the area, together with commercial and industrial development and potential, constitute the external constraints and influences on the alignment. To the extent that they are known or assumed the geometry of the line can be fixed. The principal geometrical elements are:

(1) gradient
(2) curves
(3) tunnel cross section.

8.3.1 Gradient

The ruling gradient is of fundamental importance to the planning of a line and should be fixed provisionally as early as possible, even though it may later be found advantageous to modify it. It may be fixed to be the same as that on existing lines to which the new line will be linked and thus to suit the locomotives currently in use or available, but it is possible that overall economy will be achieved by the use of new and improved locomotives which can haul full train loads up a steeper gradient.

Gradients have commonly been described as: 1 in x, or as a percentage, but there is a growing practice of describing them as per thousand, using the symbol $^0/_{00}$. A gradient of 1 in 100 may thus be displayed as 1.0% or as $10^0/_{00}$. Unfortunately this last form, which has many merits, is not sufficiently distinctive, and is all too easily reproduced as per cent. The percentage is therefore normally used here. There is no well defined limit for gradient, but rail adhesion in relation to loco weight and power imposes practical limits. A railway as a whole is planned with a ruling gradient as a normal maximum although exceptions are possible where a critical length can be overcome by 'double heading' the heaviest trains. Nevertheless, gradients exceeding about 1% are not adopted without good reason, although much steeper gradients over 2% were used in the early Sarnia tunnel, and the Appenine Pracchia tunnel gradient is 2.5%. The 12½ km Cascade Tunnel has a gradient of 1.6% which causes difficult ventilation problems for diesel locos on the ascent; in order to ensure fresh air for the loco drivers a portcullis at the upper portal is shut and a fan is used to provide a flow of fresh air against the train until it is about to emerge. 1.8% has been referred to above for the Kanmon tunnel.

The factors ultimately controlling gradient are loco power and rail adhesion. The power required is proportional to the product of weight of train, speed and gradient with the addition of frictional resistance and wind resistance. This power must be provided in the locomotive and transmitted through its driving wheels to the rail, where the adhesion available is proportional to the effective weight carried on the driving wheels and to the coefficient of friction.

A tunnel may give rise to two particularly unfavourable factors, namely, low rail adhesion because of persistently damp conditions, and high air resistance to forward speed because of the restricted space. The consequence in such tunnels is that either loco power and weight must be provided in excess of what is needed on the open line, or tunnel gradients must be limited to, perhaps, two thirds of the ruling gradient elsewhere. Icing on rails has also to be considered, but that hazard is probably less in a tunnel than in the open, except for dripping water near the portal.

At the other end of the scale gradients must be adequate to ensure drainage, for which about 0.1% is a minimum. Except in subaqueous tunnels, where pumping of drainage is unavoidable, drainage by gravity should be ensured. Something steeper than the 0.1% minimum is to be preferred if closed drains are to be self cleansing.

Fig. 8.6 – Removal of ice from a tunnel in Canadian Rockies. (*Canadian Pacific Railway*)

8.3.2 Curves

Minimum curve radius on a main line railway is commonly fixed so that the cant at maximum design speed does not exceed some arbitrary limit. The amount of cant (C) for exact balance between the inward tilt and centripetal acceleration of a coach is given by the equation.

$$\frac{C}{G} = \frac{V^2}{Rg}$$

where G is rail gauge, V is train speed, R is radius of curve and g is gravitational acceleration. The minimum radius is likely to be some hundreds of metres, up to perhaps 4000 or more for high speed and the maximum cant may be as little as 50 mm but is unlikely to exceed 150 mm. The structure gauge may have to be rotated in the same degree.

This curve limitation has little direct influence on tunnelling. The small consequent increase in width of structure gauge makes necessary a corresponding increase in excavated width, and the setting out and survey are slightly more complex than for a straight tunnel. In difficult country, however, a sharp curve in the tunnel approach may be very helpful in permitting the tunnel to be aligned more favourably and its portals to be sited in better ground.

8.4 FORM OF TUNNEL

For the railway tunnel, there is a choice to be made between one double track tunnel and twin single track tunnels. There may, of course, be cases where the traffic foreseen does not justify more than a single track, possibly with provision for later duplication.

The choice will depend on tunnelling considerations and costs and also on operational requirements. Among the factors to be considered are:

(1) Construction
- (a) Total excavated volume
- (b) Length of tunnel
- (c) Stress interactions of twin tunnel
- (d) Separation of approaches and problems of alignment
- (e) Sequence of construction and problems of access, ventilation and drainage.

(2) Operation
- (a) Aerodynamics and ventilation
- (b) Safety – collisions and fires

The difference in width is that a double track tunnel requires a minimum width of about 8.5 m and a single track about 5 m after allowing for safe clearances. The minimum height over coaches is obviously the same but the difference in maximum height at the crown depends on the curvature of the roof arch and therefore on the characteristics of the rock and the lining adopted. With the normal horseshoe section the area of the double track tunnel is likely to be rather less than twice that of the single track tunnel, but, with a full circular section, which is more typical of subaqueous tunnels, the proportionate increase in height makes the area of the larger tunnel more than double. Cross sectional area is by no means the only measure of cost because excavation and support problems in tunnelling can increase disproportionately with size as discussed in Volume 1.

The length to be tunnelled is obviously very relevant to the decision. Twin tunnels require to be separated, and if the tunnelled length is short the extra cost of spreading the approach tracks is unlikely to be justified by any saving in tunnelling costs. The special circumstances in Metro tunnelling may be compared: the tunnels are long and separation of tracks may assist station design especially at interchanges.

Duplication of a single tunnel by a parallel tunnel involves tunnelling through ground in which the stress pattern has already been distorted and, in turn, causes a change in stress on the tunnel already built. The wider the separation, the less severe are the effects, and in this respect the widest practicable separation is obviously to be sought; one diameter of tunnel may be considered as a minimum in moderately good ground. It may be necessary for the tunnels to converge at the portal if the two tracks are to be accommodated on the same formation outside, but this can make more difficult the portal construction problems in what may be weathered and disturbed surface strata. In a long tunnel, of course, it is relatively easy to effect, in the tunnelling, as wide a separation as thought necessary as distance from the portal increases.

If twin tunnels cannot be made to converge at the portal, the tracks outside must be similarly separated, and, if in cutting, may necessitate separate deep cuttings eventually converging. If on a sidelong slope the levels of the tracks may have to be different. These geometrical problems of the alignment are of most significance in rugged country and for long tunnels.

In construction of the tunnel, there are benefits to be gained from separation into twin single track tunnels. One heading will almost invariably be driven ahead of the other, and the information on the behaviour of the ground should be of great benefit. If treatment of the ground, such as grouting or rock bolting, is necessary, continuity of specialist operations can be maintained by alternating between the two headings.

If the two are interconnected, ventilating air can be circulated up one tunnel and down the other, a particular advantage if fumes from explosives, or gases such as methane, have to be dispersed. Drainage also can be more easily maintained. In the event of falls of rock occurring behind the face access is much better.

8.4.1 Operation

Among the operational problems in long tunnels is the 'piston effect', the wave of compression ahead of a train, which is most significant in a single track tunnel in which the ratio of the cross sectional area of the train to that of the tunnel is high. This is of increasing importance when operating speed is high, both in its resistance to the train and in the possible unpleasant effects to passengers from abrupt changes of pressure. It is more fully discussed in Chapter 9.

Planning of ventilation is of even greater importance where fuel burning locos are in use rather than electric traction. In single track one way tunnels of moderate length, the draught produced by train movement will usually ensure reasonable conditions by effecting a progressive change of air unless there is a persistently unfavourable wind. In a double track tunnel, on the other hand, the train movements will cancel out and change of air will be dependent on wind effects or forced ventilation, in the absence of which there is likely to be a stagnant area near the middle of a long tunnel.

Operational safety may also influence the choice. Head on collisions are obviously not to be feared in a one way tunnel; derailments on a single track are limited to the train concerned whereas with double track, both lines may be blocked. Fire likewise appears a lesser hazard in twin tunnels. There is the further advantage in any emergency of being able to use the second tunnel for evacuation of passengers and for rescue equipment, provided that there are cross connections available. Most maintenance and repair operations can also be carried out more efficiently with sole possession of a single tunnel, when traffic can be diverted to its twin.

8.4.2 Cross section

The size of tunnel to be excavated is based on the structure gauge for the line, more fully discussed in the previous chapter. It is a profile, set out in relation to the rails, within which no part of any structure or fixed equipment may encroach and is derived from the loading gauge by making appropriate allowances for such things as sway arising from springing of coaches, working tolerances and wear in suspension, wheels and rails, and then adding appropriate clearances at various points of the cross section. Where a walkway is to be provided this in turn adds to the area of the structure gauge. On curves, as explained in Chapter 7, a wider structure gauge is specified, but in tunnels designed for high speed running the curves are of such large radius that this allowance will be small. On curves also where the rails are canted, the structure gauge must be similarly canted. Another concept sometimes found useful is the 'clearance diagram' for tunnel excavation and lining, leaving room within for such fixtures as cable brackets and signals etc. and for varying cant on the track.

Area to be excavated

The excavated area will extend outside the structure gauge. Allowances must be made for:

(1) Specified construction tolerances
(2) Space for cables, ducts and brackets

(3) Overhead line installations
(4) Tunnel lining
(5) Track support
(6) Overbreak
(7) Other requirements.

1. *Construction tolerances*

The tolerance applicable to the finished lining may be as much as 150 mm for a high speed line because very little adjustment in wriggle diagrams can be accepted, and correction of the lining is likely to be difficult and time consuming and may be deleterious to its strength and integrity.

2. *Cables, ducts and brackets*

Cables will usually be accommodated in ducts or on brackets on the walls of the tunnel, preferably on both sides of the track, separating power cables from signalling and control cables.

3. *Overhead line installations*

The space necessary for high voltage (25000 V) wires and their supporting structures determines the crown height of the tunnel. Accurate and secure fixing in this difficult position must be ensured and also as far as practicable freedom from moisture. Special waterproofing work may be necessary in this area, more especially if winter conditions are such that icicles can form.

4. *Tunnel lining*

The thickness of the specified lining must be added to the excavation profile. Greater thickness may be required where bad ground is encountered. Depending on the ground, lining may or may not be required in the invert.

5. *Track*

The choice of track bed, discussed below, will determine the allowance to be made in the invert.

6. *Overbreak*

In all but the most precisely trimmed machine excavated tunnels there is bound to be some overbreak. It may be substantial locally in fissured rock and must be taken into account in the design of the lining and in calculating quantities of spoil and backfilling.

7. *Other requirements*

It is sometimes considered worth while to allow outside the normal structure gauge a further space to accommodate scaffolding for remedial work to the arch.

In seismic zones, specially large clearances may be specified at known fault breaks to simplify the correction of misalignment resulting from earthquake shocks.

Aerodynamic effects in high speed running may, in exceptional cases, justify a larger tunnel area. Refuges, at regular intervals may be specified outside the general cross section. The cross section finally adopted will have a smooth profile enclosing all the relevant allowances.

It will almost invariably have an arched crown which may be continued as a horseshoe shape down to track bed level. Straight vertical sides up to arch springing may be adopted. The invert construction will depend on the expectation of ground loading in the long term. Where there is lateral pressure the side walls must be strutted apart at the base and where uplift may develop as in clay, an arched invert lining must be allowed for. The ultimate shape for all round pressure is the circle, extensively used in subaqueous tunnels. The use of full face tunnelling machines also may impose a circular shape.

8.5 TRACK

The merits of ballasted track with sleepers on main line railways are such that other forms of track are only used in special situations.

In very many tunnels ballasted track is used, but some disadvantages arise and alternatives must be considered. The depth required for ballasted track is about 300 mm compared with 200 mm for a concrete slab. The tendency of ballast to move slightly under the impact of train loads may result in enchroachment on the clearances and make necessary more frequent track adjustment and repacking. This is a more difficult and potentially dangerous operation in the limited space of a tunnel, particularly if mechanical equipment is used, even if specially designed to work within the structure gauge. Renewal of track is likely to be necessary at more frequent intervals of time in tunnels because of the generally humid and perhaps corrosive atmosphere, and after renewal the necessarily precise setting and re-adjustment of the track is troublesome for the bedding-in period.

For a short tunnel, it is not desirable to introduce the discontinuities of a altered type of track, but in a long tunnel a continuous concrete bed laid to precise dimensions with modern slip-form machines has many merits. The track is retained securely in its required alignment, and requires a minimum of maintenance work. The rails may be secured to the concrete by resin grouted bolts if workmanship of a high standard can be ensured, and may rest on resilient pads to cushion impact forces.

A transition 'ladder', between ballasted track outside the portal and concrete slab track inside, can with advantage be provided. Good track drainage is necessary in all cases, preferably gravity drainage provided by the gradient of the tunnel.

Fig. 8.7 – Solid concrete bed for track in 5 km Heitersberg rail tunnel: Zurich–Berne. (*Swiss Federal Railways*)

8.6 SUBAQUEOUS TUNNELS

On main line railways subaqueous tunnels are few in number. In Great Britain, out of over a thousand British Rail tunnels only the Severn and Mersey tunnels, both opened in 1886, are subaqueous.

Tunnelling is not favoured for a river crossing by a main line railway if a bridge provides a feasible alternative. The fundamental reason, apart from difficulties of tunnelling, is that, to pass under a river, the railway must descend far below ground level and ascend again. This makes severe demands on locomotive power especially when a heavy train is stopped and has to restart against the gradient. Also, capital and maintenance costs are high.

The necessary descent of the tunnel is compounded of:

(1) Initial height of railway above water level

(2) Depth of water
(3) Added depth for scour or future dredging
(4) Cover over tunnel
(5) Height of tunnel from rail level to crown
(6) Depth of invert construction below rail.

These are likely to add up to at least 20 m. If a gradient of 1% is adopted the lengths of approach descent and ascent will be 2000 m each under level ground. At least half of this is likely to be in tunnel, making the tunnelled length equal to the channel width plus 2000 m.

The profile makes it necessary for all drainage to be pumped and the volume is likely to be substantial. Forced ventilation may also be necessary; the Mersey tunnel was an early example, in the age of steam.

These disadvantages do not apply in anything like the same degree to a Metro system because the general rail level is already well below ground level so that any descent will be very much less, or entirely absent. Also, Metros do not have the gradient problems imposed by heavy goods trains.

It is in estuarine conditions, with a broad alluvial plain, that tunnels are advantageous because bridges become difficult and costly to build. Shipping requirements demand long spans and high clearances, and foundation problems may be severe. It is no coincidence that where railway tunnels have been built under the Hudson river and East river in New York and the Severn and Mersey in England, all are estuarine.

For the very long crossings of straits by railways, tunnelling becomes indispensible. The Channel tunnel between England and France is an outstanding example of the concept, under discussion for over a century as a practicable project, and initiated in earnest on at least two occasions. The Seikan tunnel, to connect the Japanese islands of Honshu and Hokkaido, by crossing under the strait of Tsugaru is a railway tunnel of even greater length and in more difficult ground: it has been under construction since 1964. Both these major projects are more fully described in Volume 1.

In Japan also, the islands of Honshu and Kyushu were linked by the Kanmon tunnel, built 1936–44 as two single track tunnels for 1.07 m gauge track, 3.6 km in length. This has been paralleled in 1970–74 by a new tunnel for the high speed railway, carrying two tracks. The tunnel is 18.7 km long, although the width of the strait is less than 1 km. The deepest point is 67 m below sea level. The undersea section of the tunnel is circular, 10.06 m diameter, with concrete lining 0.9 m thick. The tunnel descends steeply at a gradient of 1.8% to pass under the sea bed with over 20 m cover, and rises more slowly on the Kyushu side at about 0.7%, where it passes under mountainous country and under the built up land between mountain and sea.

Beneath the strait a major fault, including a 30 m wide belt of clay, was encountered, and was overcome by an interesting technique. The first operation was to drill ahead horizontally some 200 holes lined with 110 mm steel tubes at

200–300 mm centres, and to inject grout round the tubes and cement fill them. The tubes constituted buried hoods for the main tunnel and for a series of headings. Excavation and concreting for the walls and roof proceeded successively upwards in these headings from the bases of the walls, the invert being left to the last.

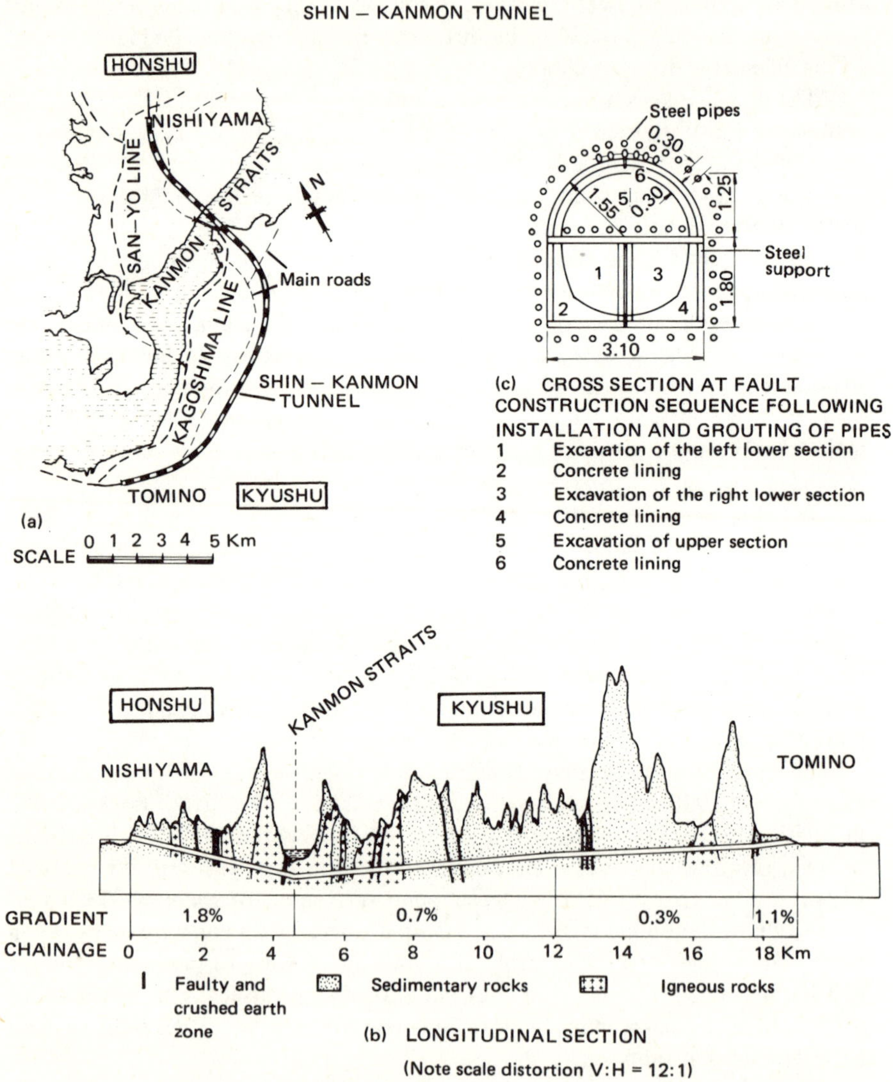

Fig. 8.8 – Shin-Kanmon Tunnel. (a) Plan, (b) Longitudinal section. (c) Special cross section for clay filled fault zone.

Subaqueous tunnels in general are aligned as nearly as practicable to cross the waterway by the shortest route, at the shallowest depth, but geology as well as geometry must be considered, and a longer, deeper route may offer more favourable tunnelling conditions, particularly in respect of favourable rock and inflow of water whether through permeable and jointed rock or at faults.

The questions of alignment, gradients, curves, cross section and lining have already been discussed. In subaqueous tunnels more emphasis is necessarily placed on waterproofing and drainage and ventilation problems.

In such long tunnels, with no intermediate shafts, yet with the necessity for the utmost economy in cross section, aerodynamic problems may be of major importance, particularly if high speed running is required.

8.7 TUNNEL LININGS

There is little about tunnel linings that is special to railways. In the great railway age of the nineteenth century brickwork was the normal lining in British practice. The 'English Method' was based on alternate excavation of a length and building of its brick lining. A merit of brickwork was its adaptability to varying conditions: an extra ring of brickwork could readily be built in the arch if found necessary.

Fig. 8.9 – Old brickwork undergoing repair using quick setting waterproof mortar in London tunnel. (*Ronacrete Ltd.*)

Adequate supply of bricks was a problem in long tunnels, sometimes solved by setting up a brickyard, utilising excavated clay from the tunnel itself or approach cuttings.

In the long term, deterioration of some brickwork linings has resulted from attack on the lime or cement mortar by percolating water, possibly containing sulphates. Acid attack has also come from the products of combustion from steam or diesel locos. Renewal of one, two or more rings of brickwork is possible.

Most other types of lining have been employed. Masonry, where suitable stone and skilled masons could be obtained, has been widely used in the past. *In situ* concrete has been, and is, used on an extensive scale. Sprayed concrete may be adopted either as an immediate lining to be followed by *in situ* concrete, or as a complete lining. In circular tunnels, cast iron segments or precast concrete segments are particularly applicable to subaqueous tunnelling. Steel arch ribs may provide immediate temporary support, but require to be subsequently embedded in *in situ* concrete or sprayed concrete.

Decorative aspects have little relevance to main line railway tunnels, but a smooth finish may be of considerable importance for the aerodynamic aspects of high speed railways.

The importance of waterproofing, particularly in the crown, has already been referred to.

8.8 CURRENT ACTIVITY

Although less in the public eye than highway tunnels there is very considerable world activity in construction and planning of railway tunnels. The Channel Tunnel and Seikan Tunnel have been dealt with in some detail in Volume 1. The Japanese rail system, as already noted, has been and is building numerous mountain tunnels and is stated to have over 3500 tunnels with a total length exceeding 1500 km.

Italy has built the 250 km 'Direttissima' high speed railway between Rome and Florence, incorporating four major tunnels of lengths 5.7, 7.4, 9.3 and 10.9 km and numerous shorter tunnels totalling a further 27 km. Techniques employed have included use of a tunnelling machine and extensive treatment with complex grouts. By way of contrast tunnels have been avoided in the new French line between Paris and Lyon, very deep cuttings being preferred to avoid the aerodynamic pulses occurring with high speed trains in tunnels: gradients up to 3.5% have been employed as being practicable in the absence of heavy goods traffic.

In New Zealand the difficult 8.8 km Kaimai Tunnel was opened in 1978, giving direct rail access to the port of Tauranga. Canada plans long new rail tunnels through the Rockies to cope with the growing volume of bulk haulage. Greece hopes to construct new railway routes through its mountainous terrain, and in Iran plans have been prepared for major works.

Fig. 8.10 – Kaimai Tunnel, New Zealand. (a) Opening September 1978. (*New Zealand Railways.*) (b) Rails in 160 m lengths being transported into tunnel. (*New Zealand Railways*)

In London proposals are under consideration for a deep tunnelled cross city link between south and north. New York's East 63rd Street crossing under the East River includes provision for main line tracks.

In Switzerland the 5 km Heitersberg tunnel has been completed on the Zurich–Berne line and plans have been made for a new St. Gotthard base tunnel which would have a length of 49 km.

Jugoslavia completed in 1976 a railway from Belgrade to the Adriatic coast, through very difficult mountain country, where a 250 km length required 254 tunnels up to 6 km in length and, to reach the 1032 m summit, a 3.2 km tunnel on a gradient of 2.5%.

BIBLIOGRAPHY

This bibliography, though brief, ranges widely in time and space, including mostly descriptions of specific tunnels of special interest.

Kossuth, F., The Mont Cenis tunnel, *Engng*, 1871, **11** pp. 347, 377, 421, 429; **12** pp. 37, 180, 193, 211, 241, 283, 367; 1872, **13** p. 391.

Fox, F., The Mersey Railway, *Min. Proc. Instn Civ. Engrs*, 1886, **86**.

Walker, T. A., *The Severn Tunnel*, Bentley, 1888 (reprinted Kingsmead, 1969).

Rogers, A. E., The location and construction of railway tunnels, *Min. Proc. Instn Civ. Engrs*, 1901, **146**.

Fox, F., The Simplon tunnel, *ibid*, 1907, **168**.

Jacobs, C. M., The Hudson River tunnels of the Hudson and Manhattan railroad, *ibid*, 1910, **181**.

Wilgus, W. J., The Detroit River tunnel, between Detroit, Michigan, and Windsor, Canada, *ibid*, 1911, **185**.

Henderson, B. H., The Transandine railway, *ibid*, 1913, **195**.

Dennis, A. C., Construction methods for Rogers Pass tunnel, *Trans. Amer. Soc. Civ. Engrs*, 1917, **81**.

Bayley, V., The Khyber railway;

Hearn, G., The survey and construction of the Khyber railway, *Min. Proc. Instn Civ. Engrs*, 1926, **222**.

Keays, R. H., Construction methods on the Moffat tunnel, *Trans. Amer. Soc. Civ. Engrs*, 1928, **92**.

Betts, C. A., Completion of Moffat tunnel of Colorado, *ibid*, 1931, **95**.

Kerr, D. J. *et al.*, The eight-mile Cascade tunnel, Great Northern railway: a symposium, *ibid*, 1932, **96**.

Robarts, H. E., Tunnel maintenance, *Instn Civ. Engrs Rly Div.*, 1947 (23).

Champion, C. L., The construction of the trans-Iranian railway, with particular reference to works of Lot 6 (South), *J. Instn Civ. Engrs*, 1947, **29** (Dec.).

Campion, F. E., Part reconstruction of Bo-Peep tunnel at St. Leonards-on-Sea, *ibid*, 1951, **36** (Mar.).

Scott, P. A. and Campbell, J. I., Woodhead new tunnel: construction of a three-mile main double-line railway tunnel, *Proc. Instn Civ. Engrs*, 1954, **3** (Sept.).

Terris, A. K. and Morgan, H. D., New tunnels near Potters Bar in the Eastern region of British Railways, *ibid*, 1961, **18** (Apr.); *see also* discussion, 1962, **21** (Apr.).

Saito, T., Construction of Shin-Kanmon Tunnel, *Civ. Engng in Japan*, 1974, **13**.

Sacrison, H., Flathead Railroad Tunnel, *Proc. Amer. Soc. Civ. Engrs, J. Constr. Divn*, 1971, **97** (Mar.).

Hermans, R. E. *et al.*, The Kaimai Tunnel (9 papers), *Trans. New Zealand Instn Engrs*, 1978, **5** (1).

9

Ventilation and Aerodynamics in Vehicle Tunnels

9.1 HISTORY

The need to provide ventilation for vehicle tunnels arises mainly from the atmospheric pollution caused by the propulsive units and the heat generated. It is only quite exceptionally, as in the proposed Channel tunnel, that the oxygen requirement of the passengers transported becomes significant.

In the 19th century, railway tunnels faced the problem of pollution by smoke from coal burning locomotives, not important in short tunnels on an open main line, but very serious in urban tunnels and long subaqueous or mountain tunnels. Electrification solved some of these problems and made possible the development of metro systems. Oil burning locos also reduced the smoke pollution but produced their own smoke and irritant fumes.

Ventilation of highway tunnels is a 20th century development, the requirement arising initially from the exhaust gases of petrol driven internal combustion engines, and later also from smoke and fumes produced from heavy diesel engined vehicles.

Pollution levels may be controlled by limiting vehicle access to the tunnel and by improvements in engine design and fuels, but tunnel design must still ensure the supply of adequate fresh air to dilute the pollutants to acceptable levels, and the extraction of the polluted air. In addition to this aspect of toxicity, the progressive accumulation of heat in any tunnel network must be examined, but more particularly in metro tunnels, where direct pollution is minimal. Another need in metros is to avoid excessive draughts in stations.

Entirely different phenomena have more recently become significant in railway tunnels, namely the aerodynamic effects of the passage of high speed trains. Such trains interact with the tunnel structures to generate air pressure pulses causing discomfort to passengers. In addition the train, within the constricted area of the tunnel, is subject to aerodynamic drag, which may reduce speed seriously in a long tunnel, or make necessary much higher loco power and give rise to consequent heat problems.

There are thus three principal aspects to consider: pollution in highway tunnels, heating effects and draught relief in metros, aerodynamic effects in high speed rail tunnels.

The need to safeguard users of long highway tunnels became apparent in design of the Holland Tunnel (1927) in New York under the Hudson river. The essence of the solution was the dilution and extraction of the toxic exhaust gases. For many years the proportion of CO in the atmosphere to which vehicle drivers were exposed was seen as the fundamental criterion of any ventilation system, and it remains in most cases the primary consideration, but the importance of other aspects has become apparent. The proportion of diesel engined traffic has since grown to very significant levels. Their emission of CO and other toxic gases is low compared with petrol engines, but they create smoke to a degree that interferes with visibility, and also gases which, although non-toxic, can be unpleasant unless adequately diluted.

From the first the problems of fire control have also been of major importance in design of ventilation systems. Another aspect, which normally requires no separate provision, is the heat energy given out by vehicles (1) in the exhaust gases, (2) from radiators and (3) from brakes. Fresh air requirements for dilution of noxious fumes are usually more than adequate to carry away the surplus heat without the temperature of the tunnel structures building up to objectionable levels.

The earlier tunnels, were usually designed to accept two-way traffic either normally or occasionally, but probably a majority of modern highway tunnels carry one-way traffic, except temporarily in emergency or for reasons of maintenance. There is therefore a traffic-induced flow of air, which may in some cases be wholly adequate for ventilation requirements, and, at least, contributes very substantially. It is in some degree self-regulating, in that more vehicles and higher speeds induce greater flow of air. Conditions with congested traffic moving slowly, or total blockages, remain to be provided for by fan installations.

The principal patterns of ventilation are longitudinal, semi-transverse and fully transverse. These are discussed in more detail below. Design has progressed from the early 'static' concept of dilution and removal of CO, by transverse air flow across the traffic space at each point where it is generated along the tunnel, to more complex concepts making use of the longitudinal flow and turbulent mixing of the air and exhaust gases, resulting from aerodynamic drag of the vehicles.

9.2 HIGHWAY TUNNEL REQUIREMENTS

Artificial ventilation for highway tunnels other than very short tunnels is normally essential, even if only for use at times of traffic congestion. While traffic in a tunnel is one-way and free flowing the traffic-induced air flow is found to be adequate up to a length of 1–2 km, and possibly longer, but if traffic may be stationary or crawling forced ventilation is likely to be essential. 'Natural' ventilation is thus only adequate for one-way tunnels not subject to congestion up to about 0.8 km and for two-way tunnels of half that length.

For the occasional use required in relatively short tunnels jet fans meet most requirements. Although not efficient in terms of energy consumption they are economically advantageous, because they do not require major structures to house them, nor tunnelled ducts for supply and exhaust air.

For longer tunnels major fan installations are necessary, and may present difficult design problems involving heavy costs.

The system to be adopted can only be selected after study of every aspect: no single factor can be optimised in isolation. Safety, amenity, aerodynamic efficiency, tunnel construction, maintenance problems, capital costs, operating costs, system of operation and traffic control and pattern and composition of traffic must all be examined in the attempt to achieve a balanced effective and economical system, contributing to the main objective of providing an unobstructed highway.

The fundamental requirements are that the ventilation system should:

(1) dilute and remove toxic gases, in particular carbon monoxide,
(2) dilute and remove smoke, generated largely by diesel engines,
(3) remove the heat energy given out by vehicles,
(4) be adaptable to the needs of fire fighting.

The evaluation of air requirements is a subject of much difficulty in detail because individual vehicles vary greatly in their output of pollutants, and the composition and density of traffic flow fluctuates over a wide range from a nearly empty road to a state of acute congestion.

In a long tunnel the air mass moved by fans is very large and there is a considerable time lag in response to any changes in fan speed. The supply of air cannot be varied with the minute to minute fluctuations of traffic flow. Automatic, or manual, regulation cannot be efficient if it simply follows an indicated level of pollution. Some degree of prediction is almost essential, but can only be based on experience of the particular tunnel.

After examination in detail of all the varying factors the ultimate installation must provide safety in the worst conditions, together with economy in normal operation, having due regard also to amenity.

Standards have to be specified for acceptable levels of pollution in terms of toxicity and visibility and for maximum acceptable air velocity. Energy consumption costs, capital costs and maintenance and supervision costs have then to be assessed in comparing alternative systems.

The most important factors in the assessment of needs are:

(1) length between shafts, and number and direction of lanes,
(2) traffic flow patterns and composition, and traffic control,
(3) environmental impact.

The normal pattern and speed of traffic and the appropriate levels of amenity will be influenced substantially by the location and consequent function of the highway tunnel as, for example: subaqueous, mountain, motorway or urban.

9.3 POLLUTANTS

The common pollutants arising in tunnel conditions from engine exhausts are: carbon monoxide, nitrogen oxides, sulphur dioxide, organic compounds (principally unburnt hydrocarbons and aldehydes), carbon particles (soot), lead compounds.

Published figures for exhaust gas composition vary over a very wide range. The figures below are typical for cruising conditions but higher values are experienced, in acceleration for some pollutants, and in deceleration for others.

Toxic and irritating exhaust gases in parts per million (ppm)

	CO	NO + NO_2	Hydrocarbons	Aldehydes
Petrol Engines	27,000	650	1,000	10
Diesel Engines	trace	240	100	10

The physiological effects of these substances are briefly described below.

9.3.1 Carbon monoxide

The major pollutant from petrol engine exhausts is carbon monoxide whose deadly character at quite low concentrations is well known. It usually constitutes about 3% of the exhaust gas volume and is very rapidly dangerous at a level of 0.1% (1,000 ppm). Dilution of the exhaust fumes by a factor of about 200 is essential, that is to say from 3% to 150 ppm.

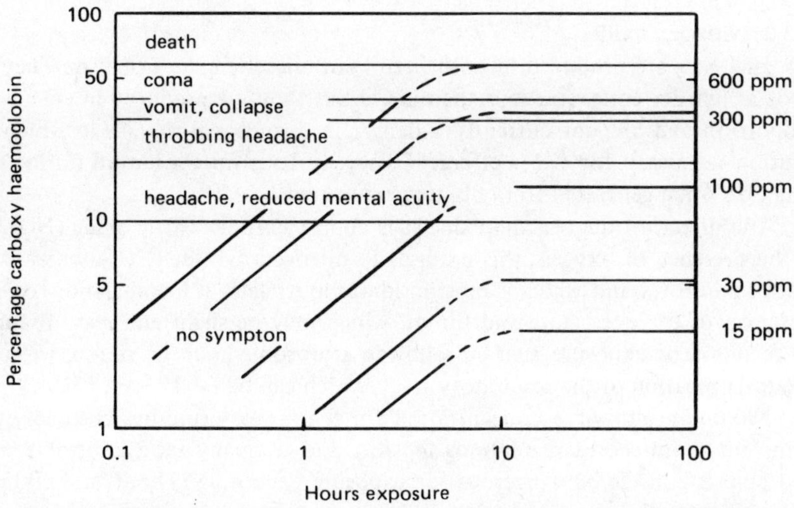

Fig. 9.1 – Carbon monoxide – physiological response of people with normal health, at light activity, to various exposures. (*Pursall and King, Aerodynamics and Ventilation of Vehicle Tunnels, publisher B.H.R.A.*)

CO acts by combining with haemoglobin in the bloodstream, thus depriving the system of oxygen. The effect of exposure can be measured in terms of the percentage of carboxy-haemoglobin in the blood. Below 10% there are normally no obvious effects, but above this level headache, fatigue and dizziness may appear and with rising concentration symptoms will become progressively more severe, with possible collapse at 30%–40%, followed by coma at higher levels and possibly death above 60%–70%

The time taken to reach any level depends on:

(1) initial conditions for the individual (3% would not be unusual)
(2) CO concentration
(3) time of exposure
(4) activity, for which a factor varying from 3 for rest to 11 for heavy work is applicable,
(5) height above sea level.

At sea level a passenger through a tunnel with a concentration as high as 250 ppm requires about two hours exposure to reach 10%, but for someone doing heavy work the time is only about 20 minutes.

At a height of 1500 m these times are reduced to about 1 hour and 10 minutes respectively while at a height of 3000 m the altitude factor alone results in a deficiency exceeding 10%. Special studies have therefore been necessary for such tunnels as the Eisenhower Memorial tunnel through the continental divide in Colorado at a height of 3350 m (11200 ft). At this height air pressure, and therefore density, is only about two thirds of that at sea level. An hourly average for CO not exceeding 100 ppm under adverse traffic conditions was specified for the ventilation system.

9.3.2 Nitrogen oxides

NO and NO_2 are produced in both petrol and diesel engine exhausts. They are toxic at smaller concentrations than is CO but they are produced in yet smaller proportion and are not currently a danger in operational tunnels in which the dilution necessary for CO is effected. They do constitute a hazard during construction when generated from blasting explosives.

The direct product in the exhaust is almost entirely Nitric oxide (NO) but, in the presence of oxygen, this oxidises to nitrogen dioxide (NO_2), which is the more toxic form and which is most insidious in its action. Its inhalation, causing irritation of the eyes, nose and throat, which may be slight and may disappear on cessation of exposure, can be followed after some hours by serious irritation and inflammation of the respiratory tract, which can be fatal.

Maximum allowable concentration for 8-hour exposure has been set at 25 ppm, but more recent regulations in USA and Germany set a limit of 5 ppm. 100 ppm is considered dangerous for exposure exceeding ½ hour, and 200 ppm is 'rapidly lethal'.

One important and dangerous difference from CO is that NO_2 symptoms cease to be reversible beyond some threshold, whereas CO symptoms at almost any stage are reversible by supply of clean air.

Nitric oxide (NO) appears to act in the same way as CO by combination with the haemoglobin in the blood, and although less toxic than NO_2, it is much more toxic that is CO.

9.3.3 Sulphur dioxide

SO_2 is the principal oxide of sulphur experienced. It is a bronchial and nasal irritant with a characteristic taste. It is unpleasant in the range 5 to 10 ppm. The production from motor vehicles is a very small proportion of the total from other sources and it does not normally present any tunnel ventilation problem.

9.3.4 Organic compounds

The production of these is extremely variable. In combination with oxygen and oxides of nitrogen unburned hydrocarbons can give rise to 'photochemical smog' in the presence of ultraviolet light. They are the cause of eye irritation.

Unburned hydrocarbons are mainly attributable to badly tuned engines.

Aldehydes, alcohols and organic acids have been identified in highway tunnels. Although unpleasant odours are observed no health risks have been found. Careful study of the health records of employees in tunnels of the New York Port Authority have shown no special identifiable risk over long periods.

9.3.5 Particulate matter

Carbon, mainly from diesel engines, but also from over-rich mixtures in petrol engines, is the principal solid particle which may act as a nucleus for droplet formation and consequent haze. Very many other solids have been recorded.

9.3.6 Lead

The use of lead to improve octane ratings constitutes a continuous source of compounds with oxygen, chlorine and sulphur. About 40% of the lead in the fuel becomes airborne. Lead is a cumulative poison, slow in action. It is not considered to be at a dangerous level in tunnels but reduction of permissible levels for lead additives in petrol is being effected in most countries, more particularly in Germany. The possibility that air exhausted from road tunnels may cause undesirable concentration of lead in areas near the discharge shaft has recently emerged as a potential hazard to be studied and avoided. At present it seems likely that the problem is limited to areas where lead is already present in quantities which make further accumulation unacceptable.

9.3.7 Acceptable standards for toxic substances

There are no statutory standards for pollution directly applicable to tunnel atmospheres. The responsible authority promoting any tunnel project will be guided by reports and studies on a world-wide basis.

Threshold Limit Values (TLV) recommended for various airborne noxious substances are published by H. M. Factory Inspectorate, based on a list adopted by the American Conference of Government Industrial Hygienists in 1968. They are more particularly related to exposure in factory conditions throughout an 8-hour day and a 5-day week and are not directly applicable to tunnel conditions, but do offer helpful guidance. Higher values are in most cases acceptable for short term and intermittent exposure.

The relative significance of the various pollutants actually experienced in road tunnels may be put in perspective by a comparison of these TLV figures with maximum figures measured in a series of Drager Tube samples in a number of UK tunnels.

Substance		TLV	Max. measured
Carbon Monoxide	CO	50 ppm	110 ppm
Nitric Oxide	NO	25 ppm	5 ppm
Nitrogen Dioxide	NO_2	5 ppm	0.25 ppm
Sulphur Dioxide	SO_2	5 ppm	0.5 ppm
Hydrocarbons		111 mg/m^3	10 mg/m^3
Carbon Dioxide	CO_2	0.5%	0.05%
Lead	Pb	0.15 mg/m^3	0.032 mg/m^3

It is clear that in practice CO is fundamental in determining the necessary degree of dilution as none of the other pollutants are observed at more than 20% of TLV. Oxides of nitrogen, however, call for a special note of caution for the reasons already discussed and in the light of suggested lower limits proposed in recent years for these substances. Questions of smoke and visibility are separately discussed.

9.3.8 Limits for carbon monoxide

Acceptable standards for CO require to be examined in more detail. The fact that the TLV of 50 ppm is substantially exceeded in normal practice is entirely proper because the exposure of drivers and passengers is for a very short time only.

In the first of the modern highway tunnels, the Holland Tunnel in New York (1927), the design maximum for CO was 400 ppm, with the assumption that the average would be much lower. Even this value is unlikely to have serious consequences for brief exposures, but an emergency maximum of 250 ppm or 200 ppm, with a working level not exceeding 100 ppm or 150 ppm is more usual now. It is of interest to note that the mean level in the Holland Tunnel has in fact remained at about 70 ppm over the half century of its use.

The Permanent International Association of Road Congresses (PIARC) through its Technical Committee Report on Road Tunnels presented at the 16th Congress in Vienna in 1979 has recommended that ventilation should be so designed that the average CO level at any time in normal conditions should not exceed 150 ppm, with an absolute limit of 250 ppm for accidental and short peaks anywhere. With longitudinal ventilation, in which there is a maximum of twice the average at the point of discharge, 250 ppm maximum is considered acceptable.

9.4 AIR SUPPLY

Whatever limit is adopted there will inevitably be great short term variation in the levels of CO actually experienced, because air supply cannot be regulated to follow immediately and closely fluctuations of traffic in density and composition.

9.4.1 Dilution of CO

The basic formula for air requirements for a single lane in terms of dilution of CO is:

$$Q = \frac{fq}{60S} \times \frac{10^6}{\delta} \text{ m}^3/\text{s.km}$$

where f = coefficient for gradient and altitude
q = unit volume (l/m) of CO emitted by a typical vehicle
S = average spacing of vehicles (m)
δ = permitted CO content of air in parts per million

Diesel engined vehicles are intially equated to cars although their percentage of CO in the exhaust is much lower, offset in part by the higher volume of exhaust from larger engines. Smoke output needs to be examined separately if the proportion of diesel vehicles is other than small.

The value for δ in this formula has already been discussed: f, q and S require examination.

9.4.1.1 Gradients and altitude

The appropriate corrective coefficients for gradient are in some degree dependent on the composition of traffic. For a mix of 90% cars and 10% heavy diesels the recommended coefficient varies from 0.7 for a 6% descending gradient to 1.3 for a 6% ascent. If the tunnel is two-way these tend to counterbalance and the modification is not of great significance, but in a one-way tunnel they must be taken into account particularly on a continuous rising gradient.

They may also be important in a subaqueous tunnel where the profile includes a descent, a mid-river level length and an ascent, the gradients being typically of the order of 4% to minimise the tunnelled length. The rising gradient may slow down heavy vehicles disproportionately and some congestion may result, contributing further to pollution. Local conditions over a short length of tunnel are more immediately relevant where the ventilation system is fully transverse, because any longitudinal element in the air flow tends to average out pollution. On the other hand, a heavy vehicle travelling at the same speed as the longitudinal air flow will tend to carry with it its own exhaust output.

At high altitudes two direct effects must be taken into account: increased CO generation because of reduced engine efficiency, and increased volume of air necessary because of reduction in density. These are combined to give a height correction coefficient ranging from 1.0 at sea level to 2.9 at 2000 m. There is also a physiological effect because of the oxygen deficiency.

9.4.1.2 CO Emission

In determining the air supply necessary to reduce to acceptable dilution carbon monoxide, or any other pollutant, it is essential to ascertain the characteristics of individual vehicles and to establish a typical basic emission. The output of CO from the average vehicle can only be an approximate estimate. It obviously depends on the engine capacity and on its rate of consumption of fuel, but also depends on the design and tuning of the engine and on its speed, and whether accelerating, decelerating or running at a steady speed. Outputs up to 40 l/min have been quoted for level roads, averaged over various speeds and cycles of operation, and up to 80 l/min for ascending gradients. Even higher figures may apply for acceleration, but these are not relevant to sustained speeds in a tunnel. A French survey in 1967 gave observed average rates varying from 15 to 28 l/min for tunnels in Holland, France and Switzerland.

The PIARC Technical Committee in 1979 published a new basic emission curve from a typical vehicle. The output rises from 8 l/min when idling to 16 l/min at 20 km/hr, remaining constant at high speeds.

The former curve (of 1975) continued to rise to a maximum of 22 l/min at 80 km/hr, but a more significant change has been the elimination of a peak of 22 l/min at 10 km/hr for very congested conditions, in which vehicles are moving intermittently and accelerating in low gear. These changes are attributed to improvements in engine design and control of emission, and also to new studies of traffic in stop/start conditions.

In using any such curve for air requirements it must be combined with an estimate of vehicle spacing

9.4.1.3 Traffic density

The concept most familiar to traffic engineers is that of vehicles per hour per lane, but this must be translated into vehicle spacing, or traffic density, for use in calculations of tunnel pollution. The quantities are related by the equations

$$S = \frac{10^3}{n} = \frac{V \times 10^3}{N}$$

where S = spacing of vehicles centre to centre (m)
n = vehicles per km = traffic density
V = average speed (km/hr)
N = traffic flow in vehicles per hour

What is found in practice is that vehicles space themselves out in accordance with the speed of the traffic flow, more particularly in tunnels in which overtaking is not permitted. When a line of traffic is stationary the minimum spacing is about 5 m for cars. In motion this opens out immediately to about 10 m or more, and increases progressively with speed, provided that a condition of free flow is established. If there is further obstruction, the spacing again decreases towards complete congestion; under these conditions there is no stable relationship between speed, spacing and pollutant output. Speed and spacing vary erratically in stop-start motion, and pollutant output fluctuates.

If traffic is flowing freely a rate of 2000 vehicle/hr or more for a single lane may be achieved, but maximum normal sustained flow is about 1750 vehicle/hr anywhere in the speed range 25 to 55 km/hr. This corresponds to spacing from 14 to 31 m. Above 55 km/hr the capacity tends to become less because of greatly increased spacing.

These generalisations need to be examined more closely for the traffic in any particular tunnel. In mountain or motorway tunnels congestion is unlikely except as an occasional emergency, but in urban tunnels unobstructed conditions of exit may be of fundamental importance if congestion is to be avoided. On the other hand if entry to the tunnel can be restricted by traffic control signals the congestions of peak hour traffic may be kept outside in free air.

In subaqueous tunnels a characteristic feature tending to cause congestion is that a descent and a level length are followed by a rising gradient which tends to reduce abruptly the speed of vehicles, particularly those heavily loaded and requiring the use of a low gear. A form of 'shock wave' in the traffic flow has been observed, but studies in New York tunnels demonstrated the possibility of improving traffic flow in those conditions by controlled admission, limited to the current capacity of the lane.

9.4.1.4 Air requirements

In respect of CO there are thus three speed ranges to examine in providing ventilation capacity: namely, stationary traffic, congested flow up to about 20 km/hr and free flow from 25 km/hr upwards.

First, when traffic is at a standstill, whether from congestion ahead or from an accident or emergency the spacing may be reduced to 5 m. With engines idling the suggested CO output is 8 l/min per vehicle. The air requirement for one lane with a limit of 150 ppm is then 178 m^3/s km but with the limit increased to 250 ppm this becomes 107 m^3/s. To the extent that emergency instructions to switch off engines can be enforced, pollution can be reduced.

The second range is that of congested traffic moving intermittently. If such a condition is allowed to continue the worst case might be supposed to be progress at 10 km/hr with a spacing of 10 m, which implies a traffic flow of 1000 vehicles/hr. The suggested CO output is about 14 l/min which gives an air requirement of 156 m^3/s.km if 150 ppm is to be achieved or 94 m^3/s.km if 250 ppm is acceptable as an occasional short term condition. These figures are rather uncertain in that the average spacing over 1 km is unlikely to be as close as 10 m in the stop-start motion but the output of 14 l/min may be exceeded locally.

The third range for steady free flow at 25 km/hr upwards requires the maximum capacity at any speed to be combined with the output of 16 l/min per vehicle. Taking this capacity as 1750 vehicle/hr the spacings at 25 and 55 km/hr are respectively 14 and 31 m and the air requirements are 127 and 57 m^3/s km for a pollution level of 150 ppm.

It is obvious that the air requirements decrease progressively and substantially as average speed increases. It must be noted that they are calculated for ordinary maximum capacity at the speed selected but that traffic density may often be much lower for long periods, and air requirements will be lower in proportion.

A supply of about 150 m^3/s km is thus likely to be adequate as a basis for a normal tunnel, provided that it is available quickly if congestion develops.

9.4.2 Diesel engined vehicles

The calculation of requirements arising from diesel engined vehicles is less direct. Carbon monoxide and other gaseous pollution from this source is adequately diluted without separate assessment, but smoke, principally comprised of solid particles of finely divided carbon, obscures visibility in the tunnel atmosphere, and its dilution to acceptable levels must be ensured. Additional provision is not likely to be necessary unless the proportion of heavy diesel vehicles is over about 15%, or on rising gradients where maximum power can only be achieved at the expense of smoke emission particularly in the case of underpowered and overloaded vehicles in poor mechanical condition.

9.4.2.1 Visibility limits

The fundamental principle is that of ensuring visibility adequate for safe driving by specifying an appropriate coefficient of absorption defining the rate of loss of light. The standard of lighting must be considered and the safe stopping distance for vehicles, which depends on speed and gradient and also on vehicle design and maintenance.

Direct measurement of loss of light along a set length of the tunnel can be used to assess visibility. In some countries control is based on the mass of airborne soot particles, but this is less direct and is complicated by the presence of particles of road dust, which may contribute a substantial proportion of the suspended mass but, being less finely divided, cause little obscuration.

The PIARC Technical Committee (1979) has published a formula closely analogous to that for carbon monoxide assessment and a series of curves relating to diesel smoke. The basic curve relates obscuration effect of exhaust smoke to vehicle speed the value shown being then multiplied by vehicle weight and by appropriate factors for gradient and altitude, and by the number of such vehicles per kilometre in order to arrive at the air supply required.

It is interesting to note the suggested dilution of exhaust fumes is of the order of 200:1, very similar to the figure for 3% CO in car exhausts, but the volume per vehicle is higher, while the number of vehicles per km is lower.

9.4.3 Total air requirements

The design must provide enough air for safety, not necessarily for comfort, in the worst conditions forseeable. This may be the volume of air necessary to keep CO pollution in an emergency traffic jam below some such level as 250 ppm. The fan capacity and power for this should be quickly and reliably available with standby provision to cover breakdowns and maintenance operations. It may be noted that CO pollution and obscuration by smoke are independent phenomena and their air requirements are not additive: the same air can do both jobs.

The ordinary hour-to-hour air requirements will be substantially less than the emergency peak, and economy in operation will require a system of control and regulation of fan use, possibly combined with traffic control.

Short tunnels in which one-way traffic flows freely are largely self-ventilating, but in urban conditions may be subject to peak hour congestion, requiring ventilating fans.

A long tunnel exceeding 1 or 2 km is unlikely to be self ventilating unless conditions are unusually favourable and must therefore be provided with fans for continuous operation. Substantial economies can be effected by regulating the fan output from zero or a minimum at night to a normal maximum at peak traffic periods and possibly a further boost in emergencies. This is more fully discussed below.

It is not unlikely that CO emissions will in the future be appreciably reduced by the influence of regulations for vehicle manufacture, but requirements for fresh air may not be correspondingly diminished, because higher standards for air purity may be applied and also because of the visibility standards. Even if perfect combustion, or electric propulsion, became practicable any great reductions in air supply would give rise to problems of the build up of heat in the tunnel.

Intake air temperature can be significant, especially a high average over a long perid. There is considerable transfer of heat between the air and the tunnel walls resulting in fairly steady temperatures at mid-tunnel. Close to the intakes, however, it may be necessary to safeguard water pipes against freezing, and in warm humid conditions condensation of moisture from the air may be objectionable. In very hot climatic conditions the tunnel temperature could become unacceptably high as the energy from fuel consumed in vehicles is all ultimately transferred as heat into the tunnel air. The heating problem is more fully discussed below in the context of metro tunnels. It has been estimated that heat balance alone might require air supply at about half the current rates determined by pollution.

Another point to be noted in determining the volume of air required is its purity. It should be drawn from a source as free as practicable from pollution. In city streets CO can exceptionally be as high as 50 ppm: twice as great a volume of such air would be required to achieve dilution to 100 ppm within the tunnel. Factory chimneys in the vicinity may also contribute dust, smoke, fumes or other pollutants particularly when the wind is in certain directions.

In practice the normal provision for major urban tunnels, and others liable to congestion, lies in the range 100 to 200 m^3/s km per lane and usually in the upper half of that range. In some, particularly where a fully transverse system is used, the supply may be varied in accordance with the gradient. Examples are: Mersey Kingsway 155 Elbtunnel (Hamburg) 165–185, Coentunnel (Antwerp) 175, Lincoln Tunnel (NY) 114–248, Vieux Port (Marseilles) 125–250. It is found in practice that the full supply is not required except for short peak periods, and that substantially lower levels of supply are adequate for much of the time.

In contrast, the long Alpine and other mountain tunnels are generally designed for traffic much below road saturation capacity. The Mont Blanc Tunnel, 11600 m long at a height of approximately 1300 m, was originally designed for a supply of only 26 m^3/s km per lane on an assumed limited traffic flow of 400 vehicles/hr. In 1970 the average flow was 2067 veh/day but with a maximum of 7652 per day and 762 in one hour. Because of increasing traffic the fresh air supply is accordingly being increased by 50%. Other Swiss mountain tunnels have been designed within the range of 47 m^3/s km per lane for a 2.5% descending gradient to 154 m^3/s km per lane for an ascending gradient of 2.6%. The Eisenhower Tunnel at a height of 3350 m (see above, p. 240) has a provision of 280 m^3/s km per lane to meet the needs of heavy traffic flows in the summer holiday period. In Austria the more recent motorway tunnels such as the Arlberg allow 135–190 m^3/s km per lane.

9.5 SYSTEMS OF AIR DISTRIBUTION

The conventional classification of systems as longitudinal, transverse, and semi-transverse is an over simplification.

The simplest form is purely longitudinal ventilation, in which the air enters at one portal and is discharged at the other. The movement may be maintained by wind, and more rarely by pressure differences, or by the drag of vehicles in a one-way tunnel, or by a fan or jet system.

Purely transverse ventilation is independent of any longitudinal flow and depends on flow across the tunnel from openings in a supply duct to openings in an extract duct, both ducts following the whole length of the tunnel and fan operated, but in one way tunnels traffic flow contributes a longitudinal element.

Semi-transverse ventilation typically comprises a fan operated supply duct as for the transverse system, but no separate exhaust duct, the whole traffic space functioning for that purpose and discharging either at a portal or through an extract shaft.

In any system of forced ventilation the energy considerations are significant, and for long tunnels can be of major importance. To supply a tunnel of length, L, through a duct of length, l, and cross-sectional area, A, one of the biggest elements of the power requirement is the duct loss, which is approximately proportional to:

$$fLl^3/A^{2.5} \quad f \text{ being a friction factor.}$$

It is obvious from this formula that for maximum economy duct lengths must be short and areas large. Tunnel length, L, cannot be altered except by dividing into sections; duct length, l, must be kept to a minimum; duct area, A is usually severely limited by tunnelling considerations.

Energy advantages for the longitudinal and semi-transverse systems arise from the use of the whole traffic space as a principal air duct.

9.5.1 Longitudinal ventilation

Very short tunnels, particularly in mountain areas where wind is usual and in which traffic is uncongested, are normally self-ventilating by longitudinal flow. An extreme example, which cannot be called short, is the Tateyama Tunnel in Japan, which is a single lane tunnel of which the main section is 2100 m in length, passing under a ridge at a height of 2400 m. Natural ventilation was found to be adequate, but the designed capacity is only 24 cars per hour.

With one-way traffic in a tunnel the vehicles can induce air flow with a velocity of the order of one third of the traffic speed. The adequacy of the air supply then depends on maintaining traffic speed and avoiding congestion, which is doubly detrimental in that air flow is reduced at the same time as traffic density is increased. Unless traffic can be very precisely regulated as to numbers and speed, any such tunnel will require at least standby mechanical ventilation for use in times of congestion.

The principal advantages of longitudinal ventilation are simplicity and minimum energy demand, both reflected in low capital and operating cost. The disadvantages appear progressively with length of tunnel and volume of traffic, as more air can only be provided by increased velocity of flow.

Up to 10 m/s in the direction of traffic flow can be accepted with one-way working, but with two-way traffic a normal limit might be about 5 m/s (18 km/hr) or exceptionally 7 m/s. If a maximum air supply is specified at the rate of 150 m^3/s.km and the traffic lane dimensions including clearances are 4 m wide × 5.5 m high the velocity and length are proportional. An air speed of

7 m/s limits the length to 1000 m. Reduced supply at 100 m^3/s.km would allow the length to be 1500 m.

Higher velocities have been adopted for some of the long alpine tunnels.

The risk that fire will spread rapidly along a congested tunnel, and difficulties in the control of smoke must be fully considered in application of the purely longitudinal system.

With longitudinal flow the degree of pollution rises uniformly along the length of the tunnel to the maximum acceptable at the point of discharge, which is normally the portal, but may be an extract shaft. As average pollution is only half the maximum there is a case for allowing a relatively high peak value, unless stationary vehicles are to be expected at the maximum point. The P.I.A.R.C. standards take account of this.

The polluted air is most readily discharged directly through the tunnel portal, but this may be unacceptable in urban conditions in the immediate vicinity of public or commercial buildings or in residential areas. In such cases an extract fan and discharge shaft may become obligatory. Another aspect may be the risk of recirculation of the polluted discharge into an adjacent tunnel entrance.

It is only for short lengths and free flowing one-way traffic that traffic-induced air flow can be adequate. Forced ventilation becomes essential either because of length of tunnel or congested traffic. The fan installation may be a boost system within the tunnel or may comprise a shaft for fresh air supply, a duct and guide vanes into the tunnel, and a corresponding extract system.

Either system may be used to supplement or replace traffic-induced air flow; with the latter it is possible to ventilate a long tunnel by providing intermediate shafts which divide it into sections.

Many short urban tunnels, as for underpasses, are likely to be self-ventilating for long periods when traffic is flowing freely but can lose almost all their self ventilating capacity at rush hours when traffic is blocked. Boost fans mounted outside the traffic space, near the roof of the tunnel can be used to provide the necessary flow. In energy consumption they are relatively inefficient, but they are in use only for limited periods and are cheap to install and maintain, normally requiring no elaborate ducting.

Tunnels of circular or horseshoe cross section provide ample overhead space for installation of small diameter fans for this purpose, but rectangular tunnels, which are the normal type in shallow cut-and-cover construction and are also a common form of submerged tube tunnel, may not lend themselves so readily to the system, although space can be provided, either above walkways, or by local enlargements.

9.5.2 Transverse ventilation

Fully transverse ventilation is probably the ideal system for the central sections of a long tunnel if costs of construction and operation do not preclude it. The necessary air is fed from a duct into the tunnel at or near road level, and the same quantity is extracted at roof level or diagonally opposite. The air movements in the traffic space thus impose no limitations on the length of tunnel and there is no progressive build-up of pollution as in the case of longitudinal flow. Local requirements, as for example on gradients, can be met by increased supply and exhaust at that section, and in the event of fire, smoke is extracted directly without longitudinal spread. The traffic flow may be in either direction, or two-way, without disrupting the pattern of ventilation.

The principal obstacle to the adoption of the system is cost, both in construction and in operation. Ducts must be provided for both supply and exhaust and, as is clear from the equation on p. 249 above, energy costs increase in proportion to the cube of the duct length, and are duplicated for supply and exhaust, while the capital cost of providing in tunnel ducts of large area can be prohibitive. There is also the problem of cleaning and maintenance of the exhaust duct in which deposits of oily soot tend to accumulate.

In a tunnel of circular section there is normally available useful but limited space both below the road level and above the height requirement, and this shape lends itself to fully transverse ventilation, which was adopted for the first of the urban tunnels designed for modern motor traffic, the New York Holland Tunnel (1927). The system has worked well and proved adaptable to developments over nearly 50 years. It performed satisfactorily in a serious fire in 1949. Similar systems were provided by the same authority for the three tubes of the Lincoln Tunnel (1937, 1945, 1957).

Fully transverse ventilation systems were also adopted in the long Swiss tunnels of St. Gotthard (1980), which is divided into five sections, and San Bernardino, and in Holland for the Maas, Velsen, and Ij Amsterdam submerged tube tunnels.

Transverse ventilation systems may be further classified as upward, lateral, etc. according to the positions of the supply and exhaust apertures. The upward transverse is possibly the best because the polluted air is hot and tends to rise and assist the circulation, but the turbulence induced by traffic flow is likely to be dominant. The lateral transverse is suitable for submerged tube tunnels where depth of construction must be kept to a minimum but extra width is more readily available. In some of the horseshoe mountain tunnels both supply and exhaust ducts are in the crown of the tunnel, making it necessary to feed fresh air by local ducts down the side of the tunnel to kerb level.

9.5.3 Semi-transverse ventilation

The semi-transverse systems are, like most engineering designs, a compromise between the ideal, the practical, and the economical. The Mersey Queensway Tunnel (1933) was an early example of such a system in modern conditions. The system was adopted after study of the Holland Tunnel system and after extensive experiments including smoke extraction from fires. Here also the system has stood the test of time and has been developed and adopted in the new Mersey Kingsway Tunnel (1971) and also in the Dartford, Tyne and new Blackwall tunnels. It has also been chosen for similar subaqueous crossings in the Hong Kong Harbour Tunnel and the Vieux Port Marseilles, and in mountain tunnels such as the Scheiteltunnel in Austria and the Mosi Baregg tunnels in Switzerland.

The typical feature is the use of the whole traffic space as the exhaust duct. In the subaqueous tunnels of circular section a ventilation station on each side of the channel supplies fresh air through the space under the roadway and this is fed into the traffic space by regulating apertures at kerb level. At mid-river there is minimal longitudinal flow and from that point back to the exhaust shafts on each side the air velocity in the traffic space increases progressively to a maximum. Maximum acceptable air velocities both in ducts and in traffic space impose limitations on length of tunnel ventilation from a single station, just as in the case of longitudinal ventilation.

The apertures must be carefully calculated to provide uniform supply as the pressure and velocity in the supply duct diminish along its length. The theoretical effect of this system is that with uniform output of pollutants and perfect mixing along the length of the tunnel there will be uniform dilution and therefore a constant level of pollution. Figure 9.2 illustrates the patterns of air flow and pollution for various systems.

By operation of exhaust fans only, it is possible to make a semi-transverse system function as a longitudinal system. It is worth noting that the semi-transverse system comprises transverse supply and longitudinal exhaust, and does not lend itself to the reverse pattern of longitudinal supply with transverse exhaust into a duct, because air exhausted near the point of supply will be almost unpolluted while the total air flow in the traffic space will be steadily diminishing in volume along its length, becoming progressively more polluted to a final point of nil flow and undiluted exhaust fumes.

9.5.4 Mixed systems

Although most tunnels fall broadly into one of the above classes there is often a considerable departure from the simple system such as the combination of semi-transverse and transverse or longitudinal and semi-transverse systems.

The former is illustrated by the initial scheme for the Mont Blanc Tunnel (1965) which is 11.6 km in length and where intermediate shafts were impracticable; the system adopted was effectively transverse at the centre of the tunnel and semi-transverse at the ends. All the ducts are below the roadway so that the crown of the arch is unobstructed and is readily accessible for inspection and treatment of any seepage of water or other maintenance problems. The ventilation design was based on a traffic flow of 415 vehicles per hour and a maximum CO content of 100 ppm.

With great increases in traffic, both tourist traffic up to 800 vehicle/hr in

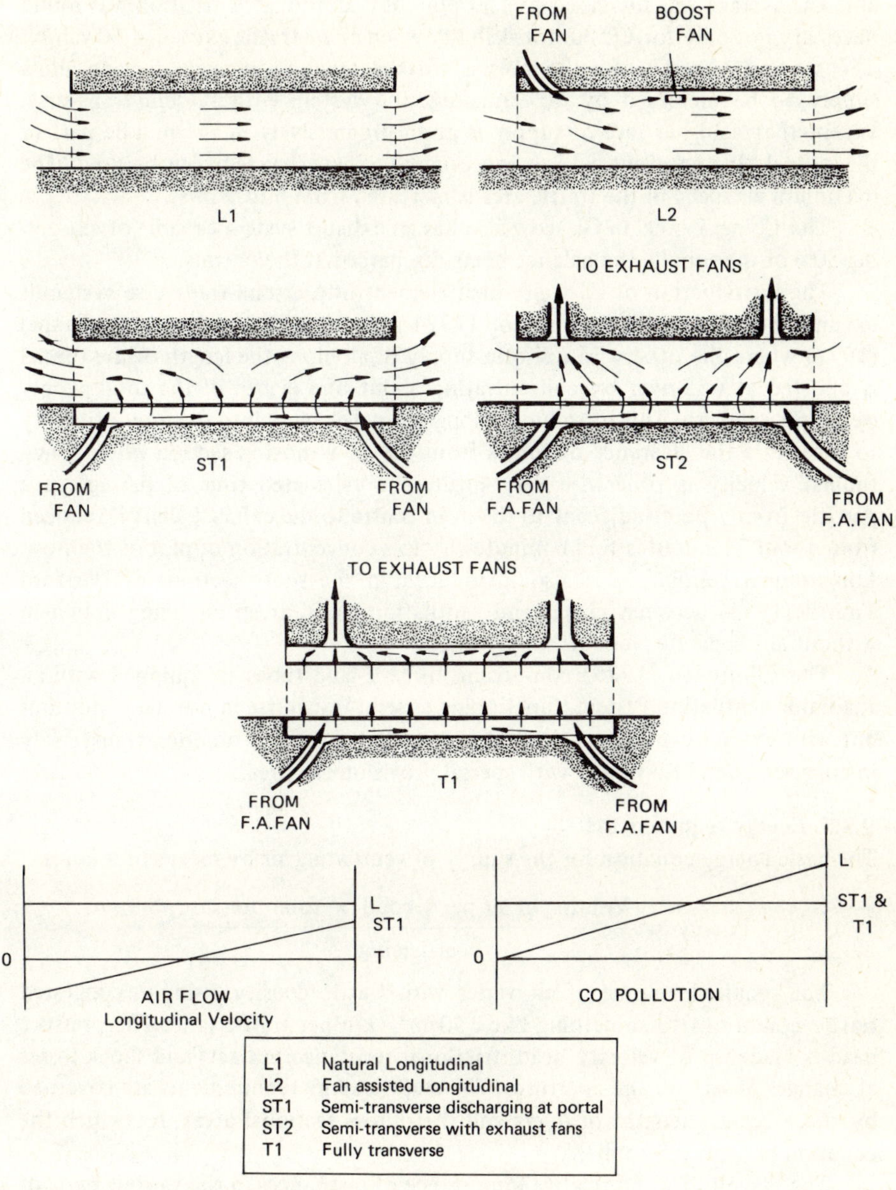

Fig. 9.2 – Tunnel ventilation – patterns of air flow.

summer, and heavy diesel traffic at other times exceeding 60 veh/hr, the acceptable CO average was increased to 150 ppm but additional ventilation was found necessary not only for CO but for visibility when diesel traffic exceeded 60 veh/hr. The scheme adopted was the reversal of the separate exhaust duct to allow supply to be increased by 50%, making the system entirely semi-transverse, except that a higher rate of supply is given progressively in the middle part of the tunnel. In the event of fire the original exhaust system can be used. The maximum air speed in the traffic area is increased from 6 to 9 m/s.

The Clyde Tunnel in Glasgow also has an exhaust system capable of extracting 50% of the supply, the balance being discharged at the portals.

The introduction of a longitudinal element into a semi-transverse system is exemplified in the mid-river section (1214 m) of the Mersey Kingsway Tunnel (1971) where the otherwise uniform supply of air along the length of the tunnel is boosted at mid-river by concentrating about one eighth of the total supply over a 60 m length. This introduces a longitudinal element into the flow, primarily to accelerate the clearance of fumes from the slow moving section where longitudinal velocity is otherwise very small. The calculated time of passage of a particle from a point adjacent to the dead centre to the exhaust shaft is reduced from about 21 minutes to 11 minutes by this concentration of part of the flow. Longitudinal ventilation is also introduced in the short sections of Dartford Tunnel (1963) between portal and ventilation shaft, fresh air being drawn in without any separate fresh air supply duct.

The Elb-tunnel (1975) comprising three 2-lane tubes is equipped with an adaptable ventilation system, functioning as semi-transverse in normal conditions but with extract capacity of 65% of the supply to make it function transversely in congested conditions, and with special provision for fires.

9.5.5 Energy requirements

The basic energy equation for the supply of ventilating air by means of a fan is

$$\text{Energy} = \frac{\text{Volume of air per second} \times \text{total pressure}}{\text{efficiency}}$$

The required volume of air varies with traffic density, rising in congested traffic conditions to something like 150 m^3/s.km per traffic lane. The pressure head is made up of velocity head, frictional resistance in ducts and shock losses at changes of section and apertures. The major energy requirements are governed by the necessary lengths of ducts and their cross sectional areas, for which the equation in para. 9.5 applies.

The magnitudes and relative importance of resistances in the various parts of a system may be illustrated by quoting the design figures for one half of one tube of the Mersey Kingsway Tunnel, where a single axial flow fan having a diameter of 5.2 m supplies 350 m^3/s to 1122 m of two-lane tunnel against a pressure of 695 N/m^2 (2.79 in w.g.)

a

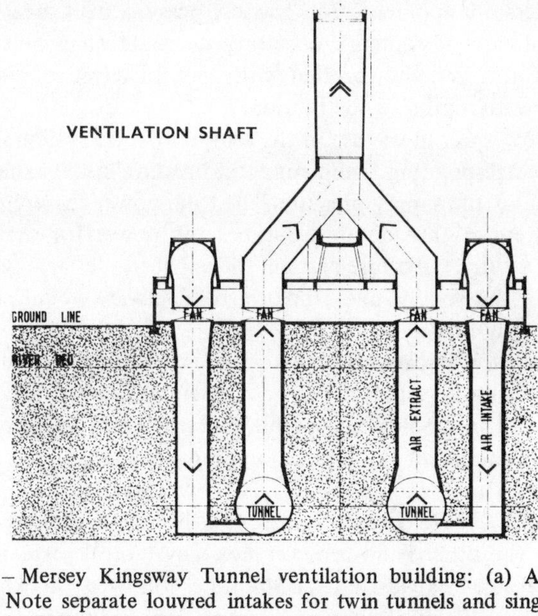

b

Fig. 9.3 – Mersey Kingsway Tunnel ventilation building: (a) As seen from R. Mersey. Note separate louvred intakes for twin tunnels and single exhaust shaft 53 m high. (b) Diagram of ventilation station and shafts.

Source of resistance	Pressure (N/m^2)	%
Intake louvres and duct, fan, safety screen, and shaft and duct into tunnel	204	28
Provision for acoustic treatment	(40)	(5)
Duct under roadway (Area 12 m^2)	419	57
Adjustable slots from duct into road space	37	5
Miscellaneous losses	35	5
	735	100

Insertion of the figure 695 into the energy equation, with an efficiency of 75% gives:

$$\frac{350 \text{ m}^3/\text{s} \times 695 \text{ N/m}^2}{0.75} = 324 \text{ kW}.$$

These resistances are for maximum air flow, but for other flows are proportional to the square of the velocity, energy being proportional to the cube of the velocity.

The largest element is the duct resistance, and the importance of length, area, and smoothness are apparent. The greatest possible duct area, uniformity of section, large radius bends aided if necessary by guide vanes must all be provided in design of the system, together with careful siting of any essential cables and pipes so as to minimise obstruction.

The economy of power in the use of the semi-transverse system is illustrated by comparing the corresponding figure for total pressure in the exhaust system, which is about 27% of the supply pressure. The total power requirement for the semi-transverse system is thus about 1.3 times the power for the supply fan alone, whereas for a fully transverse system, the exhaust duct area would be less than the 12 m^2 of the supply duct and the total power would substantially exceed twice the supply power.

Energy consumption is proportional to the cube of the rate of air supply, and it follows that in large system it becomes essential to provide for reduction when traffic is light. The simplest acceptable pattern would operate with maximum, normal and minimum air flows but more steps are likely to be advantageous in practice. Air supply rates in the ratios 1:2:3:4 will give energy consumptions of approximately 1:8:27:64. The economy of utilising the lowest practicable rate is obvious. On the other hand too complex a system of regulation of flow is not justified because traffic needs cannot be followed quickly and with precision, and too much elaboration may defeat its own ends.

9.5.6 Choice of plant

For major tunnels very large fan capacity is required and fans are normally specially designed for the duty required. A choice must be made between a few large fans and a larger number of smaller units and also between centrifugal and axial flow types. It is not possible here to do more than indicate the broad basis of choice. Local electrical and mechanical facilities are very relevant.

With large fans the desired variability of output may be obtained by multiple pole motors having a range of speeds, or by other electrical devices such as commutator motors or thyristor control, or by gearing or variable fluid couplings.

When a single large fan supplies a section of tunnel the ducting can be designed for the smoothest flow and consequently minimum losses, but it is vital that air supply to the tunnel shall not be interrupted by breakdown of the fan system, and a standby system must be provided to operate very rapidly in emergency. In the Mersey Kingsway Tunnel and in other U.K. subaqueous tunnels 100% standby is made available by provision of a duplicate fan and motor so mounted on a transfer carriage that changeover can be effected in a very brief period. There is also provision against electrical supply failure by means of standby diesel generating plant. The alternative of fixed fans and branching ducts complicates the air flow pattern and results in losses at junctions and discontinuities, as well as introducing further mechanical problems of installation, operation and maintenance.

If a number of smaller fans are installed the complexity of the ducting increases as do the losses, but it may be possible to use standard sizes and types of fans and there is also unlikely to be a need for 100% standby. The air loss resulting from breakdown of one out of three or four fans working in parallel is not catastrophic, and it is likely to be satisfactory if replacement fans and motors, held in reserve, can be installed in the course of a few hours. This basis was adopted for the Hong Kong Tunnel.

A further consideration in making the choice is the accessibility of the site for installation, maintenance and replacement of heavy plant. The differences between a high mountain tunnel and an urban subaqueous tunnel are obvious.

9.5.6.1 Fan type

The principal merits of the axial flow fan are: (1) Efficiency at low pressures, but over a limited range of pressure. (2) Economy of space when installed as a single fan directly in the intake duct.

The radial flow fan operates more efficiently at somewhat higher pressures and is tolerant of a greater range of pressure without substantial loss of efficiency. It is also claimed to be quieter, but when operated at low speed can produce a deep beat sometimes found more unpleasant. More space is required by radial flow fans of equivalent capacity, and the problem of maintaining a clean motor house is more difficult.

Direct cost of fan and motor may be somewhat less for the radial flow fan, but the extra space, and consequently larger building will more than counterbalance this particularly in urban areas.

Radial flow fans are installed in the Holland and Lincoln Tunnels in New York and also in the original Mersey Queensway Tunnel, and the Mont Blanc Tunnel but axial flow fans have been adopted in the Dartford, Tyne, and Mersey Kingsway Tunnels and also in Hong Kong.

Where longitudinal ventilation, with separate shafts for supply and exhaust by fan, is adopted the same considerations apply, but where there are no separate shafts and air enters at one portal and discharges at the other, boost fans within the main tunnel space are necessary. Jet fans are used, mounted high in the tunnel and may be required at intervals along the length.

In a study for a two-lane rectangular tube the National Engineering Laboratory at East Kilbride found that two booster fans of 1 m diameter mounted in the upper corners of the tunnel were required at about 50 m intervals to maintain an air velocity of 6 m/s necessary to ensure that CO was kept below 250 ppm with congested traffic in a tunnel 700 m long.

9.5.7 Regulation and control

From what has already been said it will be appreciated that by proper regulation of ventilation substantial savings of power can be made, but the scope for saving must not be exaggerated. In a typical subaqueous tunnel the annual average power consumption per lane for the passage of 3 million vehicles, may be about 0.1 kWh per vehicle. Obviously doubling an already adequate air supply and increasing the power cost eight times would be absurd, but on the other hand reduction of supply to a point where the public are inconvenienced and, in greater degree, the tunnel staff, is not justified by the much more limited scope for cost reduction.

'Fine tuning' of the air supply to follow a particular pollution standard by monitoring CO in the vitiated air is impossible, particularly in a long tunnel. The CO produced in the tunnel will not reach the gauges and be recorded for some minutes after it is produced. If the fans are immediately speeded up to meet the new situation further time will elapse, so that the increased air supply entering the traffic space is not effective until possibly 15 or 20 minutes after the traffic initiating the change has passed. The inertia of the system may be illustrated by the fact that maximum air supply to a two-lane tube of the Mersey Tunnel is 50 tonnes per minute. It seems clear that regulation of air supply should be in terms of anticipated and actual traffic flow.

It is of interest to note that in the New York Port Authority's Holland and Lincoln Tunnels: 'The main feature of the ventilation control systems now being developed for these tunnels is to base the ventilation decisions primarily on knowledge of actual traffic conditions throughout the various tunnel sections, rather than primarily on measurement of air quality. By estimating emissions from the known traffic, the control system should be able to anticipate the need for more or less ventilation, and to act accordingly so the effect of the decision will be felt closer in time to the need for such response'. (BHRA B3-37).

The daily pattern of traffic flow is known for most tunnels, building up in urban tunnels from zero at night to a morning peak, falling off again until an evening peak period and thereafter dropping to near zero again. The air supply can be regulated by a programmer to fit this daily pattern, preferably on a seven day cycle because weekend patterns will differ greatly from weekdays, but there must also be the ability to over-ride the programme on special occasions such as national or local holidays or sporting events creating increased traffic demand.

Manual over-ride must always be available, but whether or not in combination with automatic over-ride, there must be an alarm system bringing effectively to the notice of the controller abnormally high levels of CO polution or other dangerous conditions.

If circumstances permit ventilation and traffic flow control to be combined then economies in the maximum provision can be effected because the maximum flow of traffic through a tunnel does not make maximum demand on air. Maximum lane capacity is usually about 1750 vehicles when a speed of about 40 km/hr can be sustained. Vehicle spacing is then 23 m and pollutant output is less than half of that for the worst condition when traffic speed drops to 10 km/hr at a spacing of 10 m and flow to 1000 v.p.h.

A ventilation supply scheme regulated on the basis of the number of vehicles in the tunnel at any time, or preferably on measurement of average speed and spacing, appears to offer the best system for adquate and economical ventilation.

9.5.8 Air intake and discharge

The ideal situation for the intake is as close as possible to the point of supply to the tunnel in a position free from air pollution and protected from rain and snow. It is also important to avoid a site where noise either from the fans or from any intake screen is objectionable to neighbours.

Polluted air intake in urban conditions can arise from street traffic on the surface or from industrial processes in the vicinity, perhaps when the wind is from a particular direction. An example is at the Dartford Tunnel under the Thames where cement works adjacent to the river discharge dust into the atmosphere, which is drawn into the ventilation ducts when the wind is from one direction, and has to be cleaned out from time to time although otherwise harmless. The Tyne Tunnel likewise can suffer from adjacent industrial pollution.

Unless some settling chamber can be provided in the intake, or a sealed duct is provided, snow and fine rain can be drawn in by the fan and can affect the motors and control gear.

Noise from fans can be kept to a low level by good design, and particularly by keeping down blade tip speeds. If the environment is industrial there is little difficulty in complying with any reasonable noise standard, but in the vicinity of

Fig. 9.4 – Ventilation building for second Dartford Tunnel. Intake is through circular louvred chamber. Exhaust is through inverted conical evasee. Fans and equipment are on ground floor.

dwelling houses or hospitals it may be more difficult, and acoustic treatment may be necessary, with consequent added air resistance and cleaning problems. Noise is particularly objectionable and noticeable at night, but traffic is then normally very light and fan speeds can be minimal. It may even be possible to operate a tunnel at night on a varied circulation which dispenses with use of the fans in a sensitive area.

The problems of discharge are similar, but with the vital proviso that vitiated air should not be able to recirculate to the intake under any conditions. The ideal system in urban areas is probably discharge at high level through a tower or chimney acting as an evasée in the form of a cone whereby velocity head is transformed to suction head, contributing to economy of operation.

In the simpler systems where discharge is at a portal the questions of recirculation and local environmental pollution likewise require careful study. One possibility in a shallow tunnel is to exhaust air by means of a number of jet fans straight up through the crown of the tunnel. This may be particularly economical where cut and cover construction is used and where the surface area is reclaimed as an open space. It must be remembered that the vitiated air is not usually at a dangerous level of pollution, that it is discharged vertically and will continue to rise being hot, and that in the open air there is rapid natural dilution.

9.5.9 Construction problems

Apart from the problems faced in all tunnelling work, ventilation has a few special aspects which must be considered at an early stage in the design. If a major tunnel is to have as few ventilation stations as practicable they will be large structures whose siting is closely determined by the alignment of the tunnel and the position of the portals. There will be heavy foundation loads to be carried down in the immediate vicinity of the tunnel and large shafts will be required as air ducts for both supply and exhaust.

The openings from these ducts into the tunnel present a special problem as large smooth continuous ducts without abrupt bends are desired. In any ground this presents difficult tunnelling conditions, and in bad ground aerodynamic efficiency may have to give way to practicalities of construction.

Construction of the duct connections and a short length of the tunnel in a caisson was adopted at the Dartford tunnel and elsewhere, the bored tunnel terminating by driving the shield into this chamber. At the Mersey Kingsway Tunnel the junctions were in rock and could be constructed in a series of headings but always adapted to the boring and lining of the main tunnel by machine.

9.6 VENTILATION IN FIRES

The frequency of vehicle fires in tunnels is not very different from that on the open road. A rate of one fire per km for every 10 million vehicles is representative.

The most important special feature is that the flames, smoke, and fumes cannot ascend vertically but rise to the roof of the tunnel and spread laterally, still burning and forming a layer possibly 2 or 3 m thick which travels horizontally at a rate of a few metres per second. Although there may be a zone of fresh air below the smoke cloud, heat radiated downwards may constitute a hazard to life. The spread and behaviour of this smoke cloud is greatly influenced by the ventilation air flow, and it is therefore important that fire fighters should be familiar with the possibilities of regulating air supply and extraction by selection of the most suitable combination of fans. Longitudinal ventilation without extraction presents a particular hazard of the spread of fire along the tunnel, but at least has the merit of supplying fresh air upstream from the fire and thus allowing access. Too high a rate of flow creates turbulence and the smoke obscures the whole tunnel. Semi-transverse ventilation has also a longitudinal flow, but its average velocity is usually lower and the continuous supply of fresh air at road level may be helpful. Fully transverse ventilation helps to localise the fire hazard, but the normal extraction rate may be quite inadequate to clear the smoke generated, and longitudinal spread in both directions may result, unless emergency outlets can be brought quickly into use.

In all types the hazard from a small fire, as from a single car, is not found to be severe; it is likely that the fire will remain localised, not spreading to adjacent vehicles, even at fairly close spacing. In a larger fire, the high temperature of the smoke cloud may make access to the seat of the fire difficult or dangerous. The problems of fire in a tanker carrying liquid fuel are such as to justify strict rules about the carrying of dangerous loads.

There is not complete agreement on the control of ventilation to the best advantage, but use of exhaust fans to extract smoke and to limit the spread of smoke to a single direction is certainly helpful, provided that the fans are not liable to be burnt out. The supply of fresh air is a help to fire fighters requiring access, but in major fires possible increased rate of combustion is advanced as a reason for restricting air supply. On the other hand, incomplete combustion tends to generate lethal quantities of carbon monoxide.

The particular requirements of each tunnel must be examined in the light of controls and facilities available and general policy on the hazards. In fact there have been very few major fires in highway tunnels. Two instances are described in Chapter 6.

9.7 RAILWAY VENTILATION

Ventilation is a problem of long tunnels. In a tunnel of moderate length the passage of a train creates a strong current of air so that polluted air is carried out to the exit portal while fresh air is drawn in behind.

Smoke was troublesome from the earliest days of railways. In addition to being environmentally objectionable to passengers, and to drivers, it produced corrosive deposits within the tunnel. Although dirt and unpleasantness in some degree were tolerated as inevitable, much was done to reduce smoke pollution by means of open shafts to the surface and by extract fans.

An early example of smoke extraction was on the world's first passenger railway at Liverpool terminus, when, after 40 years, in 1870 loco working was introduced in place of rope haulage on the 1 in 97 gradient from Edge Hill down to Lime Street. A 9 m diameter extract fan running at 45 r.p.m. was installed half way along the 1850 m tunnel. The tunnel was subsequently opened out as a cutting and widened to take four tracks. Another tunnel ventilation scheme was in the nearby Mersey Railway tunnel (1886) in which there were provided ventilation headings connecting to two extract fans in shafts on each side of the river crossing. The total output was 300 m^3/s, fresh air being drawn in through the stations. The Severn railway tunnel (1886) also had to be provided with ventilating fans before it was allowed to operate. These examples are not of course typical of the average railway tunnel: the conditions were particularly unfavourable to smoke clearance in that the Lime Street station is a terminus on the approach to which the trains were moving too slowly to be effective in sweeping out the polluted air, while the Mersey and Severn railway tunnels are subaqueous and deep with no possibility of open shafts.

In the first metro, the London Metropolitan Railway (1863), coal burning steam locos were employed, the last of which were not withdrawn until 1905. Ventilation of the shallow tunnels was by 'blow holes' of area about 90 m^2 at intervals of about 1 km. Smoke was a serious deterrent to passengers until electrification of the system.

The electrically operated London 'tube' railways which followed could not be provided with blow holes for ventilation and depended on piston action of trains which fitted closely into the twin tunnels, and exchange of air through station passages. Progressively, air supply to stations by fans has been found necessary to eliminate hot spots and the build up of heat. In the Victoria Line (1968–1971) reversible fans midway between stations were incorporated in the original design and are operated as exhaust fans, fresh air being drawn in through station entrances. Draught relief shafts are also provided at stations to suit the particular conditions. As a quantitative example, one length of 1 km incorporating one station and carrying 300 trains a day at a maximum speed of 18.6 m/s involves heat gain of 14.6 MWh, for which the extract fan provision is 40.5 m^3/s of air based on a temperature rise of 10.5°C, after allowing a 15% exchange of air through draught relief shafts.

In New York the main lines from the north into the Grand Central Station passed in a shallow tunnel 3 km long under Park Avenue. Steam operation gave rise to many difficulties and complaints: signals were obscured, passengers endured smoke and heat, and the public in the street above were subjected to blasts of ash, smoke and fumes through the ventilation openings at road level. Electrification was seen as the remedy, and after much careful study was adopted and effectively introduced in 1907.

Electrical operation has abolished the smoke problem for all metros and many main line railways, but other aspects of air control have become of growing importance. For tunnelled metro systems draught relief in stations and long term removal of accumulating heat are universal problems. For main line railways high operating speeds have introduced new problems of pressure pulses and train resistance more particularly in long tunnels.

The major differences from highway tunnel ventilation lie in:

(1) the absence of toxic gases,
(2) the self ventilating piston effect of trains, which occupy a substantial proportion of the tunnel cross section,
(3) the very high speeds attainable,
(4) the possibility of sealing trains against external atmospheric conditions,
(5) closely controlled use by stock of known characteristics on a fixed path.

Ventilation and aerodynamics cannot be entirely separated, but in the railway tunnel context ventilation may be considered to comprise the functions of removal of vitiated air and heat and, perhaps, air conditioning, while aerodynamics in its simplest aspect deals with the calculation and limitation of air speeds in public areas. Where operating speeds are high more complex problems arise, of air flows and pressures and transient pulses, which can be environmentally unacceptable, and also may call for substantial increases in loco operating power.

The impact of these problems differs considerably in metros and main line railways and may be considered for each type of system under the two aspects of environmental effects on passengers and staff and operational effects on plant and equipment. The control and adequacy of ventilation in the event of fire in a tunnel may be of particular importance in metros, but emergency provisions for long main line tunnels may also be thought necessary.

9.7.1 Metros

In tunnelled metros there are four zones used successively by passengers, namely,

(1) entrances and ticket halls
(2) escalators, stairs and access passages
(3) station platforms
(4) trains.

The environment in each is determined in part by the outside climatic conditions and in part by the ventilation system adopted. The general flow of air may either be inward from station entrances to platforms and tunnels, or outwards through escalator tunnels to discharge at station entrances. The former is generally preferable in ensuring fresh uncontaminated air in public areas, but in hot or cold extremes of climate or weather the unconditioned air entering stations may not be favoured.

In metros train speeds are not high because of the short runs between stations; the more complex aerodynamic problems do not therefore obtrude.

The ventilation system is required to ensure:

(1) an adequate supply of fresh air at acceptable temperatures,
(2) control of air flows induced by movement of trains where these create objectionable draughts in passenger areas,
(3) avoidance of abrupt and excessive temperature differences,
(4) removal of excess heat from all sources, which would otherwise accumulate within the metro.

Passage of a train through a tunnel generates considerable air movements; air will be expelled in front of a train and drawn in behind it through any ventilation shafts or other openings, including station passages. The effective exchange of stale air for fresh air will obviously depend on such factors as the proportion of tunnel cross-section occupied by the train, the area and length of shafts or other openings and their siting, the frequency of the train service and – to a major extent – on whether twin single track tunnels or double track two-way tunnels are used.

Self ventilation of the system by such means has generally been found to provide adequate quantities of fresh air, but fans are likely to be required to improve and regulate distribution, and more especially to deal with the removal of excess heat, which eventually is likely to become the major problem. The contingency of fire is also a reason for provision of fans, so designed and controlled as to allow smoke to be extracted in the most suitable path which leaves clear a passenger escape route. Possible heat damage to fans and their supply cables needs to be examined.

Having ensured the adequacy of the fresh air supply, it must be so controlled as to avoid draughts in passenger areas. The surges of air as trains enter and leave underground stations necessarily create air flows on platforms, in cross passages, and in escalator shafts and other connecting passages. These flows can be calculated from a knowledge of the layout of the tunnel system, including stations and of the operating characteristics of the trains – speed, acceleration, braking and frequency of service. Air speed of 7 m/s may be acceptable as a maximum, but 4 m/s is preferable. Excessive air flows can be reduced by such means as enlargement of passage areas, and, sometimes, provision of additional draught relief openings or passages in critical places, but, usually, the most effective action is the provision of draught relief shafts to the open air from the point of junction between running tunnels and station tunnels. Where the distance between stations so requires, ventilation shafts about half way between stations may also be provided.

The climate within a tunnel system will usually fluctuate less than outside, but with an inevitable tendency for the air to heat up, and for the structures also to become somewhat warmer than the long term outside average. In temperate climates there is not often any serious immediate problem, but at station entrances there may be some discomfort if air is drawn in when the weather is abnormally cold or hot.

In tropical climates refrigeration of the air supply may be thought necessary, but the problems of regulating temperatures in the various passage areas are very great.

Any attempt to cool the air in stations is seriously impaired by the exchange of air between stations and running tunnels arising from the piston action of trains. The construction of a separating wall between platform and train, with controlled openings to allow passengers to enter trains and alight, has been proposed and would effect substantial energy savings if found safe and practicable.

Air conditioning in trains is also at a disadvantage because of the opening of doors at frequent stations, and also because extra waste heat is generated by the refrigeration process and must be removed, probably by additional ventilating air to the tunnel. The air conditioning within a train has also to be adaptable to surface operation where trains run partly in a central tunnelled area and partly at the surface in outer suburban areas.

9.7.1.1 Heat gain

The accumulation of heat within a metro system is a phenomenon that may only become apparent over a period of years, but may appear quite quickly if traffic is heavy. The principal source of heat is the energy used in driving the trains, mostly converted to heat in the braking system but some also more directly in the driving motors. The major traction power requirement is for frequent and rapid acceleration out of stations: also significant are the lighting currents in trains and in stations and the body heat of large numbers of passengers. The first effect of these heating sources is to warm up the air in the system, which in turn transfers some of the heat to the tunnel lining and other structures and carries away some when it is exhausted.

The ground behind the lining acts as a 'heat sink' absorbing heat from the lining, or returning it if the air is unusually cold. This balancing effect evens out fluctuations between day and night and between summer and winter, but in the longer term average lining temperatures build up until there is no net annual loss of heat to the ground. The ventilating air has then to carry away the whole of the heat generated and its temperature rise is determined accordingly. It is for this condition that the ventilation system must ultimately be designed: the air flow must be sufficient in quantity to carry away the heat at the specified temperature rise, and must be directed so as to avoid discomfort to passengers. The environmental conditions in continuing hot weather are likely to determine maximum requirements. Conditions of severe cold have also to be allowed for.

Heated air can be extracted through fans in vent shafts at the end of stations or midway between stations. Another possibility is to site an extract duct under station platforms, which has the advantage of direct removal of much of the heat generated in the brake linings as the train slows down to enter the station. Any such extract system must draw in the same volume of air elsewhere, probably through station entrances.

In addition to the comfort and convenience of passengers and staff, ventilation must be related to the operational efficiency of plant and equipment. In general, electric motors and cables suffer loss of efficiency as their temperature rises, and therefore cooling by adequate ventilation is to be sought in the design of the system. Fans are less effective in handling warm air than cold and appreciable allowances must be made for extract fans. Fans must also be protected against surges of air or designed to withstand them.

9.7.1.2 Dust

A problem of some importance in metros is that of dust in tunnels. The dust is likely to contain metallic particles from the wear of wheels and rails and brake shoes. Dust is also brought into stations from the streets. Metallic dust blown about by passage of trains within the tunnels can be particularly deleterious to mechanical and electrical equipment, more especially to any delicate electrical contact devices, which must therefore be carefully protected against it. Removal of the dust by extract ventilation alone is not often practicable.

It will be clear that the requirements for ventilation of a metro system are varied and complex. In systems which have been built up progressively over many years, ventilation needs have become apparent concurrently with other development. Modifications and additions have been necessary in existing lines and the experience gained has been incorporated in new construction. In newer metros ventilation systems are almost invariably part of the initial design, but have not always proved adequate in operation. Until recent years calculations of sufficient accuracy for a complex system of tunnels have not been practicable, but computer programmes have now been devised and tested giving the necessary information for design.

9.7.1.3 Environment

Environmental standards required, if public transport services are to attract passengers, have risen in recent years and at the same time train frequency and speed have also risen so that the ventilation systems become more extensive and elaborate. It is all the more necessary that their planning and design is integrated with that of the system as a whole. The costs and benefits cannot be studied in isolation. All tunnelling operations for the provision of shafts, ducts, and underground plant rooms are necessarily difficult and costly, and wherever possible must be adapted and often subordinated to the tunnel construction as a whole. In urban areas the surface sites desirable for ventilation shafts and plant may be unobtainable. Shafts necessary for construction access may serve also as ventilation shafts, and, in improvement of ventilation in existing stations, old lift shafts may be utilised.

Environmental conditions have been examined in some detail by the U.S. Department of Transportation in their Subway Environmental Design Handbook Vol. I (1975) and limits of temperature, air flow speeds and air pressure pulses have been proposed. It is correctly suggested that increased attention to these factors has become necessary for the reasons given above, namely: greater heat loads for heavier faster and more frequent trains and higher environmental standards. It is questionable how far the pursuit of comfort should be carried when its capital and operating costs and their impact on fares are fully considered. The primary function of a metro system is to provide a fast and efficient service at a low cost. Cushioned armchair comfort and protection from every vagary of wind and weather can push costs up to excessive levels.

In a new system early and comprehensive studies are, of course, essential if appropriate provisions are to be incorporated in the system at the construction stage, having regard to the greatly enhanced cost and difficulty of effecting later improvements. Development of the ventilation design requires close liaison and interaction with those responsible for all the other aspects. In this, as in other aspects it may be just as important to reject unnecessarily high standards, after proper examination, as to provide for every possible contingency of discomfort.

9.8 MAIN LINE RAILWAYS

For main line railway tunnels, other than in the vicinity of termini, problems of fresh air supply arise only in long tunnels. Aerodynamic problems of pressure pulses, and train resistance are a consequence of high speed operation in tunnels of any length, becoming significant where speeds much exceed 120 km/h, and specially so in very long tunnels.

For main line tunnels less than a few kilometres long, self ventilation by train movement is usually fully adequate both for quality of air supply and elimination of heat build up. In twin single track tunnels the one way piston effect is particularly effective. In longer tunnels vent shafts are likely to provide the required changes of air if the tunnel is not at too great depth and if the necessary surface sites are available. On electrified lines air pollution is minimal: the worst conditions are with coal burning locos hauling heavy trains up a severe gradient.

A high speed train entering a tunnel portal acts abruptly on the nearly stationary air in the tunnel and sets up a complex set of aerodynamic effects. The air ahead of the train is compressed sharply and a shock wave is set up, travelling forward through the length of the tunnel at a speed close to that of sound. At the far portal this wave is reflected back as a pulse of rarefaction and is experienced as a sudden reduction of pressure when it reaches the train. Similar, but normally smaller, rarefaction pulses are generated at any discontinuities such as open shafts or cross passages.

By the mechanism of these pulses, the column of air in the tunnel is set in motion forwards and a stream of turbulent air flows back, through the annulus between train and tunnel lining, to the rear of the train, where suction is generated.

The difference of air pressures at front and rear of train constitutes a braking force, as also does the frictional drag of the turbulent air, whose motion relative to the train is rapid, but relative to the tunnel lining much slower.

In a short tunnel these braking effects are less than in a long tunnel, because, (1) the smaller mass of stationary air is more easily projected forward, (2) the turbulent flow and resultant drag are also less, (3) the forces act against the train for the shorter time during which the train is in the tunnel. The energy loss and consequent reduced speed are unlikely to be significant. The critical aerodynamic factor for operation of high speed trains through short tunnels is environmental, namely the development of pressure pulses which, if too great, can cause discomfort to the ears of passengers.

Research on this effect by British Rail has led to the adoption of criteria of acceptability. The unit adopted for pressure change in the pulses is 1 kN/m^2, which is approximately 1% of atmospheric pressure. It is considered that rapid changes should not be allowed to exceed 3 kN/m^2, but a limit of 0.5 kN/m^2 should be set for frequently repeated pulses, such as occur in a tunnel with numerous cross passages or shafts. It may be noted that the recommendation of the U.S. Department of Transportation Handbook is that the rate of change should not exceed 0.41 kN/m^2s.

The magnitude and importance of these effects may be exemplified by considering a twin single track tunnel 3 km long between portals without intermediate openings, and trains 350 m long running at 200 km/h. (See Fig. 9.5(a)). It will take 54 seconds for any point on the train to pass through the tunnel. A pressure rise of 4 kN/m^2 occurs during the first 7 s, dropping slightly as the tail of the train enters the tunnel and then after a further 7 s dropping sharply to a deficiency of 1 kN/m^2 when the wave of rarefaction has returned from the far portal. Thereafter the pressure fluctuates through a diminishing range. The drop of 5 kN/m^2 is not acceptable, and reduction has to be effected either by construction of intermediate shafts, or cross passages to the adjacent tunnel, or speed limitation to 150 km/h, or enlargement of the cross section of the tunnel.

The case of a double track tunnel with trains in both directions is more complex. The pressure pulse at entry is much less because of the greater area of tunnel, but maximum pressure changes occur as two trains meet and pass. Since the positions of maximum pressure change will be fixed by the meeting point of the trains, pressure relief by provision of vent shafts is unlikely to be practicable because very close spacing would be required to give relief at all times. If the surges are unacceptably high and if speeds cannot be reduced, a larger area of tunnel is called for.

Where a long tunnel is to carry high speed passenger trains the balance of advantage, aerodynamically, appears to be in favour of twin single track tunnels linked by cross passages. Even in the most favourable circumstances it is difficult to ensure the primary objects of ventilation: fresh air, minimum drag, adequate cooling, and limitation of pulse pressures. This was particularly apparent in the

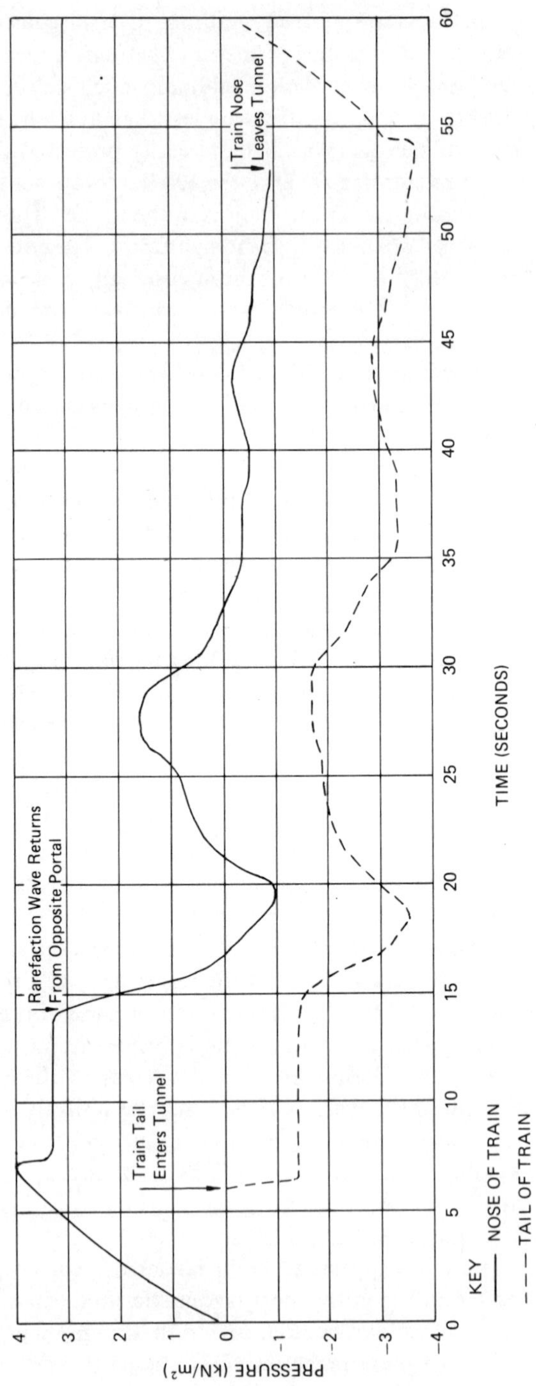

Fig. 9.5(a) – Pressure/time for train passing through single track tunnel (*Henson & Lowndes*).

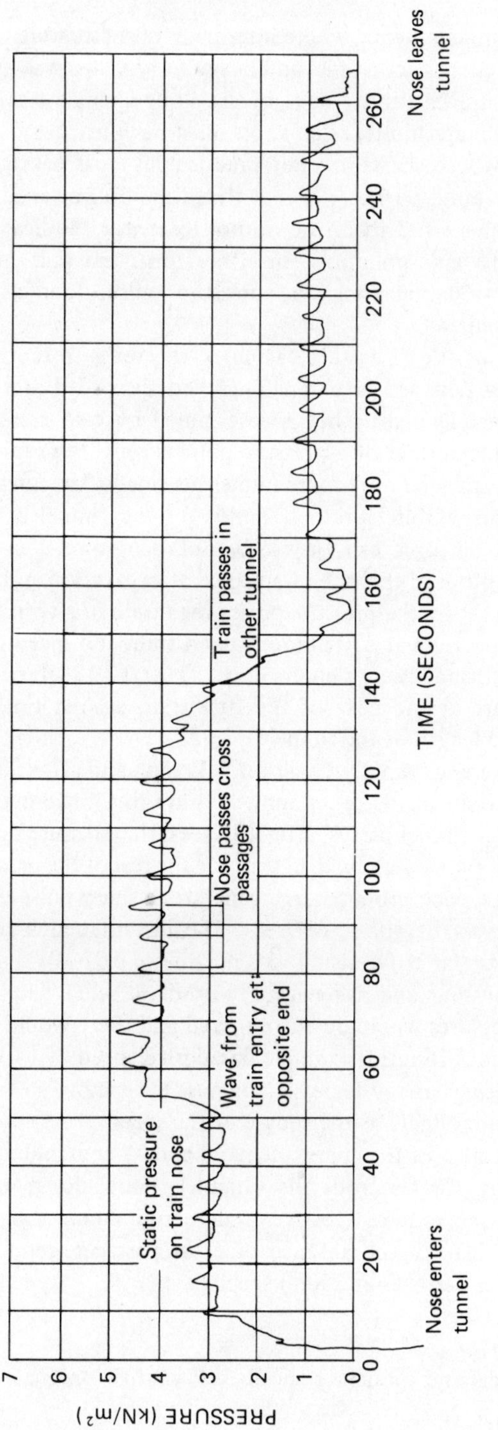

Fig. 9.5(b) – Typical static pressure on train nose in a twin bore system with cross passages (*Henson & Lowndes*).

1974 Channel tunnel scheme, where intermediate shafts were quite impracticable and a very frequent service of fast and heavy ferry trains was expected.

If the pressure and movement of air along a single track tunnel are to be kept within moderate limits, relief vents must be constructed, of sufficient cross sectional area. Where shafts are not practicable, cross passages to the adjacent tunnel serve the same purpose, except that they do not ensure exchange of air with the atmosphere and therefore do not discharge the heat building up in the system or relieve any pollution. In either form the vent openings constitute discontinuities reflecting back the pressure pulses, and as the train passes, generating new pulses.

The design of the size and spacing of the vents in terms of aerodynamics poses the choice between large openings widely spaced and smaller but more frequent openings. Draught relief is determined by vent area per unit length of tunnel but the magnitude of a pressure pulse is roughly proportional to the area of a single vent, and therefore more numerous smaller openings are advantageous in keeping pulses within specified limits. In the tunnel construction process numerous cross passages can be costly out of proportion to the direct cost because of disruption of the rate of advance of excavation and lining.

Figure 9.5b shows the pressure pulses for a train in a twin bore tunnel 10 km long with cross passages at 250 m intervals. A train at a speed of about 140 km/h enters the tunnel, and another enters at the far end 30 s later. The graph is of the static air pressure at the nose of the first train against time, the train taking 260 s to pass the length of the tunnel.

The pressure rises rapidly to about 3 kN/m^2 and after 50 secs is met by the pulse from the train in the other tunnel, rising to a little over 4 kN/m^2. As the train in the other tunnel passes, after 130 secs, the pressure drops quickly nearly to zero and only rises by about 1 kN/m^2 for the rest of the passage. Superimposed on these changes, each cross passage generates a sharp pulse of about 0.5 kN/m^2 at regular intervals of about 6½ secs. The instantaneous aerodynamic power follows a very similar curve rising to a maximum of 7 MW, falling back to 3 MW at the time of passing, and rising again to 6 MW. If relief vents were not provided much higher pressures would be encountered and there would be a severe penalty in power demand. From this example based on a speed of 140 km/h, the importance of minimising energy losses is obvious: even more so for higher speeds as pressures increase roughly as the square of the speed.

The significance of the energy demand and its control by proper design may be illustrated by figures from the Channel tunnel design of 1974 where the estimate of power required to overcome aerodynamic drag for a single ferry train was 13.5 MW in the absence of cross passages, reduced to 5.8 MW by provision of cross passages at 250 m spacing.

9.8.1 Heat gain

For short tunnels and moderate speeds self ventilation ensures that heat gain is not a serious problem, but as length of tunnel and operating speed increase, there may be an unacceptable rise in tunnel temperature. The operating power required to overcome aerodynamic drag increases as the square of the operating speed, and, once tunnel length increases beyond the capacity for self ventilation, all extra energy has to be otherwise exhausted. In the Channel tunnel the large heat input from the frequent train service could not readily be cooled by ventilation in the absence of shafts intermediate between those at the coasts, and therefore mechanical cooling systems were to be employed, fresh air being supplied normally through the service tunnel only to the extent of physiological demand, namely 50 m^3/s. In emergency, 200 m^3/s could be supplied through the interconnected servicing tunnels.

In the projected St. Gotthard Base tunnel proposed by Swiss Railways, the double track rail tunnel would be 48.67 km in length having a maximum gradient of 0.73% and with 3 deep shafts. Heat input from 282 trains a day and from fans and other plant is estimated at 25 MW, and in addition the high rock temperatures at depths of 1500 m would contribute 10 MW. Some 7 MW would be carried out by heating up of trains and half of the remainder by air supplied at a rate of 1600 m^3/s through two shafts and exhausted through the middle shaft and at the portals. It is proposed that the balance should be absorbed by vaporisation of water sprays.

BIBLIOGRAPHY

The subjects of ventilation and aerodynamics in vehicle tunnels are rapidly growing. As with compressed air working (Volume 1, Chapter 5), there are both physical and physiological aspects, the latter being a major factor in the ventilation of highway tunnels. Most of the information must be sought in papers presented at conferences and elsewhere, and in articles in the technical press. Many papers on the construction of individual vehicle tunnels, listed in the main bibliography, discuss their ventilation in some detail.

The following are considered of particular relevance. A very extensive bibliography, with abstracts, is contained in Pursall and King's book. The B.H.R.A. symposia provide comprehensive and up-to-date information.

Singstad, O., Ventilation of vehicular tunnels, *Proc. World Engng Congr., Tokyo*, 1929, **9**, 381–99.

Haldane, J. S., The ventilation of tunnels, *J. Instn Heat. Vent. Engrs*, 1936, **4** (Mar.).

Anderson, D., The Construction of the Mersey Tunnel, *J. Instn Civ. Engrs*, 1936, **2** (Apr.).

Murdock, C. W., Ventilating the Lincoln vehicular tunnel, *Trans. Amer. Soc. Heat. Vent. Engrs*, 1938, **44**, 273–88.

Daugherty, R. L., Piston effect of trains in tunnels, *Trans. Amer. Soc. Mech. Engrs*, 1942, **64** (Feb.).

Mount, S. C., Ventilation and cooling in London's tube railways, *J. Instn Heat. Vent. Engrs*, 1947, **14** (Jan.–Feb.).

Scheyer, E., Ventilation of Subways, *Mun. Engrs J.*, 1955, **41** (2).

Road Tunnels Committee:
Report, 11th World Congress, Rio de Janeiro, 1959;
Report, 12th World Congress, Rome, 1964;
Report, 13th World Congress, Tokyo, 1967;
Report, 14th World Congress, Prague, 1971;
Report, 15th World Congress, Mexico, 1975;
Report, 16th World Congress, Vienna, 1979;
Permanent International Association of Road Congresses, Paris.

Rennie, R. P. *et al.*, Effect of diesel locomotive operation on atmospheric conditions in a railway tunnel, *Can. J. Chem. Engng*, 1960, **38** (Aug.).

Ackeret, J. *et al.*, Ventilation of vehicle tunnels, *Mitt. Inst. Strassenbau, E.T.H. Zurich*, 1961 (10) (in German).

Brown, W. G., Temperature and ventilation in the Toronto subway, an experimental study, *Engr*, 1964, **218** (28 Aug.).

Brown, W. G., Basic theory of rapid-transit tunnel ventilation, *Trans. Amer. Soc. Mech. Engrs*, 1966, **88B** (Feb.).

Kennedy, E. R., *A mathematical model of tunnel ventilation*, Report 65–5, 1965.

Helly, W., *A mathematical model of tunnel ventilation*, Report 65–3, 1965.

Kennedy, E. R., *Port Authority tunnel ventilation: 1965–1966 experimental program*, Report 66–5, 1966.

Kennedy, E. R., *ditto, 1966–1967 experimental program*, Report 67–5, 1967.
Engineering Department, Port of New York Authority.

Fox, J. A. and Henson, D. A., The prediction of magnitudes of pressure transients generated by a train entering a single tunnel, *Proc. Instn Civ. Engrs*, 1971, **49** (May); *see also* discussion, 1971, **50** (Dec.).

The ventilation of road tunnels:
Constant, J., A general approach to the problem;
Cuaz, F., Mont Blanc tunnel;
Megaw, T. M., Road tunnels in U.K.;
Tunnels and Tunnelling, 1969, **4** (May and July).

The Aerodynamics and Ventilation of Vehicle Tunnels:
Proc. 1st Int. Symp., Canterbury, 1973;
Proc. 2nd Int. Symp., Cambridge, 1976;
Proc. 3rd Int. Symp., Sheffield, 1979.
British Hydromechanics Research Association.

Henson, D. A. and Fox, J. A., Transient flows in tunnel complexes of the type proposed for the Channel Tunnel, *Proc. Instn Mech. Engrs*, 1974, **188** (15/74).

Associated Engineers, *Subway Environmental Design Handbook*, U.S. Department of Transportation, 1975.

Pursall, B. R. and King, A. L., *The Aerodynamics and ventilation of vehicle tunnels: a state of the art review and bibliography*, British Hydromechanics Research Association, 1976.

Henson, D. A. and Lowndes, J. F. L., Design of a Ventilation system for an English Channel Tunnel, *ASHRAE J.*, 1978 (Feb.).

Bibliography

This is confined to publications in the English language, and is classified in six sections, the items in each section being listed in order of date of publication:

A. Books and reports
B. Institution of Civil Engineers
C. American Society of Civil Engineers
D. Other organisations
E. Conferences and Symposia
F. Journals and Periodicals

Further extensive bibliographies can be found in Pequignot (1963), Pursall and King (1976) and Hoek and Brown (1980) as listed in Section A.

A. BOOKS AND REPORTS

Simms, F. W., *Practical tunnelling*, Crosby Lockwood, 1844 (revised and enlarged, 1859, 1877, 1896).

Brunel, I., *The life of Isambard Kingdom Brunel, Civil Engineer*, Longmans, 1870.

Drinker, H. S., *Tunneling, explosive compounds and rock drills*, J. Wiley, 1878.

Gripper, C. F., *Railway tunnelling in heavy ground*, Spon, 1879.

Burr, S. D. V., *Tunnelling under the Hudson River*, J. Wiley, 1885 (reprinted A. A. Mathews, Arcadia, 1968).

Walker, T. A., *The Severn Tunnel*, Bentley, 1888 (reprinted Kingsmead, 1969).

Copperthwaite, W. C., *Tunnel shields and the use of compressed air in subaqueous work*, Constable, 1906 (reprinted A. A. Mathews, Arcadia, 1968).

Gilbert, G. H. and others, *The Subways and tunnels of New York*, J. Wiley, 1912.

Hewett, B. H. M. and Johannesson, S., *Shield and compressed air tunneling*, McGraw Hill, 1922 (reprinted A. A. Mathews, Arcadia, 1968).

The McAlpine system of reinforced concrete tunnel lining, Sir Robert McAlpine & Sons, 1935.

Noble, C., *The Brunels*, Cobden-Sanderson, 1938.

Richardson, H. W. and Mayo, R. S., *Practical tunnel driving*, McGraw-Hill, 1941 (revised edn., R. S. Mayo, 1975).

Terzaghi, K., Rock defects and loads on tunnel supports: Introduction to tunnel geology, in Proctor, R. V. and White, T. L., *Rock tunneling with steel supports*, Commercial Shearing and Stamping Co., Ohio, 1946, 2nd edn., 1968.

Terzaghi, K., Geologic aspects of soft-ground tunneling, Ch. 11 in Trask, P. D., (ed), *Applied Sedimentation*, J. Wiley, 1950.

Rolt, L. T. C., *Isambard Kingdom Brunel*, Longmans, 1957.

Slater, H. and Barnett, C., *The Channel Tunnel*, Wingate, 1958.

The Work in Compressed Air Special Regulations, SI 1958 no. 61, HMSO, 1958.

Pratt, A. W. and Daws, L. F., Heat transfer in deep underground tunnels, *National Building Studies Research Paper 26*, HMSO, 1958.

The Stockholm Underground Railway: a technical description 1958 and *A technical description of the Stockholm Underground 1964*, Public Works Dept. and Passenger Transport Co., Stockholm.

Payne, R., *The canal builders*, Macmillan, New York, 1959.

Hammond, R., *Tunnel engineering*, Heywood, 1959.

Jeffery, A. H. G., *Immersed tube tunnels*, Kemp's, 1962.

Jackson, A. A. and Croome, D. F., *Rails through the clay: a history of London's tube railways*, Allen & Unwin, 1962.

Lampe, D., *The Tunnel: the story of the world's first tunnel under a navigable river, dug beneath the Thames 1824–42*, Harrap, 1963.

Ministry of Transport, *Proposals for a fixed Channel link, Cmnd 2137*, HMSO, 1963.

Pequignot, C. A. (ed.), *Tunnels and tunnelling*, Hutchinson, 1963.

Sandström, G. E., *The history of tunnelling*, Barrie & Rockliff, 1963.

Blower, A., *British railway tunnels*, Ian Allan, 1964.

Guthrie Brown, J., *Hydroelectric engineering practice*, **1** *Civil Engineering*, Blackie, 1964.

Howson, F. H., *World's underground railways*, Ian Allan, 1964.

Reid, A. S. H., Road tunnels more than 500 feet long, *Road Research Technical Paper 78*, HMSO, 1965.

Coleman, T., *The railway navvies*, Hutchinson, 1965.

Pequignot, C. A., *Chunnel*, C. R. Books, 1965.

Havers, H. C. P., *Underground railways of the world*, Temple Press, 1966.

Terzaghi, K. and Peck, R. B., *Soil mechanics in engineering practice*, J. Wiley, 2nd edn., 1967.

McGregor, K., *The drilling of rock*, C. R. Books, 1967.

Proctor, R. V. and White, T. L., *Rock tunneling with steel supports* (incl. Terzaghi, *q.v.*), Commercial Shearing & Stamping Co., Ohio, 2nd edn., 1968.

Stagg, K. G. and Zienkiewicz, O. C., *Rock mechanics in engineering practice*, J. Wiley, 1968.

Follenfant, H. G., *Underground railway construction*, London Transport, 1968.

Mayo, R. S. and others, *Tunneling: the state of the art*, U.S. Department of Housing and Urban Development, (NTIS PB 178036), 1968.

Safety in sewers and at sewage works, Institution of Civil Engineers, 1969.

Deere, D. U. and others, *Design of tunnel liners and support systems,* U.S. Department of Transportation (NTIS PB 183799), 1969.

Peck, R. B. and others, *Some design considerations in the selection of underground support systems*, U.S. Department of Transportation (NTIS PB 190443), 1969.

Household, H., *The Thames and Severn Canal*, David & Charles, 1969.

Clements, P., *Mark Isambard Brunel*, Longmans, 1970.

Duncan, J. M. and others, *Materials handling for tunneling*, U.S. Department of Transportation Report FRA-RT-71-57 (NTIS PB 197331), 1970.

Fenix and Scisson Inc., *A systems study of soft-ground tunneling*, U.S. Department of Transportation Report DOT-FRA-OHSGT-231 (NTIS PB 194769), 1970.

Major Tunnels in Japan, 1970, Japan Society of Civil Engineers.

Howson, F. H., *The rapid transit railways of the world*, Allen & Unwin, 1971.

Parker, H. W. and others, *Innovations in tunnel support systems*, U.S. Department of Transportation Report FRA-RT-72-17 (NTIS PB 204437), 1971.

Beaver, P., *A History of tunnels*, P. Davies, 1972.

Safety in wells and boreholes, Institution of Civil Engineers, 1972.

I.C.I., *Blasting Practice*, Nobel's Explosives Co. Ltd., Scotland, 4th edn., 1972.

Underground Construction Research Council, *The use of underground space to achieve national goals*, American Society of Civil Engineers, 1972.

Department of the Environment, *The Channel Tunnel Project, Cmnd 5256*, HMSO, 1973.

Department of the Environment, *The Channel Tunnel, Cmnd 5430*, HMSO, 1973.

Haining, P., *Eurotunnel, an illustrated history of the Channel scheme*, New English Library, 1973.

Nock, C. S., *Underground railways of the world*, A. & C. Block, 1973.

Parker, H. W. and others, *Testing and evaluation of prototype tunnel support systems*, U.S. Department of Transportation Report FRA-ORDD 74-11 (NTIS PB 231912), 1973.

Paul, S. L. and others, *Research to improve tunnel support systems*, U.S. Department of Transportation Report FRA-ORDD 74-51 (NTIS PB 235762), 1973.

Semple, R. A. and others, *The effect of time-dependent properties of altered rock on tunnel support requirements*, U.S. Department of Transportation Report FRA-ORDD-74-30 (NTIS PB 230 307), 1973.

Sverdrup & Parcel and Associates, *Cut-and-cover tunneling techniques,* **1**, A *study of the state of the art,* **2**, *Appendix*, U.S. Department of Transportation Reports FWHA-RD-73-40 & 41 (NTIS PB 222997 & 8), 1973.

Szechy, K., *The art of tunnelling*, Akademiai Kiado, 2nd edn., 1973.

Wahlstrom, E. E., *Tunneling in rock*, Elsevier, 1973.

Blyth, F. G. H. and de Freitas, M. H., *A geology for engineers*, E. Arnold, 6th edn., 1974.

Jacobs Associates, *Ground support prediction model (RSR concept)*, U.S. Bureau of Mines (NTIS AD 773018), 1974.

Follenfant, H. G., *Reconstructing London's underground*, London Transport, 2nd edn., 1975.

Cording, E. J. and others, *Methods for geotechnical observations and instrumentation in tunneling,* 2 vols., Dept. of Civ. Engng, University of Illinois, 1975 (NTIS PB 252585 & 6).

The present status of tunnelling activities in Japan, Japan Tunnelling Association, 1975.

Clough, G. W., *A Report on the practice of chemical stabilisation around soft ground tunnels in England and Europe*, Dept. of Civ. Engng, Stamford University, 1975.

Associated Engineers, *Subway Environmental Design Handbook*, U.S. Department of Transportation, 1975.

Muir Wood, A. M., 'Tunnelling', ch. 30 in Blake, L. S. (ed.), *Civil Engineer's Reference Book*, Newnes-Butterworth, 3rd edn., 1975.

Bell, F. C. (ed.), *Methods of treatment of unstable ground*, Butterworths, 1975.

Cording, E. J., *Displacements around tunnels in soil*, U.S. Department of Transportation Report DOT-TST 76T-22 (NTIS PB 267356), 1976.

Glerum, A. and others, Motorway tunnels built by the immersed tube method, *Rijkswaterstaat Communications*, 1976 (25).

Manual of applied geology for engineers, Institution of Civil Engineers and HMSO, 1976.

Herndon, J. and Lenahan, J., *Grouting in Soils*, 2 vols., U.S. Federal Highway Administration Reports FHWA-RD-76-26 & 27 (NTIS PB 259043 & 4), 1976.

Harding, H. J. B., Tunnels, ch. 2 in Pugsley, A. (ed.), *The works of Isambard Kingdom Brunel*, Institution of Civil Engineers, 1976, and Cambridge U.P., 1980.

Golder Associates and James F. MacLaren Ltd., *Tunnelling technology*, Ontario Ministry of Transport and Communications, 1976.

Pursall, B. R. and King, A. L., *The aerodynamics and ventilation of vehicle tunnels: a state of the art review and bibliography*, British Hydromechanics Research Association, 1976.

Parsons, Brinckerhoff, Quade & Douglas, *Subsurface exploration methods for soft ground rapid transit tunnels*, 2 vols., U.S. Department of Transportation Report UMTA-MA-06-0025-76-1 & 2 (NTIS PB 258343 & 4), 1976.

Bechtel Inc., *Systems study of precast concrete tunnel liners*, U.S. Department of Transportation Report DOT-TSC-OST-77-7 (NTIS PB 264761), 1977.

Consulting Engineers Group, *Prefabricated structural members for cut and cover tunnels*, **1**, *Design concepts*, **2**, *Three case studies*, U.S. Department of Transportation Reports FHWA-RD-76-113 & 4 (NTIS PB 273530 & 1), 1977.

DeLeuw, Cather & Co., *Study of subway station design and construction*, U.S. Department of Transportation Report UMTA-MA-06-0025-77-6 (NTIS PB 268894), 1977.

Proctor, R. V. and White, T. L., *Earth tunneling with steel supports*, Commercial Shearing & Stamping Co., Ohio, 1977.

Brakel, J., *Submerged Tunnelling*, Technische Hogeschool, Delft, 1978.

BS 5573 Code of Practice for safety precautions in the construction of large diameter boreholes for piling and other purposes, (Revision of CP 2011), British Standards Institution, 1978.

Langefors, U. and Kihlstrom, B., *The modern technique of rock blasting*, J. Wiley, 3rd edn., 1978.

Bayliss, D. A., *The Post Office Railway London*, Turntable Pubns., 1978.

Ranken, R. E. and others, *Analysis of ground-liner interaction for tunnels*, U.S. Department of Transportation Report UMTA-IL-06-0043-78-3 (NTIS PB 294818), 1978.

Metropolitan Atlanta Rapid Transit Authority, *The Atlanta research chamber; applied research for tunnels: blasting techniques, conventional shotcrete, steel-fiber-reinforced shotcrete; monographs on the state-of-the-art of tunnelling,* U.S. Department of Transportation Report UMTA-GA-0007-81-1 (NTIS PB 81-231151), 1981.

MacPherson, H. H. and others, *Settlements around tunnels in soil: three case histories*, U.S. Department of Transportation Report UMTA-IL-06-0043-78-1 (NTIS PB 290856), 1978.

O'Rourke, T. D., *Tunneling for urban transportation: a review of European construction practice*, U.S. Department of Transportation Report UMTA-IL-06-0041-78-1, 1978.

Hadfield, C., *British canals,* David & Charles, 6th edn., 1979.

Jaeger, C., *Rock mechanics and engineering*, Cambridge U.P., 2nd edn., 1979.

Schach, R. and others, *Rock bolting, a practical handbook*, Pergamon, 1979.

Rogers, L. H. (ed.), *International statistical handbook of urban public transport*, International Union of Public Transport (U.I.T.P.), Brussels, 1979.

Hoek, E. and Brown, E. T., *Underground excavations in rock*, Institution of Mining and Metallurgy, 1980.

The Present Status of Tunnelling Activities in Japan, Japan Tunnelling Association, Tokyo, 1980.

Schlaug, R. N., *Management of air quality in and near highway tunnels*, U.S. Department of Transportation Report FHWA-RD-78-186, 1980.

BS 5930, Code of Practice for site investigations (Revision of CP 2001), British Standards Institution, 1981.

Bowen, R., *Grouting in engineering practice*, Applied Science, 2nd edn., 1981.
House of Commons, Second Report from the Transport Committee, Session 1980–81, *The Channel Link, Report & Minutes, (HoC 155–I, II & III)*, HMSO, 1981.
Harding, H. J. B., *Tunnelling history and my own involvement*, Golder Assocs., 1981.
A guide to pipe jacking design, Pipe Jacking Association, 1981.
Stack, B., *Handbook of mining and tunnelling equipment*, J. Wiley, 1982.
BS 6164, Code of Practice for Safety in Tunnelling in the construction industry, British Standards Institution, 1982.

B. INSTITUTION OF CIVIL ENGINEERS

The principal publications embodying papers and discussions are:

Minutes of Proceedings	1837–1935,	**1–240**	*MP*
Journal	1935–1951,	**1–36**	*J*
Proceedings	1952– ,	**1–**	*P*

From Proceedings 1957, **8** (Sept) discussions were published later and are separately noted.

Additional series of papers are indicated:

Selected Engineering Papers	1922–1935	*SEP*	
Engineering Division Papers	1941–1951:	*Works*	*Wks*
		Railway	*Rly*
Civil Engineer in War	1948	*CEW*	
Chartered Civil Engineer	1949–1957	*CCE*	

'Description of the works on the Netherton tunnel branch of the Birmingham canal',
Walker, J. R., *MP*, 1860, **19**
'On the construction and enlargement of the Lindal tunnel, on the Furness railway',
Stileman, F. C., *MP*, 1860, **19**
'Description of the Lydgate and of the Buckhorn Weston railway tunnels',
Fraser, J. G., *MP*, 1863, **22**
'The actual state of the works on the Mont Cenis tunnel, and description of the machinery employed',
Sopwith, T., *MP*, 1864, **23**
'On the main drainage of London, and the interception of the sewage from the River Thames',
Bazalgette, J. W., *MP*, 1865, **24**
'On the geological conditions affecting the construction of a tunnel between England and France',
Prestwich, J., *MP*, 1873, **37**
'The St. Gothard tunnel',
Clark, D. K., *MP*, 1875, **42**

'On tunnel construction, and on the Sydenham tunnel, London, Chatham and Dover railway',
Baldwin, A. E., *MP*, 1877, **49**
'The St. Gothard tunnel',
Clark, D. K., *MP*, 1879, **57**
'Tunnel outlets from storage reservoirs',
Wood, C. J., *MP*, 1879, **59**
'The main drainage of Torquay',
Chatterton, G., *MP*, 1880, **61**
'Summit-level tunnel of the Bettws and Festiniog railway',
Smith, W., *MP*, 1883, **73**
'The Metropolitan and Metropolitan District Railways',
Baker, B.,
'The City lines and extensions (Inner Circle completion) of the Metropolitan and District railways',
Barry, J. W., *MP*, 1885, **81**
'The Mersey Railway',
Fox, F., *MP*, 1886, **86**
'Yanegase Yama tunnel on the Tsuruga-Nagahama railway, Japan',
Hasegawa, K., *MP*, 1887, **90**
'Covered way as constructed on the Glasgow City and District railway',
Wilson, W. S., *MP*, 1888, **92**
'Reparation of Betchworth tunnel, Dorking, on the London, Brighton and South Coast railway',
Lopes, G., *MP*, 1888, **95**
'Alpine engineering',
Vernon-Harcourt, L. F., *MP*, 1888, **95**
'Methods adopted in constructing the Glasgow Central railway (Bridgeton and Trongate contracts)',
Barker, C. D., *MP*, 1893, **114**
'Method of underpinning and tunnelling under tenements',
Young, J., *MP*, 1893, **115**
'Tunnels on the Dore and Chinley railway',
Rickard, P., *MP*, 1894, **116**
'Timbering in the Ampthill second tunnel',
Matheson, E. E., *MP*, 1895, **120**
'Glasgow District subway',
Stewart, A. M., *MP*, 1895, **122**
'The City and South London Railway; with some remarks upon subaqueous tunnelling by shield and compressed air',
Greathead, J. H., *MP*, 1895, **123**
'Bold Street extension tunnel and Central low-level station of the Mersey railway',
Rowlandson, C. A., *MP*, 1895, **123**

'Iron tunnels',
Leitch, W. O., *MP*, 1896, **125**
'Periyar (river) tunnel',
Allen, P. R., *MP*, 1897, **128**
'Tunnels on the first division of the Mushkaf-Bolan railway, Baluchistan',
Johns, W. A., *MP*, 1897, **128**
'Tunnels on the second division of the Mushkaf-Bolan railway, Baluchistan',
Cole, C. J., *MP*, 1897, **128**
'Main drainage of London',
Worth, J. E. and Crimp, W. S., *MP*, 1897, **129**
'The Blackwall Tunnel',
Hay, D. and Fitzmaurice, M., *MP*, 1897, **130**
'Section no. 1, North Yarra main sewer',
Champion, H. V., *MP*, 1898, **133**
'The Waterloo and City Railway',
Dalrymple-Hay, H. H., *MP*, 1899, **139**
'Maintenance of railway tunnels',
Watson, A., *MP*, 1900, **140**
'The construction of the Elan aqueduct: Rhayader to Dolan',
Lapworth, H., *MP*, 1900, **140**
'Construction of the Simplon tunnel',
Fox, C. B., *MP*, 1900, **140**
'Great Central railway extension,. Northern division',
Bidder, F. W.,
'Great Central railway extension, Southern division',
Fox, F. D., *MP*, 1900, **142**
'Setting-out of two tunnels on the Elan aqueduct',
Brightmore, A. W., *MP*, 1900, **142**
'Location and construction of railway tunnels',
Rogers, A. E., *MP*, 1901, **146**
'Greenwich footway tunnel',
Copperthwaite, W. C., *MP*, 1902, **150**
'Subaqueous tunnelling through the Thames gravel: Baker Street and Waterloo railway',
Haigh, A. H., *MP*, 1902, **150**
'Notes on the construction and setting out of tunnels in the London clay',
Bartlett, H. A., *MP*, 1904, **156**
'Renewal of a portion of the invert of Netherton (canal) tunnel, 1902',
Nias, C. H., *MP*, 1904, **158**
'Alfreton tunnel',
French, E. F. C., *MP*, 1905, **161**
'Reconstruction of Moncreiffe tunnel',
McLellan, D., *MP*, 1905, **161**

'The Simplon Tunnel',
Fox, F., *MP*, 1907, **168**
'The New York Rapid-Transit Subway',
Parsons, W. B., *MP*, 1908, **173**
'The Rotherhithe Tunnel',
Tabor, E. H., *MP*, 1908, **175**
'Re-lining a portion of Cwm Cerwyn tunnel, Port Talbot railway',
Cleaver, W., *MP*, 1909, **177**
'Aberdeen main drainage works: Girdleness outfall scheme',
Conway, G. R. G., *MP*, 1909, **177**
'Hudson River tunnels of the Hudson and Manhattan Railroad Company',
Jacobs, C. M., *MP*, 1910, **181**
'The London County Council Holborn-to-Strand improvement, and tramway subway',
Humphreys, G. W., *MP*, 1910, **183**
'Electricity applied to a modern tunnelling work',
Yerbury, H. E., *MP*, 1910, **183**
'Construction and setting-out of the low-level sewer between Cannon Street and Belgrave Square',
Wilson, L. T., *MP*, 1911, **184**
'The Detroit River tunnel, between Detroit, Michigan and Windsor, Canada',
Wilgus, W. J., *MP*, 1911, **185**
'The Concorde tunnel of the Paris Metropolitan railway',
Seurot, P., *MP*, 1912, **188**
'Works for the water supply to Birmingham from mid-Wales',
Mansergh, E. L. and W. L., *MP*, 1912, **190**
'City passenger-transportation in the United States',
Snyder, G. D., *MP*, 1913, **193**
'Transandine Railway',
Henderson, B. H., *MP*, 1913, **195**
'Extensions of the Hudson River tunnels of the Hudson and Manhattan railroad company',
Davies, J. V., *MP*, 1914, **197**
'Laxaxalpam aqueduct tunnels, Mexico',
Forgie, J., *MP*, 1915, **200**
'The main drainage system of London',
Humphreys, G. W., *MP*, 1917, **204**
'Buenos Aires Western railway tunnels under the city of Buenos Aires',
Lowe-Brown, W. L., *MP*, 1917, **205**
'The Khyber Railway',
Bayley, V.,
'Survey and construction of the Khyber Railway',
Hearn, G., *MP*, 1926, **222**

'Enlargement of the City and South London Railway tunnels',
Jones, I. J. and Curry, G., *MP*, 1927, **224**
'Sub-aqueous tunnelling in compressed air, with reference to Barking power-station cable tunnel under the River Thames',
Matheson, D. S., *SEP*, 1927, (43)
'The reconstruction of Mossgiel tunnel',
McLellan, D., *SEP*, 1927, (53)
'The new Piccadilly Circus station',
Hall, H., *MP*, 1929, **228**
'The Tata Company's hydro-electric project',
Cursetjee, N. J. M., *MP*, 1929, **228**
'The Lochaber water-power scheme',
Halcrow, W. T., *MP*, 1930, **231**
'The opening-out of Cofton tunnel, London Midland and Scottish railway',
McCallum, R. T., *MP*, 1930, **231**
'The construction, testing and strengthening of a pressure tunnel for the water supply of Sydney, N.S.W.',
Haskins, G., *MP*, 1932, **234**
'Cementation in the Severn Tunnel',
Carpmael, R., *MP*, 1932, **234**
'The River Hooghly Tunnel',
Norrie, C. M., *MP*, 1933, **235**
'Power-house foundations and circulating-water tunnels at the Ford Motor Company's works, Dagenham',
Allin, R. V. and Nachshen, M., *MP*, 1933, **235**
'Circulating-water culverts for the Portishead generating station of the Bristol corporation',
Rees, G. M. G., *SEP*, 1934, (163)
'The coaling-jetty, circulating water system and cable-tunnels of the Battersea power station of the London Power Co. Ltd.',
Bartlett, V. F. and Cadwell, W. H., *MP*, 1935, **240**
'The construction of the Mersey Tunnel',
Anderson, D., *J.*, 1936, **2** (Apr.)
'The construction of infiltration galleries for the water-supply of Ndola, Northern Rhodesia',
McIlleron, R. C., *J.*, 1936, **3** (June)
'West Middlesex Main Drainage',
Watson, D. M., *J.*, 1937, **5** (Apr.)
'Reconstruction of Aldgate East station',
Harley-Mason, J. H., *J.*, 1939, **11** (Apr.)
'The strengthening and final testing of the pressure tunnel for the water supply of Sydney, N.S.W.',
Farnsworth, S. T., *J.*, 1939, **11** (Apr.)

'Notes on the station-lengthening at Shepherd's Bush station',
Filor, C. G. H., *J.*, 1940, **13** (Feb.)
'Tunnel linings with special reference to a new form of reinforced concrete lining',
Groves, G. L., *J.*, 1943, **20** (Mar.)
'The Laghouat escape tunnel',
Brickell, R. G., *J.*, 1943, **20** (Apr.)
'Tunnelling with pre-cast concrete',
Pattenden, B., *J.*, 1945, **23** (Feb.)
'Emergency repair to the Tower subway, London, after air-raid damage',
Harding, H. J. B., *J.*, 1945, **25** (Nov.)
'Particle size in silts and sands',
Glossop, R. and Skempton, A. W., *J.*, 1945, **25** (Dec.)
'The use of aluminous cement in the construction of the Mosul tunnel, Iraqi state railways',
Hagger, G. C., *J.*, 1945, **25** (Dec.)
'The Ilford Tube',
Groves, G. L., *J.*, 1946, **26** (Mar.)
'Some uses of explosives in civil engineering',
Lorimer, J., *J.*, 1946, **26** (Mar.)
'Applications of the freezing process to civil engineering works',
Mussche, H. E. and Waddington, J. C., *Wks*, 1946 (5)
'The choice of expedients in civil engineering construction',
Harding, H. J. B., *Wks*, 1946 (6)
'Tunnel maintenance',
Robarts, H. E., *Rly*, 1947 (23)
'The construction of the Trans-Iranian Railway, with particular reference to the works of "Lot 6 (South)" ',
Champion, C. L., *J.*, 1947, **29** (Dec.)
'Some special storages',
Temple-Richards, H. B., *CEW*, 1948, **3**
'Protection against flooding in the London Underground railway system',
Morgan, H. D., *CEW*, 1948, **3**
'The tunnels in Gibraltar',
Cotton, J. C., *CEW*, 1948, **3**
'Site investigations, including boring and other methods of subsurface exploration',
Harding, H. J. B., *J.*, 1949, **32** (Apr.)
'Compressed-air caisson foundations',
Wilson, W. S. and Sully, F. W., *Wks*, 1949 (13)
'The driving and lining of the Clunie tunnel on the Tummel-Garry hydro-electric project',
Grundy, C. F., *Wks*, 1950 (15)

'The Mullardoch-Fasnakyle-Affric tunnels',
Dillon, E. C., *Wks*, 1950 (16)
'Arley tunnel: remedial works following subsidence',
King, C. W., *J.*, 1951, **36** (Mar.)
'Part reconstruction of Bo-Peep tunnel at St. Leonards-on-Sea',
Campion, F. E., *J.*, 1951, **36** (Mar.)
'A 75-inch diameter water main in tunnel: a new method of tunnelling in London clay',
Scott, P. A., *P.*, 1952, **1** (May)
'Civil engineering aspects of hydro-electric development in Scotland',
Fulton, A. A., *P.*, 1952, **1** (July)
'Tunnelling plant and equipment',
Archer, G. C., *P.*, 1952, **1** (Sep.)
'Some major problems in railway civil engineering maintenance',
Cantrell, A. H., *P.*, 1953, **2** (Feb.)
'Special features of the Affric hydro-electric scheme (Scotland)',
Roberts, C. M., *P.*, 1953, **2** (Sep.)
'The trend of development in the design and equipment of underground railways',
Coombs, D. H. and Wilson, G. J., *P.*, 1953, **2** (Oct.)
'The construction of Tigres dam and Malgovert tunnel',
Pelletier, J., *P.*, 1953, **2** (Dec.)
'The construction of two concrete-lined tunnels for the Machkund hydro-electric project, South India',
Gauld, G. A. and Newport, R. J., *P.*, 1954, **3** (Mar.)
'Concerning the main drainage of London',
Warlow, W. P., *P.*, 1954, **3** (May)
'Swedish underground hydro-electric power stations',
Hagrup, J. F., *P.*, 1954, **3** (Aug.)
'Woodhead New Tunnel: construction of three-mile main double-line railway tunnel',
Scott, P. A. and Campbell, J. I., *P.*, 1954, **3** (Sep.)
'Present trends in design of pressure tunnels and shafts for underground hydro-electric power stations',
Jaeger, C., *P.*, 1955, **4** (Mar.)
'Investigations into the design of pressure tunnels in London clay',
Tattersall, F. and others, *P.*, 1955, **4** (July)
'The diversion of the Annalong River into the Silent Valley reservoir',
Colebrook, C. F., *P.*, 1956, **5** (Apr.)
'Uses of aerated cement grout and mortar in stabilization of slips in embankments, large scale tunnel repairs and other works',
Purbrick, M. C. and Ayres, D. J., *P.*, 1956, **5** (Feb.)
'The construction of Middleton connecting sewer in north Manchester',
Collier, E. H., *P.*, 1956, **5** (Dec.)

'The equipment of the Ilford tube',
Cuthbert, E. W., *Rly*, 1950 (36)

'Pretreatment of gravel for compressed air tunnelling under the River Thames at Dartford',
Kell, J., *CCE*, 1957 (Mar.)

'The introduction of tunnelling for the construction of large-diameter sewers in Kingston-upon-Hull',
Roberts, G. W. and Lucas, W. H., *P.*, 1957, **6** (Apr.)

'Shaft raising in the Scottish Highlands',
Henderson, R. S., *P.*, 1957, **6** (Apr.)

'Portishead "B" power station, with particular reference to the circulating-water works',
Morgan, H. D. and others, *P.*, 1957, **7** (Aug.)

'Preliminary planning for a new tube railway across London',
Turner, F. S. P., *P.*, 1959, **12** (Jan.), disc. 1959, **13** (Aug.)

'The use of the freezing process in the construction of a pumping station and storm-water overflow at Fleetwood, Lancs.',
Ellis, D. R. and McConnell, J., *P.*, 1959, **12** (Feb.), disc. 1959, **13** (July)

'Underground station for Western District Post Office, London: design and control',
Blackford, S. and Cuthbert, E. W.,

'Ditto: construction',
Collingridge, V. H. and Tuckwell, R. E., *P.*, 1960, **15** (Feb.), disc. 1960, **17** (Nov.)

'Zambezi hydro-electric development at Kariba: first stage',
Anderson, D. and others, *P.*, 1960, **17** (Sep.), disc. 1962, **21** (Mar.)

'The work of the Channel Tunnel Study Group, 1958–60',
Bruckshaw, J. M. and others, *P.*, 1961, **18** (Feb.), disc. 1962, **21** (Mar.)

'New tunnels near Potters Bar in the Eastern Region of British Railways',
Terris, A. K. and Morgan, H. D., *P.*, 1961, **18** (Apr.), disc. 1962, **21** (Apr.)

'The Thames to Lee tunnel water main',
Cuthbert, E. W. and Wood, F., *P.*, 1962, **21** (Feb.), disc. 1962, **23** (Dec.)

'An introduction to alluvial grouting',
Ischy, E. and Glossop, R., *P.*, 1962, **21** (Mar.), disc. 1962, **23** (Dec.)

'Special features of the civil engineering works at Aberthaw power station',
Fraenkel, P. M. and Triggs, R. L., *P.*, 1962, **23** (Sep.), disc. 1964, **27** (Mar.)

'Installation of a reinforced-concrete invert in the Gillingham (Dorset) tunnel, on the Southern Region of British Railways',
Tanner, P. W. and Burton, R. A. H., *P.*, 1963, **24** (Jan.), disc. 1964, **27** (Jan.)

'The subsidence at Fylde Street, Farnworth',
Hale, H. T. and Dyer, E. A., *P.*, 1963, **24** (Feb.), disc. 1963, **26** (Sep.)

'The Dartford tunnel',
Kell, J., *P.*, 1963, **24** (Mar.), disc. 1963, **26** (Dec.)

'Installation of trav-o-lators at the Bank Station of the Waterloo and City Railway,'
Bubbers, B. L. and Coles, D. E., 1963, **24** (Apl), disc. **26** (Nov).
'Cameron Highlands hydroelectric scheme',
Dickinson, J. C. and Gerrard, R. T., *P.*, 1963, **26** (Nov.), disc. 1964, **29** (Sep.)
'Park Lane improvement scheme: design and construction',
Granter, E., *P.*, 1964, **29** (Oct.), disc. 1966, **33** (Mar.) and 1966, **35** (Dec.)
'Clyde tunnel: design, construction and tunnel services',
Morgan, H. D. and others,
'Ditto: constructional problems',
Haxton, A. F. and Whyte, H. E., *P.*, 1965, **30** (Feb.), disc. 1967, **37** (July)
'The Bretby tunnelling machine',
Hay, J. D. and others, *P.*, 1965, **30** (Apr.), disc. 1966, **35** (Sep.)
'Victoria Line: experimentation, design, programming and early progress',
Dunton, C. E. and others, *P.*, 1965, **31** (May), disc. 1966, **34** (July)
'Hazard from blasting fumes in normal rock tunnelling, with particular reference to nitrogen dioxide',
Fogden, C. A. and Garrod, A. D., *P.*, 1965, **31** (July), disc. 1966, **34** (Aug.)
'Soft ground tunnelling for the Toronto subway',
Bartlett, J. V. and others, *P.*, 1965, **32** (Sep.), disc. 1966, **34** (May)
'Blackwall tunnel duplication',
Kell, J. and Ridley, G., *P.*, 1966, **35** (Oct.), disc. 1967, **37** (July)
'Strand underpass',
Finn, E. V. and others, *P.*, 1966, **35** (Nov.), disc. 1967, **37** (Aug.)
'Some design and construction features of the Cruachan pumped storage project',
Young, W. and Falkiner, R. H., *P.*, 1966, **35** (Nov.), disc. 1967, **37** (Aug.)
'Furnas hydro-electric scheme, Brazil: closure of diversion tunnels',
Lyra, F. H. and MacGregor, W., *P.*, 1967, **36** (Jan.), disc. 1967, **37** (Aug.)
'Horizontal earth boring',
Thomson, J. C., *P.*, 1967, **36** (Apr.)
'The Bristol regional foul water drainage scheme',
Steel, P. H., *P.*, 1967, **38** (Oct.), disc. 1968, **41** (Oct.)
'The Tyne tunnel: planning of the scheme',
Prosser, J. R. and Grant, P. A. St. C.,
'Ditto: construction of the main tunnel',
Falkiner, R. H. and Tough, S. G., *P.*, 1968, **39** (Feb.), disc. 1968, **41** (Nov.)
'Cooling water intakes at Wylfa power station',
Chapman, E. J. K. and others, *P.*, 1969, **42** (Feb.), disc. 1969, **44** (Nov.)
'Thames Cable tunnel',
Haswell, C. K., *P.*, 1969, **44** (Dec.), disc. 1970, **47** (Oct.)
'The Victoria Line',
Follenfant, H. G. and others, *P.*, 1969, **Suppl**, disc. 1970, **Suppl**

'The site investigations for a Channel tunnel 1964–65',
Grange, A. and Muir Wood, A. M., *P.*, 1970, **45** (Jan.), disc. 1970, **47** (Dec.)
'Chesapeake Bay bridge-tunnel project',
Sverdrup, L. J., *P.*, 1970, **46** (May)
'Construction of the Monmouth tunnels in soft rock',
Parry, Rj. R. and Thornton, G. D., *P.*, 1970, **47** (Sep.), disc. 1971, **49** (July)
'Design and construction of the cargo tunnel at Heathrow airport, London',
Muir Wood, A. M. and Gibb, F. R., *P.*, 1971, **48** (Jan.), disc. 1971, **50** (Oct.)
'Some considerations of submerged tunnelling',
Brakel, J., *P.*, 1971, **48** (Apr.), disc. 1972, **51** (Jan.)
'The prediction of the magnitudes of pressure transients generated by a train entering a single tunnel',
Fox, J. A. and Henson, D. A., *P.*, 1971, **49** (May), disc. 1971, **50** (Dec.)
'The Manapouri power project, New Zealand',
Langbein, J. A., *P.*, 1971, **50** (Nov.), disc. 1972, **52** (Nov.)
'Mersey Kingsway tunnel: planning and design',
Megaw, T. M. and Brown, C. D.,
'Ditto: construction',
McKenzie, J. C. and Dodds, G. S.,
'Ditto: project organization, approach roads, buildings and operating organization',
Cairncross, A. A. and Jones, S. T., *P.*, 1972, **51** (Mar.), disc. 1973, **54** (May)
'Design aspects of tunnelling in Tyneside',
Boden, J. B., *P.*, 1972, **53** (Sep.)
'Construction of the inclined penstocks at Churchill Falls power station, Labrador',
Dawson, H. J. and Lyttle, R. P. T., *P.*, 1972, **52** (Nov.), disc. 1973, **54** (Aug.)
'Shaft sinking by ground freezing: Ely Ouse-Essex scheme',
Collins, S. P. and Deacon, W. G., *P.*, 1972, **Suppl**, disc. 1972, **Suppl**
'The use of rock tunnel machines',
Pirrie, N. D., *P.*, 1973, **54** (Feb.), disc. 1973, **54** (Nov.)
'Deformation of tunnel with variable ground reaction',
Brownrigg, D. R. K. and Wood, W. L., *P.*, 1973, **55** (June)
'Tunnel under the Severn and Wye estuaries',
Haswell, C. K., *P.*, 1973, **54** (Aug.), disc. 1974, **56** (Aug.)
'The bentonite tunnelling machine',
Bartlett, J. V. and others, *P.*, 1973, **54** (Nov.), disc. 1974, **56** (Aug.)
'The Elbe tunnel, Hamburg: southern ramp and submerged tunnel',
Havno, K., *P.*, 1974, **56** (Aug.), disc. 1975, **58** (May)
'Soft ground tunnelling',
Bartlett, J. V. and King, J. R. J., *P.*, 1975, **58** (Nov.), disc. 1976, **60** (Aug.)
'Planning and design of the Hong Kong Mass Transit Railway',
Edwards, J. T., *P.*, 1976, **60** (Feb.), disc. 1976, **60** (Aug.)

'Development of Edinburgh's sewage disposal scheme',
Crockett, A. S. and Dugdale, J., *P*., 1976, **60** (Feb.), disc. 1976, **60** (Nov.)
'Ground movements caused by tunnelling in chalk',
Priest, S. D., *P*., 1976, **61** (Mar.), disc. 1976, **61** (Dec.)
'Structural performance of a temporary tunnel lined with spheroidal graphite cast iron',
Thomas, H. S. H., *P*., 1976, **61** (Mar.)
'Extension of the Piccadilly Line from Hounslow West to Heathrow Central',
Jobling, D. G. and Lyons, A. C., *P*., 1976, **60** (May), disc. 1976, **60** (Nov.)
'Pipe-jacking applied to large structures',
Clarkson, T. E. and Ropkins, J. W. T., *P*., 1977, **62** (Nov.), disc. 1978, **64** (Aug.)
'Foyers pumped storage project: planning and design',
Lander, J. H. and others,
'Ditto: construction',
Land, D. D. and Hitchings, D. C., *P*., 1978, **64** (Feb.), disc. 1978, **64** (Nov.)
'Calcutta Rapid Transit System and the Park Street underground station',
Dasgupta, K. N. and others, *P*., 1979, **66** (May), disc. 1980, **68** (Feb.)
'Tunnelling trials in chalk',
O'Reilly, M. P. and others, *P*., 1979, **67** (June)
'Tyne and Wear metro',
Howard, D. F., *P*., 1979, **66** (Aug.)
'The Jubilee Line – 1. The project',
Cuthbert, E. W.,
'Ditto – 2. Construction from Baker Street to Bond Street exclusive and from Admiralty Arch to Aldwych',
Lyons, A. C.,
'Ditto – 3. Construction from Bond Street station to Admiralty Arch',
Bubbers, B. L., *P*., 1979, **66** (Aug.), disc. 1980, **68** (Feb.)
'Design of concrete shaft linings',
Auld, F. A., *P*., 1979, **67** (Sep.)
'The Transclyde railway network: Kelvinhaugh burrowing junction',
Craige, G. B. and others, *P*., 1980, **68** (Aug.), disc. 1981, **70** (May)
'Hong Kong Mass Transit Railway Modified Initial System: system planning and multi-contract procedures',
Edwards, J. T. and others,
'Ditto: design and construction of underground stations and cut-and-cover tunnels',
McIntosh, D. F. and others,
'Ditto: design and construction of the driven tunnels and the immersed tube',
Haswell, C. K. and others, *P*., 1980, **68** (Nov.)
'Tyne and Wear Metro, Byker contract: planning, tunnels, stations and track-work',

Smyth, W. J. R. and others, *P.*, 1980, **68** (Nov.), disc. 1981, **70** (Aug.)
'Cavitation damage and the Tarbela tunnel collapse of 1974',
Kenn, M. J. and Garrod, A. D., *P.*, 1981, **70** (Feb.), disc. 1981, **70** (Nov.)
'Tyne and Wear Metro: concept, organization and operation',
Howard, D. F. and Layfield, P.,
'Ditto: management of the project',
Bartlett, J. V. and others,
'Ditto: design and construction of tunnelling works and underground stations',
Tough, S. G. and others, *P.*, 1981, **70** (Nov.)

C. AMERICAN SOCIETY OF CIVIL ENGINEERS

From 1867 to 1963 all important papers were published in full, with discussion, in *Transactions* **1–128**; subsequent volumes contain summaries and now only abstracts. Early issues of *Proceedings* were preprints of papers, most of which were subsequently reprinted in *Transactions*. From *Proceedings*, 1956, **82** onwards a series of *Journals of Divisions* has grown, each with a variable number of issues per annum; relevant *Journals* are abbreviated as follows:

Construction CO
Engineering Issues EI
Engineering Mechanics EM
Power PO
Structural ST
Surveying and Mapping SU
City Planning, later *Urban Planning CP/UP*
Pipe Line, merged in *Transportation Engineering PL/TE*
Soil Mechanics and Foundations, later *Geotechnical Engineering SM/GT*

C1 Transactions

'Arching Bergen tunnel on Erie railroad',
Houston, J., 1872, **1**
'Nesquehoning tunnel',
Steele, J. D., 1872, **1**
'Tunnels of the Pacific railroad',
Gilliss, J. R., 1872, **1**
'Reconstruction and enlargement of Cork Run tunnel on the Pittsburgh, Cincinnati and St. Louis railway',
Becker, M. J., 1877, **6**
'The location of the Chimbote tunnel',
Nichols, O. F., 1880, **9**
'The Hudson River tunnel',
Spielmann, A. and Brush, C. B., 1880, **9**
'Shaft sinking under difficulties at Dorchester Bay tunnel, Boston, Mass.',
Stauffer, D. McN., 1881, **10**
'The Hudson River tunnel',
Smith, W. S., 1882, **11**
'Tunnel surveying on division no. 6, New Croton aqueduct',
Watkins, F. W., 1890, **23**

'The Philadelphia tunnel of the Baltimore and Ohio railroad, its construction and cost',
Thayer, W. W., 1892, **26**
'A method of tunnel alignment',
Dunham, H. F., 1892, **27**
'Lining a waterworks tunnel with concrete',
Fitzgerald, D., 1894, **31**
'Brafford's Ridge tunnel',
Staniford, C. W., 1894, **32**
'Tequixquiac tunnel, Valley of Mexico',
Campbell, A. J. and Abbot, F. W., 1894, **32**
'The Pennsylvania Avenue subway and tunnel, Philadelphia, Pa.',
Webster, G. S. and Wagner, S. T., 1902, **48**
'Underground railways: the metropolitan system of Paris',
Biette, L.,
'Underground railways in Great Britain',
Mott, B. and Hay, D.,
'Underground railways in the United States',
Parsons, W. B., 1905, **54**(F)
'The Scranton tunnel of the Lackawanna and Wyoming Valley railroad',
Francis, G. B. and Dennis, W. F., 1906, **56**
'The bracing of trenches and tunnels, with practical formulas for earth pressures',
Meem, J. C., 1908, **60**
'Caisson disease and its prevention',
Japp, H., 1909, **65**
'The New York Tunnel extension of the Pennsylvania railroad: Symposium of 16 papers',
Raymond, C. W. and others, 1910, **68** and **69**
'Experiments on retaining walls and pressures on tunnels',
Cain, W., 1911, **72**
'Sinking a wet shaft',
Hogan, J. P., 1911, **73**
'The Detroit River Tunnel',
Kinnear, W. S., 1911, **74**
'Air resistance to trains in tube tunnels',
Davies, J. V., 1912, **75**
'Notes on a tunnel survey',
Noble, F. C., 1912, **75**
'Specifications for the design of bridges and subways',
Seaman, H. B., 1912, **75**
'The Laramie-Poudre tunnel',
Coy, B. G., 1912, **75**

'The Sixth Avenue subway of the Hudson and Manhattan railroad',
Burrowes, H. G., 1913, **76**
'Subaqueous highway tunnels',
Snyder, G. D., 1915, **78**
'Secure subway supports',
Lueder, A. B., 1916, **80**
'The Astoria tunnel under the East River for gas distribution in New York City',
Davies, J. V., 1916, **80**
'Construction methods for Rogers Pass tunnel',
Dennis, A. V., 1917, **81**
'Tunnel work on sections 8, 9, 10 and 11, Broadway–Lexington Avenue subway, New York City',
Werbin, I. V., 1917, **81**
'Underpinning Trinity Vestry building for subway construction',
Parsons, H. de B., 1917, **81**
'Construction problems of the Manhattan–Bronx and Lexington Avenue subway junction and Queensborough tunnel connections',
Perrine, G., 1918, **82**
'Notes on tunnel lining for soft ground',
Johannesson, S. and Hewett, B. H. M., 1920, **83**
'Grouting operations, Catskill water supply',
Sanborn, J. F. and Zipser, M. E., 1920, **83**
'The proposed New York and New Jersey vehicular tunnel',
Byrne, E. A. and others, 1920, **83**
'Bridge versus tunnel for the proposed Hudson River crossing at New York City',
Waddell, J. A., 1921, **84**
'Construction progress of the Hetch Hetchy water supply of San Francisco',
O'Shaughnessy, M. M., 1922, **85**
'Elastic stresses in the rock surrounding pressure tunnels',
Dunn, C. P., 1923, **86**
'Engineering geology of the Catskill water supply',
Berkey, C. P. and Sanborn, J. F., 1923, **86**
'Pipe tunnel under Gowanus canal, Brooklyn, New York',
Stiles, L. S., 1927, **90**
'Construction methods on the Moffat tunnel',
Keays, R. H., 1928, **92**
'The Shandakan tunnel',
Gaussmann, R. W., 1928, **92**
'Methods used in the construction of twelve pre-cast concrete segments for the Alameda County, California, estuary subway',
Horwege, A. A., 1929, **93**
'Trans-mountain water diversions: a Symposium',
Debler, E. B. and others, 1930, **94**

'Completion of the Moffat tunnel of Colorado',
Betts, C. A., 1931, **95**
'The eight-mile Cascade tunnel, Great Northern railway: a Symposium',
Kerr, D. J. and others, 1932, **96**
'Fulton Street, East River tunnels, New York',
Killmer, M. I., 1933, **98**
'Analysis of stresses in subaqueous tunnel tubes',
Eremin, A. A., 1938, **103**
'Stress distribution around a tunnel',
Mindlin, R. D., 1940, **105**
'Bridge and tunnel approaches',
Curtin, J. F., 1941, **106**
'Measuring the potential traffic of a proposed vehicular crossing',
Cherniack, N., 1941, **106**
'Tunnel construction, Sixth Avenue subway, New York',
Feld, J., 1942, **107**
'Liner-plate tunnels on the Chicago subway',
Terzaghi, K., 1943, **108**
'Earth pressure on tunnels',
Housel, W. S., 1943, **108**
'The Queen's Midtown tunnel',
Singstad, O., 1944, **109**
'Pressure on the lining of circular tunnels in plastic soils',
Krynine, D. P., 1945, **110**
'Stresses in the linings of shield-driven tunnels',
Bull, A., 1946, **111**
'Application of geology to tunneling problems',
Wahlstrom, E. E., 1948, **113**
'Friction coefficients in a large tunnel',
Peterka, A. J. and Elder, R. A., 1948, **113**
'Diversion tunnel and power conduit of Nantahala hydroelectric development',
Bleifuss, D. J., 1951, **116**
'Floating tunnel for long water crossings',
Andrew, C. E., 1951, **116**
'Construction of the Elizabeth River tunnel',
Peraino, J., 1955, **120**
'Friction measurements in the Apalachia tunnel',
Elder, R. A., 1958, **123**
'Haas hydroelectric power project',
Cooke, J. B., 1959, **124**

C2 Proceedings

'Deas Island Tunnel',
Hall, P., 1957, **83** (ST6, Nov.)

'Geological factors in tunnel construction: 3 papers',
Cleaves, A. B. and others, 1958, **84** (SM2, May)
'Evaluation of alternative subway routes',
Griffith, B. A. and von Cube, H. G., 1960, **86** (CP1, May)
'Soil freezing to reconstruct a railway tunnel',
Low, G. J., 1960, **86** (CO3, Nov.)
'Oahe dam: influence of shale on Oahe power structures design',
Johns, E. A. and Burnett, R. G., 1963, **89** (SM1, Feb.)
'Design and performance of Mammoth Pool power tunnel',
Laverty, B. R. and Ludwig, K. R., 1963, **89** (PO1, Sep.)
'Stresses at tunnel intersections',
Riley, W. F., 1964, **90** (EM2, Apr.)
'Subway tunnel construction in New York City',
Knight, G. B., 1964, **90** (CO2, Sep.)
'Tunnels for hydroelectric power in Tasmania',
Thomas, H. H. and Whitman, 1964, **90** (PO3, Oct.)
'Rock trap experience in unlined tunnels',
Mattimoe, J. J. and others, 1964, **90** (PO3, Oct.)
'Unlined tunnels of the Snowy Mountains Hydroelectric Authority, Australia',
Dann, H. E. and others, 1964, **90** (PO3, Oct.)
'Contractor's view on unlined tunnels',
Petrofsky, A. M., 1964, **90** (PO3, Oct.)
'Unlined tunnels of the Southern California Edison Company',
Spencer, R. W. and others, 1964, **90** (PO3, Oct.)
'Machine tunneling on Missouri River dams',
Underwood, L. B., 1965, **91** (CO1, May)
'British grouting practice on granular soils',
Perrott, W. E., 1965, **91** (SM6, Nov.)
'Tunnel design in massive rocks',
Adler, L., 1966, **92** (SM6, Nov.)
'Stability of clay at vertical openings',
Broms, B. T. and Bennermark, H., 1967, **93** (SM1, Jan.)
'Oroville dam diversion tunnels',
Lanning, C. C., 1967, **93** (PO2, Oct.)
'Homestake trans-divide water supply project',
Thon, G. J. and Campbell, D. E., 1967, **93** (PL3, Nov.)
'Ground freezing in construction',
Sanger, F. J. and Golder, H. Q., 1968, **94** (SM1, Jan.)
'Stresses against underground structural cylinders',
Hoeg, K., 1968, **94** (SM4, July)
'What's ahead for tunneling machines',
Hill, G., 1968, **94** (CO2, Oct.)

'Towards an optimum solution for underground rapid transit',
Caspe, M. S., 1969, **95** (TE1, Feb.)
'Overall planning of the Bay Area Rapid Transit project',
Hammond, D. G., 1969, **95** (TE2, May)
'Analysis of tunnel liner-packing systems',
Dawkins, W. P., 1969, **95** (EM3, June)
'Earthquake design criteria for subways',
Kuesal, T. R., 1969, **95** (ST6, June)
'Gravity and gradients of long tunnels',
Williams, H. S., 1969, **95** (SU1, Oct.)
'Chemical grouting for Paris rapid transit tunnels',
Janin, J. J. and Le Sciellour, G. F., 1970, **96** (CO1, June)
'River diversion for Boundary dam',
Schilling, A. A., 1970, **96** (PO3, June)
'Diversion at Portage Mountain project',
Pratt, H. K., 1970, **96** (PO3, June)
'Rock properties and steel tunnel liners',
Kruse, G. H., 1970, **96** (PO3, June)
'Planning for the Metro',
Deen, T. B., 1970, **96** (TE3, Aug.)
'Laser alignment techniques utilized in tunneling',
Cooney, A., 1970, **96** (SU2, Sep.)
'Development of tunneling methods and controls',
Armstrong, E. L., 1970, **96** (CO2, Oct.)
'Cabin Creek pumped storage hydroelectric project',
Hight, H. W., 1971, **97** (PO1, Jan.)
'Flathead railroad tunnel construction',
Sacrison, H., 1971, **97** (CO1, Mar.)
'Design of seismic joint for San Francisco Bay tunnel',
Douglas, W. S. and Warshaw, R., 1971, **97** (ST4, Apr.)
'Fourth dimension for urban environment',
Sorensen, K. E., 1971, **97** (UP1, Apr.)
'Surveys for South Shore rapid transit construction',
Pacelli, A. J., 1971, **97** (SU1, May)
'Precast concrete tunnel linings for Toronto subway',
Bartlett, J. V. and others, 1971, **97** (CO2, Nov.)
'Sixty years of rock tunneling in Chicago',
Pikarsky, M., 1971, **97** (CO2, Nov.)
'Vehicular tunnels in rock, directions for development',
Robbins, R. J., 1972, **98** (CO2, Sep.)
'Computer simulation of subway environment',
Hoover, T. E. and others, 1973, **99** (TE1, Feb.)

'Recent development of New Austrian Tunneling Method',
Nussbaum, H., 1973, **99** (CO1, July)
'Tunnel survey and tunneling machine control',
Peterson, E. W. and Frobenius, P., 1973, **99** (SU1, Sep.)
'Escalators in rapid transit stations',
O'Neill, R. S., 1974, **100** TE1, Feb.)
'BART in operation – innovations in rapid transit',
Bugge, W. A., 1974, **100** (TE2, May)
'Subsurface construction contracts, a contractor's view',
Fox, G. A., 1974, **100** (CO2, June)
'Planning economical tunnel surveys',
Thompson, B. J., 1974, **100** (SU2, Nov.)
'Subsidence over soft ground tunnel',
Butler, R. A. and Hampton, D., 1975, **100** (GT1, Jan.)
'Barjansky's shallow-tunnel solution',
Agrawal, J. S. and Richards, R., 1975, **101** (EM1, Feb.)
'Developments in trench-type tunnel construction',
Palmer, W. F. and Roberts, K. C., 1975, **101** (CO1, Mar.)
'Liner-medium interaction in tunnels',
Mohraz, B. and others, 1975, **101** (CO1, Mar.)
'Introduction to fifty years of tunneling',
Mayo, R. S., 1975, **101** (CO2, June)
'Fifty-year highlights of tunneling equipment',
Day, D. A. and Boiser, B. P., 1975, **101** (CO2, June)
'Importance of groove spacing in tunnel boring machine operations',
Rad, P. F., 1975, **101** (GT9, Sep.)
'Karl Terzaghi and the Chicago subway',
Peck, R. B., 1975, **101** (EI4, Oct.)
'Ventilated approach regions for railway tunnels',
Vardy, A. E., 1975, **101** (TE4, Nov.)
'Design of transit structures for urban areas',
Hammond, D. G. and Desai, D. B., 1976, **102** (TE1, Feb.)
'Tunnel and shaft sealing methods using polyvinyl chloride sheet',
Peduzzi, A., 1976, **102** (CO1, Mar.)
'Analysis of backpacked liners',
Dar, S. M. and Bates, R. C., 1976, **102** (GT7, July)
'Model Study of tunnel pre-reinforcement',
Korbin, G. E. and Brekke, T. L., 1976, **102** (GT9, Sep.)
'Applications of photoelasticity to structural analysis',
Jullien, J. F. and others, 1976, **102** (EM5, Oct.)
'Stresses in shallow elliptical tunnel',
Mark, R., 1976, **103** (GT11, Nov.)

'Urban tunnels – an option for the transit crisis',
Walton, M. S. and Proctor, R. J., 1976, **102** (TE4, Nov.)
'Half a century progress in soft ground tunneling',
Fox, G. A. and Nicolau, L. E., 1976, **102** (CO4, Dec.)
'Subsidence above shallow tunnels in soft ground',
Atkinson, J. H. and Potts, D. M., 1977, **103** (GT4, Apr.)
'Segmental liner for soft ground tunnels',
Tartaglione, L. C., 1977, **103** (CO2, June)
'Construction methods in tunneling under a railroad',
Soto, M. H., 1977, **103** (CO3, Sep.)
'Fire protection systems on (WMATA) Metro',
Ell, W. M., 1978, **104** (TE1, Jan.)
'Damage to rock tunnels from earthquake shaking',
Dowding, C. H. and Rozan, A., 1978, **104** (GT2, Feb.)
'Model tests for plastic response of lined tunnels',
Kennedy, T. C. and Lindberg, H. E., 1978, **104** (EM2, Apr.)
'Evaluation of alternative station spacings for rapid transit lines',
Permut, H., 1978, **104** (TE3, May)
'Field study of tunnel prereinforcement',
Korbin, G. E. and Brekke, T. L., 1978, **104** (GT8, Aug.)
'Case study of Buffalo's rail transit development',
Wilson, T. M., 1978, **104** (TE5, Sep.)
'Application of epoxy resins in tunnel lining concrete repair',
Borden, R. C. and Selander, C. E., 1978, **98** (CO2, Sep.)
'Temporary tunnel support by artificial ground freezing',
Jones, J. S. and Brown, R. E., 1978, **104** (GT10, Oct.)
'Simplified analysis for tunnel supports',
Einstein, H. H. and Schwartz, C. W., 1979, **105** (GT4, Apr.)
'Pipe jacking state-of-the-art',
Drennon, C. B., 1979, **105** (CO3, Sep.)
'Settlements of immersed tunnels',
Schmidt, B. and Grantz, W. C., 1979, **105** (GT9, Sep.)
'Park River auxiliary tunnel',
Blackey, E. A., 1979, **105** (CO4, Dec.)
'The Central Artery project, construction techniques',
Sholock, F. R., 1980, **106** (TE2, Mar.)
'Economic potential of tunnel standardisation',
Hampton, D. and McCusker, T. G., 1980, **106** (CO3, Sep.)
'Mixed face tunneling on Melbourne underground',
Petrofsky, A. M., 1980, **106** (CO3, Sep.)
'Ground control for shallow tunnels by soil grouting',
Tan, D. Y. and Clough, G. W., 1980, **106** (GT9, Sep.)

'Remote sensing for tunnel siting studies',
Russell, O. R. and others, 1980, **106** (TE5, Sep.)
'Tokyo's Dainikoro underwater tube tunnel',
Paulson, B. S., 1980, **106** (CO4, Dec.)

D. OTHER ORGANISATIONS

Four organisations having a substantial and continuing output of material directly relevant to tunnelling are listed separately:

D1 British Tunnelling Society (BTS)
D2 International Tunnelling Association (ITA)
D3 Construction Industry Research and Information Association (CIRIA)
D4 Transport and Road Research Laboratory (TRRL)

There are a growing number of other organisations, including many national societies affiliated to ITA, holding conferences and publishing reports and journals, some of which are mentioned in Sections A, E and F.

The **International Union of Public Transport (UITP)** primarily represents operators, and their quarterly *Revue* and bienneial *Congresses* contain many references to Metros. UITP also publish a quarterly *Biblio-Index* and occasional *Metro Bibliography.*

In the United States of America, the **U.S. National Committee on Tunneling Technology** issue their quarterly *Tunneling Technology Newsletter*; the **American Underground Space Association** have *Underground Space* (Pergamon Press, 6 issues per annum); while there is a steady flow of *Research Reports* from universities, consulting engineers and contractors sponsored by the various administrations in the **U.S. Department of Transportation.**

In the United Kingdom, besides the learned Institutions devoted to Engineering, Geology, Metallurgy and Mining, there are societies for enthusiasts such as the **London Underground Railway Society, Tunnel Study Society** and **Subterranea Britannica.**

In an English language bibliography there is no space to give adequate coverage to **STUVA (Studiengesellschaft fur unterirdische Verkehrsanlagen)** or West Germany whose volumes of *Forschung + Praxis U-Verkehr und Unterirdisches Bauen* have made a valuable contribution.

D1 British Tunnelling Society

This was established in 1971, with the aid of the Institution of Civil Engineers, and holds frequent meetings for presentation of papers and discussions. Its meetings have been reported in some detail in the journal *Tunnels and Tunnelling* and the following selected references are to subjects of general and continuing interest. The dates are those of publication.

'Standardisation in tunnelling',
1971, **3** (July)

'Efficient excavation (hard rock machine heads)',
1972, **4** (Jan., Mar., May)
'Ventilation of road tunnels',
1972, **4** (May, July, Sep.)
'Contractual risks in tunnelling – how should they be shared?',
1973, **5** (Nov.), 1974, **6** (Mar.)
'Expanded tunnel linings',
1974, **6** (Mar., May)
'Compressed air and the new decompression table',
1974, **6** (Nov.), 1975, **7** (Jan.)
'Geological information for underground works',
1975, **7** (Jan., Mar.)
'Extending the Paris Metro',
1975, **7** (May, July)
'Problems of shaft sinking',
1975, **7** (July, Sep.)
'Instrumentation and monitoring of tunnels',
1975, **7** (Sep., Nov.)
'Problems of tunnelling in chalk',
1975, **7** (Nov.), 1976, **8** (Jan.)
'Early history of sapper tunnelling',
1976, **8** (Jan., Mar.)
'Blasting in urban areas',
1976, **8** (Sep.), 1977, **9** (Mar.)
'Probing ahead for tunnels',
1977, **9** (Jan.)
'Tunnelling trials in chalk',
1977, **9** (Mar.)
'Tunnelling – improved contract practices',
1977, **9** (Nov.)
'U.S. and U.K. tunnelling – developments, comparisons and trends',
1978, **10** (Jan.)
'Lining and waterproofing techniques in Germany',
1978, **10** (Apr., May)
'Hydraulics versus pneumatics',
1978, **10** (July)
'Modern surveying instruments',
1978, **10** (Dec.)
'Immersed tunnels – Danish style',
1979, **11** (July)
'Urban tunnelling in the eastern U.S.',
1979, **11** (July)

'Ground classification – continental and British',
  1980, **12** (July)

D2 International Tunnelling Association – ITA

(Association Internationale des Travaux en Souterrain – AITES, 109 Av. Salvador Allende, 69500 Bron, France), was formed in 1974 as an international commission on tunnelling activities. Almost all major countries with any interest in tunnels are represented by their national Society, Institution or Authority. In recent years the most significant functions of ITA have tended to move away from its international congresses into the activities of nine working groups on the following subjects briefly described in *Tunnels and Tunnelling*:

Contractual sharing of risks	1980, **12** (Jan)
Standardization	(Mar)
Research	(Apr)
Safety in works	(May)
Catalogue of works in progress	(June)
Subsurface planning	(Oct)
Structural design models	(Nov)
Seismic effects	(Dec)
Maintenance and repair	1981, **13** (Jan)

The *Proceedings* of the first four *Annual Meetings* and the first three *Reports by the Working Groups* were published by AITES, Bron. All subsequent *Proceedings* and *Reports* and published in *Advances in Tunnelling Technology and Subsurface Use*, from 1980, **1** (1), Pergamon Press, 3 issues per annum.

D3 Construction Industry Research and Information Association (CIRIA)

Founded in 1964 as the Civil Engineering Research Association (CERA), CIRIA is an association of contractors, manufacturers, consulting engineers, universities, local and public authorities. A sectional committee on Underground Construction was formed in 1974. Results of research projects are published as *Reports* (*Rep*) and *Technical Notes* (*TN*).

'Hydraulic design of unlined and lined-invert rock tunnels',
  Wright, D. E., *Rep*. 29, 1971

'Full-scale tests to determine the hydraulic performance of a lined-invert rock tunnel',
  Wright, D. E., *TN* 16, 1971

'Medical Code of Practice for work in compressed air',
  Rep. 44, 1973, 2nd edn. 1975, 3rd edn. 1981

'The economics of vehicular tunnel ventilation and the automatic control of tunnel systems',
  Pursall, B. R. and Swann, C. D., *TN* 53, 1974

'Experiences with a new decompression table for work in compressed air',
  TN 59, 1974

'Classified index of research requirements in underground construction', 1975

'Recommended tunnel sizes – a cost benefit study', *Rep.* 66, 1977

'Lateral movement of heavy loads', *Rep.* 68, 1977

'Recording drilling performance for tunnelling site investigations', Brown, E. T. and Phillips, H. R., *TN* 81, 1977

'Down hole instrumentation – a review of tunnelling ground investigation', *TN* 90, 1977

'Tunnelling, improved contract practices', *Rep.* 79, 1978

'A review of instruments for gas and dust monitoring underground', *Rep.* 80, 1978

'Tunnel waterproofing', *Rep.* 81, 1979

'Precast concrete tunnel linings – review of current test procedures', Lance, G. A., *TN* 104, 1981

'Acoustic techniques for rock quality evaluation', *TN* in prep.

'Guide to the health and safety aspects of ground treatment materials', *Rep.* 95, 1982.

'Hydraulic roughness of segmentally-lined tunnels', *Rep.* 96, 1982.

'Rock reinforcement: a state-of-the-art review', *Rep.* in prep.

D4 Transport and Road Research Laboratory (TRRL)

Founded in 1933 as the Road Research Laboratory (RRL), the TRRL is currently responsible to the Departments of the Environment and Transport. In 1970 there was formed what has now become the Tunnels and Underground Pipes Division. Results of research are published as either *Laboratory Reports* (*LR*) or *Supplementary Reports* (*SR*).

'The classification and description of soils for engineering purposes: a suggested revision of the British system', Dumbleton, M. J., *LR* 182, 1968

'The effect of tides on subsoil pore-water pressures at a site on the proposed Shoreham by-pass', Margason, G. and others, *LR* 195, 1968

'A preliminary study of the cost of tunnel construction', Margason, G. and Pocock, R. G., *LR* 326, 1970

'Preliminary sources of information for site investigations in Britain', Dumbleton, M. J. and West, G., *LR* 403, 1971, revised 1976

'Channel Tunnel: vehicular movements in proposed ferry trains',
Ellson, P. B. and Layfield, R. E., *LR* 435, 1972
'Channel Tunnel: pedestrian movements in proposed ferry trains',
Cundill, M. A., *LR* 436, 1972
'A review of methods of ventilating road tunnels',
Hignett, H. J. and Hogbin, L., *LR* 477, 1972
'An analysis of vehicle breakdowns in the Mersey Tunnel',
Bartlett, R. S., and Chhotu, S. R., *LR* 484, 1972
'Some examples of underground development in Europe',
O'Reilly, M. P., *LR* 592, 1974
'Guidance on planning, directing and reporting site investigations',
Dumbleton, M. J. and West, G., *LR* 625, 1974
'A pilot scale machine for tunnel boring research',
Hignett, H. J. and Howard, T. R., *LR* 632, 1974
'Measurement of ground movements during a bentonite tunnelling experiment',
Boden, J. B. and McCaul, C., *LR* 653, 1974
'Techniques and equipment for detecting underground services',
Keir, W. G., *SR* 69, 1974
'Note on a visit to Italy – September 1973',
O'Reilly, M. P., *SR* 80, 1974
'A horizontal inclinometer for measuring ground movements',
Hudson, J. A. and Morgan, J. M., *SR* 92, 1974
'An assessment of geophysics in site investigation for roads in Britain',
West, G. and Dumbleton, M. J., *LR* 680, 1975
'The influence of tunnel excavation on an adjacent shaft in chalk',
Hudson, J. A. and McCaul, C., *SR* 161, 1975
'Probing ahead for tunnels: a review of present methods and recommendations for research',
SR 171, 1975
'Rock quality in the Kielder Experimental Tunnel, Co. Durham',
Priest, S. D. and Hudson, J. A., *SR* 173, 1975
'Research studies into the behaviour of tunnels and tunnel linings in soft ground',
Atkinson, J. H. and others, *SR* 176, 1975
'Measurements of ground movement and lining behaviour on the London Underground at Regents Park',
Barratt, D. A. and Tyler, R. G., *LR* 684, 1976
'An impact penetrometer for assessing the cuttability of soft rocks',
Hudson, J. A. and Drew, S. D., *LR* 685, 1976
'The particle-size distributions of debris produced during tunnelling trials',
Toombs, A. F. and others, *LR* 714, 1976
'Ultrasonic wave propagation in discontinuous rock',
New, B. M., *LR* 720, 1976

'Site investigation for tunnelling trials in chalk',
Priest, S. D., *LR* 730, 1976
'A guide to site investigation procedures for tunnels',
Dumbleton, M. J. and West, G., *LR* 740, 1976
'Noise measurements during tunnelling trials at Chinnor',
Foster, K. and others, *SR* 188, 1976
'Machine performance data recorded during the Chinnor tunnelling trials',
Snowdon, R. A., *SR* 196, 1976
'Measurement of ground movement due to excavation of a shallow tunnel in lower chalk',
McCaul, C. and others, *SR* 199, 1976
'A note on a use of stereophotography in tunnelling',
West, G., *SR* 211, 1976
'Shotcreting on chalk: some preliminary trials',
McCaul, C. and Ryley, M. D., *SR* 212, 1976
'Traffic induced ground vibration in the vicinity of road tunnels',
Bean, R. and Page, J., *SR* 218, 1976
'Site investigation aspects of the Sydenham Road sewer tunnel',
Dumbleton, M. J. and Toombs, A. F., *SR* 235, 1976
'The tunnelling system for the British section of the Channel tunnel phase II works',
Morgan, J. M. and others, *LR* 734, 1977
'Tunnelling trials in chalk: rock cutting experiments',
Hignett, H. J. and others, *LR* 796, 1977
'Site investigation aspects of the River Tyne Siphon Sewer Tunnel',
Dumbleton, M. J. and Priest, S. D., *LR* 831, 1978
'Site investigation aspects of the Empingham reservoir tunnels',
Dumbleton, M. J. and Toombs, A. F., *LR* 845, 1978
'The effects of ground vibration during bentonite shield tunnelling at Warrington',
New, B. M., *LR* 860, 1978
'A review of tunnel lining methods in the United Kingdom',
Craig, R. N. and Muir Wood, A. M., *SR* 335, 1978
'The arrangements of rock cutting tools on full face tunnel boring machines',
Hignett, H. J. and O'Reilly, M. P., *SR* 376, 1978
'Settlements caused by tunnelling in weak ground at Stockton-on-Tees',
McCaul, C., *SR* 383, 1978
'The propagation of ultrasonic waves in strong rocks',
Toombs, A. F., *SR* 384, 1978
'Site investigation aspects of the River Medway cable tunnels',
Dumbleton, M. J. and others, *SR* 451, 1978
'Site investigation aspects of part of the Tyneside South Bank interceptor sewer',
Dumbleton, M. J. and others, *SR* 454, 1978

'Site investigation and construction of the Liverpool Loop and Link Tunnels',
West, G. and Toombs, A. F., *LR* 868, 1979
'Speed/flow relationships in road tunnels',
Bampfylde, A. P. and others, *SR* 455, 1979
'Tunnel boring machine performance and ground properties. Report on the initial 1½ km of the North Wear Drive, Kielder Aqueduct',
Morgan, J. M. and others, *SR* 469, 1979
'A preliminary study of the reproducibility of joint measurements in rock',
West, G., *SR* 488, 1979
'Tunnelling machine instrumentation for use in coal mines',
Temporal, J. and Hignett, H. J., *SR* 502, 1979
'Geophysical and television borehole logging for probing ahead of tunnels',
West, G., *LR* 932, 1980
'Measurements of ground movements around three tunnels in loose cohesionless soil',
Ryley, M. D. and others, *LR* 938, 1980
'Some observations of machine tunnelling at the Kielder Aqueduct',
O'Rourke, T. D. and others, *SR* 532, 1980
'Settlement caused by tunnelling beneath a motorway embankment',
Toombs, A. F., *SR* 547, 1980
'Site investigation and construction of the Dinorwic diversion tunnel',
West, G. and Ewan, V. J., *LR* 984, 1981
'Site investigation and construction of the Cardiff cable tunnel',
West, G. and McLaren, D., *LR* 1012, 1981
'Reproducibility of joint spacing measurements in rock',
Ewan, V. J. and others, *LR* 1013, 1981
'Kielder aqueduct tunnels – predicted and actual geology',
Davis, T. P. and others, *SR* 676, 1981

E. CONFERENCES AND SYMPOSIA

The world-wide growth of such meetings presents difficulties in keeping track of the information presented. The papers vary in their importance as to permanent value, wide interest, repetition elsewhere, and inclusion or absence of discussion. The section is divided:

E1 Tunnelling Conferences
E2 Other related conferences with a significant tunnelling content.

E1 Tunnelling Conferences

Symposium on shaft sinking and tunnelling, Instn. Min. Engrs, London, 1959
Rapid excavation – problems and progress; Proc. Tunnel and Shaft Conf., A.I.M.E., Minneapolis, 1968
Large permanent underground openings, Proceedings of the International Symposium, Intnl Soc. Rock Mechs, Oslo, 1969

Deep tunnels in hard rock: a solution to combined sewer overflow and flooding problems, U.S. Environmental Protection Agency, Milwaukee, 1970

The technology and potential of tunnelling: South African Tunnelling Conference, S. African Instn Civ. Engrs, Johannesburg, 1970

6th National Tunnel Symposium, Japan Soc. Civ. Engrs, Tokyo, 1970

Advisory Conference on Tunnelling, Organisation for Economic Co-operation and Development, Washington, 1970

Seminar on Engineering problems in tunnelling, Indian Soc. Engng Geol., New Delhi, 1971

Symposium on Underground Rock Chambers, Amer. Soc. Civ. Engrs, Phoenix, 1971

Proceedings, International Symposium on Underground openings, Swiss Soc. Soil Mechs Fndtn Engng, Lucerne, 1972

Proceedings of Rapid Excavation and Tunneling Conferences, 1st, Chicago, 1972; 2nd, San Francisco, 1974; 3rd, Las Vegas, 1976; 4th, Atlanta, 1979; 5th, San Francisco, 1981, Society of Mining Engineers of A.I.M.E.

Tunnelling in Rock, a course of lectures, S. African Instn Civ. Engrs, Scientia, 1973

Proceedings of a symposium on rock mechanics in tunnelling problems, Indian Geotechnical Society, Haryana, 1973

International Symposium on the Aerodynamics and Ventilation of Vehicle Tunnels, BHRA Fluid Engineering, 1st, Canterbury, 1973; 2nd, Cambridge, 1976; 3rd, Sheffield, 1979

Use of shotcrete for underground structural support, Amer. Soc. Civ. Engrs, South Berwick, 1973

Shotcrete for ground support, Amer. Soc. Civ. Engrs, Easton, 1976

Shotcrete for underground support III, Engineering Foundation, St. Anton am Arlberg, 1978

Subsurface exploration for underground excavation and heavy construction, Amer. Soc. Civ. Engrs, Henniker, 1974

Australian Tunnelling Conferences, 1st, *Reshaping cities using underground construction*, Melbourne, 1974; 2nd, *Design and construction of tunnels and shafts*, Melbourne, 1976; 3rd, Sydney, 1978; 4th, Melbourne, 1981, Australian Tunnelling Association

Hazards in tunnelling and on falsework; Proc. 3rd Intnl Safety Conf., Instn Civ. Engrs, London, 1975

Mechanical boring or drill and blast tunnelling: 1st U.S.–Swedish underground workshop, Swedish Detonic Research Foundation, Stockholm, 1976

Tunnelling '76: Proceedings of an international Symposium, Instn. Min. Metal., London, 1976

Tunnelling '79: Proceedings of the 2nd international Sympsoium, Instn Min. Metal., London, 1979

Computer methods in tunnel design, Instn Civ. Engrs, London, 1977

Storage in excavated rock caverns, Rockstore '77, Proc. 1st Intnl Symp., Stockholm, 1977 (publ. Pergamon)

Subsurface Space, Rockstore '80, Proc. Intnl Symp., Stockholm, 1980 (publ. Pergamon)

Immersed Tunnels: Delta Tunnelling Symposium, Royal Instn Engrs, Netherlands, Amsterdam, 1978

International Symposium on tunnelling under difficult conditions, Japan Tunnelling Assoc., Tokyo, 1978

Seminar on utility services underground, Royal Society of Arts, London, 1978

Eurotunnel Conference, Basle, 1978 and 1980

Analysis of tunnel stability by the convergence-confinement method, Proc. Conference, Paris, 1978; in *Underground Space*, 1980, **4** (4, 5 & 6)

International Symposium, the safety of underground works, Brussels, 1980

E2 Other Conferences

PIARC World Road Conferences:

Road Tunnel Committee *Reports*: 11th, Rio de Janeiro, 1959; 12th, Rome, 1964; 13th, Tokyo, 1967; 14th, Prague, 1971; 15th, Mexico City, 1975; 16th, Vienna, 1979

Road Tunnel Committee *Documentation Digests*: 13th, Tokyo, 1967; 14th, Prague, 1971; 15th, Mexico City, 1975

Permanent International Association of Road Congresses

Grouts and drilling muds in engineering practice, British National Society of ISSMFE, London, 1963

In situ investigations in soils and rocks, Proceedings of the Conference, British Geotechnical Society, London, 1969

Conferences on Soil Mechanics and Foundation Engineering:

7th International, Mexico, 1969, esp. **State of the Art**, Peck, R. B., 'Deep excavations and tunneling in soft ground', **2** Session 4, **3** General report and Session 4

5th Panamerican, Buenos Aires, 1975, esp. **4**, Cording, E. J. and Hansmire, W. H., 'Displacements around soft ground tunnels'

9th International, Tokyo, 1977, esp. *Speciality Session no. 1*, and **Case History** (both published separately)

6th Asian, Singapore, 1979, esp. **2**, Watanabe, T., 'Subway tunnelling in soft ground'

6th Panamerican, Lima, 1979, esp. *Speciality Session on Soft ground tunnelling* (published separately, Balkema, Rotterdam, 1981)

10th International, Stockholm, 1981, esp. **1** Session 2, and **4** Session 2

Dynamic Rock Mechanics, 12th Symposium on Rock Mechanics, Society of Mining Engineers, AIME, Rolla, Missouri, 1970, esp. Part VI

2nd International Congress of the International Association of Engineering Geologists, Sao Paulo, Brazil, 1974, esp. **2** Theme VII

3rd International Congress of Engineering Geology, Madrid, 1978, esp. **2**, Section III-c

Design Methods in Rock Mechanics, 16th Symposium on Rock Mechanics, Amer. Soc. Civ. Engrs, Minneapolis, 1975, esp. Session 3

Proceedings of the Symposium on Exploration for Rock Engineering, S. African Instn Civ. Engrs, Johannesburg, 1976

Proceedings of a Conference on Rock Engineering, British Geotechnical Society, Newcastle-upon-Tyne, 1977

Field Measurements in Rock Mechanics, Proceedings of the International Symposium, ISETH, Zurich, 1977

Large Ground Movements and Structures, Proceedings of the Conference (esp. Session IV and State of the Art), UWIST, Cardiff, 1977

2nd Conference on Ground Movements and Structures (esp. Sessions IV and VII), UWIST, Cardiff, 1980

International Symposium on Ground Freezing, Bochum, W. Germany, 1978

2nd International Symposium on Ground Freezing, Trondheim, Norway, 1980

Mass Transportation in Asia, Concrete Society (HK) Ltd., Hong Kong, 1980

F. JOURNALS AND PERIODICALS

F1 contains a selection of papers from *Tunnels and Tunnelling*. First issued in May 1969, now a monthly, it is the premier journal on the subject and the official journal of the British Tunnelling Society; issues reporting some BTS meetings have already been listed in Section D1.

F2 contains a selection of papers from the vast range of engineering and related journals and from a few non-tunnelling conferences.

Apart from *Tunnels and Tunnelling, Tunneling Technology Newsletter* and *Underground Space* (referred to in Section D), there are many periodicals currently carrying up to date news of tunnelling projects, among the prominent are:

Civil Engineering – London
New Civil Engineer
New Civil Engineer International
International Water Power and Dam Construction
Civil Engineering – New York
Engineering News-Record

F1 *Tunnels and Tunnelling*

Selective bibliography on immersed tubes,
Pequignot, C. A., 1969, **1** (July)

Some aspects of rail tunnel maintenance (UK),
Walton, J. A., 1969, **1** (July)

Tunnel construction and maintenance on the Austrian Federal Railway system,
Ziermann, R., 1970, **2** (Mar., May)
Maintaining the tunnels of French Railways,
Bogaert, G. van den, 1970, **2** (July, Sep.)
Tunnels of Norwegian State Railways,
Hartmark, H., 1970, **2** (Nov.)
Design, construction and maintenance of tunnels of German State Railways,
Spang, J., 1971, **3** (Sep.)
Rail tunnels in Czechoslovakia,
Streit, J., 1973, **5** (May)
Air pollution in vehicular road tunnels,
Pursall, B. R. and Swann, C. D., 1972, **4** (July)
Some aspects of fire behaviour in tunnels,
Roberts, A. G., 1973, **5** (Jan.)
Foundation of a tunnel by the sandflow system,
Griffioen, A. and Van der Veen, R., 1973, **5** (July)
The generation and alleviation of air pressure transients in tunnels,
Fox, J. A. and Vardy, A. E., 1973, **5** (Nov.)
Subterrene rock melting devices,
Altseimer, J. H., 1974, **6** (Jan.)
Channel Tunnel survey,
Hulme, T. W., 1974, **6** (Sep.)
Visco-elastic tunnel analysis,
Curtis, D. J., 1974, **6** (Nov.)
The Shin Kanmon connection,
Brown, R. L., 1975, **7** (Jan.)
Immersed-tube tunnels,
Culverwell, D. R., 1976, **8** (Jan., Mar.)
Mechanised tunnelling – progress and expectations,
Robbins, R. J., 1976, **8** (May)
A view of the Australian tunnelling scene,
Neyland, A. J., 1976, **8** (July)
Survey of hydraulic drilling performance,
Bullock, R. L., 1976, **8** (Sep.)
Road tunnels in Norway,
Grønhaug, A., 1976, **8** (Nov.)
Pipe jacking case histories,
1977, **9** (July, Nov.); 1978, **10** (Jan., Mar., Apr., May, June, July)
Some road and rail tunnels in Japan,
Leeney, J. G., 1977, **9** (Nov.)
Japanese slurry shield tunnelling,
Leeney, J. G., 1978, **10** (Jan.)

Removing misconceptions on the New Austrian Tunnelling Method,
Müller, L., 1978, **10** (Oct.)
Another view of the NATM,
Golser, J., 1979, **11** (Mar.)
The German tunnelling industry – special report,
Harding, P. G., 1979, **11** (Apr.)
Ground behaviour and support for mining and tunnelling,
Muir Wood, A. M., 1979, **11** (May, June)
Assessment of tunnel boring machine performance,
McFeat-Smith, I. and Tarkoy, P. J., 1979, **11** (Dec.)
Tunnel boring machines in difficult ground,
McFeat-Smith, I. and Tarkoy, P. J., 1980, **12** (Jan.)
Site investigations for machine tunnelling contracts,
McFeat-Smith, I. and Tarkoy, P. J., 1980, **12** (Mar.)
A history of the Channel tunnel,
Muir Wood, A. M., 1980, **12** (May)
"Earthworm" system will threaten conventional tunnel jacking,
Richardson, M. and Scruby, J., 1981, **13** (Apr.)
The Unitunnel in action,
Richardson, M. and Scruby, J., 1981, **13** (May)
World list of immersed tubes,
Culverwell, D. R., 1981, **13** (Sep.)
Putting the NATM into perspective,
Brown, E. T., 1981, **13** (Nov.)

F2 Journals and Periodicals

The Mont Cenis tunnel,
Kossuth, F., *Engng*, 1871, **11** and **12**, 1872, **13**
The Channel Tunnel,
Tylden-Wright, C., *Trans. N. England Inst. Min. Mech. Engrs*, 1882, **32**, 3 and 55
On the construction of the Glasgow City and District railway,
Simpson, R., *Trans. Instn. Engrs Shipbuilders Scotland*, 1887–88, **31**
Cast-iron segments for railway and other tunnels,
Carey, E. G.,
Tunnelling in soft materials with special reference to Glasgow District subway,
Simpson, R., *Trans. Instn Engrs Shipbuilders Scotland*, 1895–96, **39** (Dec. 24, Jan. 28, Feb. 25)
Construction and demolition of tunnels,
Ross, M. G., *Trans. Liverpool Engng Soc.*, 1908, **29**
Tunnelling in water-bearing strata,
Baker, R. F., *Trans. Liverpool Engng Soc.*, 1921, **42**

Tube railway tunnelling,
Anderson, D., *Trans. Liverpool Engng Soc.*, 1924, **45**
The Holland vehicular tunnel under the Hudson River,
Skinner, F. W., *Engng*, 1927, **124** (Nov. 11 and 25, Dec. 9)
The Post Office tube railway, London,
Engng, 1928, **125** (Jan. 27, Feb. 10 and 24, Mar. 2 and 16)
Construction of a tunnel under the River Liffey,
Nicholls, H., *Trans. Instn Civ. Engrs Ireland,* 1929, **55**
Ventilation of vehicular tunnels,
Singstad, O., *Proc. World Engng Congr.*, Tokyo, 1929, **9**, 381–99
Driving a mines drainage tunnel in North Wales,
Francis, J. L. and Allan, J. C., *Trans. Instn Min. Metal.*, 1932, **41**
The ventilation of tunnels,
Haldane, J. S., *J. Instn Heat. Vent. Engrs*, 1936, **4** (Mar.)
Ventilating the Lincoln vehicular tunnel,
Murdock, W., *Trans. Amer. Soc. Heat Vent. Engrs*, 1938, **44**, 273–88
Mono Craters tunnel construction problems,
Jacques, H. L., *J. Amer. Waterworks Assoc.*, 1940, **32**, 43
The road tunnel under the River Maas at Rotterdam,
Van Bruggen, J. P., *Engng*, 1940, **150** (Aug. 9 and 30, Sep. 27)
A century of tunnelling,
Halcrow, W. T., *Proc. Instn Mech. Engrs*, 1941, **146**
Piston effect of trains in tunnels,
Dougherty, R. L., *Trans. Amer. Soc. Mech. Engrs*, 1942, **64** (Feb.)
Shield tunnels of the Chicago subway,
Terzaghi, K., *J. Boston Soc. Civ. Engrs*, 1942, **29** (July)
Ventilation and cooling in London's tube railways,
Mount, Sc C., *J. Instn Heat & Vent. Engrs,* 1947, **14** (Jan.-Feb.)
Historical development of subaqueous tunneling,
Singstad, O., *J. Boston Soc. Civ. Engrs*, 1949, **36**
Underground railway planning and construction,
Anderson, D., *Proc. Joint Engng Conf.*, London, 1951
The main access tunnel at London Airport,
Concrete Constr. Engng, 1952, **47** (Mar.)
Stresses in rocks about cavities,
Terzaghi, K. and Richart, F. E., *Geotechnique*, 1952, **3** (June)
Lining the Bowland tunnel,
Wat. Pwr., 1953, **5** (Feb.)
Geology of the Bowland Forest tunnel,
Earp, J. R., *Bull. Geol. Surv. G.B.*, 1955 (7)
Tunnel linings,
Dawson, O., *Trans. Instn Civ. Engrs Ireland*, 1956, **83**

Modern blasting practice in tunnelling operations,
Rankin, W. W. and Haslam, R., *Civ. Engng Publ. Wks Rev*., 1957, **52** (Feb., Mar. and Apr.)

Design and construction of the Hampton Roads tunnel,
Bickel, J. O., *J. Boston Soc. Civ. Engrs*, 1958, **45** (Oct.)

The construction of the Howth tunnel,
O.Shee, S. F., *Trans. Instn Civ. Engrs Ireland*, 1958, **84**, 49 and 171

Effect of diesel locomotive operation on atmospheric conditions in a railway tunnel,
Rennie, R. P. and others, *Can. J. Chem. Engng*, 1960, **38** (Aug.)

The invention and development of injection processes,
Glossop, R., *Geotechnique*, 1960, **10** (Sep.) and 1961, **11** (Dec.)

A contribution to the analysis of stress in a circular tunnel,
Morgan, H. D., *Geotechnique*, 1961, **11** (Mar.)

Theory and practice of rock bolting,
Lang, T. A., *Trans. Soc. Min. Engrs AIME*, 1961, **220**, 333

Technical description of rock cores for engineering purposes,
Deere, D. U., *Rock Mech. Engng Geol*., 1963, **1**, 18

The tunnel of Eupalinus (Samos tunnel),
Goodfield, J., *Scientific American*, 1964, **210** (June)

Temperature and ventilation in the Toronto subway, an experimental study,
Brown, W. E., *Engr*., 1964, **218** (Aug. 28)

Christchurch–Lyttleton road tunnel (4 papers),
Spooner, B. W. and others, *New Zealand Engng*, 1964, **19** (Dec.)

Prestressed concrete in sub-aqueous tunnel construction,
Hall, P. F., *J. Prestressed Concr. Inst*., 1965, **10** (Aug.)

Basic theory of rapid-transit tunnel ventilation,
Brown, W. G., *Trans. Amer. Soc. Mech. Engrs*, 1966, **88B** (Feb.)

Recent developments in the design of submerged tunnels,
Brink, A., *Struct. Engr*., 1966, **44** (Feb.); *see also* discussion 1966, **44** (Aug.)

Trench type subaqueous tunnels: design and construction,
Bickel, J. O., *Struct. Engr*., 1966, **44** (Oct.); *see also* discussion 1967, **45** (Feb.)

Hard rock tunnelling machines,
Muirhead, I. R. and Glossop, L. G., *Trans. Instn Min. Metall*., 1968, **77A**, 1; *see also* discussion, 118

Modern tunnelling methods,
Donovan, H. J., *Public Works & Municipal Services Congress*, London, 1968

The Gibraltar Hill tunnels, Monmouth (Great Britain),
Baker, C. O. and Howells, D. A., *Engng Geol. Amsterdam*, 1969, **3**, 121

Seikan undersea tunnel,
Tanaka, T., *Civil Engineering in Japan*, 1970, **9**

The Tower Subway: the first tube tunnel in the world,
Lee, C. E., *Trans. Newcomen Soc.*, 1970, **43**
Tunnels for roads and motorways,
Muir Wood, A. M., *Q. J. Engng Geol.*, 1972, **5** (1 and 2)
Shotcrete support with special reference to Mexico City drainage tunnels,
Mason, E. E. and Mason, R. E., *Rock Mechs*, 1972, **4**, 115
Seikan undersea tunnel,
Harma, K., *Japanese Railway Engineering*, 1972 (Jan.)
Structural aspects of submerged tube tunnel construction,
Williams, G. M. S. and Innes, K. W., *Struct. Engr.*, 1972, **50** (Feb.); *see also* discussion 1972, **50** (Aug.)
Long horizontal boring,
Mocheda, Y., *Permanent Way*, 1973, 14 (Apr.)
Grouting in pilot tunnel-heading,
Shinohara, H. and Akita, K., *Ibid*
Engineering classification of rock masses for the design of tunnel support,
Barton, N. and others, *Rock Mechs.*, 1974, **6**, 189
Construction of Shin-Kanmon tunnel,
Saito, T., *Civ. Engng in Japan*, 1974, **13**
Transient flows in tunnel complexes of the type proposed for the Channel tunnel,
Henson, D. A. and Fox, J. A., *Proc. Instn Mech. Engrs*, 1974, **188** (15/74)
The design of the Channel tunnel,
Gould, H. B. and others, *Struct. Engr*, 1975, **53** (Feb.); *see also* discussion, 1975, **53** (Dec.)
The circular tunnel in elastic ground,
Muir Wood, A. M., *Geotechnique*, 1975, **25** (Mar.); *see also* discussion, 1976, **26** (Mar.)
The invention and early use of compressed air to exclude water from shafts and tunnels during construction,
Glossop, R., *Geotechnique*, 1976, **26** (June)
Some developments in segmental tunnel linings designed in the United Kingdom,
Lyons, A. C., *Underground Space*, 1977, **1** (3)
Seikan undersea tunnel,
Matsuo, S. and Endo, K., *Civil Engineering in Japan*, 1977, **16**
Design of a ventilation system for an English Channel Tunnel,
Henson, D. A. and Lowndes, J. F. L., *ASHRAE J.*, 1978 (Feb.)
Geotechnical performance of a tunnel in till,
Eisenstein, Z. and Thomson, S., *Canad. Geotech. J.*, 1978, **15** (Aug.)
The Kaimai Tunnel (9 papers),
Hermans, R. E. and others, *Trans. New Zealand Instn Engrs*, 1978, **5** (1)
The 50-year history of the subways in Japan,
Watanabe, T., *Civil Engineering in Japan*, 1978, **17**

Hong Kong MTRC harbour tunnel,
Hansen, F. J., *Hong Kong Engr*, 1979, **7** (Apr.)
Tunnel design: philosophy and practice,
Hulme, T. W., *Struct. Engr.*, 1979, **57A** (May); *see also* discussion, 1981, **59A** (Apr.)
The Hong Kong mass transit immersed tube – some aspects of construction,
Kennedy, M. J., *Hong Kong Engr*, 1980, **8** (Jan.)
Dams and their tunnels,
Muir Wood, A. M. and others, *Int. Wat. Pwr Dam Constr.*, 1980, **32** (Feb., Mar., Apr. and May)
Field measurements in two tunnels in Edmonton, Alberta,
Thomson, S. and El-Nahhas, F., *Canad. Geotech. J.*, 1980, **17** (Feb.)
The stability of shallow tunnels and underground openings in cohesive material,
Davis, E. H. and others, *Geotechnique*, 1980, **30** (Dec.)
An initial appraisal of ground probing radar for site investigation in Britain,
Darracott, B. W. and Lake, M. I., *Ground Engng*, 1981, **14** (Apr.)
Recent developments in geophysical techniques for the rapid location of near-surface anomalous ground conditions,
McDowell, P., *Ground Engng*, 1981, **14** (Apr.)
Soil mechanics aspects of soft ground tunnelling,
Atkinson, J. H. and Mair, R. J., *Ground Engng*, 1981, **14** (July)
Site investigation for tunnels,
West, G. and others, *Inst. J. Rock Mech. Min. Sci.*, 1981, **18** (5)
The construction of the Ahmed Hamdi road tunnel, Suez,
Harries, D. A., *Highway Engr.*, 1981, **28** (Nov.)

Index